Scientific Computation

Optimization can be viewed as an exhaustive search, starting at some quite arbitrary place and then climbing or descending through a very rugged landscape searching for the highest mountaintop, or lowest valley (where altitude represents the quality of the current solution). We will consider landscapes at least as rugged as in the picture above, taken in the Navaho Reservation's Monument Valley (Arizona and Utah, USA) by Norman Koren. Although the problems we consider usually offer more than two dimensions in which to explore, they share the challenges of peaks which arise in clumps, separated by long stretches which one must go both up and down to cross, and valleys of contorted shape, difficult to find your way out of. Metaphors abound in this field, ranging from simulated annealing (q.v.) to millenial floods, such that the walker flees to the mountaintops to keep his feet dry. We recommend returning to this picture from time to time, not only for inspiration but also to ask just how the more advanced optimization algorithms might work in this landscape.

Johannes J. Schneider
Scott Kirkpatrick

Stochastic Optimization

With 189 Figures and 29 Tables

Springer

Johannes Josef Schneider
Johannes Gutenberg-Universität Mainz
Institut für Physik
Fachbereich 08
Physik, Mathematik und Informatik
Staudinger Weg 7
55099 Mainz, Germany
e-mail: schneidj@uni-mainz.de

Scott Kirkpatrick
The Hebrew University of Jerusalem
The Selim and Rachel Benin School
of Engineering and Computer Science
Edmond Safra Campus
91904 Jerusalem, Israel
e-mail: kirk@cs.huji.ac.il

Cover picture: See Fig. 4.4 (*left*)

Library of Congress Control Number: 2006922326

ISSN 1434-8322
ISBN-10 3-540-34559-0 Springer Berlin Heidelberg New York
ISBN-13 978-3-540-34559-6 Springer Berlin Heidelberg New York

Springer is a part of Springer Science+Business Media

springer.com

Printed in Germany

Typesetting: Camera-ready copy from the authors
Data conversion and production by LE-TEX Jelonek, Schmidt & Vöckler GbR, Leipzig, Germany
Cover design: *Erich Kirchner* Heidelberg
Printed on acid-free paper SPIN: 10887739 55/3100/YL - 5 4 3 2 1 0

To Yael and Tom Kirkpatrick
and to Josef Sebastian and Maria Schneider

Preface

Our purpose in writing this book was to provide a compendium of stochastic optimization techniques, some guides to when each is appropriate in practical situations, and a few useful ways of thinking about optimization as a process of search in some very rich configuration spaces. Each of us has come to optimization, traditionally a subject studied in applied mathematics, from a background in physics, especially the statistical physics of random mixtures or materials. One of us (SK) has used ideas developed in the study of magnetic alloys to explore the optimal placement of computer circuits subject to many conflicting constraints, while at IBM Research, in Yorktown Heights, NY. The other (JJS) while completing his studies in physics under Prof. Ingo Morgenstern in Regensburg, Germany, and working at the IBM Scientific Center Heidelberg, was exposed to optimization problems as varied as scheduling the pickup of fresh milk and planning automobile assembly line schedules. We had the opportunity to work together after SK moved from IBM to a professorship at The Hebrew University of Jerusalem, Israel, and JJS was, for a year, a postdoc there. JJS has taught a course on stochastic optimization at the University of Mainz, where his students have used portions of the present manuscript. We hope to make this material readable by undergraduates, and useful to graduate students and practitioners as well, in computer science, applied mathematics, physics, and economics.

Mainz, April 2006 — *Johannes Josef Schneider*
Jerusalem, April 2006 — *Scott Kirkpatrick*

Contents

Part I Theory

Overview of Stochastic Optimization Algorithms

Part I

Theory

Overview of Stochastic Optimization Algorithms

0 General Remarks

0.1 Why Optimize Things?

Why do we want to optimize things? Well, first of all, this is part of our nature: each race has developed and optimized strategies to survive in the struggle of life over the millions of years of evolution. Humankind, not equipped with weapons like the poisonous fangs of a snake, the fast legs of a tiger, the strong bite of a crocodile, or the size and strength of an elephant, has developed a large brain in order to think of complex strategies to avoid enemies and to obtain food. Although our animal origin is now a little bit hidden below a civilizing layer, we still act in a resourceful way: we drive the shortest way from work to home; we want to earn the most money with the least effort.

This is not a book about how to make money. However, it is a book that may bring you some other pleasure, namely, being a champion. It is a hidden desire of all of us to be the best, at least in one discipline. There are several avenues for becoming the best. Many people play some kind of sport. The best meet in the Olympic games and try to win medals. Some people derive pleasure from beating others in games such as chess. There are also many competitions to find the best solutions to problems that must be solved using a computer as a human being would not be able to find a good solution manually. Examples of these are competitions to find optima for instances of the traveling salesman problem: the traveling salesman problem is given through a set of cities, through which the traveling salesman must drive the shortest closed route possible, visiting each city exactly once. It has been found that humans can solve random instances up to 50 cities optimally by hand. With larger instances, the results are at least 10% worse than the optimum. Therefore, a number of people have developed complex strategies for computers working on problems like this in order to arrive at either the true optimum or at least a very good solution to the problem in a reasonable amount of time. One further prominent example of a successful computer algorithm was the defeat of the world chess champion Garry Kasparov by the IBM supercomputer Deep Blue in 1997.

The most frequent motivation for optimization is business: production processes are improved in order to save time, resources, and human workers and in order to increase product quality. This must be done to stay competitive with other companies in the industry. Optimization algorithms are used in various fields of industrial processes: starting with the determination of the requirements of material, workers, and machines, and the estimation of the costs and the possible gain, a company owner or executive has to decide which parts of the production processes should be outsourced to subcontractors and suppliers. Then there is a vehicle routing problem with time windows as the preproducts must be delivered to the factory in time. These preproducts have to be stored and distributed inside a factory. The distribution of the workers on individual tasks also has to be planned carefully. The main production processes themselves have to be supervised in such a way that one is faced with an information-gathering, filtering, and administration problem. Errors have to be registered and corrected such that there might arise the need to exchange information with subcontractors and customers in real time. After that the final products have to be checked for their quality and delivered to the customers. Finally, the prospective profit has to be estimated and distributed. There are decisions such as whether some portfolio selection shall be made, whether the profits shall be reinvested, or whether a dividend shall be disbursed. All these problems must not be seen as independent of one another. They must be considered as parts of a complex system with long-range interactions. This overall system can only be solved by a global optimization technique.

0.2 Moral Aspects of Optimization

Besides the fact that the topic of optimization is satisfying and might lead to some financial gain, the moral aspects of optimization have to be considered: as always in science, the question arises as to whether all that can be done should be done.

One might argue that in our world today optimization is no longer necessary. We in the western world have already reached a rather high standard of living that could be extended to the whole world if some global circumstances were changed. And even worse, optimization is said to eliminate jobs, as in the process of rationalizing work formerly done by humans and now done by machines, which forces many people into unemployment. If one has to optimize the problem of, e. g., a dairy farm, which consists of the vehicle routing problem to route trucks that collect the milk from individual farmers, one is most successful if one not only considers constraints like time windows in which the farmers would like to have the milk fetched but also saves one or more of the trucks. However, one must also think about the driver of the trucks, who might be unemployed after the real-life implementation of this successful optimization result. Similarly, an optimization process might lead to the result that some workers can be saved in the manufacturing process or that even an expensive machine, replacing a whole working group, would pay for itself quickly and have the added advantage that it works 24 hours a day without making mistakes, without getting sick, and without going on strike and demanding more money.

However, it is not so easy. Probably the most successful social system in the modern world is that of the market economy and capitalism. It makes use of the fact that people want to possess more in the future than they possess at present. Furthermore, people want to possess something more or better than their neighbor. If a company owner or executive wants to stay competitive, he cannot simply maintain the present state of affairs, as stagnation means regression, because his competitors will optimize their companies for a reduction in production costs and an increase in product quality. Globalization now leads to even greater pressure on all companies worldwide to stay competitive. Therefore, if one has to write an optimization program for a company, then it should be done as well as possible. One takes some responsibility for the large majority of the workers, for if the company cannot compete, the consequences could be closure and unemployment. This is even more the case as company owners or executives clamor for optimization tools from outside experts when they are unable to solve their problems from within.

0.3 How To Think About It

The first question to ask about any problem is whether there exists an exact algorithm to generate the mathematically proved optimum solution for a given problem in a reasonable amount of time. If this is the case, then this exact and fast algorithm should be implemented. The calculation time of such an algorithm is commonly estimated as a function of the problem size N: for example, calculating the scalar product of two vectors consisting of N elements takes $2 \times N$ operations (N additions and N multiplications). Therefore, this algorithm is linear in N, and one writes that the algorithm is of the order $\mathcal{O}(N)$. Another example is the matrix summation: let us assume two matrices A and B consisting of $N \times N$ elements; then calculating one element of $A + B$ takes only one operation [this is of the order $\mathcal{O}(1)$] such that the calculation of all elements is of the order $\mathcal{O}(N^2)$. Matrix multiplication takes even more time: calculating one element of the product matrix $C = A \times B$, $c_{ij} = \sum_{i=1}^{N} a_{ik} b_{kj}$, takes $2 \times N$ operations (N multiplications and N additions) such that the overall order of this algorithm is $\mathcal{O}(N^3)$. The time t such a fast and exact algorithm needs generally depends on the system size in a way that one can estimate $t \propto N^p$ or $t \propto N^p \ln(N)$. As the time can therefore be estimated by some polynomial, these problems are usually solved by use of an exact algorithm, except if the system size is very large and p exceeds the value of 3.

However, such algorithms do not exist for every problem. Indeed, there is no such algorithm for most of the complex problems occurring in practical applications of operations research, applied mathematics, computational physics, chemistry, and biology.

0.4 Minima, Maxima, and Extrema

When looking for the best solution, this means finding the extreme value of some proposed function, depending on many variables. This function is often called the cost function or the objective function. In physics, it is also called the Hamiltonian of the system. As physical systems tend to reduce their energy, and thus to find the minimum value of their Hamiltonian, physicists usually search for the minimum of a proposed cost function. In contrast, mathematicians usually search for the maximum, perhaps because many of them liked to climb mountains in their youth. Thus, most optimization literature in mathematics deals with finding the maximum, whereas this book, written by physicists, mostly deals with finding the minimum.

However, it usually makes no difference whether the optimization algorithm searches for a minimum or a maximum. Some problems, of course, behave differently when searching for a maximum as opposed to a minimum [181]. But usually maximizing the cost function of a problem with a minimization algorithm can simply be done by taking the negative of the cost function, and vice versa. Thus, in these cases one often speaks of the optimum such that both the minimum and maximum cases are covered by one single term.

Some algorithms that gradually improve configurations might get stuck at some specific configuration σ and are then unable to improve the system even further, although there are better configurations in the system. In this case, the configuration σ is called a local optimum of the system. This local optimum might be much worse than the overall best configuration, which is called the global optimum of the system.

0.5 What Is So Hard About Optimization?

In this book we will introduce a number of exact and approximate algorithms for optimization and apply them to problems, most generally to what is called the traveling salesman problem. Since most of the methods under discussion are heuristic, rather than exact, methods, it is important to remember that what works for one problem may not be a viable approach to another problem. As a result, we shall try to present a set of approaches for developing heuristics of steadily increasing power for any given problem. One hopes that the simplest method, which requires the least detailed study of the problem of interest, will give a result good enough for one's immediate objectives. Otherwise, one of the stronger methods will be needed. The particulars of the problem, such as the cost of calculating the objective quality or utility of each proposed solution, or the difficulty of rearranging its elements to see if an improvement results, will determine which of a group of comparably powerful methods should be used.

We shall see that the really hard optimization problems are those with many local optima of widely varying quality. It may be necessary to introduce methods that will allow getting "unstuck" from a local optimum in order to find another that is of much better quality. But how hard must one work to achieve this, and how far from the quality of a given local optimum must one explore (how high a barrier will we have to climb over?) in order to get to something much better? Is it best to wait until one has found a reasonably good local optimum before searching among inferior solutions for a path to the ultimate goal, or should one from the beginning include solutions with obvious flaws, hoping to isolate their best characteristics and ultimately combine them into one successful solution? To answer such questions we must first gain experience with a range of problems and a variety of methods.

0.6 Algorithms, Heuristics, Metaheuristics

In the literature on optimization, one often comes across the terms algorithms and heuristics and sometimes even metaheuristics. But different authors use these terms in different ways. Some authors differentiate between these terms as follows. Algorithms are approaches that solve a problem exactly and then return the global optimum solution of the problem. In addition, they also provide the mathematical proof that the returned solution is optimal. On the other hand, for these authors heuristics are approaches that only return some configuration that might be quite good and that could even be optimal if the applicant was lucky. But usually these heuristics cannot provide a proof of whether the returned configuration is optimal or at least how good the returned configuration is if compared to the optimum solution. For other authors, generally every approach leading to either the optimum or even any quite good solution of the given problem is called an algorithm. They thus regard the set of heuristic algorithms to be a subset of the set of all optimization algorithms.

Furthermore, some authors differentiate between heuristics and metaheuristics. For such authors metaheuristics use some heuristics for providing good solutions to the proposed problem, which cannot usually be achieved with the underlying heuristics alone. Some of them consider a metaheuristic to be only a framework that needs an underlying, mostly simple, heuristic to enable it to optimize the specific application to the proposed problem.

In this book, we will call any approach leading to a quite good or optimum solution to a proposed problem an optimization algorithm. If the algorithm leads always to the optimum solution of the problem and in addition provides the proof that this solution is optimal, then we will call the algorithm an exact algorithm; otherwise we will speak of a heuristic. Furthermore, we will not use the term metaheuristics in this book.

1 Exact Optimization Algorithms for Simple Problems

1.1 A Simple Example—Exact Optimization in One Dimension

We are interested in finding optimal operating points of systems with many variables and a definite cost for any operating point so that there will exist a cost function whose value we can minimize. In order to have a clear mental picture of the obstacles to be overcome, this discussion usually proceeds from one degree of freedom to a few degrees, and finally to many. Although we shall see that the problems of "many" are entirely different in nature than those of optimizing a function of a few variables, the classical literature of optimization contains very detailed treatments of the already difficult problems of optimizing functions of a few variables, where a "few" might be 50 or less. Simple pictures capture the essential obstacles to be overcome. Let us start with a function of a single variable, x.

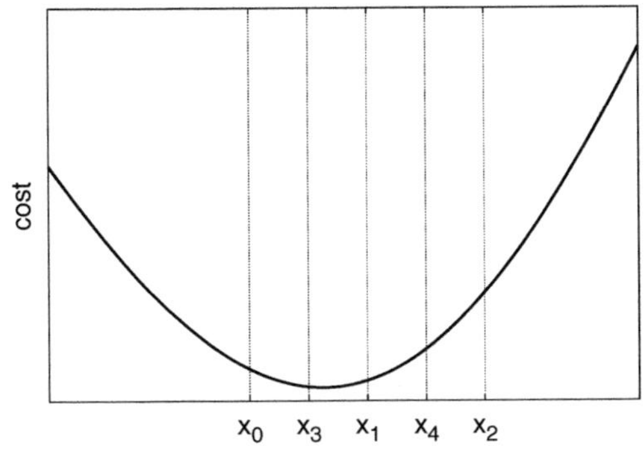

Fig. 1.1. Minimizing a function of one variable

The sketch in Fig. 1.1 shows the starting point for finding a minimum. We need to know the value of our function, $\mathcal{H}(x)$, at three points, x_0, x_1, and x_2, with the value at the middle point, x_1, less than the values at x_0 and x_2. Even if the only information available is the value of $\mathcal{H}(x)$ at selected additional points on the line, we can tighten up this bracket of the minimum by sampling at points x_3 between x_0 and x_1 and x_4 between x_1 and x_2. Now the new bracket is whichever sequence of three consecutive points has the middle point lowest, in this case x_0. x_3, and x_1. One can bisect the intervals at each step and obtain linear convergence, that is, the error term, or difference between the lowest value of $\mathcal{H}(x)$ found so far and the true minimum, will decrease by a constant ratio with each narrowing of the bracket. Press [166], Chap. 10.1, shows that an improvement over this scheme is obtained by tightening the brackets using intervals whose ratio forms a golden mean $(1+\sqrt{5})/2$, but convergence remains linear.

1.2 Newton–Raphson Method

Now suppose that it is possible and inexpensive to compute derivatives of $\mathcal{H}(x)$. In this case, if we start with a value of x not too far from a minimum, the Taylor expansion,

$$\mathcal{H}(x) = \mathcal{H}(x_0) - b(x - x_0) + A(x - x_0)^2/2\,, \tag{1.1}$$

allows us to estimate the location of the minimum (the point at which the first derivative of $\mathcal{H}$ vanishes), at the point where

$$b = A(x - x_0)\,. \tag{1.2}$$

Choosing the next approximation, x_1, for the location of the minimum to be

$$x_1 = x_0 + b/A \tag{1.3}$$

gives a quadratically convergent series of approximations to the minimum, in which the error in minimizing $\mathcal{H}$ decreases by the square of the deviation

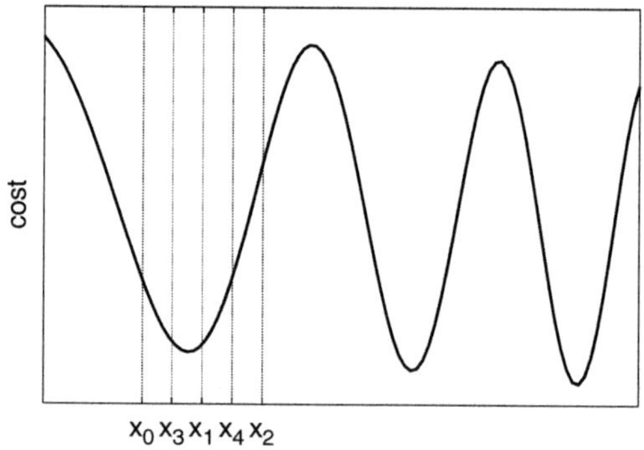

Fig. 1.2. Even functions of one variable may contain several minima

remaining in x with each iteration, at least until roundoff and other machine computation limitations halt further convergence. However, this scheme does not distinguish a maximum from a minimum and may not converge if we start too far from the minimum. And finally, what if the function, even in this limiting case, actually looks like Fig. 1.2, with many minima? And what

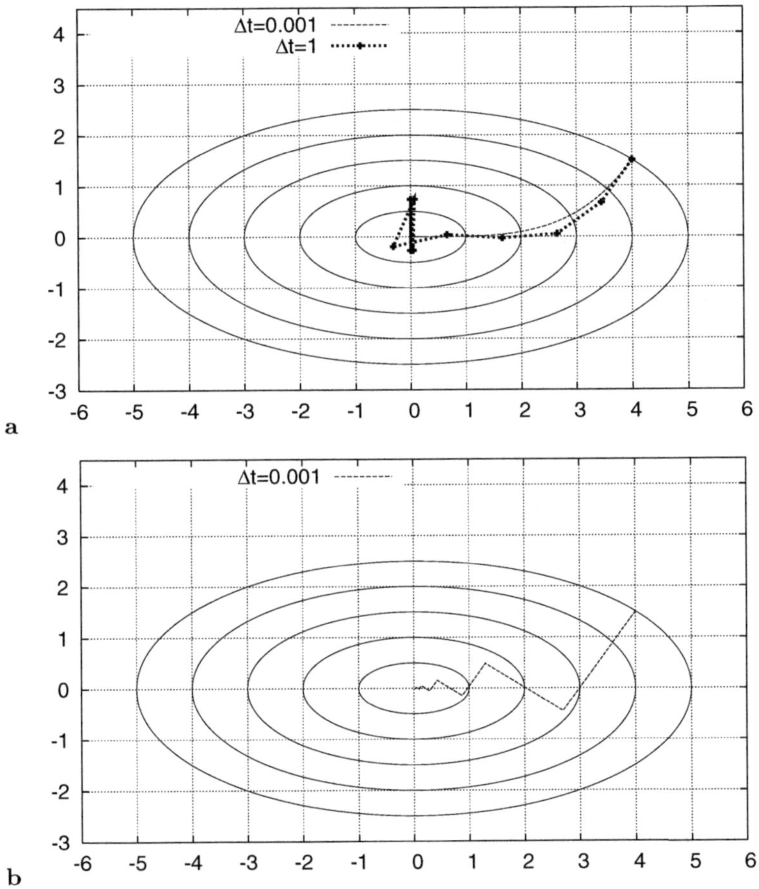

Fig. 1.3. (**a**) At each time step, the normalized gradient at the current point $\boldsymbol{r}(t)$ is determined. Then the system moves to a new point $\boldsymbol{r}(t+1) = \boldsymbol{r}(t) + \mathrm{d}\boldsymbol{r}(t)/\mathrm{d}t \times \Delta t$. The two curves are shown for $\Delta t = 1$ and $\Delta t = 0.001$. Finally it freezes near the origin or performs periodic jumps if Δt is chosen too large. A maximum of 10,000 time steps is used. (**b**) Initially, the gradient at the initial point is calculated. Then the system moves in the direction of this gradient over several time steps, thus reducing the energy value gradually until a deterioration occurs if the energy value is reduced even further. At this final point, the local gradient is evaluated. Then the system moves in the direction of this new gradient, again until some deterioration occurs if it moves even further. This approach is repeated up to 10,000 time steps

if the minimum we just lavished such care on reaching is not even the lowest of them? We will return to the problem of functions with many local minima in subsequent chapters. They are not at all rare.

These two approaches, bracketing and interpolation, extend to more dimensions, although bracketing is more difficult to use in higher dimensions than interpolation. Pictures are instructive in two dimensions, so consider the function $\mathcal{H}(x)$ whose contour lines are shown as Fig. 1.3.

1.3 Descent Methods in More Than One Dimension

The 1D optimization methods, bracketing and Newton's method, are a valuable building block for optimization of functions in more than one dimension. As before, the calculation of a local gradient may be easy, or relatively expensive. Pictures are instructive in two dimensions, although the problems are richer as the number of dimensions increases further. Consider minimizing the function $\mathcal{H}(x, y)$ with a single minimum whose contour lines are shown in Fig. 1.3:

$$\mathcal{H}(x, y) = x^2 + 4y^2 . \tag{1.4}$$

If evaluation of the local gradient of $\mathcal{H}$ is easy (as is certainly the case here, where $\boldsymbol{\nabla}\mathcal{H} = 2x\boldsymbol{e}_x + 8y\boldsymbol{e}_y$), we may find its minimum by taking many sufficiently small steps, each of them in the direction of the local gradient. This is shown in Fig. 1.3a by the solid line of steps of length $\Delta t = 0.001$: starting out at an initial point $\boldsymbol{r}(0) = (4, 1.5)$, we continue in discrete steps according to

$$\begin{aligned} \boldsymbol{r}(t+1) &= \boldsymbol{r}(t) + \frac{\mathrm{d}\boldsymbol{r}}{\mathrm{d}t}(t) \times \Delta t \\ &= \boldsymbol{r}(t) + \frac{\boldsymbol{\nabla}\mathcal{H}}{|\boldsymbol{\nabla}\mathcal{H}|}(t) \times \Delta t \\ &= \boldsymbol{r}(t) + \frac{x(t)\boldsymbol{e}_x + 4y(t)\boldsymbol{e}_y}{\sqrt{x^2(t) + 16y^2(t)}} \times \Delta t . \end{aligned} \tag{1.5}$$

While a 2D generalization of Newton's method would require far fewer steps, this approach requires absolutely minimal thought and programming effort, even with functions that are more contorted than our example. But suppose that we wish to avoid unnecessary evaluations of the gradient and as a result took longer steps. Figure 1.3a presents an extreme example. Using steps of length 1, we reach the vicinity of the minimum in six giant steps. Unfortunately, the search then oscillates helplessly across the innermost ellipse in the contour plot, so some more sophisticated routine must then be used to reduce the step length, or search more finely along our path. Our desire to speed things up has brought us back to the domain of 1D search and optimization. There are a large number of algorithms for selecting a sequence of 1D searches to find minima of a higher-dimensional function. For a thorough review, consult Press [166], Chap. 10. We will sketch here some of the basic ideas involved.

The immediate generalization of our downhill search to deal with functions in which gradient calculations are expensive is to determine a direction of search from one local gradient calculation and then search along that direction, moving in a straight line, until a local minimum is found. Then stop, calculate a new local gradient, and search along that line for a second local minimum. Repeat as needed until the gradient is as close to zero as desired. This common form of "steepest descents" is shown in the example of Fig. 1.3b. It requires roughly seven 1D searches to reach the minimum of $\mathcal{H}(x, y)$. If our function were even more anisotropic, or its valleys curved back and forth as our search went through them in descending order, we might find that the steepest descents involve many more crossings of the valley, and the method can become rather inefficient. Especially in higher dimensions, which are difficult to illustrate on 2D paper surfaces, the proliferation of 1D search directions can be quite wasteful.

1.4 Conjugate Gradients

A reasonable strategy for enforcing efficient use of 1D search in D-dimensional ($D > 1$) potentials is to require that each such search explores a previously unexplored direction. Naturally, we can explore only D orthogonal directions in a D-dimensional space, so we should take no more than D 1D search steps, or try to forget about previously explored directions in some graceful way every D searches. In two dimensions, the gradient is perpendicular to the direction of each 1D search when we reach a minimum, so that the right angle turns will constantly oscillate across the narrow valley that we are searching. In higher dimensions, the space orthogonal to a line search is much richer, so the concept of finding a "conjugate" gradient direction that is not exactly the local gradient becomes interesting. The trick in generating each search direction is to ensure that, while orthogonal to previous searches, it also points toward a reasonable estimate of the location of the minimum we are seeking. The Newton method, assuming that the function to be minimized is a quadratic form, is the basis for this. And since we choose our direction orthogonal to all previous directions, the search must stop in D iterations. The details of this approach to solving large systems of linear equations as well as minimizing multivariate functions are given in Press et al. [166] and in the original papers referenced therein.

This is a powerful method, but because it only works within one minimum, and the hard part of these problems, we believe, is finding the best minimum to work on, we will address that class of issues for the remainder of this book.

2 Exact Optimization Algorithms for Complex Problems

For complex problems, there are some rather general exact optimization methods guaranteed to lead to the optimum solution of the proposed problem. These will in general be appropriate only for modest-sized problems, but they serve to calibrate and to tune the heuristic methods that are the main subject of this book. In this chapter we shall describe several such methods, the simplex algorithm for linear cost functions over continuous and integer variables and the various branch & bound techniques for exhaustive, but pruned, search. We also refer the interested reader to several texts and articles cited for more detail.

2.1 Simplex Algorithm

The simplex algorithm is an exact method for solving linear problems of the following kind: the state of a system can be described with an N-dimensional vector $\boldsymbol{x}$ with real-valued components. The cost or objective function of such a linear problem is given by

$$\mathcal{H}(\boldsymbol{x}) = d_1 x_1 + d_2 x_2 + \cdots + d_N x_N + \vartheta = \boldsymbol{d} \circ \boldsymbol{x} + \vartheta \,, \qquad (2.1)$$

with $\boldsymbol{d} \in \mathrm{I\!R}^N$ and $\vartheta \in \mathrm{I\!R}$. As usual, this objective function is to be minimized. [With a linear function $f(\boldsymbol{x})$ to be maximized, simply set $\mathcal{H}(\boldsymbol{x}) = -f(\boldsymbol{x})$.] However, some constraints must be met, usually given as inequalities. The standard form of this algorithm demands

$$\boldsymbol{x} \geq \boldsymbol{0} \,, \qquad (2.2)$$

i. e., all components of vector $\boldsymbol{x}$ are restricted in their sign: $x_i \geq 0$. (If there is the constraint that $x_i \geq c$ and $c \neq 0$, then the variable x_i can easily be replaced by $x_i - c$ to obtain the standard form. Analogously, a variable $x_i \leq c$ can be replaced by $-x_i + c$.) According to this constraint in the sign, any positive d_i causes costs if the corresponding $x_i > 0$, while any negative d_i leads to gains. Furthermore, the following set of linear inequalities must be fulfilled:

$$\begin{aligned} a_{11}x_1 + \ a_{12}x_2 + \cdots + \ a_{1N}x_N &\leq b_1 \,, \\ a_{21}x_1 + \ a_{22}x_2 + \cdots + \ a_{2N}x_N &\leq b_2 \,, \\ &\vdots \\ a_{M1}x_1 + a_{M2}x_2 + \cdots + a_{MN}x_N &\leq b_M \,. \end{aligned} \qquad (2.3)$$

This is usually abbreviated to $A\boldsymbol{x} \leq \boldsymbol{b}$ with $A \in \mathbb{R}^{M \times N}$, $\boldsymbol{b} \in \mathbb{R}^{M}$, and $\boldsymbol{b} \geq \mathbf{0}$. One can also consider constraints given as equations, e. g., by rewriting them as two inequalities, instead of the "=" sign, once "$\leq$" and once "$\geq$" are used.

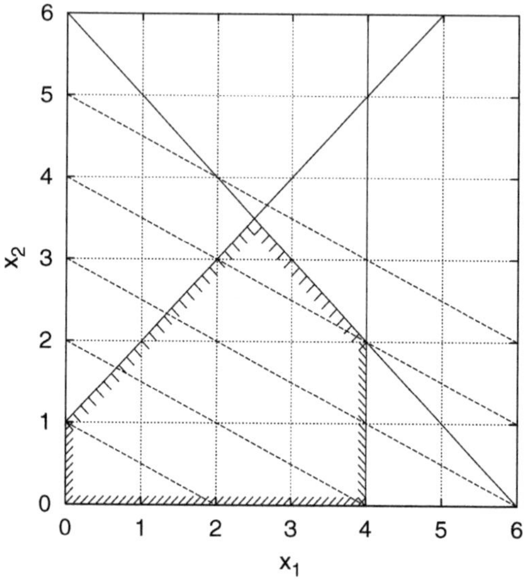

Fig. 2.1. Small example depicting an $N = 2$-dimensional linear optimization problem: the variables x_1 and x_2 must be nonnegative; furthermore the inequalities $x_1 \leq 4$, $x_1 + x_2 \leq 6$, and $-x_1 + x_2 \leq 1$ will be fulfilled. The *framed part* of the graphics shows the bounded region in which all feasible solutions lie. The cost function to be minimized shall be $\mathcal{H}(\boldsymbol{x}) = -x_1 - 2x_2$. The *parallel lines* that exhibit a gradient of $-\frac{1}{2}$ will demonstrate how this 2D problem can be solved in a geometric way by shifting this line through the convex polytope of the feasible solutions. One can easily imagine that the optimum solution must lie not inside this polytope but on one of its edges, in this example at $(x_1, x_2) = (2.5, 3.5)$. The simplex algorithm starts at the edge $(0, 0)$ in the lower left edge and jumps from edge to edge minimizing the cost function value in a way that the decrease of the cost function shall be as large as possible at each jump

In two dimensions, this problem can be solved geometrically, as demonstrated in Fig. 2.1: one first determines the polytope, i. e., the region of feasible solutions bounded by hyperplanes (which are straight lines in two dimensions) given by the inequalities. This polytope is always convex. Searching for the extreme value of the cost function, one simply shifts a straight line with the gradient $-d_1/d_2$ through the polytope of the feasible solutions. One sees that this objective function takes its extreme value on one edge of this polytope. There might be the case that the cost function has the same gradient as one of the boundary lines and that one ends up at this boundary line when searching for the optimum. In this case, the optimum is degenerate as both edges of the boundary line, and therefore all points on the boundary line, exhibit the same minimum value. The coordinates x_1 and x_2 of the optimum edge can be determined by either estimating them from the graphics or by solving the set of equations $a_{i1}x_1 + a_{i2}x_2 = b_i$ and $a_{j1}x_2 + a_{j2}x_2 = b_j$ given by the two border lines from inequalities i and j, which cut this edge. From the coordinates of the optimum solution the minimum objective value can be calculated. For higher dimensions, such a graphical solution is usually impossible. The simplex algorithm [136, 155] must be applied.

In the solution process of problems occurring in practice, one usually starts with a preprocessing phase in which inequalities that are automatically fulfilled by other inequalities are removed. Furthermore, given an inequality where a variable $x_i \leq 0$, this variable can be set to 0 due to the further constraint that it must be nonnegative. Thus, the cost function and the inequalities containing this variable can be simplified, decreasing the dimension of the problem.

The simplex algorithm itself starts with introducing so-called slack variables $x_{N+1}, \ldots, x_{N+M}$ with $x_{N+i} \geq 0 \; \forall 1 \leq i \leq M$ supplemental to the variables $x_1, \ldots, x_N$. With these slack variables, the set of inequalities can be transformed to a set of equations:

$$\begin{array}{rcrcccrcrcl} a_{11}x_1 &+& a_{12}x_2 &+& \cdots &+& a_{1N}x_N &+& x_{N+1} &=& b_1\,, \\ a_{21}x_1 &+& a_{22}x_2 &+& \cdots &+& a_{2N}x_N &+& x_{N+2} &=& b_2\,, \\ && && \vdots && && && \\ a_{M1}x_1 &+& a_{M2}x_2 &+& \cdots &+& a_{MN}x_N &+& x_{N+M} &=& b_M\,. \end{array} \tag{2.4}$$

This can also be written in the compressed form

$$\left(A | I_M \right) \begin{pmatrix} x_1 \\ \vdots \\ x_N \\ \hline x_{N+1} \\ \vdots \\ x_{N+M} \end{pmatrix} = \boldsymbol{b}, \tag{2.5}$$

with I_M being the $M \times M$ identity matrix.

Furthermore, an index array σ is introduced that is initialized with $\sigma(i) = i$ for all $1 \leq i \leq N+M$. As the column vectors of the identity matrix I_M form the basis of the $\mathbb{R}^M$, the corresponding new components $x_{N+1}, \ldots, x_{N+M}$ of vector $\boldsymbol{x}$ are called the basis variables and the old components $x_1, \ldots, x_N$ are the nonbasis variables.

Now an iterative scheme is started in which the entries of vectors $\boldsymbol{d}$ and $\boldsymbol{b}$, of the matrix A, of the constant ϑ, and of the index array σ are changed:

- If $\boldsymbol{d} \geq \boldsymbol{0}$, then the simplex algorithm has already reached the optimum such that it can be stopped.

(If this is already the case at the very beginning, then all d_i would incur costs if the corresponding x_i were larger than zero such that the point $\boldsymbol{x} = 0$ is the optimum solution.)
If $\boldsymbol{d} > \boldsymbol{0}$, then the optimum is even nonambiguous.
Otherwise, usually the so-called pivot column s of the matrix A is chosen according to

$$d_s = \min_j \{d_j\} < 0 \,. \tag{2.6}$$

- Then it is checked whether all $a_{is} \leq 0$. If so, then there exists no minimum, as the objective function is not limited to small values.
- Otherwise, the so-called pivot line r is chosen according to

$$\frac{b_r}{a_{rs}} = \min \left\{ \frac{b_i}{a_{is}} \middle| a_{is} > 0 \right\} . \tag{2.7}$$

- The following settings are made:

$$(a_{ij})_{\text{new}} = \begin{cases} a_{ij} - \dfrac{a_{is}a_{rj}}{a_{rs}} & \text{if } i \neq r \text{ and } j \neq s \,, \\ \dfrac{1}{a_{rs}} & \text{if } i = r \text{ and } j = s \,, \\ -\dfrac{a_{is}}{a_{rs}} & \text{if } i \neq r \text{ and } j = s \,, \\ \dfrac{a_{rj}}{a_{rs}} & \text{if } i = r \text{ and } j \neq s \,, \end{cases}$$

$$(b_i)_{\text{new}} = \begin{cases} b_i - \dfrac{b_r a_{is}}{a_{rs}} & \text{if } i \neq r \,, \\ \dfrac{b_r}{a_{rs}} & \text{if } i = r \,, \end{cases} \tag{2.8}$$

$$(d_j)_{\text{new}} = \begin{cases} d_j - \dfrac{d_s a_{rj}}{a_{rs}} & \text{if } j \neq s \,, \\ -\dfrac{d_s}{a_{rs}} & \text{if } j = s \,, \end{cases}$$

$$(\vartheta)_{\text{new}} = \vartheta + \frac{d_s b_r}{a_{rs}} \,.$$

- Finally, the old values are replaced by the new values and $\sigma(N + r)$ is exchanged with $\sigma(s)$ in order to indicate that the basis variable $x_{\sigma(N+r)}$ has been exchanged with the nonbasis variable $x_{\sigma(s)}$.
Then the algorithm jumps back to the first step.

Ultimately, ϑ contains the optimum value. The corresponding values for the components of $\boldsymbol{x}$ are stored in $\boldsymbol{b}$:

$$x_{\sigma(i)} = \begin{cases} 0 & \text{if } 1 \leq i \leq N \,, \\ b_{i-N} & \text{if } N + 1 \leq i \leq N + M \end{cases} . \tag{2.9}$$

The values for the slack variables determine how much space would be left according to the corresponding inequality. Incidentally, similarly one can check during execution of the simplex algorithm which edge of the polytope the current solution is at.

When one searches for the maximum instead of the minimum of a linear function $\boldsymbol{d} \circ \boldsymbol{x} + \vartheta$ under the constraints $\boldsymbol{x} \geq \boldsymbol{0}$ and $A\boldsymbol{x} \leq \boldsymbol{b}$, nearly the same algorithm can be applied. Thc above scheme must be changed only at the following point:

- If $\boldsymbol{d} \leq \boldsymbol{0}$, then the simplex algorithm has already reached the optimum; if $\boldsymbol{d} < \boldsymbol{0}$, the optimum is nonambiguous. The rule (2.6) for choosing the pivot column s is changed to

$$d_s = \max_j \{d_j\} > 0 \,. \tag{2.10}$$

The simplex algorithm is constructed in such a way that it will lead as fast as possible to the optimum value. The choice of the position of the pivot element a_{rs} is done in such a way that the greatest of all possible improvements is guaranteed. Thus, the number of steps the simplex algorithm needs is usually reduced. However, one can construct instances in which this type of pivoting leads to an endless cycling over two or more edges. If this happens, the problem can be solved by selecting not the optimum pivoting indices r and s but the minimum or maximum ones that consider the constraints of $d_s < 0$ and $a_{rs} > 0$. Furthermore, other strategies might lead to a faster convergence to the optimum. However, the simplex algorithm in its proposed form is optimal for most problems.

In practical applications, usually not all constraints are implemented explicitly. Typically, some important inequalities are chosen at the beginning. The achieved solution is checked to see if it fulfills the other constraints. Inequalities that are not fulfilled are introduced step by step. A simplex run is performed at each step. One sometimes investigates what the minimum cost of additional constraints is, which are nice to fulfill but not necessary.

In performing the simplex algorithm manually, it is quite useful to use a so-called simplex tableau in which all values are written, e. g.:

$$\begin{array}{c|ccccc|c|c}
 & \sigma(1) & \dots & \sigma(s) & \dots & \sigma(N) & & \\
\hline
\sigma(N+1) & a_{11} & \dots & a_{1s} & \dots & a_{1N} & b_1 & b_1/a_{1s} \\
\vdots & \vdots & & \vdots & & \vdots & \vdots & \vdots \\
\sigma(N+r) & a_{r1} & \dots & \boxed{a_{rs}} & \dots & a_{rN} & b_r & b_r/a_{rs} \\
\vdots & \vdots & & \vdots & & \vdots & \vdots & \vdots \\
\sigma(N+M) & a_{M1} & \dots & a_{Ms} & \dots & a_{MN} & b_M & b_M/a_{Ms} \\
\hline
 & d_1 & \dots & d_s & \dots & d_N & \vartheta &
\end{array} \tag{2.11}$$

The small example of Fig. 2.1 can be solved by this method. The initial tableau is given as

	1	2		
3	1	0	4	
4	1	1	6	6
5	−1	$\boxed{1}$	1	1
	−1	−2	0	

The pivot element, chosen with $s = 2$ and $r = 3$, is marked by a frame. In the intermediate tableau

	1	5		
3	1	0	4	4
4	$\boxed{2}$	−1	5	$\frac{5}{2}$
2	−1	1	1	
	−3	2	−2	

$\sigma(2)$ and $\sigma(2+3)$ are exchanged and the new values for a_{ij}, b_i, d_j, and ϑ are calculated. The new pivot element is again determined by selecting the minimum component of vector $\boldsymbol{d}$ and the minimum of the fractions b_i/a_{is} for all $a_{is} > 0$. With the pivot element $a_{21} = 2$, $\sigma(1)$ and $\sigma(2+2)$ are exchanged and the final tableau is calculated:

	4	5	
3	−0.5	0.5	1.5
1	0.5	−0.5	2.5
2	0.5	0.5	3.5
	1.5	0.5	−9.5

One gets the solution point $(x_1, x_2) = (b_2, b_3) = (2.5, 3.5)$, the values $x_3 = b_1 = 1.5$ and $x_4 = x_5 = 0$ for the slack variables, and the optimum value $\vartheta = -9.5$. Thus the simplex algorithm jumps from the initial point $(x_1, x_2) = (0, 0)$ via the intermediate edge point $(x_1, x_2) = (0, 1)$ to the final point $(x_1, x_2) = (2.5, 3.5)$ in only two steps, which is the minimum amount of jumps needed, as Fig. 2.1 shows.

2.2 Integer Optimization

Integer optimization problems have in common that many or even all x_i must be integers or are otherwise only allowed to take values from a discrete set of numbers. (As is common in the literature, henceforth we will only speak of integer values.) Usually, such problems can only be solved exactly with a large amount of calculation time, which increases exponentially with system size.

A special case of this class of problems is integer linear optimization problems, which, just as above, are given by a linear cost function $\mathcal{H}(\boldsymbol{x}) = \boldsymbol{d} \circ \boldsymbol{x} + \vartheta$ and an inequality set $A\boldsymbol{x} \leq \boldsymbol{b}$, $\boldsymbol{x} \geq \boldsymbol{0}$, and $\boldsymbol{b} \geq \boldsymbol{0}$. But there is the additional restriction that some or even all of the x_i are noncontinuous but can only take discrete values, e. g., must be integers. For solving this type of problem, one usually starts with a so-called relaxation of the problem in such a way that all x_i can take any real value, so that the simplex algorithm can be applied. If the simplex algorithm finds that there is no solution to this relaxed problem, then there is also no solution to the integer problem. Otherwise, the solution to the relaxed problem is examined. Sometimes one is lucky, such that all x_i meet the desired constraints and are already integers or they can at least simply be rounded up to the next highest integer or down to the next smallest integer, thus also ending up at a feasible solution. However, this rounding idea can mostly only be used if the entries in vector $\boldsymbol{x}$ are rather large; otherwise the probability for choosing a solution much worse than the optimum is too large.

The Gomory algorithm and related algorithms follow another approach for solving such integer linear problems [63, 178]: after solving the relaxed problem with the simplex algorithm, it is checked whether there are some real-valued components of the solution vector that should be integers. In this case, a further restriction is introduced in such a way that

- The resulting configuration of the simplex algorithm performed becomes a nonfeasible solution and
- Each feasible integer solution of the linear integer problem fulfills this new restriction.

The new restriction is added to the optimization problem as an inequality such that the dimension of the optimization problem is enlarged ($M \to M+1$). This enlarged problem is again solved in its relaxed form by the simplex algorithm, which leads to a new solution. This solution is again checked to verify whether it meets all integer constraints. If not, the algorithm is iterated until it ends at a feasible solution with integer components.

As further cutting hyperplanes are introduced in the N-dimensional space in this algorithm, which cut off the current nonfeasible solution, one speaks of a cutting plane algorithm. Usually, only a small amount of the polytope is cut off by an added hyperplane using this Gomory approach such that this kind of algorithm converges very slowly. However, one can prove that the algorithm ends after a finite amount of iterations is performed in a certain way.

2.3 Branch & Bound

Most combinatorial optimization problems, i. e., optimization problems in which the variables must take values from a discrete set of values, are not given as linear problems. If the simplex algorithm or other algorithms for some special problems cannot be applied (e. g., because too many variables must be introduced to formulate the problem in a linear way such that the system size is too large for the memory or the calculation time needed), the only way to arrive at the optimum solution to the proposed problem in a mathematically proved way consists in searching for this optimum through the set of feasible solutions. The simplest search strategy is an exact enumeration, i. e., all possible configurations σ are visited, their corresponding cost values $\mathcal{H}(\sigma)$ are calculated, and the configuration σ_0 with the minimum cost value is returned as the result of this search. However, this approach can only be used for usually very small system sizes, as the calculation time needed increases exponentially with system size.

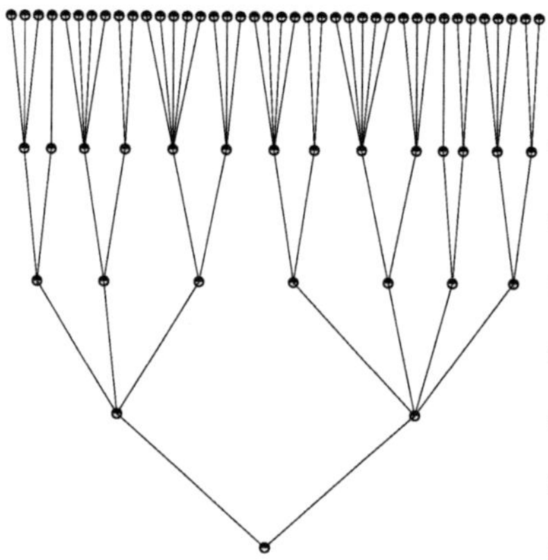

Fig. 2.2. Schematic search tree for the branch & bound method: the configurations are ordered as the leaves on top of a search tree that splits into various branches for each considered property of the configurations

The time for this search process can be drastically reduced if the search process is performed in a more intelligent way by creating an appropriate search tree: starting at a root node, the tree divides into two or more branches, which again split into several branches, which again and again split until the tree ends in the single leaves that are formed by the configurations, as shown in Fig. 2.2. The branching is performed in such a way that the leaves of one subbranch have one or more properties in common that the configurations of all other subbranches at this branching point do not have. At each branching layer, the splitting of the individual branches is usually performed according to the same property of the configurations. For example, such a splitting at a layer a can mean that a certain variable x_i is set to one of its possible values $v_{i1}, \dots, v_{in}$; these values are used instead of the variable already at the nodes in the next layer $a + 1$. Usually, not just one branch is split into subbranches from layer a to $a + 1$ according to the possible values of the variable x_i, but all branches are split to the possible values of exactly this variable between layers a and $a + 1$. However, it might be that the variable can have, e. g., five possible values in one branch and only three possible values in another branch if the number of possible values of the variable x_i depends on the values of other variables that had been set before. In such a case, one does not get a symmetric search tree.

Till now, the search process has only been ordered according to the properties used for the branching process. The nodes at the branching points already contain information about the properties of all configurations of all outgoing subbranches, such that it is often possible to estimate a lower bound for the cost function values of all these configurations by examining only the node at the branching point. Sometimes some (rather) good solution of the proposed optimization problem is already known; otherwise some heuristic is applied in order to achieve a good solution in a short time. At the very least, the objective function of the first configuration of the search process is calculated. The objective function of such a configuration provides a first upper bound for the optimum value. During the search process, better configurations might be found, such that this global upper bound can be stepwise decreased as it is given by the optimum value found so far. If the global upper bound is smaller than the local lower bound at a specific branching node, then it is impossible for any configuration in one of the subbranches going out from this node to be the optimum configuration. Therefore, all of the outgoing subbranches can be "cut off." As the configurations on them need not be visited, the search process is sped up considerably. This branch & bound method ends when the lower and upper bounds coincide at the optimum configuration.

The speedup of this method strongly depends on the properties used for branching and on the quality of the bounding functions, especially for the lower bound. Sometimes even local upper bounds are introduced to guide the search process as to which branch should be examined first. Above all, detailed mathematical insight into the proposed optimization problem is needed to determine appropriate branching properties and good lower-bound functions. Sometimes it is better to work with a "wide tree", i. e., a tree with many branches going out from each node; for other cases, it is advantageous to work with a "deep tree", i. e., a tree that splits into only a few branches at each node such that more branching layers are needed. A special case of a deep tree is the binary tree in which each branch splits into two subbranches.

A further question of this search process relates to the bound functions used. Elaborate bounding functions take some time to calculate the bounding value. Therefore, one often must make some compromise between a simple but fast bounding function that provides worse bounding values such that the nodes in the next branching layer at least partially must be visited and examined, whereas a better bounding function, which needs more calculation time, would have been able to cut off all subbranches.

The success of this exact branch & bound method furthermore depends on the quality of already known good solutions: the best known solution (usually created using good heuristics) serves as an upper bound for the optimum value. However, parts of other good solutions can also be used for local lower and upper bounds, if their properties agree with the properties of the corresponding subbranch.

Summarizing, this exact branch & bound algorithm is an intelligent search technique in which branches of the search tree can be cut off if the lower bound function value for the subtree is larger than the global upper bound. Therefore, the algorithm makes it possible to solve larger instances of a given problem exactly than is possible with a complete search. However, the calculation time of this algorithm still increases exponentially with the system size such that the problem of not being able to solve an instance exactly is simply shifted to larger instance sizes.

2.4 Branch & Cut

Currently the most powerful exact algorithm for problems that can be formulated in a linear way without increasing the system size too much and in which some or all of the variables can only take values from a discrete set is the branch & cut algorithm, which combines the branch & bound algorithm with the cutting plane idea from integer programming. Again the problem of minimizing a linear cost function $\mathcal{H}(\boldsymbol{x}) = \boldsymbol{d} \circ \boldsymbol{x} + \vartheta$ with the set of constraints in the form $A\boldsymbol{x} \leq \boldsymbol{b}$, $\boldsymbol{b} \geq \boldsymbol{0}$, and $\boldsymbol{x} \geq \boldsymbol{0}$ is considered.

Let us first restrict ourselves to the case where all x_i are only allowed to take discrete values and let us consider only the feasible points inside the polytope a formed by the inequality set $A\boldsymbol{x} \leq \boldsymbol{b}$ and $\boldsymbol{x} \geq \boldsymbol{0}$. Then the convex hull of all feasible points inside a forms a polytope c that is smaller than the polytope a and that can be described by an inequality set $C\boldsymbol{x} \leq \boldsymbol{d}$ and $\boldsymbol{x} \geq \boldsymbol{0}$. (This inequality system can be derived with the Fourier–Motzkin elimination or the double description method. See [158] or [211].) Determining the convex hull of all feasible points inside polytope a leads to a new linear optimization problem with the former cost function but with a new inequality set, which restricts the search process for the desired optimum solution much more strongly. For this reason and as polytope c is smaller than polytope a and really bounded by the extreme discrete points among which the optimum solution lies, the inequalities $C\boldsymbol{x} \leq \boldsymbol{d}$ are called strong inequalities.

However, the number of these strong inequalities is usually much larger than M. Furthermore, it usually takes too much calculation time to determine all strong inequalities, i. e., the boundaries of the convex hull of the points. Therefore, this approach is seldom used. Instead, a sequence of convex polytopes $s_0, s_1, s_2, \dots$ with $a = s_0 \supset s_1 \supset s_2 \supset \cdots \supset c$ is constructed by successively adding strong inequalities to the already existing set of inequalities, thus shrinking the original polytope a stepwise into c. For each step i, an optimization run with the simplex algorithm is performed that searches for the optimum of the relaxed linear problem in polytope s_i. If the optimum found meets all desired constraints not yet incorporated, then the algorithm stops. Otherwise one or more strong inequalities are added to matrix A and the algorithm is iterated. There is also another possibility in this iteration: instead of trying a further simplex step, a branch & bound step can be performed: one of the variables x_i that is only allowed to take some discrete values can be chosen for a branching step. For each branch, the variable is set to one of its possible values. Thus, the optimization problem can be partially rewritten by inserting the value of the variable. For each of the newly created smaller optimization problems, a simplex run is performed. This branching is especially useful if the number of possible values is very small, e. g., if x_i is a binary variable and can only take the values 0 and 1. Of course, this algorithm can easily be generalized to problems in which some of the variables are continuous. However, for such a continuous x_i no branching step in the above sense can be performed; instead one can split the interval of possible values into subintervals.

To demonstrate the usage of this branch & cut algorithm, we first reconsider our problem of Fig. 2.1, for which we now want to achieve a solution with x_1 and x_2 as integers. As the simplex algorithm ends with the solution $(x_1, x_2) = (2.5, 3.5)$, which is nonfeasible in this case, it is necessary to cut off this solution by introducing at least one further hyperplane, which is defined via an inequality and belongs to the convex hull of all integer solutions. If we determine the convex hull of all integer solutions in Fig. 2.1, we find at first glance that all of the inequalities that are already in the inequality set are strong inequalities, i. e., are part of the convex hull, and that only the inequality $x_2 \leq 3$ must be added to the inequality set in order to close the convex hull. For this extended problem, again a simplex run is performed:

	1	2		
3	1	0	4	
4	1	1	6	6
5	−1	[1]	1	1
6	0	1	3	3
	−1	−2	0	

$\longrightarrow$

	1	5		
3	1	0	4	4
4	2	−1	5	$\frac{5}{2}$
2	−1	1	1	
6	[1]	−1	2	2
	−3	2	−2	

$\longrightarrow$

	6	5		
3	−1	1	2	2
4	−2	[1]	1	1
2	1	0	3	
1	1	−1	2	
	3	−1	−8	

$\longrightarrow$

	6	4		
3	1	−1	1	
5	−2	1	1	
2	1	0	3	
1	−1	1	3	
	1	1	−9	

After three steps, we achieve the optimum integer solution $(x_1, x_2) = (3, 3)$ with the value -9. The slack variables have the values $x_3 = 1$, $x_4 = 0$, $x_5 = 1$, and $x_6 = 0$.

Finally, we consider the solution to a famous old problem, the propositio de lupo et capra et fasciculo cauli (the problem of the wolf, the goat, and the bunch of cabbage) which was, among others, introduced around 800A.D. by Flaccus Alcuinus of York, who was a royal advisor to Charlemagne at the Frankish court and head of Charlemagne's Palace School at Aachen. The problem is stated as follows: Homo quidam debebat ultra fluvium transferre lupum et capram et fasciculum cauli, et non potuit aliam navem invenire, nisi quae duos tantum ex ipsis ferre valebat. Praeceptum itaque ei fuerat, ut omnia haec ultra omnino illaesa transferret. Dicat, qui potest, quomodo eos illaesos ultra transferre potuit.

Thus, a man had to transfer a wolf, a goat, and a bunch of cabbages across a river and he could not find another boat except one that could only carry two of them. Thus, the constraint for him was how to transfer all these without any damage. However, if the wolf stays alone with the goat, then the wolf will kill and eat the goat. Similarly, the goat will eat the cabbage if not supervised. However, the wolf is not interested in the cabbage. The question is how to safely transfer all three across the river [6]. Of course, the transfer should take a minimum amount of time. Alcuin himself provided the solution to this problem No. 18: Simili namque tenore ducerem prius capram et dimitterem foris lupum et caulum. Tum deinde venirem lupumque ultra transferrem, lupoque foras misso rursus capram navi receptam ultra reducerem, capraque foras missa caulum transveherem ultra, atque iterum remigassem, capramque assumptam ultra duxissem. Sicque faciente facta erit remigatio salubris absque voragine lacerationis.

Thus, Alcuin would take the goat across first and leave the wolf and the cabbage. Then he would return and take the wolf across the river. Then, after setting the wolf down on the other side, he would take the goat back across the river to the original side. Then, with the goat on the other side, Alcuin would take the cabbage across the river. Finally, he would go back for the goat one more time and take it across the river. In this way, the wolf, the goat, and the cabbage would all be transported across the river safely. This is one of two possible optimum solutions. Alcuin arrived at this solution but was not able to prove that his solution of crossing the river seven times was optimal.

This proof was performed 1200 years later by Borndörfer, Grötschel, and Löbel [25]. They first formulated the problem introducing vectors $\boldsymbol{l}(t)$, $\boldsymbol{b}(t)$, and $\boldsymbol{r}(t)$ for the left bank of the river, the boat, and the right bank, respectively, for each time step t. These vectors are triples of binary numbers for the locations of the wolf at the first position, for the goat at the second, and for the cabbage at the third position. Thus, in the beginning at time $t = 0$, one obtains the initial vectors $\boldsymbol{l}(0) = (1,1,1)$, $\boldsymbol{b}(0) = (0,0,0)$, and $\boldsymbol{r}(0) = (0,0,0)$. On the other end, there must be some final time t_f at which all three items must be on the right bank, $\boldsymbol{r}(t_f) = (1,1,1)$. Then one must distinguish between even time steps at which the traveler starts out on the the left bank and odd time steps in which he starts out on the opposite side:

$$\left.\begin{array}{l} \boldsymbol{l}(t+1) = \boldsymbol{l}(t) - \boldsymbol{b}(t+1) \\ \boldsymbol{r}(t+1) = \boldsymbol{r}(t) + \boldsymbol{b}(t+1) \end{array}\right\} \text{ if } t \geq 0 \text{ even};$$

$$\left.\begin{array}{l} \boldsymbol{l}(t+1) = \boldsymbol{l}(t) + \boldsymbol{b}(t+1) \\ \boldsymbol{r}(t+1) = \boldsymbol{r}(t) - \boldsymbol{b}(t+1) \end{array}\right\} \text{ if } t \geq 0 \text{ odd}.$$

One must take care that the goat must not stay alone with the wolf or with the cabbage. For example, at odd time steps, when the traveler is on the right bank, vector $\boldsymbol{l}$ can only take one of the five values $(0,0,0)$, $(1,0,0)$, $(0,1,0)$, $(0,0,1)$, or $(1,0,1)$; the other three bit combinations are forbidden. From this, one can estimate an upper bound T for the last time step t_f: if there are only five possible values for $\boldsymbol{l}$ for each odd time step, there can be a maximum of only nine rowing steps, as otherwise some steps will be repeated, such that $T = 9$.

Analogously, one must restrict the possible values for $\boldsymbol{r}(t)$ for the even time steps when the traveler is at the left bank to the same values as $\boldsymbol{l}(t)$ for the odd time steps. However, one must furthermore allow the desired final value $(1,1,1)$ for both odd and even time steps. Analogously, $\boldsymbol{b}(t)$ can only take the values $(1,0,0)$, $(0,1,0)$, $(0,0,1)$ for transporting one of the three items and $(0,0,0)$ for rowing an empty boat for all time steps.

Furthermore, the scientists introduced a cost function to be minimized, namely, summing up the number of items remaining on the left bank over the time steps:

$$\mathcal{H}((\boldsymbol{l}(t), \boldsymbol{b}(t), \boldsymbol{r}(t))_{0 \leq t \leq T}) = \sum_{t=0}^{T} \sum_{i=1}^{3} l(t,i) \,.$$

Thus, the Hamiltonian does not depend on all $3 \times 3 \times 10 = 90$ variables but only on the 30 describing the states on the left bank over the time period.

The solution by means of branch & cut starts with determing the convex hull of all allowed vectors for $\boldsymbol{l}(t)$, $\boldsymbol{b}(t)$, and $\boldsymbol{r}(t)$. As can easily be checked, the cost function above must be minimized subject to the following constraints, if relaxing the linear integer problem to a linear problem [25]:

$$\begin{aligned} \boldsymbol{l}(0) &= (1,1,1)\,, \\ \boldsymbol{b}(0) &= (0,0,0)\,, \\ \boldsymbol{r}(0) &= (0,0,0)\,; \end{aligned}$$

$$\left.\begin{aligned} \boldsymbol{l}(t+1) &= \boldsymbol{l}(t) - \boldsymbol{b}(t+1) \\ \boldsymbol{r}(t+1) &= \boldsymbol{r}(t) + \boldsymbol{b}(t+1) \end{aligned}\right\} \text{ if } 0 \leq t \leq T \text{ even}\,,$$

$$\left.\begin{aligned} \boldsymbol{l}(t+1) &= \boldsymbol{l}(t) + \boldsymbol{b}(t+1) \\ \boldsymbol{r}(t+1) &= \boldsymbol{r}(t) - \boldsymbol{b}(t+1) \end{aligned}\right\} \text{ if } 0 \leq t \leq T \text{ odd}\,,$$

$$\left.\begin{aligned} l(t,1) + l(t,2) \qquad\qquad &\leq 1 \\ l(t,2) + l(t,3) &\leq 1 \end{aligned}\right\} \text{ if } 0 \leq t \leq T \text{ odd}\,,$$

$$\left.\begin{aligned} -\,r(t,1) + r(t,2) + r(t,3) &\leq 1 \\ r(t,1) + r(t,2) - r(t,3) &\leq 1 \end{aligned}\right\} \text{ if } 0 \leq t \leq T \text{ even}\,,$$

$$b(t,1) + b(t,2) + b(t,3) \leq 1 \text{ for all } 0 \leq t \leq T\,,$$

$$\boldsymbol{r}(T) = (1, 1, 1)\,,$$

$$0 \le l(t,i), b(t,i), r(t,i) \le 1 \text{ for all } 0 \le t \le T, 1 \le i \le 3\,.$$

However, in solving this relaxed linear problem one achieves the following solution: the man first transports the goat to the right bank, returns, then transfers half the wolf and half the cabbage from left to right, returns, and, finally, transports the remaining halves of the wolf and of the cabbage to the right. This solution, which only needs $t_f = 5$ times of rowing, does not violate any of the inequalities; however, the objective was to transfer all the items to the other bank in good condition.

As mentioned above, there are now two ways to overcome the problem of not achieving a desired integral solution: one can either add at least one more equation or inequality, thus introducing one or more new hyperplanes and cutting off the current solution. For this, it is necessary to determine—at least partially—the convex hull of all integer solutions of the problem and to select equations and inequalities that are not fulfilled by the current solution. Adding the equation $r(4,1)+r(4,2)+r(4,3) = 1$ and the inequality $b(3,2)+b(4,1)+b(4,3)+b(5,2)-b(6,2) \le 0$, one ends up at Alcuin's solution [25], which is thus proved to be optimal.

For this small problem, it is still a relatively easy task to determine the convex hull of the 20 integral solutions, described by only 16 inequalities and 79 equations. In general, however, this approach requires too much computational effort as the number of inequalities increases exponentially. Furthermore, there is no guarantee of ending up at an integral optimum solution in general [25].

Therefore, it is advantageous to work with the second possibility and to introduce a branching step, splitting the problem into two subproblems: the first wrong part of the solution was transporting half a wolf and half a cabbage. Thus, one forces the traveler to leave either the wolf or the cabbage behind in time step No. 3, i. e., either $l(3,1) = 1$ or $l(3,1) = 0$. Adding one of these two equations to the relaxed problem leads to two relaxed subproblems that are independently solved. The branch with the subproblem extended with $l(3,1) = 0$ actually ends up at an integer solution with a value of 12. Thus this branch need not be considered any further. As all constraints are met in this solution, the value 12 for this solution provides a global upper bound for the branch & cut algorithm.

Following the other branch could lead to a better solution such that the linear problem extended with $l(3,1) = 1$ must also be solved. In this case, one achieves a fractional solution with a value of 12. If one added further restrictions in a cutting or branching step, one would end up at a solution with a value not less than 12. Therefore, the value of 12 for this fractional solution serves as a local lower bound for all solutions in this branch of the search tree. But there is also the global upper bound of 12; thus lower and upper bounds coincide here already. As a solution with a value of 12 has already been found, one can stop the algorithm here. (Of course, if proceeding one could detect the other optimum solution.) All in all, the integer solution of the first branch is optimal and, as it is equal to Alcuin's solution, it is also proved that Alcuin's solution is optimal.

The art of using the branch & cut algorithm is to determine which are the important strong inequalities to be added first and when a branching step or a plane cutting step should be performed. For an example of an even more difficult problem see [126].

3 Monte Carlo

3.1 Pseudorandom Numbers

As most of the algorithms we introduce later include some randomization in their rules, we start with the question of how random events can be produced on a computer. After all, a computer is a machine working off a sequence of instructions in a deterministic way. A computer is therefore unable to produce any real random event. However, there has long been the need for simulating random events on a computer. (Think, e. g., of game simulations like roulette, blackjack, or poker.)

Therefore, many algorithms have been developed that produce a sequence of numbers that appears to be a sequence of random numbers. These so-called pseudorandom numbers are used instead of real random numbers in order to decide about the results of some sequence of events. For any unknowing spectator, the sequence of pseudorandom events should look like real random events.

As the city of Monte Carlo has been famous for decades for its casinos with roulette tables and many other games with random results, algorithms that make use of pseudorandom numbers are generally called Monte Carlo (MC) algorithms. The name inspired by the casino at Monte Carlo was first used by von Neumann and Ulam, when they used random numbers for simulating nuclear reactions in developing the atomic bomb. Usually, one speaks of random numbers instead of pseudorandom numbers.

It is worth mentioning here that the possible amount of generable pseudorandom numbers is always finite. After a certain number of calls the sequence of random numbers that has already been produced is repeated again. The larger this sequence length is, the better the random number generator should be. Random number generators always need at least one integer, an initial value x_0, called the seed, to get started. Different seeds do not usually lead to different sequences of random numbers, but the random number generator starts at different points in its finite sequence of random numbers.

The type of problem determines whether the knowledge of the seed, and therefore of the random numbers, is important or not. Sometimes one keeps a record of the seed in order to be able to reproduce the results of a MC program. The seed may be either encoded in the program or otherwise saved.

This is important, e. g., when trying to set new world records for some benchmark problems, because only if the seed is known can the result be reproduced, providing the proof that it was that program that generated the new record. However, in programming an application like a poker game, one would like to start every time with a new seed, one that cannot be controlled by the players. Here often the exact time of day, perhaps including even the milliseconds, is used as a seed.

Although "Monte Carlo" is commonly used to describe any randomized algorithm, some researchers distinguish between MC and Las Vegas (LV) algorithms. In contrast to MC algorithms, LV algorithms always lead to a correct result, whereas MC algorithms lead to the correct result only with a certain probability. A further difference is that MC algorithms have a deterministic running time whereas the running time of LV algorithms varies according to the random events. It it is also possible for a LV algorithm not to terminate. A subset of these LV algorithms are the Macao algorithms, which have a deterministic running time and which are guaranteed to terminate. In summary, different random seeds often lead to different results for a MC algorithm, such that its outcome is random, but a LV algorithm always produces the same result. The seed only influences the way to get there and therefore the running time [151].

3.2 Random Number Generation and Random Number Tests

Various algorithms leading to random numbers and tests checking the quality of such random numbers for their "randomness" have been developed. That is, the tests compare the results of using pseudorandom numbers instead of truly random numbers in the specific problems for which they are intended. The so-called random number generators are distinguished in different classes: "good" random numbers pass most of these elaborate tests. Some generators produce random numbers of a low quality but with low computational overhead. The resulting "quick and dirty" random numbers do not pass all tests checking their randomness. However, they should pass at least some basic tests to be useful for some MC techniques.

There are several ways to produce "quick and dirty" random numbers. A common way is to use the linear congruential method [128]: starting from the seed value x_0 the following iterative rule is applied to produce a sequence of random numbers:

$$x_{i+1} = (a \times x_i + b) \bmod c\,. \tag{3.1}$$

The parameters a, b, c, which are integers, define the random number generator. Each generated random number only depends on its predecessor. As the random numbers are calculated modulo the integer number c, they can only take the values $0, 1, \ldots, c-1$. The basic sequence of the random number generator can therefore only have a length of up to c; after that the sequence repeats. Therefore, the chosen parameter c must be rather large. (You can also get longer sequences if you keep more previous states, see, e. g., the R250 random number generator [115].) One can also accelerate the speed of the random number generator: as the modulo operation takes more time than addition or multiplication, one wants to avoid performing this modulo operation. This can be done by making use of the fact that integers are stored in a finite amount of bits on a computer: if one is added to the largest positive integer that can be represented on a computer, then the smallest negative number that can be represented is returned as the result. This overflow is used for a natural type of the modulo operation. Working with signed integers of 32 bits, the largest positive number is $2^{31} - 1$ and the smallest negative number is -2^{31} such that c is effectively set to 2^{32} and only the rule $x_{i+1} = a \times x_i + b$ is applied. Then the following values for a and b provide some good quick and dirty random numbers:

- $a = 1{,}664{,}525$, $b = 1{,}013{,}904{,}223$
- $a = 16{,}807$, $b = 0$
- $a = 65{,}539$, $b = 0$
- $a = 65{,}549$, $b = 0$

In the multiplicative congruential cases where $b = 0$, only odd seed values shall be used; otherwise correlations are apparent even in the screen pixel test. Some such random number generators are famous; for example, $7^5 = 16{,}807$ is often called magic due to some of its properties. Throughout this book, we use only random numbers generated with $a = 1{,}664{,}525$ and $b = 1{,}013{,}904{,}223$.

From integer random numbers x_i, uniformly distributed in the interval $[-2^{31}; 2^{31} - 1]$, random numbers r_i are derived by dividing by the largest possible absolute random number (in this case 2^{31}) and taking the absolute value:

$$r_i = \text{abs}\left(x_i \times 4.656612 \cdot 10^{-10}\right) . \tag{3.2}$$

These r_i are uniformly distributed in the interval $[0; 1]$.

Other methods also exist for producing random numbers very quickly, some of them leading to fairly good random numbers, for example the Kirkpatrick–Stoll random number generator [115].

More elaborate random number generators, which produce random numbers of a high quality, often use the basic generation methods, introduced above, not just once but several times and in a combined way. However, it takes more time to produce a sequence of random numbers using one of these generators [166]. One must be careful when trying this because there are ways of combining random number generators that lead to lower quality in the resulting random numbers. See [166] on this.

Therefore, one has to weigh the necessity of such good random numbers. For most purposes in optimization, the quick and dirty random numbers are quite sufficient. Furthermore, one must even be careful when using such a mathematically proved good random number generator: it has been shown that sometimes (e. g., if they are correlated with the proposed problem) these lead to worse results than the quick and dirty generators [70].

To investigate the test routines of random numbers, let us consider here only random numbers that are uniformly distributed in some interval. The simplest test is the histogram test. In this test, the interval is divided into a certain number of subintervals. A counter is assigned to each subinterval and initialized with zero. Then a long sequence of random numbers is calculated. For each random number the counter for the subinterval in which it lies is incremented. For a large amount of random numbers, these counters should exhibit roughly the same value, i. e., they should be roughly equal to the overall number of random numbers divided by the number of subintervals, as seen in Fig. 3.1.

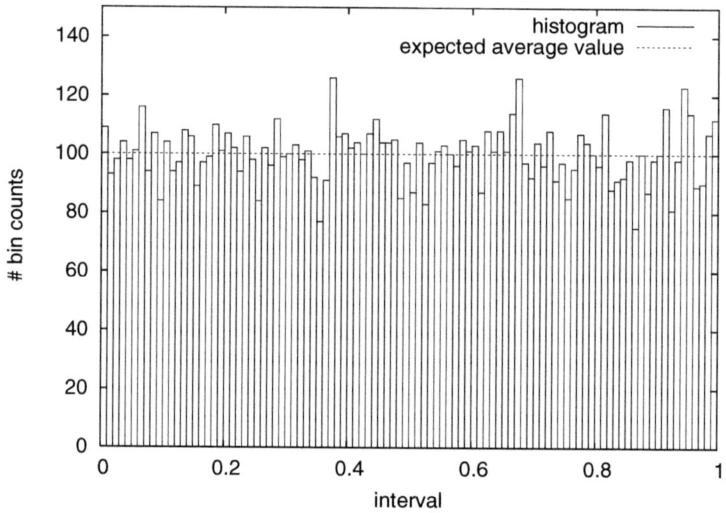

Fig. 3.1. Example of a histogram generated from 10,000 random numbers uniformly distributed in the unit interval. The unit interval is divided into 100 subintervals, also called bins, with a width of 0.01 each. As the random numbers are chosen uniformly, we expect to get on average 100 random numbers in each bin

However, even for very many random numbers, the heights of the histogram bars differ significantly. Actually they must differ with a certain variation if the random numbers are independent as well as random. Let us denote the overall number of random numbers as N, the number of bins as N_{b}, the probability that a random number gets sorted in bin No. i as p_i, and the expected number of random numbers in bin i as $N_{\text{exp}}(i) = N \times p_i$. As our random numbers are uniformly chosen, p_i is given as $p_i = 1/N_\mathrm{b}$ such that $N_\mathrm{exp} = N_\mathrm{exp}(i) \equiv N/N_\mathrm{b}$. The variations should be of order $\mathcal{O}(\sqrt{N})$.

The χ^2-test is a good means for determining whether the random numbers are truly random: the normalized χ^2-measure is given as

$$\chi^2 = \frac{1}{N_\mathrm{b}-1} \sum_{i=1}^{N_\mathrm{b}} \frac{(N_i - N \times p_i)^2}{N \times p_i} , \tag{3.3}$$

with N_i being the actual number of random numbers falling in bin i in the test. If N_i are identical to N_exp, the normalized χ^2-measure would give a value of 0. But with the usual fluctuations, the normalized χ^2-measure should have a value of roughly 1. For our example in Fig. 3.1, we get $\chi^2 = 0.909$.

We performed many more tests for various values of N_b and N_exp. The results are shown in Fig. 3.2. For example, we get $\chi^2 = 1.0019697 \pm 0.0045$ for $N_\mathrm{exp} = 100$ and $N_\mathrm{b} = 100$, averaged over 1000 sequences. The minimum value we find here is 0.687676768, and the maximum value is 1.7. Generally, we get that χ^2 is roughly 1, independent of the number of bins and the expected number of counts per bin. Moreover, the variation of our normalized χ^2-measure around 1 decreases with an increasing number of bins. For the precise values of the probability distribution of χ^2 without our normalization by $1/(N_\mathrm{b}-1)$, see [1]. It can also be evaluated in popular math packages, such as Mathematica.

Now let us consider normalized χ^2-values of distributions tampered with by people thinking, e. g., that each bin should have the same number of

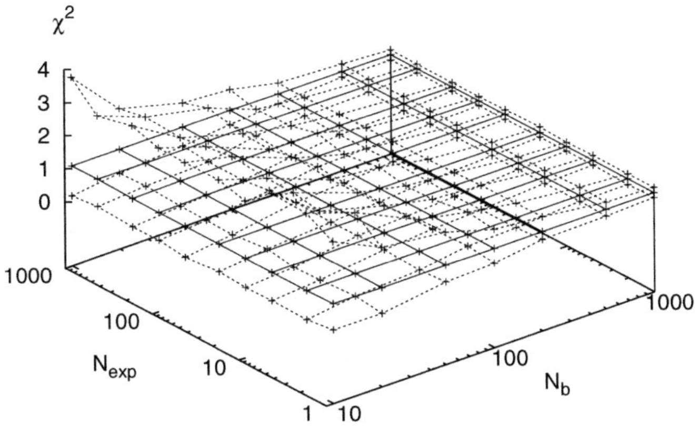

Fig. 3.2. χ^2-values for random numbers generated with a quick and dirty random number generator: for various numbers N_b of bins and for various expected values N_exp of counts per bin, we generated 1000 sequences containing $N_\mathrm{b} \times N_\mathrm{exp}$ random numbers each. The graphic shows the minimum, average, and maximum χ^2-values taken from the 1000 sequences each. The deviation of the minimum χ^2-value and the maximum χ^2-value from the average χ^2-value decreases with increasing N_b

counts. As already mentioned, the ideal normalized χ^2-value for large N_b is 1. If each bin had exactly the same number of counts, such that it was equal to N_exp, then the nominator in Eq. (3.3) would vanish, such that $\chi^2 = 0$. Thus, people showing you a "perfect" looking histogram with bins of the same size created by an "excellent" newly invented random number generator are simply trying to fool you. Then there are those who mess around with histograms trying to reduce the visible amount of randomness in order to impress their bosses, who do not have a clue as to what randomness is about.

There are approaches in which some people simply change N_i to $N_i + (N_\mathrm{exp}(i) - N_i)/2$, thus reducing the "distance" between the actual and the desired value by a factor of 2. Applying this approach to the instance in Fig. 3.1, χ^2 decreases to 0.245. If the distance were further reduced, e. g., even by a factor of 4, then χ^2 would be only 0.06 for this instance. But using the numbers from our experiment of Fig. 3.2, the values of χ^2 lie well outside the range we encountered in 1000 trials. Thus, their likelihood of coming from real independent random variables is less than one in 1000. In fact, much, much less.

Another widely used way of tampering with histograms is setting some of the histogram counts to their expected values. But if we, e. g., select randomly half of the bins and set their counters N_i to the expected number $N_\mathrm{exp} = 100$ for our example instance above, we get $\chi^2 = 0.531$; if we do this for even three quarters of the bins, then we get $\chi^2 = 0.276$.

One could also think of other strategies to reduce the randomness in a histogram, but most of this removing of true randomness can be detected with the χ^2-test, as shown above, as the value of χ^2 is not roughly 1 then, as it should be, as our results for thousands of runs in Fig. 3.2 show.

One might also wonder whether to add additional randomness to such a histogram. This can be done, e. g., by changing N_i into $N_i + 2(N_\mathrm{exp} - N_i)$. Applying this change to our histogram instance, we get $\chi^2 = 2.026$. Thus, values much larger than 1 are achieved when adding randomness. It is basically impossible to get such a large value of χ^2 if the number of bins is sufficiently large.

A more elaborate test, one that checks for correlations between the random numbers, is the screen pixel test: the successive random numbers x_i are considered to be the coordinates of the points $p_j = (x_{2j-1}, x_{2j})$. These points are printed as pixels on the screen. If the pixels fill the screen completely and smoothly, then also the requirement of the uniform distribution in the whole interval is fulfilled. Furthermore, the pixels should not form any patterns when printed on the screen, such that the correlation between successive random numbers is not too large. One can proceed further and look in three and even higher dimensions at pixels with the coordinates $(x_{dj-d+1}, \ldots, x_{dj})$ [133], with d denoting the number of dimensions. Looking from the right angle, one finds that the random points lie in only a small amount of (hyper)planes instead of being distributed randomly over the whole space. This Marsaglia effect reveals bad instances of the multiplicative random number generator. These three eye checks only show whether these basic requirements are fulfilled, but such checks are, of course, no replacement for a real test of "randomness".

3.3 Transformation of Random Numbers

Nonuniform distributions of random numbers may be necessary. The simplest case is a uniform distribution over a different random interval, say, $[a;b]$. In this case, the following linear transformation is applied to the random numbers:

$$\tilde{r}_i = a + (b-a) \times r_i \,. \qquad (3.4)$$

The new random numbers $\tilde{r}_i$ are uniformly distributed in the interval $[a;b]$.

Sometimes one wants to work with uniformly distributed integer random numbers $1, \ldots, N$. For this purpose, the following rule is used: let r_i again be random numbers that are uniformly distributed in the interval $[0;1]$; then the integer random numbers n_i are created according to

$$n_i = [1 + N \times r_i] \,, \qquad (3.5)$$

with $[x]$ denoting the Gaussian brackets, which simply take the integer part of x. One must be careful when applying this rule, as r_i could be exactly 1, so that $n_i = N+1$. In this marginal case, either a new random number is calculated or n_i is simply set to one of the integer numbers in its range. Of course, if the real-valued random numbers r_i are derived from integer-value random numbers x_i, as in Sect. 3.2, it is faster to turn the x_i directly into the n_i by

$$n_i = |x_i \bmod N| + 1 \,. \qquad (3.6)$$

However, if N is rather large, then an error occurs: simply assume that N is $\frac{2}{3}$ of the maximum possible random number. Then the first half of the numbers are generated twice as often as the second half of the numbers.

So far, we have only generated uniformly distributed random numbers. But often also other distributions are needed. For generating random numbers that are distributed in a different way, one usually starts out with a random number generator as discussed above that creates random numbers r_i that are uniformly distributed in the unit interval $[0;1]$. Now if one needs a distribution of the random numbers according to some distribution function P, which for simplicity we assume to lie in the range between 0 and 1, there is always the possibility of using the von Neumann rejection principle: first, one determines the interval in which the desired random numbers will be found; let this interval be $[a;b]$. Then the random numbers r_i are transformed in a linear way, as discussed above, such that they are uniformly distributed in the interval $[a;b]$; one sets $\tilde{r}_i = a + (b-a) \times r_i$. Then one chooses a second random number s_i for each $\tilde{r}_i$; these s_i are also uniformly distributed in the interval $[0;1]$. Now there are two cases, either $s_i \le P(\tilde{r}_i)$ or $s_i > P(\tilde{r}_i)$. As the s_i are uniformly distributed in the range $[0; 1]$, the case $s_i \leq P(\tilde{r}_i)$ occurs with the exact probability $P(\tilde{r}_i)$. Thus, the von Neumann rejection principle works as follows. Create pairs (r_i, s_i) of random numbers. Then, if required, transform the r_i into $\tilde{r}_i$. For each i, determine whether $s_i \leq P(\tilde{r}_i)$. If this is the case, then keep $\tilde{r}_i$; otherwise delete it from the sequence of random numbers. The remaining $\tilde{r}_i$ are then distributed according to the distribution function P.

The von Neumann method can be used for creating any desired probability distribution. However, it often consumes a lot of computing time for generating random numbers that are thrown away later on, especially if the desired distribution peaks in some areas of the interval and nearly vanishes in other areas. Furthermore, it converges rather slowly toward the desired distribution function. Therefore, other techniques for creating a certain distribution must be given preference.

In another widely used approach, uniformly distributed random numbers are transformed directly into random numbers that are distributed according to the desired distribution function P. Let (r_i) be a set of uniformly distributed random numbers. Then the task is to transform this set into a set (s_i) of random numbers distributed according to the probability distribution P. Here one makes use of the fundamental transformation law of probabilities [166], according to which one gets

$$|\,\mathrm{d}r| = |P(s)\,\mathrm{d}s|\,. \tag{3.7}$$

This equation can be transformed into

$$P(s) = \left|\frac{\mathrm{d}r}{\mathrm{d}s}\right|\,. \tag{3.8}$$

Now we want to derive the transformation for the widely used exponential distribution

$$P(s) = \exp(-s)\,. \tag{3.9}$$

Thus, one gets $r(s) = \exp(-s)$, i. e., $s(r) = -\ln(r)$. Therefore, one can transform the random numbers r_i, which are uniformly distributed in the unit interval, into the random numbers $s_i = -\ln(r_i)$, which are exponentially distributed over the interval $[0; +\infty[$. One has only to address the fact that the random number generator for the r_i might create $r_i = 0$. This random number is dropped, and another random number r_i is then generated, from which the s_i can be derived.

Another widely used distribution of random numbers is the Gaussian or normal distribution. The Gaussian probability distribution P is given by the formula

$$P(x) = \frac{1}{\sqrt{2\pi}\sigma}\exp\left(-\frac{(x-\mu)^2}{2\sigma^2}\right)\,. \tag{3.10}$$

A simple way of generating such a distribution would be to generate first a set of random numbers $(r_i)_{i=1,\ldots,N}$ that are uniformly distributed in the interval $[\mu - \alpha\sigma; \mu + \alpha\sigma]$. α has to be chosen large enough such that $P(\mu + \alpha\sigma)$ is sufficiently small. Usually, α is chosen to be ≥ 3. Then a second set of random numbers $(s_i)_{i=1,\dots,N}$ is generated; these random numbers are uniformly distributed in the interval $[0; 1/\sqrt{2\pi}\sigma]$. Now the following rule is applied: if $s_i \leq P(r_i)$, then the random number r_i is accepted, otherwise it is rejected. The remaining accepted random numbers are roughly Gaussian distributed. However, the tails of a real Gaussian distribution are missing in this distribution. Therefore, one has to find some compromise: if α is chosen very large, then most of the random numbers are rejected. However, the tails are at least partially represented. If α is small, then the tails are completely missing but little calculation time is wasted for calculating random numbers that are rejected afterwards. As was already mentioned, other techniques for creating a certain distribution must be given preference.

For the Gaussian distribution, a very simple approach based on the central limit theorem can be used for getting quick and dirty Gaussian random numbers: according to the central limit theorem, the sum over several probability distributions (which are either bounded or have a finite variance) always tends toward the Gaussian distribution. Therefore, summing up several uniformly distributed random numbers leads to Gaussian random numbers:

$$g_i = -\frac{N}{2} + \sum_{j=1}^{N} r_{i,j} \,. \tag{3.11}$$

$N/2$ has to be subtracted as $\frac{1}{2}$ is the expectation value of a $[0; 1]$ uniformly distributed random number, such that the resulting g_is are centered at 0. The larger N is, the better is the approximation to the Gaussian distribution. A natural choice is $N = 12$, as then the standard deviation of the resulting Gaussian distribution is 1: the expectation value of a uniformly distributed random number r_i in the interval $[0; 1]$ is $\langle r_i \rangle = \frac{1}{2}$, $\langle r_i^2 \rangle = \frac{1}{3}$, $\langle r_i r_j \rangle = \frac{1}{4}$ for $i \neq j$. Therefore,

$$\begin{aligned}
\sigma^2(g_i) &= \left\langle \left(-6 + \textstyle\sum_{j=1}^{12} r_{i,j}\right)^2 \right\rangle - \left\langle -6 + \textstyle\sum_{j=1}^{12} r_{i,j} \right\rangle^2 \\
&= \left\langle \left(\textstyle\sum_{j=1}^{12} r_{i,j}\right)^2 \right\rangle - 12 \left\langle \textstyle\sum_{j=1}^{12} r_{i,j} \right\rangle + 36 - 0 \\
&= \left\langle (r_{i,1} + r_{i,2} + \ldots + r_{i,12})^2 \right\rangle - 12 \times 6 + 36 \qquad (3.12) \\
&= 12 \left\langle {r_{i,j}}^2 \right\rangle + 2 \times (11 + 10 + \ldots + 1) \times \langle r_{i,j} \times r_{i,k\neq j} \rangle - 36 \\
&= 12 \times \tfrac{1}{3} + 2 \times \tfrac{11 \times 12}{2} \times \tfrac{1}{4} - 36 \\
&= 4 + 33 - 36 = 1 \,.
\end{aligned}$$

Note that the disadvantage of summing up only 12 uniformly distributed random numbers is that only Gaussian random numbers in the interval $[-6; 6]$ can be obtained. Beyond that, the tails of the Gaussian distribution are completely missing.

There is another method leading to very good Gaussian random numbers with zero mean and unit variance, the Box–Muller method: the trick of this method consists of creating two Gaussian random numbers a and b from two uniformly distributed random numbers u and v in the interval $[0; 1]$ at the same time. Let

$$f(x,y) = \frac{1}{2\pi} \exp\left(-\frac{x^2+y^2}{2}\right) \tag{3.13}$$

be the density distribution and

$$\varrho^2 = x^2 + y^2 \,, \quad \tan(\varphi) = \frac{y}{x} \tag{3.14}$$

be the transformation to polar coordinates. Then the relation between the area elements is given by

$$\mathrm{d}x\,\mathrm{d}y = \varrho\,\mathrm{d}\varrho\,\mathrm{d}\varphi\,. \tag{3.15}$$

The local distribution has to be equal for both coordinate systems:

$$f(x,y)\,\mathrm{d}x\,\mathrm{d}y = f(\varrho,\varphi)\,\mathrm{d}\varrho\,\mathrm{d}\varphi \tag{3.16}$$

such that

$$f(\varrho,\varphi) = \frac{1}{2\pi}\varrho \exp\left(-\frac{\varrho^2}{2}\right)\,. \tag{3.17}$$

The goal is to generate the radius ϱ according to the distribution $\varrho\exp(-\varrho^2/2)$ by using the first random number u and φ being homogenously distributed in $[0; 2\pi]$ by using the second random number v. Therefore,

$$F(\varrho) = \frac{1}{2\pi}\int_0^{2\pi} \mathrm{d}\varphi \int_0^{\varrho} \varrho' \exp\left(-\frac{\varrho'^2}{2}\right) \mathrm{d}\varrho' = 1 - \exp\left(-\frac{\varrho^2}{2}\right) \tag{3.18}$$

can be equalized with some uniformly distributed random number z in the interval $[0; 1]$. The radius ϱ is therefore given by

$$\varrho = \sqrt{-2\ln(1-z)}\,. \tag{3.19}$$

$u = 1 - z$ is also a uniformly distributed random number from the unit interval. Let v be the second uniformly distributed random number and set $\varphi = 2\pi v$. With the relations $x = \varrho\cos(\varphi)$ and $y = \varrho\sin(\varphi)$, one obtains the two Gaussian distributed random numbers

$$a = \cos(2\pi v)\sqrt{-2\ln(u)} \tag{3.20}$$

and

$$b = \sin(2\pi v)\sqrt{-2\ln(u)}\,. \tag{3.21}$$

However, this method of calculating Gaussian distributed random numbers is very time consuming, as the functions natural logarithm, sine, cosine, and the square root must be calculated. Furthermore, it is recommended that v should not depend strongly on u, as then a and b are not truly independently normally distributed.

Of course, this distribution can simply be transferred to a Gaussian distribution with another mean value μ or another standard deviation σ.

The Box–Muller method was improved by Marsaglia and Bray in 1964 [134]. Their improved method, which eliminates the calculation of the sine and the cosine, has become known as the polar method. As in the Box–Muller method, first two uniformly distributed random numbers u and v are chosen. Then they are linearly transformed to the interval $[-1;1]$:

$$\tilde{u} = 2u - 1 \text{ and } \tilde{v} = 2v - 1 \,. \tag{3.22}$$

Then

$$w = \tilde{u}^2 + \tilde{v}^2 \tag{3.23}$$

is calculated. If $w > 1$, then the algorithm jumps back to its start. Otherwise, let

$$t = \sqrt{\frac{-2\ln w}{w}} \,. \tag{3.24}$$

Then the Gaussian random numbers a and b are given by

$$a = t \times \tilde{u} \quad \text{and} \quad b = t \times \tilde{v} \,. \tag{3.25}$$

The savings from eliminating the calculation of one sine and one cosine is balanced by having to apply a rejection rule such that two new random numbers might have to be chosen. Note that in contrast to the Box–Muller method, in which u was used to calculate the radius and v was used to determine the phase, here u and v are used together. The rejection principle is applied in order to accept only those pairs of (u, v) for which this trick can be used, namely, those inside the unit circle.

Finally, we compare the four methods for creating Gaussian distributed random numbers. Table 3.1 shows the calculation times of the various methods. It turns out that the rejection method is the slowest method, although only $\alpha = 3$ was used for creating the random numbers, whereas the polar method is the fastest method.

Table 3.1. Comparison of calculation times of four methods creating Gaussian random numbers: in each case, an array containing 1 million Gaussian distributed random numbers was created 100 times. The times are given for a 400-MHz Pentium II

Method	Calculation time
Rejection method	200 s
Summing up 12	81 s
Box–Muller method	74 s
Polar method	55 s

3.4 Example: Calculation of π with MC

A famous example for the usage of the MC technique is the calculation of the number π by means of MC. Figure 3.3 shows the geometry used for this calculation: one shoots randomly into a square of edge length r and counts the hits in the quarter circle of a radius r. The position of the random point, i. e., both its x- and y-coordinates, is determined by two successive calls of a random number generator. It is simply checked with $x^2 + y^2 \leq r^2$ to see whether the point is inside the quarter circle or not. The area of the square is given by r^2, and the area of the quarter circle is given by $\frac{1}{4}\pi r^2$. Therefore, the ratio of the number of hits divided by the number of shots approximates $\frac{1}{4}\pi$, if the number of shots is chosen large enough.

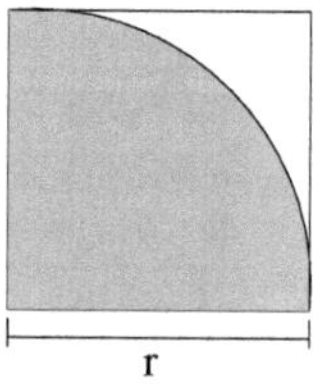

Fig. 3.3. A quarter circle in a square

Table 3.2. For each number of shots, 100 runs were performed. The minimum and maximum approximations for π for the corresponding number of shots are given, as are the mean value and the error bar of these approximations

Shots	π_{min}	π_{max}	$\pi_{\text{mean}} \pm \Delta\pi$
10	1.6	4.0	3.148 ± 0.06
100	2.64	3.40	3.1408 ± 0.015
1000	3.032	3.284	3.14484 ± 0.0052
10,000	3.1044	3.1852	3.143668 ± 0.0016
100,000	3.12500	3.15344	3.1415816 ± 0.00057
1,000,000	3.138464	3.14534	3.1415648 ± 0.00015
10,000,000	3.1408864	3.142138	3.14159355 ± 0.000023

Table 3.2 shows results for various numbers of shots. As can be easily seen, it is much better to calculate π with the formula

$$\pi = 4 \times \arctan(1) \tag{3.26}$$

or to memorize the first 15 digits of π. These can be remembered as the lengths of the words in the sentence "Now I want a drink: alcoholic, of course, after the heavy lectures involving quantum mechanics!"

4 Overview of Optimization Heuristics

4.1 Necessity of Heuristics

After reading Chap. 2, one might ask why this book does not end after that chapter, as every simple as well as every complex problem can be solved by some exact optimization algorithm. However, depending on the particular problem and on the size of the problem, it may take a large amount of calculation time to reach the global optimum solution. Often years of research time must be invested in order to improve these general approaches and to tailor them for the proposed problem such that the optimum solution can be achieved after weeks or months of calculation time. Furthermore, in real-life problems, further constraints often appear that were not mentioned or not clear at the beginning of the research or that appear due to new requirements in the company or due to new laws. In this case, often a completely new algorithmic structure must be developed, and most of the previous research might end in the waste basket. Additionally, due to this tailoring and customizing of the algorithm to the proposed problem, it is often the case that the algorithm cannot simply be applied to another problem that is identical to the first problem but has other parameter values, as the calculation time might be too long.

Therefore, it is very often necessary to use heuristic algorithms instead of exact algorithms. These heuristic algorithms do not usually guarantee reaching the global optimum of the problem, giving instead a locally optimum solution, which may be the global optimum, but most of the time is slightly worse. Whether or not a difference of 1% or less between the qualities of a near-optimum solution and the global optimum can be accepted depends on the business at hand. Usually, such a heuristic can be tuned to lead to even better results. However, this tuning can be done in minutes, hours, or days, not the weeks or months that may be required for development of new exact methods. Sometimes it is even impossible to speak of a global optimum solution for a business decision, as there are disturbances from the world outside like traffic jams and uncertainties in the data. Then the usage of heuristics is even more justified, as they lead rather quickly to such a quasioptimum solution. Besides the advantage of only needing a relatively small amount of calculation time, many of these algorithms are very easy and fast to implement and to tune after one has gotten some feeling about the interaction between the algorithm and the proposed problem. Additional constraints can usually be implemented rather easily. A heuristic algorithm can also be transferred to other identical problems with other parameter values by adjusting some parameters of the heuristics. Often such adjustments can be made automatically, as we will show later.

Of course, mathematicians are not happy with these heuristics as they do not provide any proof of having reached the global optimum solution. You may also hear that many heuristics have funny names, and already for this reason heuristics are not to be trusted. Yet, these same mathematicians often use heuristics as a source of bounds for their exact algorithms and to speed up the search process. Therefore, there is nobody really claiming that heuristics are not useful and there is nobody in the business of optimization who does not need to make use of heuristic algorithms.

4.2 Construction Heuristics

Suppose one has a problem to solve optimally and one does not know any algorithm for solving this problem. The natural way to solve this problem, then, is the same way as children put a puzzle together. One determines the single pieces of the problem, starts with a tabula rasa, and includes step by step all parts of the system in an at least locally optimal way. This approach can easily be transferred to a computer algorithm, which puts one piece after the other into the system until a solution for the complete problem is reached at the end. From the point of view of the individual pieces, each piece looks in an egoistic manner for the best possible place in the already existing torso system. This approach usually leads to rather good solutions.

Of course, such egoistic construction heuristics are not or only partly able to use synergy effects between different parts of the system. Therefore, global approaches try to keep the whole system in view and not only to use but also to control the egoistic tendencies of the individual parts. Mostly, all parts of the system are introduced in the system already at the start of the construction heuristics. For that reason, the system usually has to be extended by some auxiliary variables in order to meet the constraints of the problem. According to some set of rules, these auxiliary variables are removed, with the sequence and the way of their removal being determined by the size of the improvements according to some objective function.

In order to clarify this approach, let us consider a production problem in the chemical industry: some products have to be made in several steps in a production system with several production lines and finite storage having finite capacities. In this storage, either the preproducts or some intermediate products can be stored. The production process can continue only if the number of preproducts or intermediate products in the storage is sufficiently large. But there may not be an overflow of this storage. When creating a solution with such a construction heuristic, one might want to add some production lines or storage in order to get a feasible solution at the beginning. Then one will try to reduce the number of these nonexisting production elements, which are the auxiliary variables in this system, until they vanish at last.

4.3 Markovian Improvement Heuristics

Heuristics that try to make improvements by changing the actual configuration step by step with a sequence of moves are widely used. Each configuration σ has a certain set of neighboring configurations (τ_i) to which the search process can jump from the configuration σ by applying one of the possible moves. The set (τ_i) is called the neighborhood $\mathcal{N}(\sigma)$ of the configuration σ. Depending on the implemented move set, different neighborhood structures are possible. Usually, moves are used that change the configurations only slightly, a principle which is called a local search. The set of configurations together with the neighborhood structure between them constitutes the search space of the proposed problem. If the cost function values of the individual configurations are drawn up in an extra dimension and if this extra dimension is coupled to the search space, one gets the so-called energy landscape of the system to be optimized. By applying a series of moves in a random way, one starts a Monte Carlo (MC) process, in which a MC walker moves through this energy landscape, climbing up hills and walking down valleys. The task is to lead the walker in an optimum way.

Often only gradient methods are used, i. e., one searches in the local surroundings of the actual configuration according to a gradient rule, which determines the width of the step and the kind of change due to the largest possible improvement. This type of search is usually called steepest descent, as such a change applied to σ leads to the best neighboring configuration in $\mathcal{N}(\sigma)$, whose energy difference is largest. Mostly, the whole configuration or one of its variables is used for the change. Sometimes, a different system is created and leads to the optimization of the system until all requirements for a feasible solution are fulfilled, an approach that is used by, e. g., neural networks.

By contrast, the greedy algorithm chooses a neighboring configuration at random and accepts it in a greedy way if it is better than or at least as good as the current solution. (Note that some authors call the method that we call steepest descent the greedy algorithm. Other authors distinguish two scenarios for the greedy algorithm, the "first improvement scenario" with a neighboring configuration, which is better than σ, chosen at random and the "best improvement scenario" with a complete search through the neighborhood of σ for the best neighboring solution.) Methods like steepest descent and the greedy algorithm usually do not lead to the global optimum solution in the energy landscape, i. e., to the configuration $\sigma_{\text{global optimum}}$ with

$$\mathcal{H}(\sigma_{\text{global optimum}}) \leq \mathcal{H}(\sigma) \tag{4.1}$$

for all configurations σ in the configuration space, but only to a local minimum configuration $\sigma_{\text{local minimum}}$ with

$$\mathcal{H}(\sigma_{\text{local minimum}}) \leq \mathcal{H}(\sigma) \tag{4.2}$$

for all configurations σ neighboring $\sigma_{\text{local minimum}}$, i. e., $\sigma \in \mathcal{N}(\sigma_{\text{local minimum}})$.

Some related algorithms, like simulated annealing (SA), threshold accepting, and the great deluge algorithm, also allow for deteriorations. The probability with which a deterioration of a certain size is accepted is governed by some control parameter, which is changed stepwise during the optimization run, until only improvements are accepted at the end. The hope is not to get stuck in bad local minima, as may occur with simpler optimization heuristics.

There are also other algorithms that change the energy values $\mathcal{H}(\sigma)$ of the configurations σ, thus deforming the energy landscape. Mostly, they only allow for improvements in either a Steepest Descent or greedy-type way. The ansatz is to remove barriers in the energy landscape that a greedy method in the original landscape cannot overcome and thus to arrive at better local minima or even the global optimum.

As these algorithms jump from a configuration σ to a configuration τ and do not use any knowledge of formerly visited configurations for the decision on whether to accept or reject this move, these algorithms belong to the class of Markovian processes that create a new state only depending on the previous state and on some random event. Therefore, we review this type of improvement heuristic under the title Markovian improvement heuristics.

4.4 Set-Based Improvement Heuristics

Whereas SA and related algorithms have in common that a new configuration is created from the current one without using further information, genetic algorithms (GAs) and evolution strategies (ESs) use a large set of configurations as individuals in one or more populations. GAs use—besides moves that change single individuals and that are called mutations in this context—mainly different types of crossover operators that create children from parental configurations, whereas ESs concentrate on mutations that change a member of the population. In both types of algorithms, which are incidentally closely related to each other, new configurations are always created. According to Darwin's principle of natural selection, only the individuals with the higher fitness, i. e., with the lower cost function value for the minimization problem, survive. The best individuals usually get more chances to reproduce than worse individuals. Various implementations of these algorithms differ only in the choice of mutations and of crossover operators and in addition in the selection of those configurations that are allowed to reproduce themselves, to have offspring with a suitable partner, or that must commit suicide.

Tabu search (TS), on the other hand, stores information about previously visited configurations in a tabu list and therefore also works on a set of configurations. TS is a memory-based search strategy by which the system to be optimized is led away from the parts of the search space that have already been investigated. This can be achieved by forbidding solutions that were visited formerly or by forbidding structures common to former configurations such that they are no longer allowed. These solutions or structures are stored in a tabu list that is updated after each move according to a specific rule set. These rules attempt to guarantee that the optimization run does not lead to a solution that was already visited formerly, that the tabu list size does not diverge, and that a good solution is reached at the end.

Thus, TS is basically a gradient-descent search with memory. The memory preserves a number of previously visited states along with a number of states that might be considered unwanted. This information is stored in a tabu list. The definition of a state, the area around it, and the length of the tabu list are critical design parameters. In addition to these tabu parameters, two extra parameters are often used: aspiration and diversification. Aspiration is used when all the neighboring states of the current state are also included in the tabu list. In that case, the tabu obstacle is overridden by selecting a new state. Diversification adds randomness to this otherwise deterministic search. If the tabu search is not converging, the search is reset randomly.

5 Implementation of Constraints

5.1 Moves, Constraints, Deadlines

Most practical optimization problems are significantly complicated by constraints—limitations on the allowed values of the degrees of freedom of the system. If the optimization requires assigning locations to objects, then either constraints typically are boundaries outside which the locations will be invalid, or the requirements that certain pairs of objects will not be permitted to overlap. If the degrees of freedom include the times at which events occur, then deadlines or precedence relationships form the most common temporal constraints. The simplest of these are easily expressed as linear inequalities. When coupled with a sufficiently simple description of the problem's objective function as a linear or perhaps quadratic function of the arguments, exact methods of solution may be available. These are, of course, the techniques of linear programming and quadratic programming, respectively.

But the objective function may be nonlinear in some more complex way. It may be the result of a table lookup, introducing discontinuities or nonanalyticities, or it may involve a complex and costly calculation each time it is evaluated for a new configuration and not permit numerical differentiation. In those cases, heuristics organized around local rearrangement moves, in which only a few degrees of freedom change at each step, are appealing. The single most important step in developing such heuristics is usually the choice of a set of moves with which to explore the design space of the problem and with which to improve upon reasonable trial solutions.

5.2 Incorporation into the Configurations

Most complex optimization problems suffer from one or more constraints that have to be considered in the optimization process in order to achieve a feasible solution. Two classes of constraints are distinguished: hard constraints have to be met without fail; soft constraints may be violated, but the degree of violation shall be as small as possible.

The best way to deal with a constraint is to integrate it into the configurations and to allow only for moves which turn a feasible configuration into an other feasible configuration. Then if we start out satisfying the constraint(s), each subsequent configuration satisfies the constraint automatically. Still the moves between the configurations have to be general and powerful enough that each feasible state can be reached from any initial configuration in a random walk. There must not be any moves leading to a forbidden configuration.

This approach is highly advantageous especially for hard constraints, as they are fulfilled automatically. It is sometimes called a restriction to feasible configurations.

5.3 Consideration of Feasible Solutions Only

If such an integration of a constraint is not possible, other ways of dealing with the constraint have to be chosen. If feasible solutions can be found rather easily, then the method of choice often consists of starting the optimization run at a feasible solution and only accepting new solutions if they are likewise feasible. Therefore, moves leading to configurations in which a hard constraint is not met or a soft constraint violated too much are rejected.

This approach seldom leads to good results because often the set of feasible configurations is split into many islands in the energy landscape, such that the search by the Monte Carlo walker for a good solution is restricted to the island containing the initial configuration. Therefore, he/she can find the best local minimum inside this island but fails to reach better solutions or even the global optimum on other islands.

5.4 Penalty Functions

A way out of this dilemma is provided by the principle of the penalty functions [155, 110]. A solution that was forbidden according to the approach above is no longer forbidden by some kind of barrier function, but it is "punished". The size of the penalty depends on the extent of the violation of the constraints. The nonfeasible solutions are thus lifted up in the energy landscape as the penalties are added to their energies. As the Monte Carlo walker searches for a (quasi) optimum configuration, he is more likely to end up in a feasible solution, in which all constraints are met, such that it can be used in practice.

The Hamiltonian $\mathcal{H}$ therefore not only contains the real costs $\mathcal{H}_0$ but also some additional pseudocosts $\mathcal{H}_i = \lambda_i \times f_i$, $i \geq 1$, consisting of Lagrange multipliers $\lambda_i > 0$ and penalty functions f_i. The penalty functions and the Lagrange multipliers, which are used to balance the impact of the additional terms, have to be chosen in such a way that all f_i are zero at the end of the optimization run, i. e., that all constraints are met.

The penalty terms can have various forms (Fig. 5.1): if the constraint is met, the penalty is usually 0. If it is violated, the penalty increases linearly

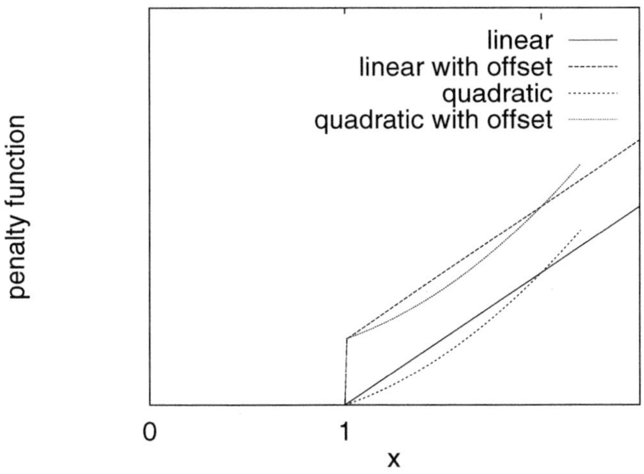

Fig. 5.1. The most common penalty functions are either linear or quadratic with the degree of violation of the constraint. If the constraint is met (here symbolic in the interval [0; 1]), then the penalty vanishes; otherwise it increases monotonously or quadratically with or without an additional offset. This offset is sometimes used to avoid small violations of a constraint. Summarizing, the more constraints of a configuration are violated, the more that configuration is punished.

Of course, one can imagine other forms of a penalty function, even some special-purpose penalty functions. Usually, the standard forms mentioned above do a good job as they do not rise too much with the amount of violation; thus they are penalty but not barrier functions. The Monte Carlo walker is able to walk on a path containing some nonfeasible solutions. However, sometimes one faces a problem in which some constraints can hardly be met or for which one has very detailed desires for what a good solution will look like. One has to be careful when incorporating a large number of sophisticated penalty functions because they may lead to a totally frustrated system in which the Monte Carlo walker is unable to find any feasible solution. If a solution is found, it might not be at all what is expected or desired.

6 Parallelization Strategies

6.1 Parallelization Models and Computer Architectures

In this chapter, the main parallelization strategies used for optimization are introduced. Each of these strategies can be implemented in one of the two major parallelization models: either one can use the cubix model, in which there is exactly one program which is loaded on all used processors. The other extreme is the master-slave model in which a host program, the so-called master, starts the other nodes, which are called slaves. Usually, each slave gets the same program, whereas the master often has a special program with completely different tasks. The master-slave model may extend from simply starting the slaves and collecting their results to the total control of the slaves. This model is also called the farm model due to the picture of a farm with a main house, in which the owner of the farm, the master, lives, and with the huts of the workers, the slaves. Furthermore, a tree model, which is an extended master-slave model, is used. Again there is a master who starts some slaves, but now the slaves themselves also start slaves. This feudal model can be repeated in several iterations. Often a binary tree is built in which each branch splits into two subbranches.

Parallel calculation can be done on a network of workstations or personal computers, but also on special parallel computers with a large number of processors. The topology of the network between the processors is usually a lattice architecture or a special tree topology. A simple lattice architecture is the hypercube. One can imagine each processor being located at a corner of a d-dimensional hypercube. Therefore, the hypercube contains 2^d processors. This architecture is advantageous if each processor mainly communicates with its d neighboring processors; the communication to more distinct processors takes $d-1$ hops at a maximum. In more complicated lattice architectures, more than two processors are in each row and each column of the lattice. Therefore, one usually needs more hops to deliver information to a distinct processor, at a maximum $(l-1) \times d$ hops, with l being the number of processors in each row or column. If the borders of the lattice are connected, such that a d-dimensional torus is formed, there are at a maximum only $[d \times l/2]$ hops. This architecture is better suited for those parallel programs in which an area is divided analogously to the lattice structure of the parallel computer.

Due to the most important applications, usually 2D and 3D lattices are used.

In the 1D case, such a lattice architecture is identical to the standard tree without branching, on which all processors are located in an open chain. If this chain is closed by introducing a connection between the first and the last processor, one gets the token ring model, in which all processors are logically located on a circle and in which each processor can directly talk to only its left and right neighbor. The maximum number of hops for transferring the information between two processors is $[l/2]$. Mostly, a tree topology contains regular branchings; for example, in the case of the binary tree each parental processor is connected with two son processors. Except for the ancestor and the offspring of the last generation, each of the $2^d - 1$ processors (assuming d generations) is connected to three neighbors, whereas the ancestor has only two connections and the youngest offspring only one connection. Furthermore, there are group models; for example, processors of different generations in the binary tree can be composed to a logical group in order to solve a partial problem.

Generally, one has to consider for each problem which parallelization model is most suitable for the given problem; then one must choose the best possible machine architecture that is available for carrying out the production runs. Furthermore, only information exchanges that are really necessary between the single processors should be performed.

6.2 Running Several Copies

The simplest approach to parallelizing a stochastic algorithm, which is sadly the most widely used type of algorithm, is simply to start the same program on each node and to use a different seed (e. g., the number of the node) for the initialization of the random number generator such that a set of different solutions is created. These solutions can be used for obtaining some statistics on the problem. However, often only the best solution is selected and returned as output of this so-called parallel algorithm. Even parallel computers, which are designed for allowing a large number of communications between the single processors, are often abused by this type of parallelization.

6.3 Divide et Impera

If a problem is so large that it cannot run on a single processor due to its requirement of storage or due to the calculation time needed, then the parallelization strategy "divide & conquer" is applied, by which this problem is divided and distributed over p processors such that it becomes calculable.

This division is especially not difficult if the problem can be split into partial problems that are completely independent of each other. The standard example for this case is the multiplication of a matrix with a vector that can be split into the multiplication of each row of the matrix with the vector. However, usually some communication between the processors is necessary. If for example some area is divided into stripes, then one has to make sure to send messages with changes at the borders to the "neighboring" processors if they are involved. One example for this scenario is a cellular automaton, which consists of a lattice of identical cells: each state of a cell is given as a discrete value out of a finite number of values. The system is evolved over several time steps. In each time step, a parallel update of all cells is performed. The new state of a cell usually only depends on its previous state and the previous states of the neighboring cells. Such a cellular automaton can easily be split up over several processors. The new states of the cells at the borders must be transmitted to the neighboring processors.

Therefore, the total time for running a parallel program consists not only of the calculation time but additionally of the communication time, which is the sum of the latency, i. e., the time needed for establishing the connection between two processors, and the transfer time, which depends linearly on the number of bytes to send. Furthermore, there is often the problem that some part of the program code cannot be parallelized completely such that there remains a part that would take some fraction α of the calculation time in a serial program, which has to be calculated by one processor in a serial way or by all processors identically. In this context, two definitions are very useful: the speedup $S(p)$ is defined as

$$S(p) = \frac{C_1}{C_p}, \tag{6.1}$$

with C_i being the calculation time on i processors. The speedup measures the acceleration achieved by splitting the problem on p processors. The efficiency

$$E(p) = \frac{S(p)}{p} \leq 1 \tag{6.2}$$

denotes how efficient the parallelization strategy is in contrast to a serial program.

Amdahl's Law

$$S(p) = \frac{1}{\alpha + \frac{1-\alpha}{p}} \leq \frac{1}{\alpha} \tag{6.3}$$

shows whether a parallelization makes sense and how many processors would be optimal. α should be as small as possible. However, the time for communication is not considered in this formula. An often used extension of Amdahl's Law considers the communication time as being proportional to the number of processors p [195]. Let therefore α be again the fraction for the serial part and β the fraction for the communication time in the overall time. Then the speedup is given by

$$S(p) = \frac{1}{\alpha + p\beta + \frac{1-\alpha-\beta}{p}}. \tag{6.4}$$

Therefore, the speedup can decrease for large numbers of processors p. The position of the maximum value, along with the optimum processor number, is determined by β, and the amount of the speedup is determined both by α and β.

6.4 Information Exchange

However, the divide et impera method is of no use for many optimization problems: due to strong correlations between parts of the system that do not neighbor each other, this approach often leads to a counterproductive divide-and-lose effect. Therefore, besides this method, which is sometimes necessary for solving a problem of a larger size, the best use of a parallel computer is to exchange information between the single processors in order to save calculation time or to improve the results. Depending on the underlying optimization algorithm, there are different bits of information that can be transferred during the optimization run: using simulated annealing and related algorithms, some information can be exchanged for determining the value of the control parameter in the next optimization step or the best-so-far solution can be transmitted to all processors. Some of these approaches are summarized by the term multichain Monte Carlo algorithms, or $(\mathrm{MC})^2$. There are also different versions of $(\mathrm{MC})^n$ algorithms with $n \geq 3$, e. g., a Metropolis coupled multichain MC algorithm. [Note that the term $(\mathrm{MC})^2$ is also used as an abbreviation for Markov chain Monte Carlo.] Working with genetic algorithms, migration processes can be simulated: a parallel implementation is ideal for dividing up all individuals among several subpopulations, with each of these simulated on one processor. After some time, some individuals are moved to another subpopulation. Whether the fitnesses of these indivuals are low (thus simulating the pressure in the time of the emigration of nations on weaker tribes to find new land) or high (thus simulating the expansion of a militarily superior tribe) is determined in different ways. A parallelization of tabu search allows for a parallel update of the tabu list. But there is an information exchange not only during the optimization run; information can also be exchanged after the end of an optimization run in order to obtain more insight into the proposed problem and in order to use this information for further parallel optimization runs.

In this context, the traditional definitions of speedup and efficiency are no longer applicable. Only problem-specific parameters can be defined with which the gain of a parallelization can be measured. This gain can consist of the improvement of the results that was possible due to the parallelization or of the savings of calculation time for achieving results that are at least as good as results from serial runs.

In the business of optimization, mostly one has to deal with problems that cannot be divided into separate subproblems, as then synergy effects are lost such that the composition of optimum results of subproblems is worse than the optimum solution of the whole problem. Therefore, we will concentrate on information-exchange strategies later in the book and give a few examples of them.

7 Construction Heuristics

7.1 General Outline of Construction Heuristics

Construction heuristics or augmentation heuristics are usually the fastest way to achieve feasible solutions for a proposed problem. Such solutions are usually not as good as solutions provided by other heuristics. Because of that, construction heuristics are mainly used if a reasonably good solution to a problem has to be found promptly or to provide initial solutions for improvement heuristics, which will be introduced in the next chapter. Construction heuristics may also be used inside improvement heuristics, as we will see later.

Generally, a construction heuristic can be split into different phases, starting with an initialization phase, continuing with a loop over selection and placement phases until some conditions are met like, e. g., having achieved a proper solution for the proposed problem, and sometimes concluding with a cleaning phase:

- The first phase is the initialization phase:
  - Some construction heuristics start with a tabula rasa, i. e., with an empty blackboard. This is the same approach as solving a jigsaw puzzle. Closely related to this approach is the idea of starting with exactly one piece of the problem, which might be specially selected, like a corner piece.
  - Another initialization method consists of starting with the optimum solution of a small subsystem of the problem. Again the parts of the subsystem might already be specially selected such that the construction heuristics start with the final result of another heuristic or an exact algorithm for the subsystems.
  - It is also possible to introduce all pieces into the system at the very beginning. To meet all the problem's constraints, the system usually has to be extended at least in one dimension or further dimensions have to be added.
- Next is the selection phase: If starting with a tabula rasa or a small subsystem, the question occurs as to which part or parts of the system will be introduced in this step of the algorithm. In some heuristics, the selection is done randomly; other heuristics have a rule set for selecting an appropriate piece. If all items were already inserted at the beginning, the question is

where to change the solution in order to reduce the length in the additional dimensions.

- Connected to that is the placement phase: the selected items have to be placed somewhere in the system according to some rules.
- The selection and the placement phases are repeated until all parts are introduced in the system or until the additional dimensions can be removed.
- Sometimes there is a cleaning phase at the end of the construction heuristics by which the already achieved solution shall be improved according to some rules.

7.2 Insertion Heuristics

Now the insertion heuristics that start with a tabula rasa, one piece, or a small subsystem will be described more thoroughly. If starting with a tabula rasa, a piece is selected either randomly or according to its relevance or according to some rule set and placed in the system. If there is already one piece, often some nearest-neighbor strategy is used for the selection phase. In order to be able to apply this strategy, a neighborhood relation or even a quantified ordered neighborhood measure must exist or has to be defined between the individual pieces of the problem. The "nearest" item is introduced next in the system. One can even start with two items that are as far apart as possible.

Now there is often already a subsystem that might be a complete solution for a smaller instance of the considered optimization problem. If the nearest neighbor strategy is applied again, the question arises as to whether only the neighborhood of the first piece, only the neighborhood of the second piece, the neighborhoods of both pieces, or the neighborhood of the connection between the two pieces will be considered. The various insertion heuristics differ mainly in this point. Some definitions of a neighborhood lead in the next few steps to better results than others, but in the end the last remaining items might have to be introduced in some awkward way, whereas other insertion heuristics that are not as good in the intermediate stage of construction pay back at the end of the algorithm.

There are also geometric insertion heuristics: the pieces can be successively inserted from top to bottom, from left to right, from the center of the system to the borders of the system, or vice versa. If the algorithm starts, for example, at the center or at the borders of the system, then it can make use of the neighborhood relation and introduce all pieces due to their "distance" to the center or it can introduce them in a clockwise or counterclockwise manner. Of course, such heuristics can only be applied to problems that exhibit some geometry at all.

7.3 Savings Heuristics

As already mentioned above, there is a different type of construction heuristic in which all parts of the system are already introduced at the beginning. One standard example for this approach is the savings algorithm for the vehicle routing problem, which was developed by Clarke and Wright [36]. Therefore, we shall describe all of these heuristics under the label savings heuristics.

The main idea of these algorithms is that after introducing all of the items independently of each other, the items are rearranged and merged in such a way that the quality of the solution increases. Usually, a rearrangement is chosen that leads to the largest savings. Instead of this approach, a randomly chosen rearrangement is simply accepted if it leads to some savings.

7.4 More Intelligent Ways of Construction

Till now, only those construction heuristics were mentioned in which the individual pieces, once they are selected and placed in the system, stay put. However, sometimes it is advantageous for an item to retreat from its good insertion and either to be removed from the system, such that it has to be introduced again later, or to be moved to another place in the system in order to allow a better overall solution. Various types of backtracking methods can be combined with the heuristics described above. Of course, unrestricted backtracking guarantees that the optimum solution will be achieved, but this can only be achieved with exponential costs in the system size such that a backtracking here is usually only used in one to three iterations in order to achieve the greatest improvements at minimum additional computational costs.

For some problems, one can make use of deeper insight into the problem if creating a construction heuristic. One can make use of geometric properties, start with subsystems that are already optimally solved, add optimally solved subsystems, or restrict the heuristic to a smaller amount of insertion or savings possibilities because of some exact or heuristic estimations of upper or lower bounds.

However, one has to be careful when extending the rule set. As we will show later, sometimes it is better to leave some space for random choices of the heuristic and not provide a too strict rule set. A good balance must be found between some degree of randomness and some otherwise stiff rule set. This holds true also for improvement heuristics.

8 Markovian Improvement Heuristics

8.1 Constructing a Markov Chain

Autoregressive processes of the order n (abbreviated AR[n]) are given by

$$X_t = \alpha_1 X_{t-1} + \alpha_2 X_{t-2} + \ldots + \alpha_n X_{t-n} + \varepsilon_t \,. \tag{8.1}$$

The system evolves therefore in a way that the state X_t in time t depends in a linear way on the states at the n previous time steps and some random element ε_t at time t. The special AR[1]-case of the autoregressive processes,

$$X_t = \alpha X_{t-1} + \varepsilon_t \,, \tag{8.2}$$

is also called a Markov process. Each Markov process therefore depends only on the previous state in time. Starting at some initial state X_0 at time $t = 0$, a Markov chain is built by successively applying formula (8.2).

Analogously, a Markovian improvement heuristic starts at some initial configuration, which might be random or already preoptimized as a result of, e. g., some construction heuristic, and constructs a Markov chain by changing the configuration successively. Thus, one starts at an initial configuration σ_0. Then one changes this configuration according to a set of rules into another configuation σ_1. Now the question is whether to accept or reject the tentative new configuration σ_1. If the move $\sigma_0 \to \sigma_1$ is accepted, then one keeps the new configuration σ_1; otherwise one sets $\sigma_1 \equiv \sigma_0$. Then one changes the configuration σ_1 into some configuration σ_2 and again decides whether to accept or reject this move $\sigma_1 \to \sigma_2$. Iterating this approach, one constructs a sequence of configurations σ_i by applying moves $\sigma_{i-1} \to \sigma_i$. As each move only depends on two configurations, i. e., the current configuration and the tentative new configuration, and as the past configurations are forgotten, one constructs a Markov process when applying an iterative improvement heuristic.

There are usually many ways to change a configuration. The various possibilities for changing a configuration are implemented in one or more move routines. A move is a description of how a configuration can be changed. Usually, these move routines contain some random elements, such that the number of configurations τ that can be reached by starting from configuration σ and applying a move $\sigma \to \tau$ is rather large.

Not every move need be accepted. Each improvement heuristic includes some rule according to which a tentative new configuration τ is accepted or not if compared to the actual configuration σ. If the move is accepted, the new configuration is τ, from which we move in the next time step. If it is rejected, σ remains the actual configuration and a new move is applied to it in the next time step. The various Markovian improvement heuristics mainly differ in the choice of the acceptance rule. This choice is completely defined by a transition probability function $p(\sigma \to \tau)$.

8.2 Trivial Acceptance Functions

The most trivial acceptance rule is used in the random walk (RW). In a RW, one of the neighboring configurations τ of the current configuration σ is chosen at random. Each move from a configuration σ to a neighboring configuration τ is accepted such that the transition probability function is simply given by

$$p(\sigma \to \tau) = 1\,. \tag{8.3}$$

The RW is mainly used in cases where there are no large structures in the energy landscape (which might be otherwise used by a more elaborate optimization technique). As a result, one can only hope to pick up a good solution while walking around at random. This is especially the case if the energy landscape is shaped like a golf course, i. e., if there are only a few isolated states of a high quality, and there is no sign of their existence in their neighborhood.

If one deals otherwise with a nicely shaped energy landscape that contains no other local minimum than the global optimum, the best choice is to use a greedy algorithm, which accepts only those moves that lead to either a better or an equally good configuration. As in the RW, a configuration τ that is a neighbor of the current configuration σ is chosen at random. Let

$$\Delta\mathcal{H} = \mathcal{H}(\tau) - \mathcal{H}(\sigma) \tag{8.4}$$

be the difference between the energies of two participating configurations; then the transition probability is simply given by

$$p(\sigma \to \tau) = \begin{cases} 1 & \text{if } \Delta\mathcal{H} \leq 0 \\ 0 & \text{otherwise} \end{cases} \tag{8.5}$$

such that each move that leads to a better or equally good configuration is accepted, but each move leading to a worse configuration is rejected. If a move is rejected, the system remains in the present configuration σ.

Sometimes one finds in the literature slightly changed acceptance rules for the greedy algorithm: either only improvements are accepted, i. e.,

$$p(\sigma \to \tau) = \begin{cases} 1 & \text{if } \Delta\mathcal{H} < 0 \\ 0 & \text{otherwise} \end{cases}\,, \tag{8.6}$$

or the case $\Delta\mathcal{H} = 0$ is accepted with a specific probability between 0 and 1, usually 0.5. According to our experience [186], acceptance rule (8.5) leads to better results than not accepting every $\Delta\mathcal{H} = 0$-move, if the construction of the moves exhibits enough randomness. If it does not, then sometimes one might end up in an endless loop between equally good solutions if using (8.5) such that one has to accept this so-called "trivial" type of move with a specific probability only in order to introduce additional randomness into the system.

There exist some problems that can be solved with this greedy algorithm optimally, but usually complex problems have energy landscapes that exhibit a large number of local minima and extended hill and valley structures. Here the greedy algorithm ofen ends up in a bad local minimum, which lies close to the starting point of the optimization run. Therefore, one has to use more elaborate transition probabilities that also accept some deteriorations and that enable the Monte Carlo walker to climb over barriers in the energy landscape.

8.3 Introduction of a Control Parameter

Since the RW and the greedy are thus not suitable to deal with complex energy landscapes, one has to think of more elaborate transition probabilities p. These transition probabilities should allow for deteriorations; however, they shall lead the Monte Carlo walker to a (quasi) optimum solution at the end of the optimization run.

Not every deterioration can be accepted, as the algorithm would then be identical to the RW. Furthermore, large deteriorations should not be accepted with the same probability as small deteriorations. Therefore, the transition probability should be a function of the energy difference $\Delta\mathcal{H}$.

The possible energy differences might be of different orders of magnitude if the energy landscape is multifractal such that there are high mountains and deep valleys on a large scale, but zooming in one finds a similar mountain-valley structure. This multifractality might be repeated in several steps of zooming in. At the beginning of the optimization run, one is mostly concerned about the largest-scale barriers and wants the Monte Carlo walker to be able to climb over these barriers in order to get to a deep wide valley. Then, however, the Monte Carlo walker should stay in this valley with a large probability and search more locally for a nice subvalley by climbing over barriers of the next smallest order of magnitude. Then an ever stronger restriction should be applied, such that the Monte Carlo walker searches inside the subvalley only. Finally, at the end of the optimization run, the greedy approach would be preferable in order to let the system climb down the local valley and to let it get stuck at a nearby (quasi) optimum solution.

Obviously, some transition from the RW mode to the greedy mode is necessary in order to fulfill these wishes. For that purpose, a control parameter has to be introduced that governs the transition: at high values of the control parameter, the system accepts (nearly) every deterioration and is therefore in the RW mode, such that the Monte Carlo walker visits unordered high-energy configurations. Reducing the control parameter, more and more deteriorations are rejected, such that the Monte Carlo walker is driven downwards in the energy landscape toward better configurations. As the control parameter reaches its final value, for example zero, only improvements are allowed, such that the Monte Carlo walker climbs down the local valley and gets stuck in a (quasi) optimum solution. In a physical context, such a control parameter is the temperature T: decreasing the temperature, a physical system loses energy to a large extent and is transferred from a high-energy unordered regime to a low-energy ordered configuration.

Besides such a global control parameter, one might wish to work locally with another control parameter. This local control parameter might be useful because there are different orders of magnitude of the energy differences in some local area and in the remaining part of the system. This approach can even be extended to defining several local control parameters for various parts of the system. But a local control parameter can also be used in a different way: while the global control parameter remains constant, the local control parameter governing a small area of the system might be decreased from a large value to zero, such that a complete optimization run is performed locally. The result of this local optimization can be accepted or rejected according to the global control parameter.

8.4 Heat Bath Approach

Till now, we have only considered the standard approach making a trial move from a current solution σ to a neighboring, tentative new solution τ. The move is either accepted, such that the system jumps from σ to τ, or rejected, in which case the system stays in σ.

In contrast, the heat bath approach works according to the following recipe: all neighboring configurations τ_i of the current solution σ are considered (or at least a specific subset of them, in which at least one of the variables of σ is changed). The transition probabilities $p(\sigma \rightarrow \tau_i)$ are calculated for all of these. Let P be their sum:

$$P = \sum_{i=1}^{\mathrm{NN}(\sigma)} p(\sigma \rightarrow \tau_i) \,, \tag{8.7}$$

with $\mathrm{NN}(\sigma)$ being the number of the neighboring configurations. A uniformly distributed random number r between 0 and 1 is generated. Then that configuration $\tau_{\tilde{i}}$ is chosen as the new configuration, for which the following expression holds true:

$$\sum_{i=1}^{\tilde{i}-1} p(\sigma \rightarrow \tau_i) < r \times P \leq \sum_{i=1}^{\tilde{i}} p(\sigma \rightarrow \tau_i) \,. \tag{8.8}$$

This heat bath approach has the advantage that the system always jumps to a new configuration, such that the dynamics is faster than that of the standard Markovian approach. However, there is the disadvantage that if the system is already in a local optimum, the heat bath approach forces it to get out of it if there are at least very small transition probabilities to neighboring configurations. Furthermore, if there are many ways to get to a worse configuration and only a few that lead to a better configuration, there is a large probability that this approach will choose one of the worse ways. In such a case, the standard approach would be preferable where the system stays in the current solution with the usually very large probability $1 - p$. There is another problem that occurs if the algorithm is in a quasigreedy mode: if the transition probabilities of all of the neighboring configurations investigated are zero, then the system must remain in the current configuration. However, if all possible neighbors are taken into account, this scenario marks the end of the algorithm, and the system is frozen in a local optimum.

According to our experience, this heat bath approach leads to worse results than the classic approach of randomly selecting one neighboring configuration and checking whether the move will be performed or whether the Monte Carlo walker should stay with the current configuration. In the heat bath approach, the Monte Carlo walker always has to leave his/her current solution. If now all neighboring solutions are far worse than the current solution, the Monte Carlo walker stays with a high probability with the current solution in the classic approach, whereas the move to one of the much worse configurations must be performed if using the heat bath approach. This problem can be solved partially by extending the neighborhood of the actual configuration with the actual configuration itself such that there is a comparatively large addend to P for staying in the actual configuration. However, if the number of neighboring configurations is very large, the weight of this addend relative to the sum over the transition probabilities to the worse neighbors might be smaller than the probability of staying in the classic approach.

9 Local Search

9.1 Classic Local Search Approach

Now the question arises of how best to define a move between various configurations. A first approach would be to let the system jump from the current configuration to any other randomly chosen configuration, as this approach provides the largest degree of freedom to the system. This type of move might be appropriate when using the random walk (RW). However, using the greedy algorithm, the acceptance rate, i. e., the number of accepted moves divided by the overall number of move trials, may soon be very small: most randomly built configurations have a much larger cost function value than the rather good solutions that we hope to reach. Furthermore, for complex problems, the number of these rather good solutions is extremely small compared to the overall number of configurations. Therefore, if some rather good solution has been found, there will be a long series of move trials until the system is able to jump to a better configuration. It would be simpler and even faster to perform an exact enumeration of the whole system by means of some intelligent search heuristics. Furthermore, the overall number of configurations for a larger-size problem is usually so large that a computer cannot visit all of them during the lifetime of the programmer.

Therefore, one is not interested in visiting all configurations. Instead one wants to get the Monte Carlo walker quickly to good solutions. In the classic local search approach, one expects a configuration that is rather similar to another good solution to be of roughly the same quality. Therefore, one chooses so-called small moves, which change the actual configuration only slightly. This way, one finds better solutions than the current one with a much larger probability than by moving around at random.

The question that remains is how to define the similarity between different configurations of a specific problem. Usually it is rather trivial to see whether or not two solutions are similar to each other, and we can define measures that quantify the similarity between two randomly selected configurations. The most common measure, sometimes called the Hamming distance, is simply the number of degrees of freedom that must change in a move.

9.2 Problems of the Local Search Approach

However, this local search approach inherits a problem: if the greedy algorithm is used in combination with this local search approach, the system soon gets stuck in a high-lying local minimum if the energy landscape is complex and contains many local minima. Furthermore, the resulting configuration strongly depends on the initial state at which the optimization process was started, as the Monte Carlo walker simply climbs down the local valley as far as possible. These problems can be partially overcome by the use of a control parameter allowing uphill moves. However, at the end of the optimization run, when the control parameter is already rather small, the system is no longer able to leave a bad local valley in a finite amount of time. In summary, if one implements the smallest move possible that performs the smallest possible change on a configuration, one then has to invest much more time in the development of a control parameter and a strategy for its use.

Secondly, one always has to ensure that the implemented move or moves lead to an ergodic system. Ergodicity simply means that, starting at a randomly chosen configuration and using the RW acceptance criterion, every other configuration must be reachable. If the system is continuous, then the algorithm must be able to get arbitrarily near any configuration in an arbitrarily large amount of time.

Finding ergodic moves is usually possible; however, one has to work on not getting stuck in order to get really good results.

9.3 Larger Moves

Often the local search approach either has to be abandoned or at least weakened to get very good solutions. An obvious idea for weakening the local search approach is simply to implement moves that change a configuration slightly more than the smallest possible moves, to take bigger steps. Often, these "next larger moves" can be composed by successively applying two or three smaller moves, but sometimes the next larger move can only be described as a long sequence of smaller moves. One might argue that the configuration that might be reached with this next larger move could therefore have been reached by applying the corresponding sequence of smaller moves. However, the sequence might have broken when one of the smaller moves is rejected. So, by implementing not only the smallest but also the next larger moves, the neighborhood of a configuration is enlarged. Depending on the considered problem, the number of neighbors may at least double, even gain an order of magnitude or more, as we will see later.

This approach can be extended by not only implementing the smallest move(s) and the next larger move(s), but also the next next larger move(s), and so on. Summarizing, one can define a move of size s that can be composed by moves of sizes $1, \ldots, s-1$. However, this weakening of the local search approach should not be extended too far: simply imagine reaching a large amount of configurations within one move with the number of these configurations being comparable to the overall number of configurations. Then one has returned to the random approach mentioned above. Usually, one restricts oneself to moves of size $s = 1$, 2, and, at a maximum, 3 to 5. Sometimes, some special cases of larger moves turn out to be very effective. Usually, they get special names in this case and are implemented together with the smaller moves.

If therefore several move routines are implemented, there is always the question as to which move to choose. Usually, there is a general move function that calls one of the various move routines with some probability. The probabilities can be chosen equal and constant during the optimization process as a first simple approach. However, usually move routines of larger sizes s should be called more often than move routines of smaller sizes, as the larger s is, the more improbable it is at an already rather good solution that the move routine will lead to a better solution. However, the acceptance of one larger move might lead to further improvements by smaller moves that would not have been possible without the acceptance of the larger move. This behavior is summarized in the saying that after a large improvement there might be a rain of small improvements.

9.4 Jumping Between Different Move Sizes

There are also other approaches to working with moves of different sizes in combination with the greedy algorithm or an algorithm in a mode similar to the greedy: first of all, moves of size $s = 1$ are applied to the system, for some time or until no further improvement can be found. Then the various approaches differ:

- Some of them then apply moves of size $s = 2$ to the system.
  - After one such move trial is accepted, some heuristics switch immediately back to moves of size $s = 1$ and search for improvements. If, finally, no more improvements are found, they switch back to size $s = 2$ and repeat this approach until no further improvement by a move of size $s = 2$ is found. Then they recursively perform the same procedure with moves of larger sizes, only differing in how and whether they switch from a larger s back to either
    - $s - 1$ or
    - $s = 1$.
  - Other heuristics stay longer in the $s = 2$ mode and search for a fixed time or until no further improvement is known to exist. Then they switch
    - either back to $s = 1$, then
      - if improvements are found there, again to $s = 2$,
      - otherwise to $s = 3$,

  - or forward to $s = 3$.

  Again this behavior is recursively repeated.
- Other approaches then apply both moves of sizes $s = 1$ and $s = 2$, then moves of sizes $s = 1, 2$, and 3, and so on.

Usually, these approaches are considered improvement heuristics, as they try to improve the current configuration. Sometimes, however, they are classified as construction heuristics or as the cleaning phase of a construction heuristic.

10 Ruin & Recreate

10.1 The Philosophy of Building One's Own Castle

The classic local search approach sometimes fails to find (quasi) optimum solutions for very complex problems when the system cannot leave a bad local minimum because of the small size of the move and the small size of the neighborhood related to it. The view of the global system is lost such that synergy between different parts of the system cannot be used during the optimization process. Therefore, a new view has to be found that considers the whole system as a unit and not as a composition of local elements.

To relate this to everyday life, think of local search as the small changes one makes in order to get home or to work faster, to work more efficiently, to get a tastier meal by adding some pepper, and so on. However, life also offers more complex problems. If one thinks of building one's own castle, one would surely not use small moves, such as changing slightly the color of a brick in the wall. Instead, one would consider moving a tower to another point and changing its appearance completely or adding a second drawbridge. One would therefore apply moves of a large size, which always lead to a rather good solution. Furthermore, if one considered the tastes of various wines, one would find that changing the chemical composition slightly might lead to a dramatic change in taste. Very complex optimization problems are sometimes like this—the local search approach must fail for them.

10.2 Outline of Approach

All in all, the system must be changed not locally but over a macroscopic scale. The question arises as to how this can be achieved. On the one hand, one should not use the local search approach. On the other hand, the system will not be destroyed completely, because then construction heuristics have to be applied in order to start over. Therefore, the approach can only be to destroy a large part of the system, then reintroduce the removed items in a specific way, and, finally, ask the underlying improvement heuristic whether this move will be accepted. Such a large move therefore consists of two parts, a destruction, which is called ruin, and a rebuild routine, which is called recreate [194]. The destroyed part of the system will be large enough such that the impact of the "bomb" that is thrown on the system will be noticeable not only locally but in the whole system. On the other hand, it should be small enough so that at least a small basis of the system, a skeleton, remains with which an efficient recreate can be performed. Both parts of the move must influence the whole system. Usually, some percentage $\mathcal{P}$ can be defined as the maximum percentage of items that are removed from the system by the ruin method. Each ruin & recreate (R & R) move starts by generating a random number in the interval $[0; \mathcal{P}]$ in order to determine the fraction of the system to be removed.

However, not every possible ruin and not every possible recreate method is suited for creating optimum or quasioptimum results. For each specific problem, one must look at the whole system from various viewpoints, think in a global manner, and then introduce a few large moves to reach the optimum. As an example of a bad recreate, consider the possibility that each item that was removed from the system is reintroduced in a random way. At first glance, this seems to be a good choice, as it recalls the way in which configurations were changed by small moves; furthermore, the system seems to achieve a large degree of freedom of movement. However, moves containing such a recreate part will seldom lead to important improvements. The acceptance rate of such moves vanishes already when the control parameter is of medium size, much earlier than if using small-sized moves, as here the size of the possible deteriorations is much larger.

Therefore, one must pay attention to which possibilities are used for a ruin and which for a recreate method and how these two routines are combined in a move. Both the ruin and the recreate parts will at least partially be of algorithmic character and partially work with rules that leave room for random choices. The moves will inherit a mixture between a certain degree of randomness and a proposed set of strict rules. Furthermore, after various possibilities for ruins and recreates have been found, under some circumstances only relatively few combinations of ruins and recreates are suitable, i. e., one has to find pairs of ruins and recreates that can work together well. Let us exemplify this with the result of a historic bad R & R move (this story is still told with a sad smile by the guides who lead tourists through the castle of Heidelberg): elector Duke Friedrich V, who reigned the Palatinate from 1610 to 1632, had some moats and furthermore a valley behind the castle of Heidelberg filled in, thereby reducing the fortifications of the castle, in order to have enough space for creating the famous castle garden Hortus Palatinus, a present to his beloved wife Elisabeth. In the Thirty Years War (1618–1648), the French general Tilly was able to capture the castle in 1622: before this change, the only way up to the castle on the mountain was from the Neckar valley, on which the cannons of the castle were aimed. One can still imagine this by looking at Fig. 10.1. But Tilly's troops climbed the back side of the Königsstuhl mountain and walked over the top of the mountain, down to and through the Hortus Palatinus, and inside the castle. In the Palatine wars

Fig. 10.1. The castle of Heidelberg, the former capital of Palatinate [83]

of 1688/89 and 1693, the castle was again captured, looted, set on fire, and blown up, so that only ruins are left.

After this historical excursion, let us return to some standard possibilities for ruins: first of all, one could imagine selecting items from the system randomly and removing them. A skeleton system remains that is the initial point for the recreate routine. This type of a ruin has an impact on all parts of the system and is not local at all. Even more than 50% of the whole system can be removed for some problems. There is also the possibility of making use of some neighborhood or similarity relation between the individual items of the system. First, one item c is randomly selected and removed from the system. Then all items that are in some relationship to c are removed. This ruin type might have a more local impact, such that only 1 to 20% of the system, depending on the problem, should be removed.

There are also several possibilities for a recreate: a standard way to devise a recreate would be to take a suitable construction heuristic of the insertion type and alter it in such a way that it recreates only the destroyed part, leaving the remaining system unchanged [194]. One can compare this approach somewhat with the work of an archaeologist, anthropologist, or paleontologist, who has a half-complete skeleton and a box full of bones in front of him and has to put them together like the pieces of a mosaic. There is also another approach that can be used, i. e., starting a local optimization run based on an improvement heuristic, but only on the ruined part of the system, such that the remaining part of the system stays constant. This can be done very nicely using the local control parameter approach mentioned in Sect. 8.3.

10.3 Discussion of Ruin & Recreate

Looking closer at moves of the R & R type, one finds that their neighborhood structure is not symmetric as is usually the case with small moves. Most often, either there is no move leading back from the new configuration to the previous one or, due to the large amount of possibilities, returning to it it is quite improbable.

Furthermore, R & R has the property that it leads from a bad configuration to a much better one very fast, sometimes by only applying one move. Especially in the case where only one item is removed from the system, the recreate part of the move leads to either a better solution or at least back to the current solution. In this special case and similarly if only removing a small number of items from the system, control parameters have no effect, or only a small one, as the move does not lead to worse configurations, such that one ends up in the greedy mode. Connected to this is the fact that the recreate part leads to a system that can only move through a subset of the configuration space [a random walk (RW) using these moves is no longer a real RW]; therefore, the ergodicity condition is violated such that one has to take special care when choosing R & R routines for getting an ergodicity at least in the set of (quasi) optimum solutions. However, R & R moves have the advantage that they create a much larger neighborhood than moves of a small size s, and it is a neighborhood between locally optimum solutions when considering the former energy landscape formed by small moves.

Additionally, the question arises as to whether this approach is not too slow. The answer is that, first of all, one does not start at a random configuration at the beginning, as is mostly done when using small moves. Instead, one can easily start with a preoptimized solution, which is created by throwing a bomb on the whole system, removing all items, and calling the recreate routine. Therefore, one does not have to climb down step by step from the range of random configurations to the range of rather good solutions. Instead, the system can concentrate on finding the last percentages of improvement. Then, if working with a control parameter, the range of change for this control parameter can be much smaller, as the large moves can easily jump over the barriers of the energy landscape of the small moves. Furthermore, as a large move leads to larger changes, one does not have to call the move routine as often as if working with small moves.

Finally, it must be mentioned that this R & R approach is implicitly suited for parallel enablement: most of the time will be spent on the recreate process such that a few R & R moves can be performed in parallel. Furthermore, when working with improvement heuristics, several runs can be performed and the best result chosen.

10.4 Ruin & Recreate as a Self-Contained Optimization Algorithm

R & R can also be used as a self-contained optimization algorithm, as it inherits a control parameter, namely, the maximum percentage $\mathcal{P}$ up to which a system is allowed to be destroyed. By this percentage, the maximum number of items that may be removed and then reintroduced by the recreate part is given. This percentage can now be used as a control parameter when working, for example, with the RW as the underlying optimization heuristic: initially, the percentage $\mathcal{P}$ is set to 100% such that a ruin part is allowed to destroy up to the whole system. At the end, only one item will be removed from the system and reinserted in the best possible way by the recreate part of the move such that here the RW and greedy modes coincide. By decreasing $\mathcal{P}$, a transition is performed from the range of the rather good solutions, which can be created by construction heuristics or improvement heuristics, to achieve better and better solutions, until hopefully the algorithm stops at the global optimum.

11 Simulated Annealing

11.1 Physical and Historical Background

Annealing is a family of techniques for creating metals with desireable mechanical properties. It is an ancient technique: even in the Bronze and Iron ages, swordsmiths and other artisans had learned that the ductility of a metal tool and the hardness of its cutting edges were strongly affected by heat treatment. These might involve heating the tool to a temperature below the melting temperature of the alloy but high enough so that the crystalline grains that make up the substance begin to change their shape. After a period of such heating the metal can be slowly cooled, resulting in a softer, more ductile material, or rapidly quenched to a lower temperature, resulting in uneven cooling with a harder surface layer. Considerable skill is needed to develop the annealing schedule of temperatures and times to which the tool should be subjected, and of course the chemical composition of the alloys being used also needs careful control, but still the method has developed independently in many different cultures and remains in use today.

An extreme form of annealing came into fashion with the need to obtain very pure single-crystal semiconductors, first for making individual transistors and subsequently for refining the wafers out of which VLSI chips are made. In this method, called zone refining, a boule or cylindrical slab of very pure Si or Ge is suspended in a ring-shaped heater and a narrow band within the cylinder is heated to just above its melting point. This barely molten zone is passed very slowly through the cylinder, from one end to the other. As the Si melts it extracts impurities from the surrounding crystal, and as it recrystallizes these impurities are expelled from the reforming crystal, remaining in the molten zone. The result is a dramatic decrease in impurity concentrations, with most of them segregated in one end of the cylinder.

While neither process is exactly comparable to the problem we have set ourselves of finding optimal solutions to complex problems, there are definite analogies between annealing and optimization. Annealing thus provides a fruitful metaphor to use in seeking new and powerful heuristics for problem solving. Note that with metals the important changes take place at temperatures high enough that significant rearrangements of the structure are possible in relatively short times. Once the materials have undergone the hoped-for medium-scale rearrangements, cooling from temperatures at which significant rearrangements no longer occur to temperatures at which one can hold the tool in one's hand safely can proceed rapidly. We shall see that this, too, has its parallel in optimization heuristics through a process that has come to be called simulated annealing (SA).

SA emerged simultaneously in the work of at least three authors known to us and undoubtedly was obvious to others as well. The optimizers to whom this was obvious were physicists exposed to computer simulation of small molecules or of magnetic systems evolving while coupled to a thermal bath that provides a source of energy fluctuations. These included one of us (S.K.), C. Daniel Gelatt [110], Ken Wilson (private comm.), and Vlado Cerny [32]. Others who must have employed the technique in some form include Nick Metropolis and Stanislaus Ulam. However, the main thrust of work in the applied mathematics of optimization at that time (mid to late 1970s) was on improving the strength and rapidity of convergence of optimization searches in problems with a single minimum but very complex attraction basins. The idea of searching stochastically, and the view that interesting large-scale problems are likely to have many minima of essentially equal quality, was not accepted in the mathematics or operations research literature (and is still rare).

Gelatt and S.K., working at IBM, had studied for some years the hard optimization problems that occur in computer-aided design of computer components—placement of circuits, routing of wires, and the like. Since these always seem to involve tradeoffs between incompatible objectives, like high-speed, low-power consumption, and the constraints of limited space for wires and restricted freedom to location-specific components, Gelatt and S.K. found it perfectly reasonable to proceed with techniques that were being employed (by S.K.) at the same time to study spin glasses. Spin glasses are disordered magnetic systems in which frustration effects occur due to a mixture of ferromagnetic and antiferromagnetic interactions. Ferromagnetic interactions cause the elementary spins to prefer to align parallel with one another; antiferromagnetic interactions insist on the opposite. Naturally, the spins become "frustrated" while attempting to find the best compromise between these conflicting demands. The term "frustration" for this situation became an accepted technical term.

Spin glasses got their name because at low temperatures, the actual materials exhibited very sluggish changes in their magnetic ordering, just as actual glasses do not freeze at a well-defined temperature but instead get steadily more viscous as the temperature is lowered. Experiments on spin glasses in the 1970s often explored the different states that the material could reach at a given temperature by different thermal sequences of slow cooling with and without applied magnetic fields. Computer experiments, performed in simulation by one of us (S.K.), used annealing cycles to explore these states in simple model spin glass systems.

Wilson was writing a compiler for high-performance VLIW (very long instruction word) computing and needed to optimally pack parts of computer instructions together to use the fewest machine cycles. Cerny was interested in solving the traveling salesman problem (TSP), or at least in achieving better solutions than existing heuristics could accomplish. They all found it perfectly natural to add noise to increase the power and flexibility of the local rearrangement search that was appropriate to each problem and chose the Metropolis framework as a natural and easy-to-control way to introduce this noise. Gelatt and S.K. were blessed with mathematician colleagues who found this whole approach counterintuitive, inelegant, and vaguely disrespectful of the algorithm gods. They demanded explanations. So Gelatt and S.K. started thinking more conscientiously about why this was a sound overall approach, why it did guarantee a solution even if that solution might be costly, and how it could be framed so generally that it would appear obvious and applicable to a very wide range of problems. The result of their efforts to convince their colleagues was the long article in the journal *Science* [110] (and a popular colloquium talk that S.K. gave very widely).

11.2 Derivation of Simulated Annealing

SA is *the* classic algorithm in physics for finding low-energy or even optimum configurations for complex physical problems that cannot be solved analytically. It simulates the cooling process of a physical system, taking advantage of the fact that if this cooling procedure is performed slowly enough, the system will end up in the optimum state (e. g., a flawless crystal). On the other hand, it only reaches a less desirable local minimum in the energy landscape (e. g., a crystal with many defects) if the system is rapidly quenched down. Therefore, starting at a very high temperature, a series of temperature steps is performed between which the temperature is slowly reduced. With decreasing temperature, the system undergoes a transition from a high-energy, unordered regime to a (relatively) low-energy, (at least partially) ordered regime. The optimization process ends when the system is frozen in a (quasi) optimum state at a low temperature. Depending on the temperature range considered and on the problem, a sharp transition at (at least) one temperature can often be observed: the system changes its behavior completely at this so-called critical or Curie temperature T_C. Examples of this are melting and boiling temperatures and the Curie temperature at which some materials become ferromagnetic.

A sequence of moves is performed at each temperature step. The question arises as to how to choose a transition probability according to this physical background. Metropolis et al. [137] first proposed to choose successive states not by means of a random walk in a search space Γ but by another Markov process in which each new state σ_{i+1} is reached from the previous, neighboring state σ_i by a move with a certain transition probability $p(\sigma_i \to \sigma_{i+1})$.

Metropolis et al. [137] considered classical physical systems whose canonical equilibrium distribution is given by the Boltzmann distribution. The probability of the system being in a state σ is therefore given as

$$\pi_{\text{equ}}(\sigma) = \frac{1}{Z} \exp\left(-\frac{\mathcal{H}(\sigma)}{k_{\text{B}}T}\right) , \tag{11.1}$$

with the energy function $\mathcal{H}$, Boltzmann constant $k_{\text{B}} = 1.3807 \cdot 10^{-23}$ J/K, temperature T, and partition sum

$$Z = \sum_{\sigma \in \Gamma} \exp\left(-\frac{\mathcal{H}(\sigma)}{k_{\text{B}}T}\right) . \tag{11.2}$$

Instead of the temperature, one often uses the inverse temperature $\beta = 1/(kT)$ in formulas. Metropolis et al. observed that it is possible to choose the transition probability p in such a way that even if starting at any arbitrary distribution, in the limit $i \to \infty$ the distribution of the states, which are created in this Markov process, tends toward the desired Boltzmann distribution for any fixed temperature $T > 0$. If the system has already reached a stationary equilibrium, the master equation

$$\pi_{\text{equ}}(\sigma) \sum_{\tau \in \Gamma} p(\sigma \to \tau) = \sum_{\tau \in \Gamma} \pi_{\text{equ}}(\tau) p(\tau \to \sigma) \quad \forall \sigma \tag{11.3}$$

is fulfilled, i. e., $\mathrm{d}\pi_{\text{equ}}(\sigma) = 0$, and therefore $\pi_{\text{equ}}(\sigma) = \text{const}$. A sufficient condition for fulfilling the master equation is the stricter principle of detailed balance. It assumes that there exists an equilibrium between each pair (σ, τ) of configurations. Let $N_{\text{equ}}(\sigma \to \tau)$ be the expected number of Monte Carlo walkers that move from state σ to state τ in equilibrium and $N_{\text{equ}}(\tau \to \sigma)$ the number that move in the opposite direction. A detailed balance demands $N_{\text{equ}}(\sigma \to \tau) = N_{\text{equ}}(\tau \to \sigma)$. The amount $N_{\text{equ}}(\sigma \to \tau)$ can be written as the product of the amount $N_{\text{equ}}(\sigma)$ being in σ and the transition probability $p(\sigma \to \tau)$. Dividing $N_{\text{equ}}(\sigma)$ and $N_{\text{equ}}(\tau)$ by $\sum_{\upsilon} N_{\text{equ}}(\upsilon)$, one receives the common form of the detailed balance equation:

$$\pi_{\text{equ}}(\sigma) p(\sigma \to \tau) = \pi_{\text{equ}}(\tau) p(\tau \to \sigma) . \tag{11.4}$$

Detailed balance is a sufficient but not a necessary condition for fulfilling the master equation. Inserting Eq. (11.1) into Eq. (11.4), one gets the combined condition that the ratio between the transition probabilities for the move $\sigma \to \tau$ and for its inverse move $\tau \to \sigma$ only depend on the temperature T and the energy difference $\Delta\mathcal{H}$ between the two participating configurations but not on the actual values of their energies or other parameters:

$$\frac{p(\sigma \to \tau)}{p(\tau \to \sigma)} = \exp\left(-\frac{\Delta\mathcal{H}}{k_{\text{B}}T}\right) . \tag{11.5}$$

Obviously, the transition probability $p(\sigma \to \tau)$ has not been uniquely defined yet; there is still some arbitrariness in the explicit choice of p [23]. Most often, one chooses either the Metropolis criterion

$$p(\sigma \to \tau) = \min\left\{1, \exp\left(-\frac{\Delta\mathcal{H}}{k_{\mathrm{B}}T}\right)\right\} = \begin{cases} \exp\left(-\frac{\Delta\mathcal{H}}{k_{\mathrm{B}}T}\right) & \text{if } \Delta\mathcal{H} > 0\,, \\ 1 & \text{otherwise} \end{cases} \tag{11.6}$$

(here the larger of the two transition probabilities is simply set to 1) or the heat bath condition

$$p(\sigma \to \tau) = \frac{1}{1 + \exp\left(\frac{\Delta\mathcal{H}}{k_{\mathrm{B}}T}\right)}\,, \tag{11.7}$$

which is often also written in the form

$$p(\sigma \to \tau) = \frac{1}{2}\left(1 - \tanh\left(\frac{\Delta\mathcal{H}}{2k_{\mathrm{B}}T}\right)\right) \tag{11.8}$$

and obtained by the ansatz

$$p(\sigma \to \tau) = \frac{\pi_{\mathrm{equ}}(\tau)}{\pi_{\mathrm{equ}}(\sigma) + \pi_{\mathrm{equ}}(\tau)}\,. \tag{11.9}$$

These ansatzes fulfills the condition of detailed balance as easily can be shown.

A series of states $\sigma \to \tau \to \upsilon \to \ldots$ created with one of these transition probabilities tends toward the Boltzmann equilibrium distribution. This can be checked by the following argument [23]. Assume that the system is not in equilibrium such that the states occur according to some probability distribution $\pi \neq \pi_{\mathrm{equ}}$. The probability $\pi(\sigma)$ changes in time according to

$$\mathrm{d}\pi(\sigma) = \sum_{\varrho\neq\sigma} \mathrm{d}\pi_{\varrho}(\sigma) = -\pi(\sigma)\times\sum_{\varrho\neq\sigma} p(\sigma \to \varrho) + \sum_{\varrho\neq\sigma} \pi(\varrho)\times p(\varrho \to \sigma)\,. \tag{11.10}$$

If this probability distribution was stationary, then $\mathrm{d}\pi(\sigma) = 0$ for all configurations σ. Now consider a pair (σ, τ) of states and let—without restriction—$\mathcal{H}(\sigma) < \mathcal{H}(\tau)$. Using the Metropolis criterion, one gets the transition probabilities $p(\sigma \to \tau) = \exp(-\Delta\mathcal{H}/(k_{\mathrm{B}}T))$ and $p(\tau \to \sigma) = 1$. Therefore, one gets

$$\begin{aligned} \mathrm{d}\pi_{\tau}(\sigma) &= -\pi(\sigma) \times p(\sigma \to \tau) + \pi(\tau) \times p(\tau \to \sigma) \\ &= \pi(\sigma) \times \left(-\exp(-\Delta\mathcal{H}/(k_{\mathrm{B}}T)) + \frac{\pi(\tau)}{\pi(\sigma)} \times 1\right) \\ &= \pi(\sigma) \times \left(\frac{\pi(\tau)}{\pi(\sigma)} - \frac{\pi_{\mathrm{equ}}(\tau)}{\pi_{\mathrm{equ}}(\sigma)}\right)\,. \end{aligned} \tag{11.11}$$

Now, if $\pi(\tau)/\pi(\sigma) > \pi_{\mathrm{equ}}(\tau)/\pi_{\mathrm{equ}}(\sigma)$, then there is some probability transfer from τ to σ; if it is smaller, then the reverse is true. Therefore, a simulation working with the Metropolis criterion at a fixed temperature leads the system to the Boltzmann equilibrium. The same holds true if using the heat bath criterion.

Using the Metropolis criterion, the proposed new state is always accepted if it is either better than or as good as the actual state. It is still accepted with a certain probability if it is worse. Comparing the Metropolis criterion with the heat bath condition, one finds that the Metropolis criterion leads to a faster dynamics, as $p_{\text{Metropolis}}$ is larger than $p_{\text{heat bath}}$ for all energy differences. As a result, the Metropolis criterion is used more often. The heat bath condition is often used for cases in which one does not want to accept every trivial move (i. e., a move with an energy difference $\Delta\mathcal{H} = 0$). Sometimes the dynamics is also slowed down by a certain factor by which the transition probabilities $p_{\text{Metropolis}}$ and $p_{\text{heat bath}}$ are divided. In simulations, the Boltzmann constant k is usually set to 1, so that the temperature is only a control parameter in the same unit as the energy.

The question of whether a certain move is actually accepted or not, when the derived transition probability only provides a probability value for accepting the move, is decided by means of a rendom number generator. If the random number returned by the generator is smaller than the transition probability, then the move is accepted, otherwise it is rejected.

SA fulfills the requirement of ergodicity. According to Ehrenfest, a system is called ergodic if the trajectory in the phase space Γ passes arbitrarily near to each point in Γ as time approaches infinity. If Γ is discrete, then the phase space trajectory actually touches each point in Γ. Ergodicity is especially important for the calculation of the expectation values of observables. Only if the condition of ergodicity is fulfilled will the average taken over several independent simulations give the same results as a time average.

Since its introduction by S.K. et al. [110] SA has become one of the standard tools for finding quasioptimum configurations of complex problems not only in physics but also in computational chemistry, biology, and even operations research. In such fields the cost or objective function of the given problem to be minimized is simply considered as the Hamiltonian of a classical physical system; then both the energy and the temperature are measured in dollars or euros. SA has often even produced optimum results for NP-complete problems, for which the calculation time for finding the optimum solution with exact methods increases exponentially with system size. A prominent example of these problems is the TSP, which will be discussed later.

11.3 Thermal Expectation Values

The thermal expectation value of an observable $\mathcal{A}$, according to the Boltzmann distribution, is defined in a discrete system as

$$\langle \mathcal{A} \rangle = \sum_{\sigma \in \Gamma} \mathcal{A}(\sigma) \pi_{\text{equ}}(\sigma) = \frac{\sum_{\sigma \in \Gamma} \mathcal{A}(\sigma) \exp\left(-\frac{\mathcal{H}(\sigma)}{k_{\text{B}}T}\right)}{\sum_{\sigma \in \Gamma} \exp\left(-\frac{\mathcal{H}(\sigma)}{k_{\text{B}}T}\right)} . \tag{11.12}$$

The expression in the denominator is simply the partition function Z [see also Eq. (11.2)]. In particular, the expectation value of the Hamiltonian $\mathcal{H}$ is given by

$$\langle \mathcal{H} \rangle = \frac{1}{Z} \sum_{\sigma \in \Gamma} \mathcal{H}(\sigma) \exp\left(-\frac{\mathcal{H}(\sigma)}{k_{\text{B}}T}\right) . \tag{11.13}$$

Formally, $\langle \mathcal{H} \rangle$ can be expressed as the logarithmic derivative of Z with respect to the inverse temperature $\beta = 1/(k_{\text{B}}T)$:

$$\begin{aligned} -\frac{\partial}{\partial \beta} \ln Z &= -\frac{1}{Z} \sum_{\sigma \in \Gamma} \frac{\partial}{\partial \beta} \exp\left(-\beta \mathcal{H}(\sigma)\right) \\ &= \frac{1}{Z} \sum_{\sigma \in \Gamma} \mathcal{H}(\sigma) \exp\left(-\beta \mathcal{H}(\sigma)\right) = \langle \mathcal{H} \rangle . \end{aligned} \tag{11.14}$$

The specific heat C, which plays a main role in considering an optimization process from a physical point of view, is defined as the derivative of the expectation value of the Hamiltonian with respect to the temperature T:

$$C = \frac{\partial \langle \mathcal{H} \rangle}{\partial T} . \tag{11.15}$$

It can be rewritten as

$$\begin{aligned} C = {} & \frac{1}{Z^2} Z \frac{1}{k_{\text{B}}T^2} \sum_{\sigma \in \Gamma} \mathcal{H}^2(\sigma) \exp\left(-\frac{\mathcal{H}(\sigma)}{k_{\text{B}}T}\right) \\ & -\frac{1}{Z^2} \sum_{\sigma \in \Gamma} \mathcal{H}(\sigma) \exp\left(-\frac{\mathcal{H}(\sigma)}{k_{\text{B}}T}\right) \frac{1}{k_{\text{B}}T^2} \sum_{\sigma \in \Gamma} \mathcal{H}(\sigma) \exp\left(-\frac{\mathcal{H}(\sigma)}{k_{\text{B}}T}\right) \end{aligned}$$

$$
\begin{aligned}
&= \frac{1}{k_\mathrm{B}T^2}\left\{\frac{1}{Z}\sum_{\sigma\in\Gamma}\mathcal{H}^2(\sigma)\exp\left(-\frac{\mathcal{H}(\sigma)}{k_\mathrm{B}T}\right)\right.\\
&\qquad\left.-\frac{1}{Z^2}\left(\sum_{\sigma\in\Gamma}\mathcal{H}(\sigma)\exp\left(-\frac{\mathcal{H}(\sigma)}{k_\mathrm{B}T}\right)\right)^2\right\}\\
&= \frac{1}{k_\mathrm{B}T^2}\left\{\langle\mathcal{H}^2\rangle-\langle\mathcal{H}\rangle^2\right\}\\
&= \frac{1}{k_\mathrm{B}T^2}\mathrm{Var}(\mathcal{H})\,,
\end{aligned}
\tag{11.16}
$$

to give a numerically more stable form. The specific heat usually exhibits a wide peak at a so-called freezing temperature T_f that indicates the temperature ranges in which most rearrangements of the system take place. Therefore, the specific heat is sometimes used for distributing the available calculation time over various temperature ranges. Note that the terms specific heat and heat capacity usually denote the same observable in the field of optimization.

One often considers composed Hamiltonians looking like

$$
\mathcal{H} = \mathcal{H}_0 + \mathcal{H}_1 + \mathcal{H}_2 + \cdots\,, \tag{11.17}
$$

with $\mathcal{H}_0$ being the Hamiltonian part of the basic problem and the further $\mathcal{H}_i$, $i \geq 1$, being additional perturbations, external forces, or penalty terms. Such terms are often of the form

$$
\mathcal{H}_i = -\lambda N \mathcal{M}\,, \tag{11.18}
$$

with the system size N, a flexible control parameter (Lagrange parameter) λ, which steers, e. g., the strength of the external force, and $\mathcal{M}$ being a function that might also be an interesting observable of the problem. Let us consider the Ising model, which is a classic model for studying spin interactions and therefore the basics of magnetism. Its Hamiltonian is given as

$$
\begin{aligned}
\mathcal{H} &= \qquad \mathcal{H}_0 \qquad + \mathcal{H}_1\\
&= -\sum_{\langle i,j\rangle} J_{ij}S_iS_j - HN\mathcal{M}\,.
\end{aligned}
\tag{11.19}
$$

The spins S_i can only take the values $+1$ and -1, which are considered as the directions “up” and “down”. The first term of the Hamiltonian considers the interactions between the spins. J_{ij} is the interaction matrix between these spins. The term $\langle i, j\rangle$ below the sum sign denotes that there is only an interaction between neighboring spins. Let us now consider the purely ferromagnetic case: here all entries in J_{ij} are positive. In order to minimize their energy, the spins are in parallel to each other; there are only the two ground states "all up" and "all down", which are degenerate with each other, i. e., they have the same energy value, if there is no $\mathcal{H}_1$. This second part of the Hamiltonian, which is often called the Zeeman term, describes the influence of an externally applied magnetic field on the N spins of the system. The absolute value of H denotes as a Lagrange parameter the strength of the magnetic field, its sign the direction of the field. As the magnetization $\mathcal{M}$ is given by

$$\mathcal{M} = \frac{1}{N} \sum_{i=1}^{N} S_i \,, \tag{11.20}$$

one sees that the magnetic field couples directly to the spins, trying to force each of them in its direction. The "answer" of the system to this external field can be calculated by means of the susceptibility

$$\chi = \frac{\partial \langle \mathcal{M} \rangle}{\partial H} \,, \tag{11.21}$$

which can be rewritten

$$\begin{aligned}
\chi &= \frac{\partial}{\partial H} \left\{ \frac{1}{Z} \sum_{\sigma \in \Gamma} \mathcal{M}(\sigma) \exp\left[-\beta \mathcal{H}_0(\sigma) + \beta H N \mathcal{M}(\sigma)\right] \right\} \\
&= \frac{\beta N}{Z} \sum_{\sigma \in \Gamma} \mathcal{M}^2(\sigma) \exp\left(-\beta \mathcal{H}(\sigma)\right) \\
&\quad - \frac{\beta N}{Z^2} \left(\sum_{\sigma \in \Gamma} \mathcal{M}(\sigma) \exp\left(-\beta \mathcal{H}(\sigma)\right) \right)^2 \\
&= \frac{N}{k_{\mathrm{B}} T} \left\{ \langle \mathcal{M}^2 \rangle - \langle \mathcal{M} \rangle^2 \right\} \\
&= \frac{N}{k_{\mathrm{B}} T} \mathrm{Var}(\mathcal{M}) \,.
\end{aligned} \tag{11.22}$$

This structure is also used for the susceptibilities of other parts of the Hamiltonian, replacing the magnetic field H by the corresponding Lagrange mulitplier λ. The susceptibilities usually exhibit a sharp peak in a narrow temperature range in which the system tries to optimize itself according to the external field. If a system contains more than one such additional field, the Lagrange multipliers of the additional terms in the Hamiltonian should be adjusted in such a way that the corresponding $\mathcal{H}_i$ are all minimized to their minimum possible value. This can often be achieved by choosing the Lagrange multipliers in such a way that the peaks of the susceptibilities are in the same temperature range. The reason for this is that it has been shown that the results are better if the system tries to optimize itself according to all requirements at the same time. On the contrary, if it would optimize itself first according to the first term, then according to the second term, and so on, then it would often destroy the formerly found good adjustment according to a previous field, such that the results get worse. Furthermore, a previous field could restrict the system to a part of the energy landscape in which there are no good solutions for adjusting to the new field such that there cannot be a good compensation between the two fields.

If a continuous system instead of a discrete one is considered, the sums have to be replaced by integrals over the phase space coordinates. The results, however, stay the same.

11.4 Inverse Simulated Annealing

There is a further approach involving a different use of SA, the so-called inverse simulated annealing. It is used for continuous problems in which each state consists of some continuous variables x_i. These variables can only take values in specific intervals, $x_i \in [a_i; b_i]$. The dynamics of the system is speeded up by performing only moves that set a variable to its expectation value. Therefore, in performing a move, first one of the variables x_i is chosen at random. Then the integral

$$I(x_i) = \frac{\displaystyle\int_a^{x_i} \exp\left(-\beta\mathcal{H}(\ldots, x_i', \ldots)\right) \mathrm{d}x_i'}{\displaystyle\int_{a_i}^{b_i} \exp\left(-\beta\mathcal{H}(\ldots, x_i', \ldots)\right) \mathrm{d}x_i'} \tag{11.23}$$

is calculated and a uniformly distributed random number r between 0 and 1 is chosen. Then the variable x_i is set to $x_i := I^{-1}(r)$.

Using this kind of move, the system converges rather fast to the equilibrium distribution. Of course, it is necessary for the application of this method that the integral be solvable and invertable. This method can even be used for problems in which (some of) the variables x_i can only take some discrete values in an interval. In this case, the single variables are "softened" such that they can take every value between their minimum and maximum values. Usually, one finds that the variable freezes in one of its extreme values at the end of the optimization run such that this approach is justified.

12 Threshold Accepting and Other Algorithms Related to Simulated Annealing

12.1 Threshold Accepting

An algorithm closely related to simulated annealing (SA) was proposed by Dueck and Scheuer [53, 54] and independently by Moscato [150]. This algorithm, which was called threshold accepting (TA) by Dueck and Scheuer, is given by the following deterministic transition probability:

$$p(\sigma \to \tau) = \begin{cases} 1 & \text{if } \Delta\mathcal{H} \leq Th\,, \\ 0 & \text{otherwise}\,. \end{cases} \tag{12.1}$$

A move is therefore accepted if it leads to either a better or equally good solution or to a solution that is maximally a given threshold worse than the actual solution. The threshold Th, up to which a move is accepted in TA, takes the role of the temperature T in (SA): this pseudotemperature is lowered gradually from a large initial value to zero, leading the system from a high-energy unordered regime to a low-energy ordered state.

TA has no physical analogy. Furthermore, it does not fulfill the condition of ergodicity as not every configuration in the phase space can be reached. Summarizing, the system cannot converge to a thermal equilibrium, so TA has to be considered a nonequilibrium algorithm in a strict physical sense. The largest difference between SA and TA can be observed at small systems inheriting energy landscapes containing some "golf holes", i. e., states that have a much lower energy than all of their neighbors, as shown in Fig. 12.1.

If the golf hole is so deep compared to all of its neighbors that the energy difference is larger than the threshold, then the system cannot leave the golf hole anymore. It is stuck in it. Therefore, it cannot find a better local optimum, much less the global optimum. If the system size gets larger, the density of such golf holes decreases, as the system has more ways to travel through an energy landscape of a higher dimension. It is less likely to get stuck in a golf hole, thus finding better solutions.

As SA needs a very large amount of time to really reach equilibrium, one should invest all calculation time available in one single run. Contrarily, as TA converges faster than SA [53] but is not ergodic, several short optimization runs should be performed instead of one long one. One then chooses the best result.

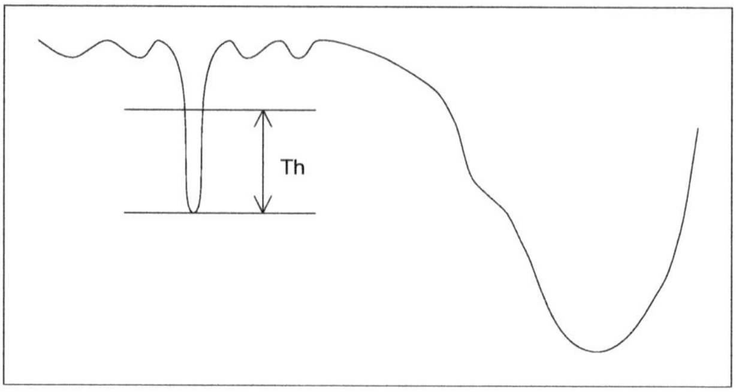

Fig. 12.1. Golf hole in energy landscape: often one finds in the energy landscape of small or specially constructed problems so-called "golf holes". These are configurations that have a much smaller energy than all of their neighboring configurations. Furthermore, there is no special evidence for them in their neighborhood. With TA, the system gets stuck in deep golf holes if the threshold is smaller than the energy difference between the golf hole state and all of its neighboring configurations. Contrarily, SA is able to leave golf holes, although the time for leaving it is rather large. The time is proportional to $\exp(\Delta/(k_\mathrm{B}T))$, with Δ being the smallest of the absolute energy differences between the golf hole and its neighboring states. When dealing with energy functions to be maximized, a similar problem occurs, but the problem of the golf holes is replaced by the problem of the so-called sugar loaves

TA can be considered as the simplest approximation of SA, fulfilling the following requirements:

- The transition probability p only depends on the energy difference $\Delta\mathcal{H}$ and a further control parameter T. It is a piecewise steady and monotonous function in both parameters:

$$p = p(\Delta\mathcal{H}, T)\,. \tag{12.2}$$

- Comparing only the Metropolis criterion and the TA transition probability, one can furthermore state that the limiting values are conserved:

$$p(\Delta\mathcal{H} \leq 0, T) = 1\,, \tag{12.3}$$

$$\lim_{\Delta\mathcal{H}\to+\infty} p(\Delta\mathcal{H}, T) = 0\,. \tag{12.4}$$

- Their integrals can be identified:
  As indicated in Fig. 12.2, the integrals over the transition probabilities of Metropolis,

$$I_\mathrm{SA} = \int\limits_0^\infty \exp\left(-\frac{\Delta\mathcal{H}}{k_\mathrm{B}T}\right)\,\mathrm{d}(\Delta\mathcal{H}) = \int\limits_0^\infty k_\mathrm{B}T\exp(-x)\,\mathrm{d}x = k_\mathrm{B}T\,, \tag{12.5}$$

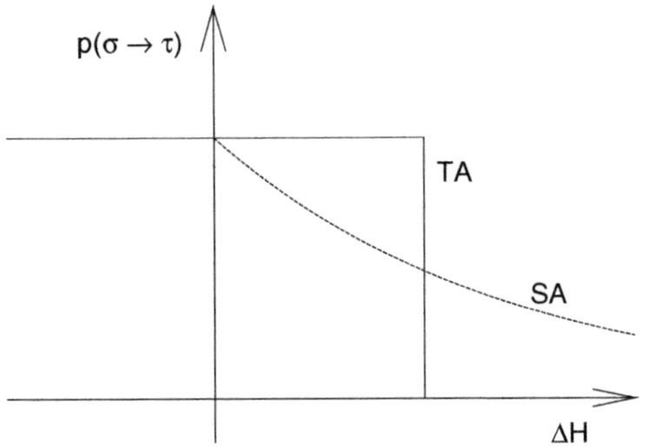

Fig. 12.2. Transition probabilities for SA (Metropolis criterion) and TA as function of energy difference $\Delta\mathcal{H}$ for fixed temperature T and threshold Th, respectively

and of TA,

$$I_{\mathrm{TA}} = \int_0^\infty \Theta(\mathit{Th} - \Delta\mathcal{H})\, \mathrm{d}(\Delta\mathcal{H}) = \int_0^{\mathit{Th}} \mathrm{d}x = \mathit{Th}\,, \tag{12.6}$$

will be equalized with each other: By setting $I_{SA} = I_{\mathrm{TA}}$, TA can be understood as the "−1"st approximation of SA, as their integrals are identified. Thus, one gets a connection between the temperature and the threshold. Of course, the calculation above assumes the energy differences are uniformly distributed and continuous between 0 and ∞. Therefore, the equivalence above holds true only for a somewhat ideal system. However, this equivalence is still approximately fulfilled in a real combinatoric optimization problem with a maximum energy difference and only a finite number of energy differences. In particular, the critical temperature and the critical threshold between the unordered high-energy regime and the ordered low-energy regime of a system are in the same order of magnitude.

In light of the above thoughts, TA seems to be the obvious choice for an approximation of SA. It is even faster than SA as there is no need for calculating an exponential value and a random number between 0 and 1, as there is no continuous probability between 0 and 1 but only the two discrete values of 0 (rejection) and 1 (acceptance).

12.2 The Steady-State Equilibrium Characteristics of TA

The transition probability of TA [Eq. (12.1)] violates the condition of detailed balance for each system consisting of at least three nondegenerate states. Let us consider a system consisting of the three states σ_1, σ_2, and σ_3 with the properties

$$\begin{aligned}\mathcal{H}(\sigma_1) < \mathcal{H}(\sigma_2) &< \mathcal{H}(\sigma_3),\\ \mathcal{H}(\sigma_2) - \mathcal{H}(\sigma_1) &< Th\,,\\ \mathcal{H}(\sigma_3) - \mathcal{H}(\sigma_2) &< Th\,,\\ \mathcal{H}(\sigma_3) - \mathcal{H}(\sigma_1) &> Th\,.\end{aligned} \tag{12.7}$$

These properties lead to the following transition probabilities:

$$\begin{aligned}p(\sigma_1 \to \sigma_2) &= 1\,,\ p(\sigma_2 \to \sigma_1) = 1\,,\\ p(\sigma_2 \to \sigma_3) &= 1\,,\ p(\sigma_3 \to \sigma_2) = 1\,,\\ p(\sigma_1 \to \sigma_3) &= 0\,,\ p(\sigma_3 \to \sigma_1) = 1\,.\end{aligned} \tag{12.8}$$

If detailed balance is fulfilled, then the following equations must hold true:

$$\begin{aligned}\pi_{\text{equ}}(\sigma_1)p(\sigma_1 \to \sigma_2) &= \pi_{\text{equ}}(\sigma_2)p(\sigma_2 \to \sigma_1)\\ \Rightarrow \pi_{\text{equ}}(\sigma_1) &= \pi_{\text{equ}}(\sigma_2)\,,\\[1em] \pi_{\text{equ}}(\sigma_2)p(\sigma_2 \to \sigma_3) &= \pi_{\text{equ}}(\sigma_3)p(\sigma_3 \to \sigma_2)\\ \Rightarrow \pi_{\text{equ}}(\sigma_2) &= \pi_{\text{equ}}(\sigma_3)\,,\\[1em] \pi_{\text{equ}}(\sigma_1)p(\sigma_1 \to \sigma_3) &= \pi_{\text{equ}}(\sigma_3)p(\sigma_3 \to \sigma_1)\\ \Rightarrow \pi_{\text{equ}}(\sigma_3) &= 0\,.\end{aligned} \tag{12.9}$$

$$\Rightarrow \pi_{\text{equ}}(\sigma_1) = \pi_{\text{equ}}(\sigma_2) = \pi_{\text{equ}}(\sigma_3) = 0\,. \tag{12.10}$$

Obviously, this disagrees with the normalization of the probabilities:

$$\sum_i \pi_{\text{equ}}(\sigma_i) = 1\,. \tag{12.11}$$

Therefore, detailed balance is not fulfilled.

The next question that arises is whether the master equation is fulfilled for TA. Let us again consider our three-state system $(\sigma_1, \sigma_2, \sigma_3)$ with

$$\mathcal{H}(\sigma_1) \le \mathcal{H}(\sigma_2) \le \mathcal{H}(\sigma_3) \tag{12.12}$$

in the case of TA. The master equation leads, in combination with the acceptance rule of TA and the energetic order of the states independently of the size of the threshold and the energy values, to the following simplified equations:

$$\begin{aligned}\pi_{\text{equ}}(\sigma_1)(p(\sigma_1 \to \sigma_2) + p(\sigma_1 \to \sigma_3)) &= \pi_{\text{equ}}(\sigma_2) + \pi_{\text{equ}}(\sigma_3)\,,\\ \pi_{\text{equ}}(\sigma_2)(1 + p(\sigma_2 \to \sigma_3)) &= \pi_{\text{equ}}(\sigma_1)p(\sigma_1 \to \sigma_2) + \pi_{\text{equ}}(\sigma_3)\,,\\ \pi_{\text{equ}}(\sigma_3) \cdot 2 &= \pi_{\text{equ}}(\sigma_1)p(\sigma_1 \to \sigma_3) + \pi_{\text{equ}}(\sigma_2)\\ &\quad \cdot p(\sigma_2 \to \sigma_3)\,.\end{aligned} \tag{12.13}$$

There are only five scenarios, depending on the relative energies of the states and of the size of the threshold:

Scenario A:

$$\mathcal{H}(\sigma_2) - \mathcal{H}(\sigma_1) < Th\,,$$
$$\mathcal{H}(\sigma_3) - \mathcal{H}(\sigma_2) < Th\,,$$
$$\mathcal{H}(\sigma_3) - \mathcal{H}(\sigma_1) > Th\,.$$

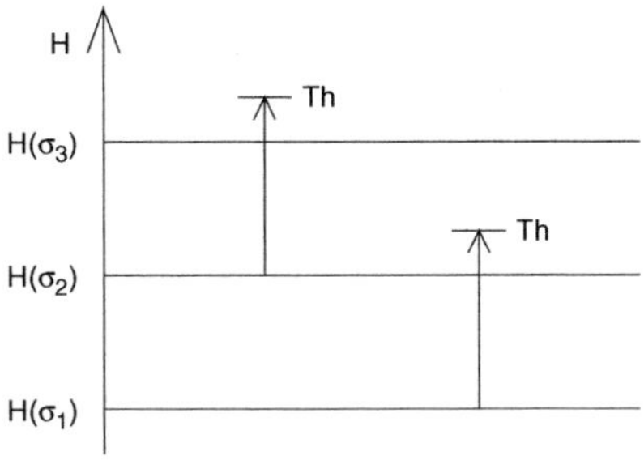

From Eq. (12.13) it follows that

$$\pi(\sigma_1) = \pi(\sigma_2) + \pi(\sigma_3)\,,$$
$$2 \cdot \pi(\sigma_2) = \pi(\sigma_1) + \pi(\sigma_3)\,,$$
$$2 \cdot \pi(\sigma_3) = \pi(\sigma_2)\,,$$

and, with the normalization of the sum of all probabilities to one, that

$$\pi(\sigma_1) = \frac{1}{2}\,, \quad \pi(\sigma_2) = \frac{1}{3}\,, \quad \pi(\sigma_3) = \frac{1}{6}\,.$$

The following results are achieved for these other scenarios:

Scenario B:

$$\mathcal{H}(\sigma_1) < \mathcal{H}(\sigma_2) < \mathcal{H}(\sigma_3)\,,$$

$$\mathcal{H}(\sigma_2) - \mathcal{H}(\sigma_1) < Th\,,$$
$$\mathcal{H}(\sigma_3) - \mathcal{H}(\sigma_2) < Th\,,$$
$$\mathcal{H}(\sigma_3) - \mathcal{H}(\sigma_1) < Th\,.$$

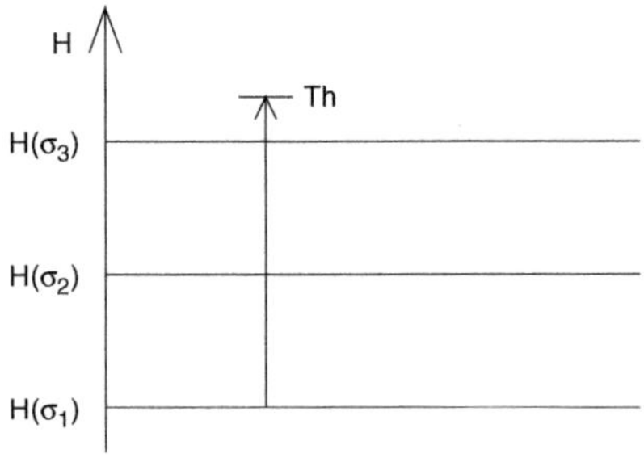

One gets the following probabilities:

$$\left.\begin{aligned} 2 \cdot \pi(\sigma_1) &= \pi(\sigma_2) + \pi(\sigma_3) \\ 2 \cdot \pi(\sigma_2) &= \pi(\sigma_1) + \pi(\sigma_3) \\ 2 \cdot \pi(\sigma_3) &= \pi(\sigma_1) + \pi(\sigma_2) \end{aligned}\right\} \Longrightarrow \quad \pi(\sigma_1) = \pi(\sigma_2) = \pi(\sigma_3) = \tfrac{1}{3}\,.$$

Scenario C:

$$\mathcal{H}(\sigma_1) < \mathcal{H}(\sigma_2) < \mathcal{H}(\sigma_3)\,,$$

$$\mathcal{H}(\sigma_2) - \mathcal{H}(\sigma_1) > Th\,,$$
$$\mathcal{H}(\sigma_3) - \mathcal{H}(\sigma_2) > Th\,,$$
$$\mathcal{H}(\sigma_3) - \mathcal{H}(\sigma_1) > Th\,.$$

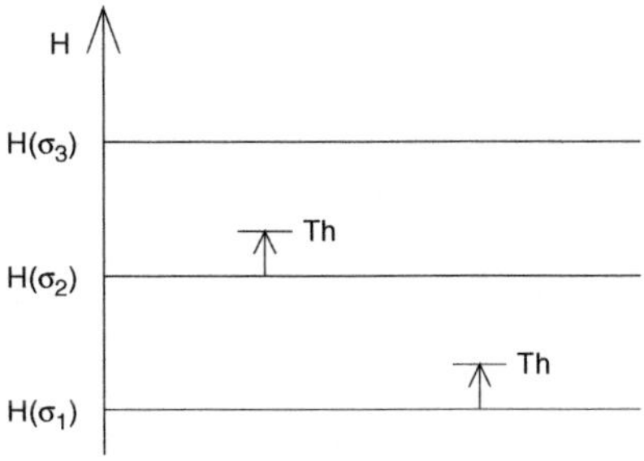

One gets the following probabilities:

$$\left.\begin{aligned} 0 &= \pi(\sigma_2) + \pi(\sigma_3) \\ \pi(\sigma_2) &= \pi(\sigma_3) \\ 2 \cdot \pi(\sigma_3) &= 0 \end{aligned}\right\} \Longrightarrow \quad \pi(\sigma_2) = \pi(\sigma_3) = 0, \pi(\sigma_1) = 1\,.$$

Scenario D:

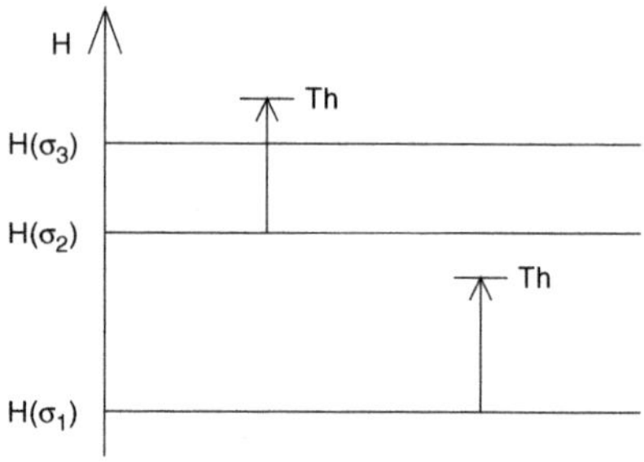

$$\mathcal{H}(\sigma_1) < \mathcal{H}(\sigma_2) < \mathcal{H}(\sigma_3)\,,$$

$$\begin{aligned} \mathcal{H}(\sigma_2) - \mathcal{H}(\sigma_1) &> Th\,, \\ \mathcal{H}(\sigma_3) - \mathcal{H}(\sigma_2) &< Th\,, \\ \mathcal{H}(\sigma_3) - \mathcal{H}(\sigma_1) &> Th\,. \end{aligned}$$

One gets the following probabilities:

$$\left.\begin{aligned} 0 &= \pi(\sigma_2) + \pi(\sigma_3) \\ 2 \cdot \pi(\sigma_2) &= \pi(\sigma_3) \\ 2 \cdot \pi(\sigma_3) &= \pi(\sigma_2) \end{aligned}\right\} \Longrightarrow \quad \pi(\sigma_2) = \pi(\sigma_3) = 0\,, \quad \pi(\sigma_1) = 1\,.$$

Scenario E:

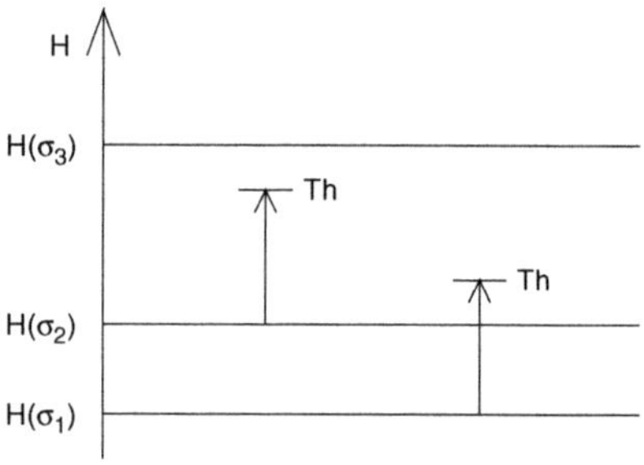

$$\mathcal{H}(\sigma_1) < \mathcal{H}(\sigma_2) < \mathcal{H}(\sigma_3)\,,$$

$$\begin{aligned} \mathcal{H}(\sigma_2) - \mathcal{H}(\sigma_1) &< Th\,, \\ \mathcal{H}(\sigma_3) - \mathcal{H}(\sigma_2) &> Th\,, \\ \mathcal{H}(\sigma_3) - \mathcal{H}(\sigma_1) &> Th\,. \end{aligned}$$

One gets the following probabilities:

$$\left.\begin{aligned} \pi(\sigma_1) &= \pi(\sigma_2) + \pi(\sigma_3) \\ \pi(\sigma_2) &= \pi(\sigma_1) + \pi(\sigma_3) \\ 2 \cdot \pi(\sigma_3) &= 0 \end{aligned}\right\} \Longrightarrow \quad \pi(\sigma_1) = \pi(\sigma_2) = \tfrac{1}{2}\,, \quad \pi(\sigma_3) = 0\,.$$

Summarizing, equilibrium distributions can be found for all scenarios without contradictions. Detailed balance is violated, but the master equation can still be fulfilled such that there can be some sort of steady state. Such nontrivial equilibrium probability distributions can also be found for all other combinatoric systems as the application of the master equation of a system with N states leads to N equations with N variables, which is always solvable in a nontrivial way.

An equilibrium probability distribution can be determined in several ways, e. g., by simulations. However, this time we do not consider the standard Monte Carlo simulation in which one Monte Carlo walker starts at a random configuration and tries to find a way through the energy landscape. Instead, one can also determine this distribution in a direct manner by calculating the transition probability matrix $(p(\sigma \to \tau))$ explicitly. These transition probabilities have to be normalized such that

$$\sum_{\tau} p(\sigma \to \tau) = 1 \quad \forall \sigma \,. \tag{12.14}$$

Starting with a uniform distribution, in which all of the N states have the same weight $\pi = \frac{1}{N}$, the following step has to be repeated again and again:

$$\pi_{\text{new}}(\sigma) = \sum_{\tau} \pi_{\text{old}}(\tau) p(\tau \to \sigma) \tag{12.15}$$

until the probability distribution $(\pi(\sigma))$ does not change anymore. Then the equilibrium probability distribution $(\pi_{\text{equ}}(\sigma))$ is reached.

Let us consider as an example the problem of the energy ladder with N equidistant steps. Each of these steps is a configuration. As the steps are numbered consecutively, we can simply set $\mathcal{H}(\sigma) = \sigma$, and the energy difference between neighboring steps is 1. We want to use a move allowing us to climb to both neighboring steps but also to jump to the two next nearest steps. One of these four steps is chosen randomly if a move is tried [57]. The temperature and the threshold are set to $\frac{3}{2}$, so that the Monte Carlo walker cannot jump to the next highest step if the basic algorithm is TA. As Fig. 12.3

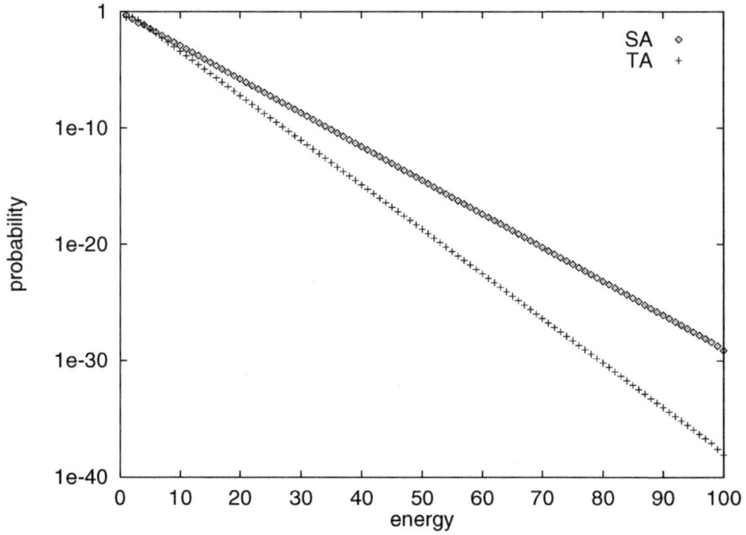

Fig. 12.3. Equilibrium distribution of energy ladder problem consisting of $N = 100$ steps for SA and TA

shows, an exponential decrease in the probability function with $\exp(-\sigma/a)$ is achieved for both SA and TA. There are only some small deviations from this exponential curve for the steps at the top and at the bottom of the ladder due to the smaller number of neighbors. For SA one finds $a = 1.5 \equiv T$ as expected; however, $a = 1.14$ is the result for TA. Therefore, the probability distribution converges to a Boltzmann-like distribution for this problem with the parameters as mentioned above for TA. Summarizing, one can state that it is justified to assume that the dynamics of TA is rather similar to that of SA as long as the system is ergodic.

However, for TA, there is no analytic expression similar to those on which one can rely when working with SA. One cannot write down a partition sum for this algorithm, as no equilibrium distribution function of this nonequilibrium algorithm is known a priori. However, one still calculates the expectation values of the observables and the derived measures specific heat and susceptibility as in SA, only replacing the temperature T by the threshold Th.

12.3 Methods Based on the Tsallis Statistics

Tsallis and Stariolo [210] introduced a method based on the generalized Tsallis entropy

$$S_q = k \frac{1 - \sum_\sigma \pi^q(\sigma)}{q-1} \tag{12.16}$$

for each real number q. If the Tsallis entropy is maximized with respect to the normalization constraint

$$\sum_\sigma \pi(\sigma) = 1 \tag{12.17}$$

and the constraint

$$\langle \mathcal{H} \rangle = \sum_\sigma \mathcal{H}(\sigma)\pi^q(\sigma) = \text{const.}, \tag{12.18}$$

then the generalized probability for the system being in the state σ is given by

$$\pi(\sigma) = \frac{\left(1 - (1-q)\frac{\mathcal{H}(\sigma)}{kT}\right)^{\frac{1}{1-q}}}{Z_q}, \tag{12.19}$$

with Z_q being the generalized partition sum

$$Z_q = \sum_\sigma \left(1 - (1-q)\frac{\mathcal{H}(\sigma)}{kT}\right)^{\frac{q}{1-q}}. \tag{12.20}$$

Note that in this generalization of the statistical mechanics, the expectation value of an observable $\mathcal{M}$ is given by

$$\langle \mathcal{M} \rangle = \sum_{\sigma} \mathcal{M}(\sigma)\pi^q(\sigma) . \tag{12.21}$$

Thus, also the detailed balance condition must be written as

$$\pi^q(\sigma)p(\sigma \to \tau) = \pi^q(\tau)p(\tau \to \sigma) . \tag{12.22}$$

Like the Boltzmann constant k_{B} in SA, the constant k is set to 1 in simulations.

In the limit $q \to 1$, the Boltzmann–Gibbs statistics are revealed, and S_q recovers the Shannon or information entropy

$$\begin{aligned} S_1 &= \lim_{q \to 1} S_q = k \lim_{q \to 1} \frac{1 - \sum_{\sigma} \pi(\sigma) \exp((q-1)\ln(\pi(\sigma)))}{q-1} \\ &= -k \sum_{\sigma} \pi(\sigma) \ln(\pi(\sigma)) . \end{aligned} \tag{12.23}$$

Tsallis and Stariolo proposed the following generalized acceptance probability:

$$p(\sigma \to \tau) = \begin{cases} 1 & \text{if } \Delta\mathcal{H} \le 0 , \\ \dfrac{1}{\left(1 + (q-1)\dfrac{\Delta\mathcal{H}}{kT}\right)^{\frac{1}{q-1}}} & \text{otherwise} . \end{cases} \tag{12.24}$$

This transition probability shows the same $T \to 0$ behavior as SA and also converges to some equilibrium distribution. However, this distribution is quite different from (12.19).

Penna also made an ansatz to generalize the Metropolis acceptance probability to the Tsallis statistics in order to solve a traveling salesman problem [163, 117]. His ansatz is given by

$$p(\sigma \to \tau) = \min \left\{ 1, \left(1 - (1-q)\frac{\Delta\mathcal{H}}{kT}\right)^{\frac{1}{1-q}} \right\} \tag{12.25}$$

and provides a quality of the results similar to SA. The temperature range between the high-energy and the low-energy regimes becomes narrower for decreasing q such that Penna suggests showing preference for negative values of q. This acceptance rule does not fulfill the condition of detailed balance; furthermore, the distribution of the states does not converge to the Tsallis equilibrium distribution. It is simply motivated by the formulas for the Tsallis statistics and by the transition to the Metropolis criterion for $q \to 1$, which can be easily checked by rewriting the acceptance probability function as the exponential of a logarithm

$$\left(1-(1-q)\frac{\Delta\mathcal{H}}{kT}\right)^{\frac{1}{1-q}} = \exp\left(\frac{1}{1-q}\ln\left(1-(1-q)\frac{\Delta\mathcal{H}}{kT}\right)\right)$$
$$\to \exp\left(\frac{-(1-q)\frac{\Delta\mathcal{H}}{kT}}{1-q}\right) = \exp\left(-\frac{\Delta\mathcal{H}}{kT}\right) \qquad (12.26)$$

for $q \to 1$. Thus, the Penna criterion is a generalization of the Metropolis criterion (11.6). Similarly, the Tsallis–Stariolo criterion is a generalization of the heat bath condition (11.7).

Investigating the Penna criterion even closer, one finds that one has to be careful when using it in its original form. Figure 12.4 shows the curves of the Penna acceptance probability function for three q-parameters with $0 < q < 1$. The Penna criterion is equal to 1 for $\Delta\mathcal{H} \leq 0$ for all values of q and T. It vanishes for

$$\Delta\mathcal{H} = \frac{kT}{1-q}\,. \qquad (12.27)$$

As Fig. 12.4 shows, the function can increase afterwards again to 1 if $1/(1-q)$ is an even number, as for $q = 0.5$ and $q = 0.9$. Here at

$$\Delta\mathcal{H} = \frac{2kT}{1-q}\,, \qquad (12.28)$$

the term $1-(1-q)\Delta\mathcal{H}/(kT)$ equals -1.

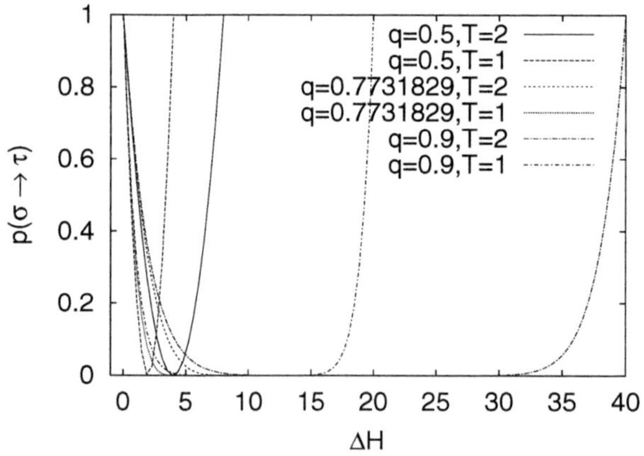

Fig. 12.4. Penna acceptance probability function for various parameters q and two temperatures $T = 2$ and $T = 1$

Thus, one should use the Penna criterion in the following form when working with $0 < q < 1$:

$$p(\sigma \to \tau) = \begin{cases} 1 & \text{if } \Delta\mathcal{H} \leq 0 \\ 0 & \text{if } \Delta\mathcal{H} \geq \dfrac{kT}{1-q}\,, \\ \left(1 - (1-q)\dfrac{\Delta\mathcal{H}}{kT}\right)^{\dfrac{1}{1-q}} & \text{otherwise}\,. \end{cases} \tag{12.29}$$

Analogously, one runs into problems if $q \leq 0$: in this case, the Penna function is negative or not defined for $\Delta\mathcal{H} > kT/(1-q)$. Here also the criterion (12.29) should be used. As this criterion contains a zero probability if the energy difference exceeds some threshold value, this algorithm inherits the properties of TA, i. e., the Monte Carlo walker is not able to leave deep golf holes and both ergodicity and detailed balance are violated.

Otherwise, if $q > 1$, one does not face any problems for $\Delta\mathcal{H} \geq 0$: in this regime, the functions are nicely decreasing with an increasing $\Delta\mathcal{H}$. Here one must be careful for $\Delta\mathcal{H} < 0$ and to modify the criterion as follows in order not to get negative probabilities:

$$p(\sigma \to \tau) = \begin{cases} 1 & \text{if } \Delta\mathcal{H} \leq 0\,, \\ \left(1 - (1-q)\dfrac{\Delta\mathcal{H}}{kT}\right)^{\dfrac{1}{1-q}} & \text{otherwise}\,. \end{cases} \tag{12.30}$$

A special case occurs for $q = 0$: in this case, the acceptance probability reduces to

$$p(\sigma \to \tau) = \begin{cases} 1 & \text{if } \Delta\mathcal{H} \leq 0\,, \\ 0 & \text{if } \Delta\mathcal{H} \geq kT\,, \\ 1 - \dfrac{\Delta\mathcal{H}}{kT} & \text{otherwise}\,. \end{cases} \tag{12.31}$$

This acceptance probability function was formerly sometimes used as a linear approximation of the Metropolis criterion in order to save calculation time.

Other Tsallis-inspired acceptance functions were subsequently generated. For example, Andricioaei and Straub [9, 10] developed from the generalized acceptance function

$$p(\sigma \to \tau) = \min\left\{1, \left(\frac{\pi(\tau)}{\pi(\sigma)}\right)^q\right\} \tag{12.32}$$

the following transition rule:

$$p(\sigma \to \tau) = \min\left\{1, \left(\frac{1 - (1-q)\dfrac{\mathcal{H}(\tau)}{kT}}{1 - (1-q)\dfrac{\mathcal{H}(\sigma)}{kT}}\right)^{\frac{q}{1-q}}\right\}\,. \tag{12.33}$$

This transition probability fulfills the criterion of detailed balance. Furthermore, it can be shown that it leads to the Tsallis equilibrium probability [Eq. (12.19)]. Furthermore, Andricioaei and Straub consider the parameter q not to be constant but to be a monotonous function $q(T)$ [9] such that the parameter q decreases with decreasing T and

$$\lim_{T \to 0} q(T) = 1 \tag{12.34}$$

such that this transition probability converges like the Metropolis criterion to the greedy acceptance probability. If the temperature T is decreased exponentially, q will as well be decreased exponentially from an appropriate starting value q_0, e. g., $q_0 = 2$.

In contrast to the TA acceptance function and the Penna criterion, this Andricioaei–Straub criterion has the disadvantage that one really must compute the new energy value $\mathcal{H}(\tau)$ and not only the energy difference $\Delta\mathcal{H}$, which is for many problems much faster to compute.

12.4 The Great Deluge Algorithm

The great deluge algorithm (GDA), which was introduced by Dueck [51], performs a random walk through a subset $\Gamma_{\mathcal{T}}$ of the configuration space Γ. All of the energy values of the configurations in $\Gamma_{\mathcal{T}}$ are smaller than a certain level $\mathcal{T}$. The transition probability is given by

$$p(\sigma \to \tau) = \begin{cases} 1 & \text{if } \mathcal{H}(\tau) \leq \mathcal{T}\,, \\ 0 & \text{otherwise}\,. \end{cases} \tag{12.35}$$

Therefore, the GDA accepts with probability 1 every configuration whose energy is smaller than the level $\mathcal{T}$. $\mathcal{T}$ assumes the part of the temperature here. In contrast to the other algorithms discussed above, the transition probability does not depend upon the energy difference but only on the energy of the tentative new configuration. The level $\mathcal{T}$ is stepwise decreasing until the system gets stuck in a local minimum. This algorithm recalls the Great Deluge or Flood legend in the Bible and is named after the same. If one wants to find the maximum in the energy landscape, the water level must rise gradually and the Monte Carlo walker is only allowed to stay on dry configurations [52]. Simply imagine that the rain starts pouring heavily out of the dark clouds in the color picture of the Monument Valley, such that the walker in this landscape is forced to climb up some mountain. In contrast to the story in the Bible, there is no Noah building an ark that is able to float on the water. When looking for the minimum in the energy landscape, however, another picture is to be preferred: a fish looks for the deepest configuration while the water level is decreasing, such that we can also speak in this case of a great drought algorithm (GDA).

The condition of detailed balance is fulfilled when using the GDA if the basic equilibrium distribution is considered as

$$\pi(\sigma) = \begin{cases} \dfrac{1}{|\Gamma_{\mathcal{T}}|} & \text{if } \sigma \in \Gamma_{\mathcal{T}}\,, \\ 0 & \text{otherwise}\,. \end{cases} \tag{12.36}$$

At a certain $\mathcal{T}$, the subset $\Gamma_{\mathcal{T}}$ will be split into several islands between which no allowed moves exist. In contrast, the whole configuration space stays connected for all temperatures $T > 0$, working with SA. There, one can always find a way between two randomly chosen configurations. The condition of ergodicity is violated by the GDA, as configurations above the water level and also on other islands can no longer be reached. Therefore, there is no thermal equilibrium with the GDA.

One might wonder—considering the picture of a valley-hill landscape partially flooded by water—why this GDA can lead to rather good results, as the allowed configuration space splits into subsets, such that one might end up on an island that does not contain any good configurations. On the surface of the earth, the movement would be restricted to one of the continents or islands; one might walk through Australia in order to end up at Mount Everest. However, this picture is only partially true. The energy landscape of a large complex optimization problem is high-dimensional, so that each state has a large number of neighbors. The Monte Carlo walker has many directions in which he/she can avoid the rising water, so the algorithm mostly leads to rather good results. The quality of the results depends on the structure of the energy landscape, on whether there are high passes between various quasioptimum configurations, and on whether $\Gamma_{\mathcal{T}}$ still percolates when $\mathcal{T}$ is already rather small.

Alternatively, one can interprete the GDA in a microcanonical way [120]: in the microcanonical ensemble, a thin energy shell $[\mathcal{H}_{\max} - \Delta\mathcal{H}, \mathcal{H}_{\max}]$ is considered. In the thermodynamic limit, the dependence upon the thickness $\Delta\mathcal{H}$ of the shell vanishes such that the GDA and a microcanonical algorithm correspond to each other.

The main disadvantage of the GDA is its slow convergence. As with SA, there is a proof showing that the global optimum of a given problem can be achieved with the GDA in an infinite amount of time. However, in practice one usually has to speed up the algorithm. Then the problem arises that the Monte Carlo walker is above the water level and he/she is not able to get below if this level is decreased too much, as the energies of all the neighboring configurations are also larger than $\mathcal{T}$. Therefore, the following acceptance function is applied for a more rapid GDA:

$$p(\sigma \to \tau) = \begin{cases} 1 & \text{if } \mathcal{H}(\tau) \le \mathcal{T}\,, \\ 1 & \text{if } \Delta\mathcal{H} \le 0\,, \\ 0 & \text{otherwise}\,. \end{cases} \tag{12.37}$$

This acceptance probability is a combination of the acceptance probability of the GDA and of the greedy algorithm. The Monte Carlo walker, when below the water level, can perform a restricted random walk below this level. By contrast, when above the level, the walker is allowed to perform greedy moves and therefore to climb downhill until he/she reaches the water level again.

Working with the GDA, the partition function of a continuous system can be written as

$$Z = \int_{\Gamma} \Theta\left(\mathcal{T} - \mathcal{H}(\sigma)\right) \, d\sigma \tag{12.38}$$

and of a discrete system as

$$Z = \sum_{\sigma \in \Gamma} \Theta\left(\mathcal{T} - \mathcal{H}(\sigma)\right) = |\Gamma_{\mathcal{T}}| \,. \tag{12.39}$$

The expectation value of an observable $\mathcal{A}$ is therefore given by

$$\langle \mathcal{A} \rangle = \frac{\int_{\Gamma} \mathcal{A}(\sigma) \Theta(\mathcal{T} - \mathcal{H}(\sigma)) \, d\sigma}{\int_{\Gamma} \Theta(\mathcal{T} - \mathcal{H}(\sigma)) \, d\sigma} \tag{12.40}$$

for the continuous system and by

$$\langle \mathcal{A} \rangle = \frac{\sum_{\sigma \in \Gamma} \mathcal{A}(\sigma) \Theta(\mathcal{T} - \mathcal{H}(\sigma))}{\sum_{\sigma \in \Gamma} \Theta(\mathcal{T} - \mathcal{H}(\sigma))} \tag{12.41}$$

for the discrete system. The expectation value is therefore the arithmetic mean value averaged over all configurations in $\Gamma_{\mathcal{T}}$.

As the decrease of $\mathcal{T}$ in the GDA leads, just as the decrease of the temperature T in SA, to a decrease in energy, a parameter like the specific heat can also be defined here:

$$C = \frac{\partial \langle \mathcal{H} \rangle}{\partial \mathcal{T}} \,. \tag{12.42}$$

13 Changing the Energy Landscape

13.1 Search Space Smoothing

Simulated annealing (SA) and related optimization algorithms use a temperaturelike control parameter by which the system to be optimized is led from an unordered high-energy regime to an ordered low-energy configuration. All these control parameters have in common that they allow not only for improvements but also for some deteriorations during the optimization runs, either with a certain probability or to a certain extent. Therefore, they give the Monte Carlo walker the possibility to climb over barriers in the energy landscape and by that means to escape bad local minima. One can imagine that the Monte Carlo walker is fed with sufficient energy to climb such that he can climb over the barriers and pull himself/herself out of a deep hole like Baron Münchhausen; then the additional energy is removed from him/her gradually.

A different way not to get trapped in bad local minima would be to introduce additional moves and thereby to create a larger neighborhood for each configuration such that more ways exist to walk around an energy barrier. This means in the picture of the lower-dimensional neighborhood that there are ways through mountains such that the Monte Carlo walker is able to tunnel through barriers. But increasing the number of ways through the energy landscape also increases the possibility of missing the global optimum if a fixed number of move trials is used.

But there is also a third way one could imagine, namely, to smooth the energy landscape in such a way that the Monte Carlo walker can easily jump over barriers or, better, to remove the energy barriers completely. An ideal way of smoothing would mean that the number of minima is reduced to one, such that only the global optimum is left. There are generally four ways to smooth the energy landscape:

- The cost function $\mathcal{H}$ can be changed for every state σ and can be smoothed by some function f such that each state gets a new objective value $f(\mathcal{H}(\sigma))$.
- On the other hand, one could apply the smoothing function f to the energy differences $\Delta\mathcal{H} = \mathcal{H}(\tau) - \mathcal{H}(\sigma)$ between pairs of neighboring configurations (σ, τ), so that the smoothed energy difference between these configurations is given as $f(\Delta\mathcal{H})$.

- If the Hamiltonian of the system is more complex and is given by, e. g., $\mathcal{H}(\sigma) = \sum_i \mathcal{H}_i(\sigma)$ (i. e., if the whole Hamiltonian $\mathcal{H}$ consists of various terms, e. g., penalty or correlation terms), then it could be necessary to smooth the different addends to the Hamiltonian with different smoothing functions f_i. This could lead to the Hamiltonian $\sum_i f_i(\mathcal{H}_i(\sigma))$ for the smoothed system.
- Similarly, the energy difference $\Delta\mathcal{H}$ can also be replaced by, e. g., $\sum_i f_i$ $(\Delta\mathcal{H}_i)$ if the Hamiltonian is composed of various terms.

Although this approach will smooth the energy landscape, it has been common to use the (strictly speaking) false term search space smoothing (SSS). However, the situation is not as easy as Fig. 13.1 suggests: if one were able to smooth the energy landscape in this way, one would already know it perfectly. If one knew it perfectly, one could spare the optimization run, as one would

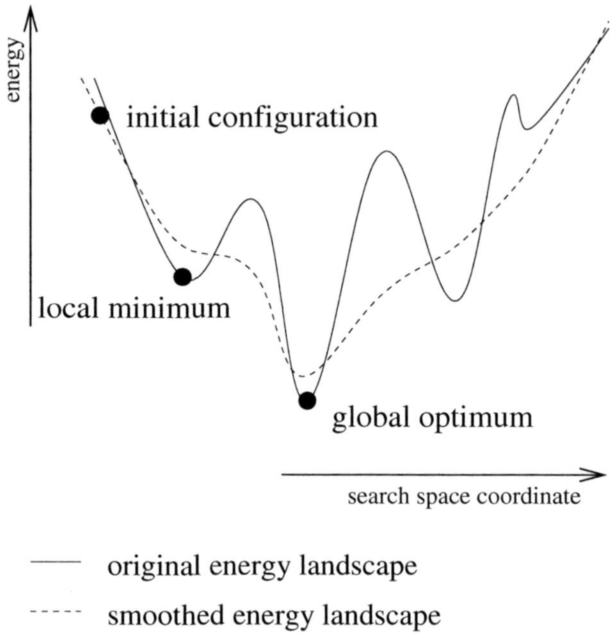

Fig. 13.1. Motivation for SSS: the graphics shows a schematic cut through the energy landscape of a simple problem. If the greedy was used as a local search algorithm in the original, i. e., unsmoothed, energy landscape, the Monte Carlo walker would get stuck in a local minimum near the starting point of his walk. Even if using more elaborate optimization algorithms like SA, TA, and GDA, one cannot be quite sure to end up at the global optimum at the end of the optimization run. However, if the energy landscape is smoothed in a way that only one minimum, identical to the global optimum, remains, the global optimum is always reached, independently of the initial configuration

know the global optimum already. In general, however, one has to deal with problems for which one does not know the energy landscape and therefore one does not know a priori where the energy barriers lie. The only way out of this dilemma is to try to smooth the energy landscape in an indirect way. This smoothing has to be done by a nonlinear formula; a linear formula f would preserve the general appearance of the energy landscape, and so it would only resize it in its height, but microscopic energy barriers are also a problem for a Monte Carlo walker who can change the height of his/her position only microscopically. An example of a nonmonotonous smoothing function, which would decrease very high energy barriers rather nicely, is the logarithm. However, applying $f(x) = \ln(x)$ to all objective values of the individual configurations would only decrease the heights of the barriers, but it would not remove them.

Therefore, one must split the Hamiltonian into many parts and apply a smoothing function f_i to each of these parts i. In the most simple way, the smoothing function is the same for all parts i. However, when smoothing the individual addends of the Hamiltonian separately by a nonlinear function like the logarithm, a further problem occurs, as the following example shows. Let σ be a local minimum and τ a neighboring configuration of σ. If these neighboring configurations only differ in one part, such that their cost function only differs in one addend, everything is fine as long as the smoothing function f is strictly monotonously increasing with the addend: let a_σ be the addend for the configuration σ and a_τ be the addend for τ. As $a_\sigma < a_\tau$, therefore $f(a_\sigma) < f(a_\tau)$. However, a difficulty occurs if neighboring configurations differ at least in two addends from the Hamiltonian: let us consider an example in which the configurations σ and τ differ in exactly two addends. Let us denote the values of these addends as a_σ and b_σ for the configuration σ and as a_τ and b_τ for the configuration τ. Of course, $a_\sigma + b_\sigma < a_\tau + b_\tau$. However, it is not necessarily the case that $f(a_\sigma) + f(b_\sigma) < f(a_\tau) + f(b_\tau)$. As an example, let us again use the logarithm as a smoothing function and choose $a_\sigma = \mathrm{e}^3$, $b_\sigma = \mathrm{e}^3$, $a_\tau = \mathrm{e}^4$, and $b_\tau = \mathrm{e}$. It is $\mathrm{e}^3 + \mathrm{e}^3 < \mathrm{e}^4 + \mathrm{e}^1$, but it is $3 + 3 > 4 + 1$. Therefore, in the smoothed landscape σ is no longer a local minimum. This type of transfer of local minima to other configurations can also happen for every other nonlinear smoothing formula. Even the global optimum might be displaced.

Therefore, one cannot simply perform an optimization run in the smoothed energy landscape, as the system will usually end up in a configuration that is a local minimum or even the global optimum in the smoothed landscape but not a minimum in the original energy landscape. The way to overcome this problem is to introduce a smoothness-control parameter α [76] by which the smoothness of the changed energy landscape can be governed: as Fig. 13.2 shows, the optimization run starts at a, for example, very large value of the smoothness-control parameter, at which the energy landscape hopefully only contains one local minimum, which is therefore the global optimum. At this

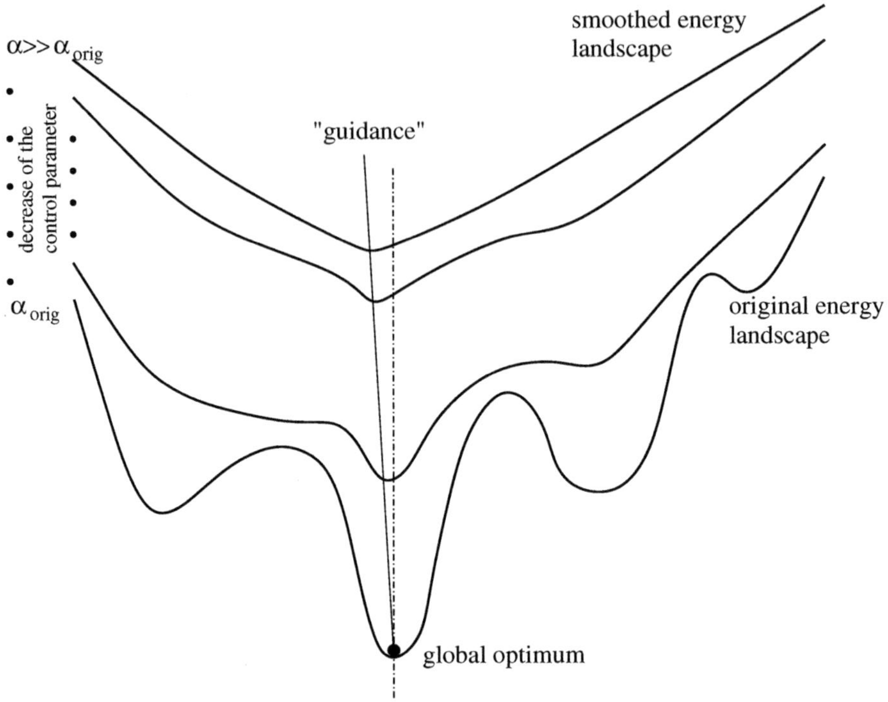

Fig. 13.2. Introduction of a smoothness control parameter: the smoothness of the energy landscape is reduced step by step by reducing some smoothness control parameter α until the original energy landscape is retrieved at the end for $\alpha = \alpha_{\text{orig}}$. After the optimum solution has been found for the smoothed landscape, this solution must be adapted in each desmoothing step such that the optimum, or at least a rather good solution, is achieved at the end (graphics constructed analogously to graphics in [76])

large value of α, a greedy optimization run is performed in order to let the system jump into the global optimum of this smoothed energy landscape. After that, the smoothness-control parameter is reduced by some small amount such that the energy landscape is slightly desmoothed. This might lead to some displacement of the minimum of the energy landscape. Therefore, a new greedy optimization run must be performed in order to get from the former minimum in the energy landscape to the minimum of the new energy landscape. Then α is decreased again and a further greedy optimization run is performed in the again slightly desmoothed energy landscape. This approach is iterated until some final value of $\alpha = \alpha_{\text{orig}}$ is reached that restores the original energy landscape.

Of course, the desmoothing steps must be very small so that one may trust in a guidance effect by which the optimum solution for the smoothed landscape is transferred step by step to the optimum or at least a quasioptimum configuration in the original landscape. By changing α too drastically, one might be moved far away from the former valley and therefore be caught in some bad local minimum. There is some analogy here between SA, in which the temperature has to be decreased slowly, and this guidance effect, which occurs only if the smoothness-control parameter is decreased slightly. Otherwise, one also finds here some quenching effect.

This analogy has been stressed further: both types of optimization heuristics, SA and relatives and SSS, rely heavily on a control parameter; furthermore, a greedy run is performed on the original energy landscape at the end of the optimization run. Therefore, the question naturally arises as to whether this SSS approach could be combined with SA or other algorithms related to it. Of course it makes no sense at all performing a complete optimization run with, e. g., SA at each α step because at large values of T, the system jumps out of the local valley in which it was already caught at small α and performs a quasi-random walk (RW) such that the guidance effect of SSS is lost. However, using the greedy, i. e., the $T = 0$, case of SA and its relatives must be considered the extreme on the other side.

Probably the most natural way of combining SSS with a more elaborate acceptance function is to use the great deluge algorithm (GDA): at each step of α, the water level $\mathcal{T}$ is set to the new value of the initial configuration of the new α step [186]. Now the Monte Carlo walker is restricted to the valley where he/she is already caught, but can search for the best local minimum inside this valley as there might be more than one possible local minimum between which new energy barriers might have arisen. Therefore, at each α step a GDA optimization run is performed, with a starting value equal to the energy of the initial configuration.

The SSS approach of removing energy barriers between neighboring local optima strongly resembles a chemical catalytic process by which reactions are enabled that would not happen in the absence of the catalyst. These reactions also take place at very low temperatures. Therefore, the approach combining SA and related algorithms with SSS would be to work with a very small and constant value of the control parameter T at which the system would be (nearly) frozen in the original energy landscape. However, due to the smoothing process, the energy differences between neighboring configurations can be much smaller in the smoothed energy landscape than in the original energy landscape. Therefore, the optimization process might start in a (quasi)-RW mode at large α. For each α step, a single optimization run shall be performed. The moves shall be accepted with the acceptance probability using the small value of T. With decreasing α, the energy differences get larger such that both a SSS and a SA/TA transition are performed by reducing α. However, when working, e. g., with threshold accepting, it is not necessary to choose $\mathit{Th} < \min\{\Delta\mathcal{H} | \Delta\mathcal{H} > 0\}$, i. e., smaller than the smallest possible energy difference in the original system. Although the system does not freeze completely, one gets roughly the same process in the simulation. The original SSS is by the way a special case, that is, the $T = 0$ case of this combined method.

13.2 Ant Lion Heuristics and Activation Relaxation Technique

The ant lion (Fig. 13.3) is an insect with an interesting behavior: in order to obtain food, which mainly consists of ants, it alters the landscape by making funnels. If an ant or another small insect comes by, it often gets trapped in one of these funnels. Additionally, the ant lion throws sand to the insect in order to increase the probability that the insect will sink into its funnel. Figure 13.4 shows such a landscape of ant lion funnels.

In summary, ant lions alter the landscape by introducing appropriate local minima in order to maximize their food. This approach is also used as an optimization algorithm, the so-called ant lion algorithm [205], which is mostly used for problems in which an objective function can only partially be written down, either because it is too difficult to quantify all constraints or because it is simply impossible. Therefore, one starts several independent optimization runs based on a restricted Hamiltonian $\mathcal{H}$, each of them leading to some solution. Then the individual solutions are analyzed. If one likes a specific solution due to fulfilling some constraints that are not represented in the Hamiltonian, then one would like to make the corresponding funnel

Fig. 13.3. An ant lion [92]

Fig. 13.4. Landscape with holes formed by ant lions [92]

in which this local minimum solution sits deeper, just like an ant lion. Similarly, if one does not like a given solution, one would like to make the funnel less deep. Thus it is not a good idea to simply change the overall objective value of a local minimum, as then the surrounding funnel would stay unchanged.

There are several ways to solve this problem: by checking all neighboring configurations of a "good" local minimum, one might be able to find out what makes this local minimum good, so that the addend to the Hamiltonian of a certain part of the configuration is changed. A second choice consists of comparing several good local minima; in this way one might find that some of them exhibit common parts. Again the values for these common parts are changed. On the other hand, when comparing "bad" solutions and their neighborhood, one will also find some structures that make these solutions bad. The corresponding addends are then increased. If now some further optimization runs are performed in this changed landscape, one will hopefully no longer end up in one of the former "bad" solutions and one might be led to "good" solutions that contain the parts one likes. However, the problem usually arises that not only are some local valleys filled up or turned into a hill, such that the system is less likely to get stuck there, and other local valleys deepened, but additional valley-hill structures occur leading to a rougher landscape than before. However, these new local minima might exhibit several nice structures and might therefore be a superposition of the individual changed addends for the "good" properties. However, it could also happen that the combination of several good properties does not lead to any good solution. This is strongly system dependent.

Another prominent example of algorithms that change the energy landscape in a constructive way is the activation relaxation technique (ART) [15, 152, 153, 208], which performs moves between local minima in the energy landscape of a continuous problem using the conjugate gradient method. These moves are accepted according to the Metropolis criterion. This method is intended to overcome the difficulty of SA to climb over barriers in the energy landscape at low temperatures.

Thus, this algorithm starts at a randomly created configuration and moves to a local minimum nearby with the Conjugate Gradient method. Other methods searching for local minima could be used instead. The idea is now to leave this local minimum and get to another local minimum by following a path of minimum energy. Therefore, the highest possible point in this path must be a saddle point of minimum order, usually of order 1. The construction of this path leading from one local minimum to another consists of three parts:

- Leaving the so-called harmonic area around the current local minimum,
- Propagating the system to a saddle point nearby, and
- Relaxing the system in a new local minimum in the energy landscape.

For the initial local minimum, a local search method is performed using the force field $\boldsymbol{F} = -\boldsymbol{\nabla}\mathcal{H}$. In order to escape the current local minimum, first a small move is performed at random. Then one continues with the force field

$$\tilde{\boldsymbol{F}} = \boldsymbol{F} - \boldsymbol{G} \tag{13.1}$$

with

$$\boldsymbol{G} = (1+\alpha)(\boldsymbol{F} \cdot \boldsymbol{\Delta r})\boldsymbol{\Delta r} \tag{13.2}$$

with

$$\boldsymbol{\Delta r} = \boldsymbol{r}(\sigma_{\text{current}}) - \boldsymbol{r}(\sigma_{\text{last local minimum}}) . \tag{13.3}$$

Thus, that part of the force field leading to the last local minimum is removed. The control parameter α determines how fast the system propagates toward the saddle point. There are several ways to develop adaptive formulas for α, e. g.,

$$\alpha = -\frac{\alpha_0}{1+\Delta r} , \tag{13.4}$$

with $\alpha_0 = 0.15$ [15, 208]. Thus, the more the system has already moved from the optimum, the faster it can move. One performs conjugate gradient steps iteratively with the force field $\tilde{\boldsymbol{F}}$, which changes after each step until $\boldsymbol{F}$ decreases again. Due to the finite step size of the moves, one will not touch the saddle point accurately. But one can check whether or not the local valley of the last local minimum has already been left: if $\boldsymbol{F} \cdot \boldsymbol{\Delta r} < 0$, then one is still in the local valley and thus in the attractor region of the former local minimum. Thus, one verifies with the condition $\boldsymbol{F} \cdot \boldsymbol{\Delta r} > 0$ that the saddle point has already been trespassed. In order to ensure that the system does not fall back in the previous local minimum, one adds some conjugate gradient steps with the modified force field and then relaxes the system in a new local minimum by applying the conjugate gradient method based on the original force field $\boldsymbol{F}$. The new local minimum is then accepted with the Metropolis criterion

$$p(\sigma_{\text{old local minimum}} \to \sigma_{\text{new local minimum}}) = \min\{1, \exp(-\Delta\mathcal{H}/(k_{\text{B}}T))\} , \tag{13.5}$$

with $\Delta\mathcal{H}$ being the energy difference between these two local minima. If this move is accepted, one performs a new move trial starting from the new local minimum. Otherwise, one tries a new move starting from the previous local minimum.

Please note that individual local minima are not found in a process with moves chosen at random but with an elaborate construction heuristic. Just as with ruin & recreate, the individual states do not occur according to their Boltzmann weights. Thus, although the Metropolis criterion is applied, no thermal euqilibrium is reached. Furthermore, when investigating the saddle points more closely, one finds that many of them are not first order but of a higher order. Up to half of the assumed saddle points are in reality local minima.

There are various ways to apply this method to combinatorial optimization problems with a discrete energy landscape. Starting at a random initial configuration, one would first perform a greedy run or the steepest descent method in order to reach a local minimum. Then one performs a few moves in the RW mode in order to leave the local minimum and the area around it. Then one switches back to the greedy mode and only accepts moves that do not increase the energy and do not increase the overlap to the previous local minimum, in order not to end up at the old local minimum again. This restriction is then removed again after a few moves so that the greedy search can lead to a true local minimum. Of course, one will do this in accordance with how one defines a Hamiltonian

$$\tilde{\mathcal{H}}(\sigma) = \mathcal{H}(\sigma) + \lambda \mathcal{O}(\sigma, \sigma_{\text{old local minimum}}) , \tag{13.6}$$

with $\mathcal{O}(\sigma, \sigma_{\text{old local minimum}})$ being the overlap between the current configuration σ and the old local minimum. If the greedy method then leads too often to the previous local minimum, then the number of moves in the RW mode must be enlarged.

This ART algorithm can be extended for both continuous and discrete problems in such a way that several former local minima get part of the force field or of the Hamiltonian in order to avoid them all.

13.3 Noising or Permutation of System Parts

Noising or perturbation heuristics also belong to the class of energy-landscape-changing algorithms but are derived from a completely different approach [206, 34, 39]: usually, one expects that if one changes an instance of a given optimization problem slightly, then the optimum solution for this changed instance should be roughly the same as for the original instance. Analogously, local minima in the energy landscape will either be the same or only slightly displaced. The extent of these changes depending on the size of the changes in the input parameters can be measured by observables like the sensitivity that are used in the analysis of computational problems in order to find out how stable a solution is if some input parameters of the problem instance are changed slightly.

From this point of view the idea arises to move from the given instance to a slightly changed instance of the proposed problem. The outline is rather similar to the SSS techniques but with a different starting point: here one starts with the original problem instance, and therefore in the original energy landscape, and performs an optimization run leading to a local optimum solution. Then the problem instance is slightly perturbed such that one now transfers the local optimum solution for the original problem to a configuration of the perturbed problem instance. There, this configuration is not necessarily a local optimum solution, so that an optimization run is performed with the greedy algorithm in order to reach a local optimum of the perturbed problem instance. Now the various approaches differ:

- One can either return to the original problem instance by transferring the solution of the perturbed instance to a configuration of the original instance. There, the current configuration might again not be locally optimal, so that a greedy optimization run is performed. After that, one either returns to the former perturbed instance or creates another perturbed instance. This scheme is iterated several times.
- On the other hand, one can also jump directly to another instance that was also created by slightly perturbing the original instance. Therefore, the local optimum solution of the first perturbed instance is transferred to a configuration of the second perturbed instance. This configuration serves as a starting point for a greedy optimization run, which is performed on this second performed instance. Also, at this point the scheme is iterated several times.

In the end, one usually returns to the original instance and performs a final greedy optimization run there. Various implementations of these algorithms furthermore differ in the amount of noise added to the original instance. Usually, this amount is decreased in some way in order only to consider instances more similar to the original instance. Note that this amount serves therefore as a control parameter like the smoothness-control parameter α of SSS or the temperature T of SA. Note again the close relation between these control parameters.

There are usually several ways of perturbing a problem instance, depending on the underlying problem: one might think of, e. g., displacing some parts of the problem, changing the size of the parts, and removing some parts of the instance or, on the other hand, introducing further parts in the instance.

13.4 Weight Annealing

Closely related to these permutation heuristics is another approach called weight annealing [156]: here one assigns weights w_i to the different parts of the problem. These weights are introduced into the cost function $\mathcal{H}$ of the problem, leading to a cost function $\mathcal{H}_{\boldsymbol{w}}$. In this way, the impact of the individual parts on the cost function is reweighted by the weight vector $\boldsymbol{w}$. There are various reasons for introducing such weights:

- One wants to represent the importance of some parts for the whole system. According to some subjective perception or objective a priori considerations, it might be that some parts have to be solved in a better way than other parts of the problem.
- After having already performed several optimization runs on the specific problem instance, one finds out that some parts are not solved or not solved very well. Therefore, one wants to give some further weight to these parts so that they are considered by the optimization process in a better way. These approaches can be summarized in the term a posteriori approaches.

- The easiest way, however, is a random approach, which simply assigns random weights to the individual parts of the system. This approach is essentially identical to the perturbation method of the last section.

By adding or multiplying this further weight, one wants to favor configurations in which the corresponding part of the problem is solved (rather) well. As the optimization process looks for the optimum in the energy landscape, it is then more probable that the final configuration will contain a better-solved part.

However, from this picture we see already the dangers of this approach: if a weight for a certain part is rather large compared to other weights, then the optimization process might concentrate too much on solving this part correctly, thus leading to an overall worse solution. The energy landscape might in this case look like a rather flat landscape, but containing a canyon system. Therefore, one must always check the final configuration to see whether it is good for the original cost function $\mathcal{H}$.

Instead of assigning weights only once, performing an optimization run based on the weighted Hamiltonian $\mathcal{H}_{\boldsymbol{w}}$, and printing the result, one could also think of changing the weights in various steps: one might either start at weights according to one's own a priori considerations or with each weight $w_i = 1$. After the first optimization run, one checks which parts of the system were solved rather badly. According to the importance of the individual parts and of the difference between the desired local solution and the current one, one changes the weights, so that those parts that are to be solved in a better way get a relatively larger weight. This approach is iterated several times or until the weights no longer change. According to the underlying optimization algorithm, one starts either with the final configuration of the previous iteration or with a random solution in the current reweighting iteration.

This approach can be used in different ways:

- It can be used like the perturbation approach of the last section. Therefore, the weights should converge to 1 for all parts.
- One can use it to solve some parts of a problem better than they are solved when using the original cost function.
- One can use it to find out which parts of a system are more or less important than other parts. Therefore, one is interested also in the final values of the weights in order to find a good cost function for the considered problem.

14 Estimation of Expectation Values

14.1 Simple Sampling

Usually, one cannot sum over all states $\sigma \in \Gamma$ in a simulation, especially if considering an NP-complete problem, for which the size of Γ increases exponentially with the system size. Therefore, the calculation of the expectation value $\langle \mathcal{A} \rangle$ of an observable $\mathcal{A}$ is restricted to its estimation by calculating a mean value $\bar{\mathcal{A}}$, averaging over some number of configurations. In the simplest approach (simple sampling), M configurations are randomly chosen and their values for the observable and their energies calculated. The average over the measurements leads to

$$\bar{\mathcal{A}} = \frac{\sum\limits_{i=1}^{M} \mathcal{A}(\sigma_i)\pi_{\text{equ}}(\sigma_i)}{\sum\limits_{i=1}^{M} \pi_{\text{equ}}(\sigma_i)} , \tag{14.1}$$

with $\pi_{\text{equ}}(\sigma)$ being the probability of the state σ for occurring in equilibrium. For a classic physical system, π_{equ} is the Boltzmann distribution function. As M approaches infinity, $\bar{\mathcal{A}}$ converges to $\langle \mathcal{A} \rangle$.

This sampling method fails in those cases where the weights of the various configurations strongly differ, especially if there are only a few configurations with a large weight. If the overall number of configurations is rather large compared to the number of these "important" configurations, then these important configurations, which would give a large addend to the sum due to their large weight, are sampled only with a small probability. This can lead to completely false estimates of the real expectation value. Therefore, some biased sampling or importance sampling is preferable.

14.2 Biased Sampling

In biased sampling, the states over which the mean value is averaged are no longer randomly chosen. Instead, they are chosen according to some distribution function $\pi(\sigma_i)$. The possible values of $\pi(\sigma_i)$ are between 0 and 1.

The process of choosing states can be done according to the von Neumann rejection principle; as in simple sampling, first a configuration σ_i is randomly chosen and its weight $\pi(\sigma_i)$ calculated. Then a uniformly distributed random number between 0 and 1 is calculated. If this random number is smaller than or equals the weight $\pi(\sigma_i)$, then the state σ_i is accepted for the calculation of the average. This way of finding configurations for the averaging process is limited to problems where configurations with weights significantly larger than zero can be found easily by random selection. Otherwise, it is preferable to perform a Markov process with acceptance probabilities leading to the distribution π. After the probability distribution converges to π, one can start taking configurations for the averaging. Most important, one should not take a successive sequence of configurations, as successive configurations are usually strongly correlated with each other. One has to take the correlation time of the process into account, which gives the number of time steps one has to wait before taking another—reasonably independent—configuration.

This distribution function π will of course include some knowledge about the real distribution function π_{equ} and will approximate it. In particular, it should be small where π_{equ} is small and large where π_{equ} is large. The mean value when configurations are chosen with bias is given by

$$\bar{\mathcal{A}} = \frac{\sum_{i=1}^{M} \mathcal{A}(\sigma_i)\pi_{\mathrm{equ}}(\sigma_i)/\pi(\sigma_i)}{\sum_{i=1}^{M} \pi_{\mathrm{equ}}(\sigma_i)/\pi(\sigma_i)} . \tag{14.2}$$

The best guess distribution function π would be the original equilibrium distribution function π_{equ}. However, sometimes it is better to work with a biased function if it is too time consuming or otherwise too difficult to generate a distribution according to π_{equ}. If it was already too time consuming to generate an approximate distribution, one restricts oneself to simple sampling.

14.3 Importance Sampling

Importance sampling is a special case of biased sampling, in which the chosen distribution π is identical to the desired distribution π_{equ}. The expression (14.2) reduces to

$$\bar{\mathcal{A}} = \frac{\sum_{i=1}^{M} \mathcal{A}(\sigma_i)}{M} , \tag{14.3}$$

an arithmetic average over the values of the chosen configurations for the observable $\mathcal{A}$.

Simulated annealing generates states distributed according to the Boltzmann distribution. In estimating the expectation value of some observable, one must wait at each new temperature until so many move trials have been performed that the system is in equilibrium again. Then one takes one measurement. After the first measurement one must again perform several move trials before taking the second measurement in order to ensure that these measurements are independent of each other. After taking a certain number of measurements, one can proceed with the next temperature step. Finally, one has a curve in a temperature-mean value diagram.

14.4 Parallel Sampling

Till now, only one Markov chain has been used for retrieving various samples of an observable in order to approximate its expectation value. This approach is usually sufficient, because if the system is ergodic, the time average over samples chosen from one Markov chain is identical to the ensemble average from samples selected from independent Markov chains. However, sometimes a parallel approach is advantageous or even necessary in which samples are derived in parallel from different Markov chains. The reasons for this include the following:

- One wants to save calculation time by using a parallel computer: in this case, at the beginning of each temperature step, one must wait until so many move trials have been performed that the system is in equilibrium again. Then one takes one measurement from each ensemble member. The expectation value is simply the mean value of these measurements if the Monte Carlo chain was created just as for the importance sampling method. After that the temperature is decreased again. In this way, one gets independent measurements from independent Markov chains. In contrast, if working with one Markov chain only, one would have to wait before taking the second measurement in order to get uncorrelated configurations.
- When dealing with quantum systems in which the ground state is usually given as an overlap over several (quasi) optimum configurations, it is necessary to work with this parallel approach in order to get correct values for the energy and other observables.
- Sometimes the low-temperature behavior of the proposed system is of interest. As each simulation is performed in a finite time only, the system is not really ergodic anymore. It cannot leave the local valley at low temperatures in a reasonable amount of time. Therefore, the average is taken over a large ensemble of Monte Carlo walkers, each of them trapped in some local valley.
- Working with threshold accepting (TA) and the great deluge algorithm (GDA), one must consider the fact that neither algorithm is ergodic, so that the time average is not identical to the ensemble average. Only the ensemble

average can be used for getting reliable estimates of the expectation value of some observables.
As we are only interested in finding optima, this does not really matter to us. Although TA and the GDA are nonequilibrium algorithms, we simply adopt Eq. (14.3) for calculating mean values of the observables of interest by using the time average and use the threshold $\mathcal{T}h$ and the level $\mathcal{T}$ instead of the temperature T in Eq. (11.16) for the specific heat and Eq. (11.22) for the susceptibility.

15 Cooling Techniques

15.1 Standard Cooling Schedules

Using the theory of Markov processes, several authors (see, e. g., [64, 142, 78, 65]) have proved the general existence of cooling schedules for simulated annealing (SA) with which the simulation ends up in the global optimum of the considered problem, however, after an infinite amount of time. Geman and Geman [65] showed for the classical case that it is necessary and sufficient for having a probability of one of ending in a global optimum that the temperature decreases like

$$T = \frac{a}{b + \log(t)}\,, \tag{15.1}$$

with a and b being positive constants that depend on the specific problem. t is the time elapsed since the start of the simulation and is usually measured in Monte Carlo steps. There are also cooling functions $T(t)$ for the other algorithms introduced in the previous chapters that lead to the globally optimum solution for some specific problem.

In practical applications, the available amount of time to produce a solution is finite. Therefore, faster ways of cooling the system down were developed that could be applied generally and that lead to very good solutions. These empirically found cooling schedules let the temperature decrease much faster to zero. However, they do not guarantee a convergence to the global optimum of the problem. Mostly two main cooling schedules are used:

- Linear/arithmetic cooling:

  $$T = a - b \times t\,, \tag{15.2}$$

  where a is the initial temperature and b is the decrement by which the temperature is decreased. Usually, b is chosen in the interval $[0.01; 0.2]$. The initial temperature strongly depends on the problem considered.
- Exponential cooling:

  $$T = a \times b^t\,. \tag{15.3}$$

  Again a is the initial temperature and b is a cooling factor, usually in the interval $[0.8; 0.999]$. In the literature, this cooling schedule is called by many names: "logarithmic", "geometric", or "exponential". We refer to it as exponential cooling.

The best cooling schedule will depend on the problem and the computing resources available (see, e. g., [207], where an optimal annealing schedule with a fixed number of steps was created for a particular problem). Much more complex cooling schedules than Eqs. (15.2) and (15.3) have been employed, with the decrease in temperature adapting to evidence of rapid or slow equilibration. The temperature need not even decrease monotonically. An example of these approaches is the following consideration for SA. One considers the relative weight $W_T(\sigma)$ of a configuration σ in relation to the weight of the ground state configuration with minimum energy $\mathcal{H}_0$, i. e.,

$$W_T(\sigma) = \frac{\pi_{\text{equ}}(\sigma)}{\pi_{\text{equ}}(\sigma_0)} = \exp\left(-\frac{\mathcal{H}(\sigma) - \mathcal{H}_0}{k_\text{B}T}\right) \tag{15.4}$$

and demands that $W_T(\sigma)$ not change too much if the temperature T is decreased from a value T_k to a new value T_{k+1}; thus, one demands

$$\frac{1}{1+\delta} < \frac{W_{T_k}(\sigma)}{W_{T_{k+1}}(\sigma)} < 1 + \delta\,, \tag{15.5}$$

with some small constant $\delta > 0$. For $T_{k+1} < T_k$, the left inequality is always fulfilled. Solving the right inequality for T_{k+1}, one gets

$$T_{k+1} > \frac{T_k}{1 + \dfrac{T_k \log(1+\delta)}{\mathcal{H}(\sigma) - \mathcal{H}_0}}\,. \tag{15.6}$$

Of course, one will not check how to fulfill this inequality for all states σ. Instead, one considers either the thermal average at temperature T_k and replaces the deviation $\mathcal{H}(\sigma) - \mathcal{H}_0$ in the last formula by $\langle \mathcal{H} \rangle_{T_k} - \mathcal{H}_0$. However, usually the optimum value for a specific problem instance is not known. In these cases, one considers the variance $\text{Var}_{T_k}(\mathcal{H})$ and approximates the mean deviation from the minimum energy value with $3\sqrt{\text{Var}_{T_k}(\mathcal{H})}$, as the individual configurations are supposed to occur according to the Gaussian distribution, in which 99.7% of all occurrences are within three times of the standard deviation range around the mean value. Thus, the formula reduces in its applicable version to

$$T_{k+1} > \frac{T_k}{1 + \dfrac{T_k \log(1+\delta)}{3\sqrt{\text{Var}_{T_k}(\mathcal{H})}}}\,. \tag{15.7}$$

But if there is no time for implementing and testing such a tuned cooling schedule, which is usually based on additional measurements that also require calculation time, the two schedules of linear and exponential cooling are the obvious choice. To choose between them, consider the characteristics of the specific heat: if $C(T)$ is more or less symmetric when plotted on a linear temperature scale, then linear cooling is preferable. If, however, the peak only becomes rather symmetric when plotted against $\log(T)$, then the system organizes itself over several orders of magnitude in temperature, and the exponential schedule is preferred.

It is rather straightforward to find a proper initial value of the temperature. At the beginning of the optimization run, a random walk is performed and $|\Delta\mathcal{H}|_{\max}$, the largest absolute value of the energy differences occurring between successive configurations, is saved. Then, if using SA, the initial value of the temperature is chosen in such a way that the acceptance rate of all moves exceeds a chosen value, e. g., it should be at least 90% at the beginning. Then the rule of thumb

$$\mathcal{T}_{\text{initial}} = -\frac{|\Delta\mathcal{H}|_{\max}}{\ln(0.9)} \approx 10 \times |\Delta\mathcal{H}|_{\max} \tag{15.8}$$

is applied. It is even simpler with threshold accepting:

$$\mathcal{T}h_{\text{initial}} = |\Delta\mathcal{H}|_{\max}\,. \tag{15.9}$$

In contrast, the maximum occurring energy value $\mathcal{H}_{\max}$ has to be saved when using the great deluge algorithm (GDA). One simply sets

$$\mathcal{T}_{\text{initial}} = \mathcal{H}_{\max}\,. \tag{15.10}$$

Instead of the maximum energy difference (or the maximum energy in the case of the GDA), often the mean value of the measurements is considered because the choice above might lead to too large initial temperatures and therefore some waste of calculation time. However, if the various moves used show different ranges of energy differences, then the problem occurs that if the initial temperature is chosen too small, the system might explore only a restricted area of the energy landscape.

Of course, other possibilities for choosing the initial value of the temperature are imaginable. In summary, if one starts at too low temperatures, the problem arises that the quality of the results decreases as the system is restricted to the local valley of the initial configuration. The Monte Carlo walker can only climb down to the local minimum and fails to reach better solutions for which he/she would have to leave the local valley and climb over barriers. Alternatively, too much calculation time might be wasted at high temperatures if the initial temperature was chosen too large. Therefore, sometimes the following procedure is chosen: a rather fast run with an initial temperature chosen as described above is performed. The decrease in the energy and the progression of the specific heat are plotted vs. the temperature T. Then a smaller initial temperature is chosen for the production runs. Besides guessing intuitively a good value for the initial temperature, one can set the initial temperature in the range of the peak of the specific heat C: let τ be the temperature at which C is maximum and $\Delta\tau$ be the width of the peak. Then one usually chooses the initial temperature as

$$\mathcal{T}_{\text{initial}} = \tau \pm \Delta\tau\,. \tag{15.11}$$

The three points $(\tau - \Delta\tau, 0)$, $(\tau, C_{\max})$, and $(\tau + \Delta\tau, 0)$ in the specific heat vs. temperature diagram form a "magic triangle", which indicates both the temperature range in which most rearrangements of the system take place and nearly the whole amount of energy the system loses while being cooled down. The final temperature has to be chosen in such a way that the system is really frozen or only trivial moves with $\Delta\mathcal{H} = 0$ are accepted. Therefore, the acceptance rate of the nontrivial moves has to vanish at the end of the optimization run. However, there is no good criterion for how long one has to wait to be absolutely sure that no nontrivial move would no longer be accepted, as this depends on the underlying energy landscape. (Of course, all possible moves from the actual solution to neighboring configurations could be considered. However, how many neighbors a configuration exhibits depends on the move set and the size of the system. The number of neighbors might be so large that it is impossible to check them all in order to prove that a certain state is really a local minimum in the energy landscape.) Therefore, one usually performs a few greedy steps at the end of each optimization run, which last long enough to be quite sure to be in a local minimum. One simply sets $T = 0$, $Th = 0$, and $\mathcal{T} =$ "best result so far".

15.2 Nonmonotonic Cooling Schedules

So far, we have only considered monotonic cooling schedules by which the control parameter, e. g., the temperature is not increased during the optimization run. This approach is motivated by the physical picture of a molten metal block: if it is cooled down very fast, i. e., quenched down, then only a polycrystalline structure can develop, in contrast to the nice results one gets if the melting is cooled down very slowly.

This physical picture can be extended even further by looking at the work of a blacksmith in former times: after cooling down the molten metal rather fast in a special form, the blacksmith treats it with a certain iterated reheating-cooling down strategy: first of all, the metal is reheated, but only up to a certain temperature at which it, e. g., glows red but does not start to melt. While the block cools down again, the blacksmith works on the block in order to improve it. Sometimes he also quenches the block down again with, e. g., water. Usually, he only works on the block during the cooling-down phase and not during the reheating phase. Due to this special treatment, the blacksmith can perform larger structural rearrangements on the considered block without having to melt his already partially completed work and therefore to start again. The temperature up to which the block is reheated usually determines the amount by which the block can be changed.

This approach can be transferred to SA and related optimization algorithms in various ways in order to get an optimum cooling schedule. However, it is usually impossible to determine an optimum cooling schedule, which is an optimization problem on its own, but there are still many ways to develop nonmonotonic cooling schedules that perform well if one considers the basic finding of Romeo and Sangiovanni–Vincentelli [173] that a result nearly as good as the global optimum can only be reached if there is always a probability large enough for leaving any configuration, i. e., also any local optimum.

Strenski and Kirkpatrick showed in their early work [207] on this subject, in which they optimized the annealing schedule for a very simple problem using a fixed number of iterations, that if the amount of calculation time available is very small, then the optimum annealing schedule is nonmonotonic: their results for various systems indicate that after starting at a random solution, one should proceed with the greedy algorithm, i. e., the $T = 0$ mode of SA and its relatives. By this approach, the system is quenched down in the fastest possible way, until it freezes in a local optimum. After that the system is reheated up to a certain problem-dependent temperature. At this temperature, the system is able to cross smaller barriers in the energy landscape and therefore to leave a local valley in order to get to a better local optimum. Depending on the calculation time available, the system should either immediately switch to the greedy mode again in order to get stuck in a better local optimum nearby or go through a more or less pronounced cooling procedure (this "more or less" depends on the computing time available), which monotonically decreases the control parameter from the reheating temperature and ends up in the greedy mode.

Another approach is called bouncing [114, 190]: after a first conventional optimization run, in which the control parameter is reduced monotonically from a large initial value T_0 to 0, such that the system is transferred from a random configuration to a locally optimum solution, the final configuration is used as the initial configuration of a further optimization run. This second optimization run starts at a value $T_B < T_0$ of the control parameter in order to correspond to the blacksmiths approach, who holds the metal block in the fire while not working on it. (This approach could be called an inverse quench.) Now the system is again at least partially able to move through the energy landscape. The amount of rearrangement possible is determined by the value of the bouncing temperature T_B. In this bouncing iteration, a standard optimization run is performed, i. e., the control parameter is decreased step by step from T_B to 0. Then the new final configuration is used as the initial configuration for the next bouncing iteration in which the control parameter is again reduced from T_B to 0. This approach is iterated several times.

The main advantage of this bouncing approach compared to the monotonic cooling approach is that the standard monotonic optimization run starts at a totally random system, whereas the bouncing scheme can start from preoptimized solutions in the bouncing iterations. These preoptimized solutions already inherit some structure, i. e., some information about the system. The question is now up to which value T_B of the control parameter will the system be reheated, i. e., which T_B is appropriate for the cooling process $T_B \geq T \geq 0$? On the one hand, local information will be kept and only a small number of structures shall be destroyed. On the other hand, the optimization process will be able to leave a bad local optimum and get to a better one.

One can distinguish three regimes for T_{B}:

- Bouncing as "slightly warming up":
  After the first conventional cooling process, the reheating is done only very slightly, i. e., in a temperature range below the critical temperature T_{C}, at which the ordering transition of the system happens. This critical temperature can be determined for systems for which an order parameter can be defined. In that case, one can measure the susceptibility, which peaks at the critical temperature.
  In this regime, the energy values of the final configurations of successive bouncing iterations first decrease monotonically and then stay nearly constant [190]. Only seldom does the system jump to a worse solution, and then it is able to jump back in the next bouncing iteration. In this regime, bouncing more than ten iterations under the same external constraints seldom gives any further significant improvement.
- Bouncing up to the maximum of the specific heat:
  The system is reheated nearly to the freezing temperature T_{f}, i. e., $T_{\mathrm{C}} < T_{\mathrm{B}} < T_{\mathrm{f}}$. This temperature range corresponds to a partially ordered phase of the system. A transition to the totally unordered range is not performed, similar to the approach of the blacksmith, who reheats the metal block in order to perform some structural rearrangements but does not melt it. The results of previous reheating iterations of the considered item do not get lost completely.
  In this temperature regime, the energy values of the final configurations of successive bouncing iterations also decrease monotonically, but only on average: there are now large fluctuations; sometimes a bouncing iteration leads to an improvement, sometimes it leads to a deterioration. But all in all, in this regime, it is possible to arrive at much better solutions than in the small T_{B} regime. The system is able to climb also over larger barriers in the energy landscape and therefore to leave a suboptimal local area.
  Generally, the optimum T_{B} value depends on the calculation time spent in each bouncing iteration: if the amount of calculation time is increased, the system has a greater ability to climb over barriers such that a smaller T_{B} value can be used than if spending less calculation time in order to get the same number of improvements [190].
- Bouncing above T_{f}:
  Finally, one can also consider the behavior of a bouncing process with $T_{\mathrm{B}} > T_{\mathrm{f}}$: the system melts such that no qualitative change of the results compared to a standard monotonically cooled optimization can exist, as the system is kicked up into the unordered high-energy regime.

Summarizing, one must work along these lines: one performs a conventional optimization run in which the specific heat C and, if possible, the susceptibility χ are measured during the cooling process. The peak of the specific heat marks the freezing temperature T_f, the peak of the susceptibility the critical temperature T_C. One can define a gain that can be achieved with bouncing by

$$g_i = \frac{\langle \mathcal{H}_0 \rangle_\mathrm{E} - \langle \mathcal{H}_i \rangle_\mathrm{E}}{\langle \mathcal{H}_0 \rangle_\mathrm{E} - \mathcal{H}_\mathrm{opt}}, \tag{15.12}$$

with i being the number of the bouncing iteration and $\langle \mathcal{H}_i \rangle_\mathrm{E}$ the ensemble average of the energy values of the final configurations of the bouncing iteration i (where $i = 0$ denotes the initial monotonically cooled optimization run). This gain g_i is normalized in such a way that it vanishes if no improvements can be achieved and that it reaches a value of 1 if the optimum is reached.

The extent of this gain g_i depends on the value of T_B; it is largest for T_B slightly smaller than T_f if only a small amount of calculation time is used in each bouncing iteration. When spending more calculation time, one should reduce this T_B. If $T_\mathrm{B} > T_\mathrm{f}$, then there is no gain at all. Similarly, the gain vanishes in the limit $T_\mathrm{B} \to 0$ as expected as the Monte Carlo walker is stuck in the local valley in the energy landscape, and cannot leave. A lower bound for T_B is the critical temperature T_C and also the width τ of the peak of the specific heat, such that one should not choose a T_B smaller than T_C or smaller than $T_\mathrm{f} - \tau/2$.

However, the question remains why this bouncing approach works: the algorithm works with random moves that could also lead to deterioration. However, the algorithm obviously can make use of the information stored in the initial configuration, i. e., the final result of the previous optimization run. By not melting the system, many structures of the previous solution are retained—they need not be found again—such that the optimization run has some preoptimized background on which it works, which leads to better solutions if T_B is chosen correctly.

A further explanation for the success of this bouncing idea can be obtained by measuring the Hamming distance between successive results of bouncing iterations: first of all, one must consider the mean Hamming distance between quasioptimum solutions produced by independent optimization runs. Then one will find that there are really three T_B regimes that can be distinguished also in the Hamming distance graphics: if $T_\mathrm{B} > T_\mathrm{f}$, then the Hamming distance between successive results of bouncing iterations is of the same size as that of independent results. On the contrary, if T_B is very small, the Hamming distance is (nearly) zero, as successive bouncing iterations lead to (roughly) the same result. One should therefore plot the graphic Hamming distance vs. bouncing iteration number for $T_\mathrm{B} = T_\mathrm{f}$ and $T_\mathrm{B} = T_\mathrm{C}$ and choose a T_B in such a way that the curve in this graphic is between these extreme curves but nearer to the curve for $T_\mathrm{B} = T_\mathrm{f}$, e. g., the distance to the curve for $T_\mathrm{B} = T_\mathrm{C}$ should be one to three times larger than the distance to the curve for $T_\mathrm{B} = T_\mathrm{f}$ [190].

Instead of using a fixed value for T_B, one can also think of a variable T_B. One can even go so far as to introduce a cooling schedule for T_B; for example, one could use the same type of cooling schedulte for T_B as for the control parameter itself, thus bouncing the system more and more gently, such that it is at first enabled to climb over larger barriers in the energy landscape and then only able to cross smaller barriers. This approach also leads to (quasi) optimum results [190].

There are also other opportunities to bounce: for example, one can perform a conventional optimization run, in which the control parameter is decreased monotonically. However, after a certain number of control parameter steps, one or more greedy steps are introduced in order to quench the system in a local optimum. After these greedy steps, the conventional optimization run is continued at the next value of the control parameter. After one or more additional control parameter steps, again some greedy steps are performed, and the system is again led to a local optimum. This approach is repeated until the value of the control parameter is so small that the system is mostly in the greedy mode. At the end of the optimization run, the best local optimum solution found is returned as the result of the optimization procedure.

15.3 Ensemble Based Schedules

There are also ways to find either good or fast cooling schedules by using parallel computers. In this case, usually the same instance of a problem is treated simultaneously on several processors. The number of the processor or the process number often serves as a seed for the random number generator, such that all processors start with different initial solutions or depart from a common initial solution in different directions.

These schedules have to avoid wasting too much calculation time in large values of the control parameter, in which the system performs a quasi-RW. On the other hand, the schedule should not start at too small values and should not decrease the control parameter too fast in important ranges as then only bad results will be achieved. The ensemble based simulated annealing (EBSA) approach tries to fulfill these requirements: if the system to be optimized is equilibrated at a certain temperature, then the energy values of the successive configurations fluctuate in a certain range around the mean value of the energy at this temperature. If the temperature T is decreased, then the energy values decrease until they oscillate around the new and smaller mean value of the energy at this smaller temperature. This decrease in the energy values of the successive configurations does not occur monotonically, but it is quite noisy. If the temperature is only decreased very slightly, it is hard to detect this decrease in an energy value vs. time diagram. However, if one averages over many optimization runs, then the fluctuations during the decrease of the mean energy cancel out, such that one usually gets a strongly monotonic decreasing curve until the new equilibrium value is reached. Small oscillations remain in the equilibrium phase, as the number of parallel optimization runs is finite. This property of such an averaging is used on parallel computers for determining for an ensemble of optimization runs whether the proposed amount of calculation time was sufficient to reach the equilibrium at the new temperature.

Therefore, EBSA starts on an ensemble of p processors with one SA step, using different initial configurations at a large value of the temperature. At the end of this step, the ensemble average $\langle \mathcal{H}_0 \rangle_{\mathrm{E}}$ over the energy values of the p configurations is calculated. Then the following steps are iterated in a loop (i denotes the number of the iteration):

- The p slaves perform one SA step at the current value of the temperature T.
- Then they send the energy values of their final configurations to the master processor.
- This master processor calculates the new ensemble average $\langle \mathcal{H}_{i+1} \rangle_{\mathrm{E}}$ and the corresponding variance $\mathrm{Var}_{\mathrm{E}}(\mathcal{H}_{i+1})$.
- If the condition

$$\langle \mathcal{H}_{i+1} \rangle_{\mathrm{E}} - \langle \mathcal{H}_i \rangle_{\mathrm{E}} \geq \gamma \left(\frac{\mathrm{Var}_{\mathrm{E}}(\mathcal{H}_{i+1})}{p-1} \right)^{\nu} \tag{15.13}$$

  (containing two parameters γ and ν) is fulfilled (often a $>$ sign is used instead of $\geq$ in the literature, but according to our experience $\geq$ is better), then the temperature is decreased; otherwise the old value of the control parameter is kept.
- The master processor overwrites the old ensemble average with the new one and sends the temperature value back to the slaves.

This way, the communication time between the master and the slaves can be kept to a minimum. Even the decision to finish the run can be easily transferred from the master to the slaves, i. e., by sending a negative value for the control parameter from the master to the slave processors, by which the slaves can identify the end of the simulation, at which they simply have to send their final configuration to the master, which chooses the best of these.

The main parameter of this parallel approach is the factor γ, which can be chosen in various ways:

- $\gamma \to -\infty$
  In this case, one gets the so-called exponential cooling scheme, because Eq. (15.13) reduces to

$$\langle \mathcal{H}_{i+1} \rangle_{\mathrm{E}} \geq -\infty \,. \tag{15.14}$$

This condition is always fulfilled, leading to an automatic decrease of the control parameter after each step.

- $\gamma = 0$
  Here one speaks of the simple adaptive cooling scheme. Equation (15.13) is simplified to
$$\langle \mathcal{H}_{i+1} \rangle_{\mathrm{E}} \geq \langle \mathcal{H}_i \rangle_{\mathrm{E}} \,. \tag{15.15}$$
  The current value of the temperature is kept constant as long as the ensemble average of the energy decreases in a monotonic way and is then decreased if this average increases or stays the same.
- $\gamma > 0$
  This case is called the weakened adaptive scheme. For this scheme, only small values for γ, e. g., $\gamma = 0.5$, are used. If one looks again at a simplified version of Eq. (15.13),
$$\langle \mathcal{H}_{i+1} \rangle_{\mathrm{E}} \geq \langle \mathcal{H}_i \rangle_{\mathrm{E}} + \varepsilon \,, \tag{15.16}$$
  with
$$\varepsilon = \gamma \left(\frac{\mathrm{Var}_{\mathrm{E}}(\mathcal{H}_{i+1})}{p-1} \right)^{\nu} \geq 0 \,, \tag{15.17}$$
  then one finds that the temperature is decreased only if the deterioration of the ensemble average of the energy values exhibits this ε.
  This approach must be used instead of the adaptive cooling scheme if the number p of available processors is rather small, e. g., $p < 100$, as in this case the fluctuations during the decrease of the energy are too large, such that the system thinks that it is already equilibrated at the new temperature whereas there was only a fluctuation. This would lead to too rapid a decrease in the temperature. However, one must be careful in the choice of ν in this case: in the literature, ν is usually chosen as 1. However, how large the ensemble variance can become depends on the problem. If it is too large, then the condition is never fulfilled and the algorithm is in an endless loop at some value of the control parameter. For every problem, the appropriateness of the ν value has to be tested. For example, $\nu = \frac{1}{2}$ is a good value for the traveling salesman problem, for which $\nu = 1$ already leads to endless loops at rather high values of the control parameter.

Summarizing, the basic thought of this ensemble based approach is to let the ensemble average of the energy values drop again and again at a fixed temperature. If it does not decrease any further, then the system is believed to have reached equilibrium at the given new temperature. Therefore, the temperature can be decreased again. At the end of the optimization run, all ensemble members get stuck as the control parameter is very small.

This ensemble based approach is defined in the context of the equilibrium properties of SA. However, this approach can also be transferred to other control strategies: for example, ensemble based threshold accepting (EBTA) works in the same way as EBSA. The temperature is simply replaced by the threshold. Analogously, this approach can be transferred to the temperature in the Tsallis-based methods or to the water level in the GDA.

The bouncing approach is also suited for parallel enablement. For example, one could simply use the best configuration of bouncing iteration i as an input for the next iteration on all p processors. But one could also define an ensemble based bouncing (EBB): in this case, the bouncing temperature T_B, up to which the temperature T is increased at the beginning of each bouncing iteration, is not held constant but decreased according to the ensemble based rule, i. e., if using the adaptive scheme, T_B is decreased if $\langle \mathcal{H}_{i+1} \rangle_\mathrm{E} \geq \langle \mathcal{H}_i \rangle_\mathrm{E}$. However, here the ensemble average $\langle \mathcal{H}_i \rangle_\mathrm{E}$ is the average over the energies of the final configurations of bouncing iteration i. Note that in contrast to EBSA and EBTA, it is not the temperature and the threshold that are decreased inside a bouncing iteration but the bouncing start temperature T_B is decreased according to the ensemble-based rule. Of course, it is also possible to use the ensemble-based approach twice for this bouncing approach: first, it is used for decreasing T_B, but secondly one can also decrease the control parameter inside a bouncing iteration according to the ensemble-based rules. This combines EBB and EBSA/EBTA.

It is also interesting to combine the ensemble-based approach with search space smoothing (SSS). Let us use the greedy algorithm as the underlying local search technique inside SSS. As therefore no energy deteriorations are allowed during one α step, the adaptive rule (15.15) of EBSA and EBTA is modified as follows: if $\langle \mathcal{H}_{i+1} \rangle_\mathrm{E} = \langle \mathcal{H}_i \rangle_\mathrm{E}$, then decrease α, otherwise keep the current value of α. However, the question arises as to which Hamiltonian will be used in this case, such that one can define three different rules for ensemble based search space smoothing (EBSSS):

- "Smoothed" rule:
  The smoothed Hamiltonian $\mathcal{H}^\alpha$ is used for calculating the ensemble average $\langle \mathcal{H}_i^\alpha \rangle_\mathrm{E}$. After changing α, the smoothed Hamiltonian changes, such that sometimes ensemble averages of different Hamiltonians are compared with each other.
- "Original" rule:
  The ensemble average is calculated with the original Hamiltonian $\mathcal{H}^0$.
- "Both" rule:
  One can also consider both Hamiltonians and must therefore change the condition above: α is decreased only if both $\langle \mathcal{H}_{i+1}^\alpha \rangle_\mathrm{E} = \langle \mathcal{H}_i^\alpha \rangle_\mathrm{E}$ and $\langle \mathcal{H}_{i+1}^0 \rangle_\mathrm{E} = \langle \mathcal{H}_i^0 \rangle_\mathrm{E}$.

15.4 Simulated Tempering and Parallel Tempering

Simulated tempering (ST) was introduced by Marinari and Parisi [132]. It works like SA, i. e., the simulation starts with an initial configuration and applies a series of moves that are accepted or rejected according to the Metropolis criterion. In contrast to SA, ST does not automatically change the temperature according to a proposed cooling schedule. Instead, the temperature is also seen as a configuration variable and is changed according to the Metropolis criterion. One considers a set of M temperatures T_i with $T_1 < T_2 < \cdots < T_M$, which play the role of the various temperatures in SA. Thus, T_1 must be chosen so small that the system is already frozen, whereas T_M has to be large enough so that the system can explore the whole configuration space. One wants to apply a move $T_i \to T_j$. The question is now how to choose an appropriate transition probability.

For this purpose, one extends the configuration space by a further dimension, in which a variable i takes the discrete values $1, 2, \ldots, M$, denoting the individual temperature steps. The configuration in this joint configuration space has to be considered as a pair $(\sigma, i) \in \Gamma \times \{1, \ldots, M\}$, with σ being the configuration of the original space Γ and i a number between 1 and M. Then one considers the partition sum of this joint system, for which a new parameter g_i is introduced:

$$Z(T_i) = \sum_{\sigma \in \Gamma} \exp(-\beta_i \mathcal{H}(\sigma) + g_i) \,, \tag{15.18}$$

with $\beta_i = 1/(k_B T_i)$. The g_i parameters are a function of parameter i and thus of the corresponding temperature T_i. They have to be chosen in a way such that

$$Z(T_i) = \text{const} = Z \,, \tag{15.19}$$

i. e., such that the partition sum of the joint system does not depend on the temperature value T_i, as this dependency will cancel out with the dependency of g_i. The equilibrium probability for the state (σ, i) is thus given by

$$\pi(\sigma, i) = \frac{1}{Z} \exp\left(-\frac{\mathcal{H}(\sigma)}{k_B T_i} + g_i \right) . \tag{15.20}$$

There are now two types of moves:

- First, one can apply as usual a move $\sigma \to \tau$, which is now the move $(\sigma, i) \to (\tau, i)$. The detailed balance condition leads to

$$\frac{p((\sigma, i) \to (\tau, i))}{p((\tau, i) \to (\sigma, i))} = \frac{\pi(\tau, i)}{\pi(\sigma, i)} = \exp\left(-\frac{\mathcal{H}(\tau) - \mathcal{H}(\sigma)}{k_B T_i} \right) . \tag{15.21}$$

Thus, this move type can be accepted with the Metropolis criterion as usual at the given temperature T_i.

- Secondly, one can apply a move to change the temperature $T_i \to T_j$. Applying detailed balance leads to

$$\frac{p((\sigma,i)\to(\sigma,j))}{p((\sigma,j)\to(\sigma,i))} = \frac{\pi(\sigma,j)}{\pi(\sigma,i)}$$

$$= \frac{\exp\left(-\mathcal{H}(\sigma)/(k_{\mathrm{B}}T_j)+g_j\right)}{\exp\left(-\mathcal{H}(\sigma)/(k_{\mathrm{B}}T_i)+g_i\right)} \tag{15.22}$$

$$= \exp(-(\beta_j-\beta_i)\mathcal{H}(\sigma)+(g_j-g_i))\,.$$

Thus, the transition probability can be chosen in a Metropolis-like criterion as

$$p((\sigma,i)\to(\sigma,j)) = \begin{cases} 1 & \text{if } \Delta \le 0\,, \\ \exp(-\Delta) & \text{otherwise}\,, \end{cases} \tag{15.23}$$

with

$$\Delta = (\beta_j-\beta_i)\mathcal{H}(\sigma)-(g_j-g_i)\,. \tag{15.24}$$

As the acceptance probability decreases exponentially with the difference between the inverse temperatures β_i and β_j, one usually only chooses moves $T_i \to T_{j=i\pm1}$, i. e., to the neighboring temperature values.

- Of course, these two moves can also be combined in one move $(\sigma,i)\to(\tau,j)$. One can derive a Metropolis-like criterion as (15.23) but with

$$\Delta = (\beta_j\mathcal{H}(\tau)-\beta_i\mathcal{H}(\sigma))-(g_j-g_i)\,. \tag{15.25}$$

Usually, however, one only uses the two moves above.

However, the main question still remains as to how to choose the g_i parameters. They are not a priori known and are usually determined by iterations of simulations that can be rather difficult for complex systems. One of the simplest approaches is first to determine the thermal expectation values $\langle\mathcal{H}\rangle(T_i)$ for each temperature by an ordinary SA run. Then the differences $g_{i\pm1}-g_i$ are given by the differences $\langle\mathcal{H}\rangle(T_{i\pm1})-\langle\mathcal{H}\rangle(T_i)$.

Thus, the outline of ST is as follows:

1. First, determine the g_i parameters.
2. Then start the simulation at some temperature T_i, preferably a large one.
3. Perform a few Monte Carlo sweeps at the given temperature T_i.
4. Then, try a move $T_i \to T_{i\pm1}$ with the transition probability Eq. (15.23).
5. If some final condition is not fulfilled, return to step 3.

Note that in contrast to SA, the system of ST does not need any time to equilibrate at the new temperature as the temperature is also changed with a Metropolis-like criterion.

A related method to overcome the difficulty of determining the g_i parameters is the replica-exchange method (REM) [98, 99], which is also called parallel tempering (PT) [41]. It was developed as an extension of ST. Again a set of M temperatures T_i with $T_1 < T_2 < \cdots < T_M$ is considered. In contrast to SA and ST, one has M different Markov chains, one at each temperature T_i. At each of these constant temperatures, a simulation as with SA at the corresponding temperature T_i is performed; thus each move is accepted according to the Metropolis criterion. The most important point of PT is that these simulations are performed independently of each other.

Thus, after some time, one has M configurations σ_i. The probability $\pi(\sigma)$ is given by the usual Boltzmann weight $\exp(-\mathcal{H}(\sigma_i)/(k_\mathrm{B}T_i))/Z(T_i)$. In contrast to ST, PT considers not the joint configuration space $\Gamma \times \{1, \ldots, M\}$ but the configuration space $\Gamma \times \Gamma \times \cdots \times \Gamma = \Gamma^M$. The probability $\mathcal{P}$ of the product state $\mathcal{S} = \sigma_1 \times \sigma_2 \times \cdots \times \sigma_M$ is given as the product of the Boltzmann weights of the single states due to the independence of the simulations, i. e.,

$$\begin{aligned} \mathcal{P}(\mathcal{S}) &= \mathcal{P}(\sigma_1, \sigma_2, \ldots, \sigma_M) \\ &= \prod_{i=1}^{M} \pi_\mathrm{equ}(\sigma_i) \\ &= \frac{\exp(-\mathcal{H}(\sigma_1)/(k_\mathrm{B}T_1))}{Z(T_1)} \times \cdots \times \frac{\exp(-\mathcal{H}(\sigma_M)/(k_\mathrm{B}T_M))}{Z(T_M)} \end{aligned} \tag{15.26}$$

with $Z(T_i)$ being the partition sum at the temperature T_i.

The new point of PT now is that it introduces a varied type of the move of ST to change the temperature: the temperatures at which two configurations σ_i and σ_j are exchanged by exchanging these two configurations between the corresponding Markov chains at the temperatures T_i and T_j. This means: after having applied a sequence of moves in process 1 at temperature T_1 with the final configuration σ_1, a sequence of moves in process 2 at T_2 with the final σ_2, and analogously in the other processes, one randomly selects two processes i and j and wants to move the configuration σ_i to process j while shifting the configuration σ_j to process i. The question is now with what probability this move

$$\begin{aligned} \mathcal{S}_1 &= (\sigma_1, \ldots, \sigma_i, \ldots, \sigma_j, \ldots, \sigma_M) \\ \rightarrow \mathcal{S}_2 &= (\sigma_1, \ldots, \sigma_j, \ldots, \sigma_i, \ldots, \sigma_M) \end{aligned} \tag{15.27}$$

will be accepted. Of course, again the detailed balance condition will be fulfilled, such that

$$\mathcal{P}(\mathcal{S}_1) \times p(\mathcal{S}_1 \rightarrow \mathcal{S}_2) = \mathcal{P}(\mathcal{S}_2) \times p(\mathcal{S}_2 \rightarrow \mathcal{S}_1) \,. \tag{15.28}$$

Thus, one gets the relation

$$
\begin{aligned}
\frac{p(\mathcal{S}_1 \to \mathcal{S}_2)}{p(\mathcal{S}_2 \to \mathcal{S}_1)} &= \frac{\mathcal{P}(\mathcal{S}_2)}{\mathcal{P}(\mathcal{S}_1)} \\
&= \frac{\pi_{\text{equ}}(\sigma_j, T_i) \times \pi_{\text{equ}}(\sigma_i, T_j)}{\pi_{\text{equ}}(\sigma_i, T_i) \times \pi_{\text{equ}}(\sigma_j, T_j)} \\
&= \frac{\exp(-\mathcal{H}(\sigma_j)/(k_{\text{B}}T_i)) \times \exp(-\mathcal{H}(\sigma_i)/(k_{\text{B}}T_j))}{\exp(-\mathcal{H}(\sigma_i)/(k_{\text{B}}T_i)) \times \exp(-\mathcal{H}(\sigma_j)/(k_{\text{B}}T_j))} \\
&= \exp(-\Delta\mathcal{H}/(k_{\text{B}}T_i)) \times \exp(\Delta\mathcal{H}/(k_{\text{B}}T_j)) \\
&= \exp(-\beta_i \Delta\mathcal{H}) \times \exp(\beta_j \Delta\mathcal{H}) \\
&= \exp(-(\beta_i - \beta_j)(\mathcal{H}(\sigma_j) - \mathcal{H}(\sigma_i))) \\
&= \exp(-\Delta\beta \times \Delta\mathcal{H}) \,,
\end{aligned}
\tag{15.29}
$$

with $\Delta\mathcal{H} = \mathcal{H}(\sigma_j) - \mathcal{H}(\sigma_i)$ and $\Delta\beta = \beta_i - \beta_j$.

As in the derivation of the Metropolis criterion, there is some arbitrariness in the explicit choice of the transition probability. Here one can use a Metropolis-like acceptance criterion:

$$
p(\mathcal{S}_1 \to \mathcal{S}_2) = \begin{cases} 1 & \text{if } \Delta\beta \times \Delta\mathcal{H} \le 0\,, \\ \exp(-\Delta\beta \times \Delta\mathcal{H}) & \text{otherwise}\,. \end{cases} \tag{15.30}
$$

Thus, if σ_j is better than σ_i, i. e., if $\mathcal{H}(\sigma_j) < \mathcal{H}(\sigma_i)$, these two configurations are always exchanged if $\beta_i > \beta_j$ and thus if $T_i < T_j$. The energetically lower configuration is thus with larger probability put to the smaller temperature and thus cooled down. The various temperatures T_i can stay constant all the time as the cooling can be achieved by the PT exchange mechanism, but one can also lower the M temperatures T_i during the simulation run.

Usually, a restriction similar to the one in ST is applied to this exchange process: as the transition probability decreases exponentially with the difference between the inverse temperatures, only exchanges of configurations are performed between neighboring temperature pairs (T_i, T_{i+1}) [98]. The individual temperature values have to be distributed in such a way that the smallest temperature is in a range where the system is already frozen, whereas the highest temperature should be well above the peak of the specific heat.

Like the problem of determining a good cooling schedule for the ordinary SA, one now has, both for ST and PT, the problem of what temperatures to use. Good results can be achieved by concentrating the temperature values around the peak of the specific heat of the problem.

There are various elaborate ways to determine these temperature values. An iterative way was proposed, e. g., by Kerler and Rehberg [109]: first, they fix a minimum β_1 and a maximum β_M. Then they start with β_i, which are equally distributed between these marginal values and measure the stay time τ_i at each β_i, which is the time between two accepted exchange moves at the corresponding β_i. The time for β_1 and β_M is divided by two, as these processes have only one neighboring process. Next they calculate the auxiliary variables

$$a_i = (\beta_{i+1} - \beta_i)/(\tau_{i+1} + \tau_i) \tag{15.31}$$

and their sum $A = \sum a_i$. Then they set the β_i to new values,

$$\beta_i^{\text{new}} = \begin{cases} \beta_1 & \text{if } i = 1\,, \\ \beta_{i-1}^{\text{new}} + \dfrac{a_{i-1}}{A}(\beta_M - \beta_1) & \text{otherwise}\,. \end{cases} \tag{15.32}$$

This procedure is iteratively repeated until the values of β_i have converged to some fixed points.

The individual implementations of these algorithms differ not only in the techniques for choosing these inverse temperature values but also in whether the individual values are kept constant during the optimization run or decreased. A cooling schedule for these temperature values can be either rather simple, e. g., the temperatures can be decreased exponentially, but they can also make use of acceptance criteria like those above [98].

The advantage of PT compared with ST is that there are no parameters g_i that have to be determined either beforehand or during the simulation process. Furthermore, one can explore the configuration space in one run much more thoroughly as there are several Markov chains. However, simulating M Monte Carlo walkers walking around obviously costs M times the compute effort of simulating only one. However, this is not necessarily a disadvantage as PT is very well suited for parallel enablement: the single processes can be run on the single nodes without interaction, except for the exchange moves that are sometimes performed.

16 Estimation of Calculation Time Needed

16.1 Exponentially Growing Space Size

The main aim in optimization is to find the ground state or at least a quasioptimum state of very low energy. A major aspect of reaching this aim is the question of how much time is needed to either achieve the global optimum solution (and in addition to prove that it is really globally optimal) when working with an exact algorithm, or to achieve a solution of at least a certain quality when working with a heuristic algorithm.

With NP-complete problems, one finds that the overall number of configurations in the configuration space explodes exponentially with the system size N, e. g., the number of configurations might be proportional to 2^N or to $N!$, which is roughly given by $N! \approx \exp(N \ln(N) - N)$ according to Stirling's formula. Furthermore, the number of quasioptimum solutions can be exponentially small compared with this overall number of solutions (for a heuristic proof see, e. g., [186]). Therefore, the time it takes to achieve the global optimum increases exponentially for an exact search algorithm, at least in the worst case.

16.2 Polynomial Approach

However, it has been shown already for several problems that one can reduce the calculation time needed if one does not rely on a proof for the global optimality of a solution. Then the calculation time needed is a polynomial function of the system size, with the parameters of the polynomial depending on the probability with which the optimizer wants to get a result of at least a required quality [180].

16.3 Grest Hypothesis

For simulated annealing and the great deluge algorithm there exist proofs that they converge to the ground state if some constraints are met, but only after an infinite amount of time [122]. There is also a construction of a finite nonmonotonic cooling schedule for a specific class of problems, leading to the global optimum with threshold accepting. In practice, however, one performs an optimization run and receives a solution not knowing how good or bad it is if the true optimum is unknown.

Generally, there is no analytic criterion proving that the solution found is the global optimum. However, in simulations of spin glasses, a relation was found between the calculation time used and the mean deviation of the solution to the true optimum [71]. This difference between the mean value $\langle \mathcal{H} \rangle_t$ of energies of final configurations of optimization runs taking a time of t and the energy value of the ground state $\mathcal{H}(\sigma_{\text{opt}})$ depends as follows on the calculation time t:

$$\langle \mathcal{H} \rangle_t - \mathcal{H}(\sigma_{\text{opt}}) \propto \frac{1}{(\ln t)^\zeta} \quad \text{with} \quad \zeta \approx 1 \, . \tag{16.1}$$

This Grest hypothesis has not yet been proved generally but has also been shown to be valid for other NP-complete problems (see, e. g., [168, 100]) and can therefore be generally used as an approximation for how much the quality of the results can be increased if all of the available computation time is used and what the true ground state energy is [77].

17 Weakening the Pure Markovian Approach

17.1 Saving the Best-So-Far Solution and Spinoffs at Good Solutions

So far, we have only considered a pure Markovian approach, i. e., a tentative new configuration is either accepted or rejected by considering the value of a control parameter and by considering the current configuration. No information about previously visited configurations or other information about the former past is used. However, this pure Markovian approach exhibits some disadvantages: sometimes it happens that the optimization run is in a good local valley but leaves this valley before reaching good solutions. After that the optimization run might end up in a not so deep valley. Of course, it is impossible to find out how deep each local valley is during the Monte Carlo walk through the energy landscape. However, there are some approaches that are still based on the Markov chain approach but that try to keep information about formerly visited good configurations in order to end up at better solutions or to speed up the optimization process.

The simplest of these approaches is to always save the best configuration visited so far. Therefore, at each move that is accepted it is checked whether the energy value of the new configuration is smaller than the energy value of the best solution so far. In this case, the "best-so-far solution" is overwritten by the new solution. Depending on the considered problem, this can be very time consuming, such that the "save the best solution" routine is sometimes only called at already rather small values of the control parameter, at which these configurations are really pretty good already, and new "best configurations" come along only rarely.

Often the optimization run ends by taking the best-so-far solution and performing some greedy steps on it in order to get quite safely to the bottom of the local valley. However, there are also more complicated ways of using the best-so-far solution.

An additional approach that can be used on these best-so-far solutions is to make a spinoff every time a new best solution is detected. This means that, besides the current optimization process, further processes are started with these best-so-far solutions as initial configurations of these optimization processes. In the simplest approach, a greedy run is performed in order to quench down each best-so-far solution to a locally optimum solution. However, one could also think of using the Bouncing technique described in Sect. 15.2. The best-so-far solution serves as an input solution, either only for the first bouncing iteration or even for all bouncing iterations. If new best-so-far solutions are found in these bouncing iterations, more spinoff processes are started. Of course, this approach will not be effective if all processes run on one processor at the same time. Ideally, it is used on a large workstation cluster with load balancing, when one can easily detect whether a spinoff process has already finished its work, so that the processor is free for a new process.

17.2 Record-to-Record Travel

Dueck altered the threshold accepting (TA) algorithm, described in 12.1, by comparing the tentative new configuration τ not with the current configuration σ but with the best-so-far solution $\hat{\sigma}$ [51]. In this record-to-record travel (R2R) algorithm, a threshold Th is used much like in TA, but the acceptance criterion is given as

$$p(\sigma \to \tau) = \begin{cases} 1 & \text{if } \mathcal{H}(\tau) - \mathcal{H}(\hat{\sigma}) \leq \mathit{Th}\,, \\ 0 & \text{otherwise}\,, \end{cases} \tag{17.1}$$

so the difference $\mathcal{H}(\tau) - \mathcal{H}(\hat{\sigma})$ is used instead of the energy difference of the successive configurations as in Eq. (12.1). It is claimed that this approach converges more rapidly to a good solution than TA [51].

Although this approach reminds one of the TA algorithm due to its acceptance criterion, it is more related to the great deluge algorithm (GDA), described in Sect. 12.4. The sum of the energy of the best-so-far solution and the threshold, $\mathcal{H}(\hat{\sigma})+\mathit{Th}$, serve as a water level $\mathcal{T}$ for the system, above which no configuration can be accepted. While TA at a small threshold might still be able to climb up an energy barrier in several steps, this is impossible with the R2R algorithm, as the comparison is always performed with a configuration at the bottom. If the threshold Th is already rather small, only seldom can a solution better than the best-so-far solution be found, so that the decrease in the water level is mainly governed by the decrease in the threshold. Without knowledge of any bounds for a good or even the optimum solution of the proposed optimization problem, this R2R algorithm might be superior to the GDA, as it is able to concentrate the calculation time automatically to more important parts of the cooling schedule. This effect can be strengthened in the following way: if a new best-so-far solution $\hat{\sigma}_{\text{new}}$ is found, the threshold Th is set to

$$\mathit{Th} = (\hat{\sigma}_{\text{old}} - \hat{\sigma}_{\text{new}}) + \mathit{Th} \tag{17.2}$$

in order to keep the water level constant in this case. The next value of the threshold is then determined according to the prescription in its cooling schedule.

Of course, this R2R approach of using the best-so-far solution instead of the current one for comparison with the tentative new solution can be transferred to algorithms with transition probabilities similar to TA, like simulated annealing (SA), in which the Metropolis criterion [Eq. (11.6)] and the heat bath condition [Eq. (11.7)] can be changed accordingly, or to the transition probabilities based on the Tsallis statistics. However, this R2R approach has the same disadvantage as the GDA in that it is necessary to compute the whole energy of the tentative new solution. In contrast, often much calculation time can be saved by calculating the energy difference between the current and the tentative new configuration locally when using SA, TA, or one of the acceptance probabilities that are based on the Tsallis statistics.

17.3 Stochastic Tunneling

Another algorithm that makes use of the best-so-far energy value and for which good results are reported is the stochastic tunneling (STUN) method. This approach belongs to the class of algorithms that change the energy landscape, but, in contrast to algorithms like search space smoothing (SSS), it tries to smooth the energy landscape directly: as already mentioned in Sect. 13.1, SSS smooths the energy landscape in an indirect way by smoothing the values of terms that are added up in the Hamiltonian. Contrarily, STUN changes the cost function values of all configurations directly by some smoothing function in which the cost function value of the best solution found so far plays a dominant role. Of course, such a smoothing function must be strictly monotonous such that local minima remain locally minimum. The outline of the smoothed energy landscape depends strongly on the best configuration found so far, as shown in Fig. 17.1.

17.4 Changing the Cooling Schedule Due to Intermediate Results

Instead of changing the acceptance criterion, one can also make use of this best-so-far solution for developing an adaptive cooling schedule.

If one achieves configurations that are much worse than the best-so-far solution, one can interpret this in two different ways:

- Obviously, the system wants to climb up some energy barriers in order to get to better solutions. In this case, one should make the situation for the system easier by increasing the control parameter. This increase could be done in a rather simple approach by simply adding some fixed amount to the control parameter or multiplying a factor to it, but it could also be done in an adaptive way by comparing the last configurations with the best-so-far solution and by determining a new value for the control parameter in an

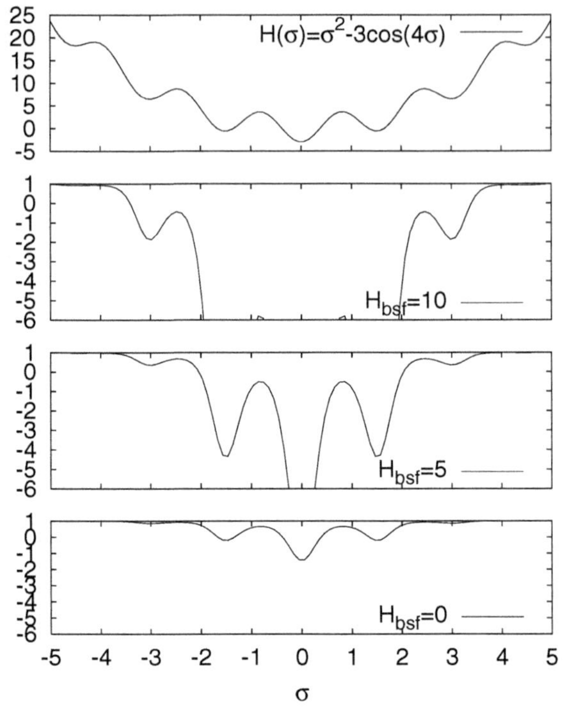

Fig. 17.1. Motivation for STUN: consider a simple toy model, in which each configuration is simply given as a real number σ. The cost function for this problem, which is to be minimized, will be $\mathcal{H}(\sigma) = \sigma^2 - 3\cos(4\sigma)$. This simple function has, of course, many local minima. The greedy algorithm, which only allows for improvements, would roll down the hill from the starting configuration to the next local minimum. Also other improvement algorithms would often fail to reach the global optimum, as there is not that standard structure in the energy landscape, that is, there are barriers of tremendous height, then barriers that are also large, but much smaller in height than the barriers of the giant class, and so on. Here the heights of the barriers are comparable for several neighboring local minima. As introduced by the inventors of STUN[215, 79], one replaces the original cost function $\mathcal{H}(\sigma)$ with the altered cost function $\mathcal{H}(\sigma, \mathcal{H}_{\rm bsf}) = 1 - \exp(-\gamma \times (\mathcal{H}(\sigma) - \mathcal{H}_{\rm bsf}))$. As can be nicely seen at the *curves* for $\gamma = 0.3$ and various values of $\mathcal{H}_{\rm bsf}$, the local minima and maxima stay at the same configurations, but the shape of the energy landscape is changed to a large extent: local minima that are deeper than $\mathcal{H}_{\rm bsf}$ are deepened in the transformed energy landscape in order to drive the search process in one of these deeper valleys. On the other hand, the structures lying above $\mathcal{H}_{\rm bsf}$ are smoothed out

elaborate way, making use of this comparison. So, generally, this approach favors more freedom for the system by increasing the control parameter if the energy difference between the last configurations visited and the best-so-far solution gets larger.

- On the other hand, one could also argue in the opposite way: the system seems to have too much freedom if it moves to configurations much worse than the best-so-far solution. Therefore, it needs a push to descend to better solutions. This can be done by reducing the control parameter in a stronger way than usual.

Both philosophies of modifying the cooling schedule in response to the energy difference between the last configurations visited and the best-so-far solution have their merits. The authors' experience suggests that the superiority of one or the other of these philosophies depends on the specific problem to which it is being applied.

But there are also other methods to use previous configurations to determine the next control-parameter values. For example, the values for the mean energy and the specific heat can be measured at the current value of the control parameter. A new value for the control parameter is determined depending on these results. One approach that works analogously to these approaches has already been discussed in Sect. 15.3.

Another approach to a possibly nonmonotonic cooling schedule is acceptance simulated annealing (ASA), which was developed by Markus Puchta [167]: starting from the random walk, i. e., from the initial temperature $T = \infty$, one always stores the energy differences $\Delta\mathcal{H}_i$ with $1 \leq i \leq m$ of the last m moves that led to an improvement. Let $\overline{\Delta\mathcal{H}}$ be the mean value of these negative energy differences. After every new improvement found, the new $\Delta\mathcal{H}_i$ is added to the list of energy differences, the oldest entry in the list is deleted if the list already contains m entries, and the mean value is updated. Then the temperature of the system is set to

$$T = \frac{\overline{\Delta\mathcal{H}}}{\log(p)}\,, \tag{17.3}$$

with p being the probability with which a move leading to a deterioration of size $|\overline{\Delta\mathcal{H}}|$ will be accepted. Instead of decreasing the temperature T, the probability p is decreased from an initial value ≤ 1 to a final value slightly larger than 0. This ASA technique has the advantage over the original SA algorithm that one does not need to determine both the start and end values for the cooling schedule for every new system. Nevertheless, one must still determine an appropriate cooling schedule for the probability p at least once.

18 Neural Networks

Homo sapiens sapiens and other animals have delevoped very complex brains during the evolutionary process. For centuries, the brain and the nerves in general have been investigated. Now, besides the general scientific interest, the aim is to understand this biological network in such a way that an artificial neural network (NN) of roughly the same or even a better quality can be created on a computer system. To date, this goal is far from being met. However, there are many implementations of artificial NNs that, while very small compared to their biological counterparts, have partially proved to work rather well. As the literature on NNs is vast (see, e. g., [8, 81, 86, 119, 172, 179, 191, 216]), and as we are only interested in using them for optimization purposes, we will give here only a short introduction to NNs in order to have the key ingredients for putting them to our own uses.

Please note that this approach to create an artificial NN represents methodogically a completely different ansatz from the artificial intelligence (AI) approach: in the school of AI, the aim is to create some intelligent system by proposing a very complex rule set such that the system reacts in the desired way or is able to solve the desired task or is even able to behave like a human only because of this rule set. (One could consider the construction heuristics as AI systems creating a good solution to a given problem.) In the NN approach, however, the aim is to create a natural intelligence system, i. e., some artificial system that learns like a biological creature from examples and that is able to generalize what it has learned such that it can also give correct answers to questions that it did not learn beforehand.

18.1 Biological Motivation

All in all, a human being has roughly 10^{11} neurons. Each of them is connected with on average 10^3 to 10^4 other neurons [125]. Sensory processing, the processing of information received from outside the network, and motor control, the creation and distribution of commands to the muscles, is not done sequentially as in a conventional computer, but in a highly parallel way. The complexity of the brain mechanisms is thus given by this huge number of connections.

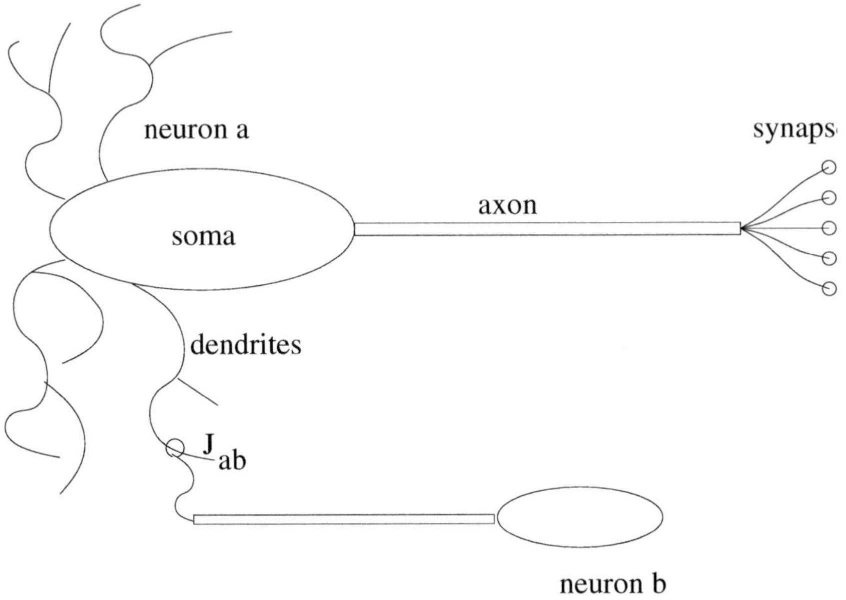

Fig. 18.1. Schematic picture of a neuron: like every cell, a neuron consists of a cell body called a soma. Furthermore, it has extensions called dendrites that collect the signals of other neurons. Additinally, a neuron also has a long string called an axon that splits and ends in the synapses. These synapses are connected to the dendrites of other neurons. Thus, if a neuron fires, the electrical peak moves from the soma via the axon to the synapses, which then influence neurons with which they are connected. The influence of neuron b on neuron a is of a certain size J_{ab}

A neuron, shown schematically in Fig. 18.1, consists of a cell body called a soma. Furthermore, it contains many so-called dendrites, which are protuberances of the cell membrane. Furthermore, each neuron has a long "tail" called an axon that can even be 1 m long. This axon splits into 10^3 to 10^4 parts that end in so-called synapses. These synapses are connected to the dendrites of other neurons.

Due to the semipermeable membrane of the neuron, which acts differently on sodium, potassium, and chloride ions, which are partially allowed to pass through ion channels in the membrane, there is usually a negative potential on the surface of a neuron called the rest potential. However, if there is a strong external influence or even several of them (for example, by preceptors or by other firing neurons) and if this influence is larger than a certain threshold, then the sodium channels open and the membrane potential becomes positive. This positive action potential runs over the surface of the neuron via the axon to the synapses—the neuron fires. After that, the neuron returns to its rest potential. When the action potential reaches the synapses, the synapses emit some transmitter molecules, which influence the neurons to which they are connected. There are two types of influences: roughly 85% of all synapses are excitatory, i. e., the firing of one neuron encourages the second to fire as well, but 15% are inhibitory, i. e., the firing of the first neuron suppresses the tendency of the second neuron to fire.

18.2 Artificial Neural Networks

Artificial NNs are computer algorithms that somehow simulate a brain, but, of course, in a strongly simplified way. They are used for two purposes: they can either lead to a better understanding of brain functions, the evolution of the brain, and brain illnesses. On the other hand, they can also be used as new approaches for solving complex problems.

The first question to be answered with regard to the creation of an artificial NN is whether the real biological behavior has to be emulated completely on a computer. First, these biological processes are not completely known and understood, so that a complete emulation is impossible. Furthermore, as the action potential always looks the same, one breaks down a biological neuron to some variable S_i that can have two states: either the neuron fires or does not fire. Computer scientists thus write

$$S_i = \begin{cases} 1 & \text{if the neuron } i \text{ fires}\,, \\ 0 & \text{otherwise}\,. \end{cases} \tag{18.1}$$

They simply imagine the neuron as an information bit. (Of course, there are also implementations of NNs in which the neurons can take more than two values, e. g., continuous values between 0 and 1, values from a discrete set of numbers, or any integer or real value.) For physicists, however, the most prominent two-state system is the Ising system, in which each spin S_i can take the values -1 and $+1$, so that they usually write

$$S_i = \begin{cases} +1 & \text{if the neuron } i \text{ fires}\,, \\ -1 & \text{otherwise}\,. \end{cases} \tag{18.2}$$

Furthermore, the synapses are represented as weight values: $J_{ji} = J_{j \leftarrow i}$ denotes the size of the influence of neuron i on neuron j. If J_{ji} is positive, then the synapse of neuron i that is connected to a dendrite of neuron j is excitatory, otherwise it is inhibitory. Thus, one adds up all influences by adding up these weight values. Analogously, the threshold that these influences have to exceed in order to make neuron j fire is also represented as some real value θ_j. Whether neuron j fires or not is then represented as some function g_j of these parameters, i. e.,

$$S_j = g_j \left(\sum_i J_{ji} S_i - \theta_j \right) . \tag{18.3}$$

Often the threshold θ_j is set to zero. This can always be done by introducing an additional neuron k that always fires, i. e., $S_k \equiv 1$, and that is connected to neuron j via the weight $J_{jk} = -\theta_j$.

The function g_j determines whether neuron j will fire or not. Often some deterministic functions are used like the slightly altered signum function, $g(x) = 1$ if $x > 0$ and $g(x) = -1$ otherwise, by physicists and the slightly altered Heaviside function, $g(x) = 1$ if $x > 0$ and $g(x) = 0$ otherwise, by computer scientists. Using such a function allows one to determine exactly whether the neuron fires. However, often also functions with some probabilistic elements are chosen, like the hyperbolic tangent function. Then the neuron fires with a probability which depends on external influences and its threshold value. Another approach adds some random number to the sum of the influences in order to randomize the firing process of the neuron.

However, the various implementations of NNs differ not only in this function g, which is mostly the same for all neurons, but much more in their architectures. Each NN contains input neurons and one or more output neurons. There may also be further hidden neurons, i. e., neurons that are neither input nor output neurons and that are thus hidden in the network. The main distinction between various NNs is whether the network contains some feedback. In this case, the output of a neuron, which is sent to other neurons as the input in the next step, influences its own future input, as this is the output of the other neurons that it had influenced before. Such networks containing loops of neurons influencing their successive neurons are called recurrent networks. NNs without feedback are called feedforward networks. In feedforward networks, the neurons can be written to be organized in single layers. Each layer of neurons then influences only the neurons in the next layer. Of course, the dynamics of such a network always reaches a stationary state. In contrast, recurrent networks do not necessarily reach such a stationary state.

Each NN must first learn some examples before it can be put to actual use. Learning mostly means that the weights J_{ij} are changed (sometimes the values of the neurons are also/instead changed) so that information about the example is stored in a distributed way in the NN. Three questions arise when working on a specific architecture with a specific learning rule:

- What can the NN learn? Is it really able to solve the proposed problem?
- How much can be stored? If the number of learned examples exceeds the capacity of the NN, then the answers the NN gives might be random.
- Is the NN able to generalize what it has learned, i. e., is it able to detect the rule behind the examples, or does it simply learn the examples by heart?

There are two types of learning algorithms: in the supervised learning, the NN gets not only the input of the example to be learned but also the desired output. Thus, it readjusts its weights if its actual output differs from the desired output. Related to this, the reinforcement learning technique simply informs the NN whether its output was correct or not, but not about the output itself. In applications using an unsupervised learning process, the learning rule only depends on the activities of the neuron; the network must learn in a self-organized way. An example of this is Hebb's learning rule [82]: if a firing neuron a influences a neuron b to fire, then the coupling J_{ba} between them is enlarged.

Thus, each NN is described by three parts, the properties of the individual neuron, the network architecture, and the learning rule. There is no general criterion regarding which NN is best suited for a given problem. One must find a good NN empirically.

Probably the simplest NN is a double layer perceptron. One layer of this feedforward network contains the input neurons. These are connected to the output neurons in the second layer. But there are no connections between the neurons in the input layer and no connections between the neurons in the output layer, either [174, 175]. Thus, each output neuron gets only inputs of the input neurons, so that the network can be split into several perceptrons with each of them containing the input neurons and one output neuron. Let $S_i = \pm 1$ be the values of the N input neurons and ζ the value of the output neuron. The N weights then simply form a vector $\boldsymbol{J}$, such that one writes

$$\zeta = g\left(\sum_{i=1}^{N} J_i S_i\right) = \mathrm{sgn}\left(\boldsymbol{J} \circ \boldsymbol{S}\right) \tag{18.4}$$

using the signum function. The supervised learning rule is applied as follows. After an initialization of the J_i with random numbers or with zeroes, some pattern $\boldsymbol{S}^\nu$ is presented to the perceptron. If its output ζ^ν is identical to the desired output z^ν, then the weights J_i do not need to be adjusted. Otherwise, the weights J_i are changed via the rule

$$J_i^{\mathrm{new}} = J_i^{\mathrm{old}} + \eta \times (z^\nu - \zeta^\nu) \times S_i^\nu\,, \tag{18.5}$$

with η being the so-called learning rate that determines how strongly the pattern is stored in the interactions. Of course, η has to be chosen between 0 and 1. Then the next pattern is learned in the same way, and so on, until all patterns are used in the learning process. It might be that one or more previously learned patterns are then partially not correctly represented, so that they have to be learned again. η is gradually reduced during the iterative application of this learning process.

This perceptron by Rosenblatt is not able to work correctly on all problems. Instead, one can only treat problems with it in which the good patterns with $z^\nu = 1$ are separated by an $N-1$-dimensional hyperplane from the bad patterns with $z^\nu = -1$ (or $z^\nu = 0$) in the N-dimensional space containing all possible patterns. Applying the learning rule above, this hyperplane is simply turned and shifted. As Fig. 18.2 shows, the perceptron can, e. g., learn

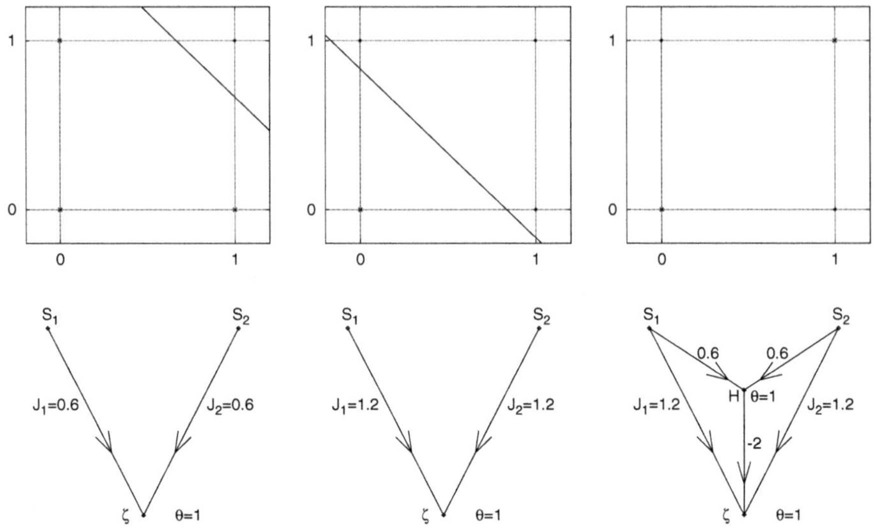

Fig. 18.2. Learning three logical operators with a perceptron. *Top row*: results for a logical AND (*left*), a logical OR (*middle*), and a logical XOR (= exclusive OR, *right*) of two logical input variables being either 0 (= false) or 1 (= true). In the case of the AND, one gets only true as a result if both input variables are true. For the OR, one gets true if at least one of the two input variables is true. For the XOR, one gets true if exactly one of the two input variables is true. In the cases of both AND and OR, one can draw a line between the solution points that are true (*filled squares*) and those that are false (*crosses*). But it is impossible to draw such a line in the XOR case. Below these graphics, some corresponding perceptrons are shown that have learned these rules: the output neuron always has a threshold $\theta = 1$. The input neurons S_1 and S_2 are connected via the weights J_1 and J_2, respectively, to the output neuron. As one can easily check for both the AND and the OR rule, the standard perceptron with only one input layer of neurons and one output neuron can solve all inputs correctly. For the case of the XOR, it is necessary to introduce at least one further hidden neuron. It would be impossible for a perceptron without hidden neurons to learn the XOR rule. The input neurons are also connected to the hidden neuron H, which also has a threshold of 1. It fires only if both input variables are true, thus forcing the output neuron ζ not to fire. Please note that the trick here is that the hidden neuron has to learn the AND rule and that the output neuron is connected with the input neurons as in the OR rule. Furthermore, the hidden neuron H is connected with a strong negative weight to the output neuron. A negative weight always inserts some NOT boolean operator. Summarizing, due to the strength of this connection, the output neuron has learned the XOR rule as (S_1 OR S_2) AND (NOT (S_1 AND S_2))

the AND rule and also the OR rule, as in both cases one can draw one line between the points leading to an overall true and the points leading to an overall false. But it is unable to learn the XOR rule, as Minsky and Papert demonstrated [140]. However, this difficulty can be overcome by introducing further hidden layers of neurons: thus, the input layer is directly connected to neurons in the first hidden layer; these are then connected to neurons in a second hidden layer, and so on, until the neurons of the last hidden layer are connected to the neurons in the output layer. For many-layer perceptrons, often the error backpropagation algorithm is used for learning: Here, first of all, an applied pattern is propagated through the layers of the network to the output neurons. If the outputs of one or more of these output neurons differ from the desired output values, then the error is calculated backwards, starting with the errors of the output neurons, then the errors of the neurons in the last hidden layer based on the errors of the neurons in the output layer, then the errors of the neurons in the second last hidden layer based on the errors of the neurons in the last hidden layer, and so on. The weights are then changed similar to the learning rule for the original perceptron.

It was shown that the introduction of two additional hidden layers is sufficient for any separation of the individual input patterns [43]. However, the question remains, for this and also for any other NN, of how many neurons and what architecture is optimal for first solving the desired task at all and second for the maximum learning velocity. Too many neurons lead to overfitting, i. e., the individual examples are simply learned by heart, but the NN cannot generalize what it has learned. On the other hand, a too small number of neurons makes it impossible for the NN to understand the basic rule as well.

As this perceptron example already demonstrates, some basic biological properties are strongly violated: a connection can change its sign, in contrast to biological synapses, which stay either excitatory or inhibitory. Summarizing, in this business of NNs, one uses some findings of biology to build up an artificial network but neglects many other findings. In the following sections, two examples of recurrent networks that are also used for optimization purposes are introduced.

18.3 The Hopfield Model

The Hopfield network consists of N binary neurons $S_1, \ldots, S_N$, which can only take the values $+1$ and -1 (again in the physicists' language). Each neuron is connected with all neurons except itself, such that there are $N \times (N-1)/2$ links between the neurons in this completely connected layer of neurons. Each neuron serves as an input and as an output neuron. A real number J_{ij} is associated to each connection $S_i \leftrightarrow S_j$. This strength of the connection is symmetric, i. e., $J_{ij} = J_{ji}$. (Of course, this is in contradiction to findings for biological systems, in which there is usually $J_{ij} \neq J_{ji}$.) As there is no self interaction in the Hopfield model, one sets $J_{ii} = 0$.

This type of network is mainly used for storing several patterns that will be recognized later. A possible application is that it will recognize the ZIP code of cities from handwritten numbers on envelopes based on some stored handwriting styles. For each number, a few example patterns are stored. These overall p patterns $\boldsymbol{\xi}^\alpha$ are given as bitstrings, e. g., $\xi_i^\alpha = +1$ if the pattern α is black in area i or -1 if it is white there. Hebb's learning rule is applied in the following way:

$$J_{ij} = \frac{1}{N} \sum_{\alpha=1}^{p} \xi_i^\alpha \xi_j^\alpha \tag{18.6}$$

and

$$J_{ii} = 0\,. \tag{18.7}$$

Thus the information about the patterns is stored in a distributed way as in a real biological network. Please note that Hebb's learning rule is altered here: the connection between two neurons becomes not only stronger if they both do fire at the same time but also if they both do not fire at the same time.

The network should be able to recognize a learned pattern. Furthermore, if a slightly altered pattern is presented to the Hopfield network, then the network should be able to reconstruct the original pattern and thus to recognize the original pattern behind the proposed altered pattern. At the beginning of this recognition algorithm at time $t = 0$, a pattern vector $\boldsymbol{S}(0)$ is presented to the network, i. e., all spins are initially set to the proposed input values. In the successive time steps, the following rule is iteratively applied to each neuron (usually in a parallel update of all neurons in the synchronous Hopfield network, sometimes in a random sequential update in the asynchronous Hopfield network):

$$S_i(t+1) = \operatorname{sgn}\left(\sum_{j=1}^{N} J_{ij} S_j(t) \right) . \tag{18.8}$$

In the case where the right side is exactly zero, $S_i(t+1)$ is randomly set to either $+1$ or -1. After a finite number of time steps, $\boldsymbol{S}$ is converged to one of the stored patterns $\boldsymbol{\xi}^\alpha$.

Let us first consider the special case that only one pattern $\boldsymbol{\xi}$ is learned. Then the coupling values are given as $J_{ij} = \xi_i \xi_j / N$ for $i \neq j$ and $J_{ii} = 0$. If $\boldsymbol{S}(0) = \boldsymbol{\xi}$, then

$$
\begin{aligned}
S_i(1) &= \operatorname{sgn}\left(\sum_{j=1}^{N} J_{ij}\xi_j\right) \\
&= \operatorname{sgn}\left(\sum_{\substack{j=1\\ j\neq i}}^{N} \frac{1}{N}\xi_i\xi_j\xi_j\right) \\
&= \operatorname{sgn}\left(\frac{1}{N}\xi_i \sum_{\substack{j=1\\ j\neq i}}^{N} 1\right) \\
&= \operatorname{sgn}\left(\frac{N-1}{N}\xi_i\right) = \xi_i \,,
\end{aligned}
\tag{18.9}
$$

i. e., $\boldsymbol{S}(1) = \boldsymbol{\xi}$, the pattern is stable in this iteration and is therefore recognized. Usually, only large NNs with large N are considered. Thus, mostly the small difference due to J_{ii} is omitted in a theoretical analysis [149].

Let us now consider the case where a pattern $\boldsymbol{S}$ is presented to the Hopfield Network that is nearly identical to the stored pattern $\boldsymbol{\xi}$ but some of the entries are changed, i. e., $S_i(0) = \xi_i$ for n out of N entries and $S_i(0) = -\xi_i$ for the remaining $N - n$ entries. Then

$$
\begin{aligned}
S_i(1) &= \operatorname{sgn}\left(\sum_{j=1}^{N} \frac{1}{N}\xi_i\xi_j S_j(0)\right) \\
&= \operatorname{sgn}\left(\frac{1}{N}\sum_{\substack{j=1\\ S_j(0)=\xi_j}}^{N} \xi_i\xi_j\xi_j - \frac{1}{N}\sum_{\substack{j=1\\ S_j(0)=-\xi_j}}^{N} \xi_i\xi_j\xi_j\right) \\
&= \operatorname{sgn}\left(\frac{n}{N}\xi_i - \frac{N-n}{N}\xi_i\right) \\
&= \operatorname{sgn}\left(\frac{2n-N}{N}\xi_i\right) \\
&= \begin{cases} \xi_i & \text{if } n > N/2\,, \\ -\xi_i & \text{if } n < N/2\,, \end{cases}
\end{aligned}
\tag{18.10}
$$

i. e., the proposed input pattern converges to the stored pattern $\boldsymbol{\xi}$ if less than half of the bits were changed, otherwise it converges to $-\boldsymbol{\xi}$ if more than half of the bits were changed. Therefore, if a pattern $\boldsymbol{\xi}$ is stored in a Hopfield network, then both this pattern and its inverse pattern $-\boldsymbol{\xi}$ serve as attractors in this network.

Let us now move to the general case where several patterns are stored in the network according to Eq. (18.6). The question is whether a stored pattern $\boldsymbol{\xi}^\alpha$ can be recognized if given as an input in the NN:

$$\begin{aligned} S_i(1) &= \operatorname{sgn}\left(\sum_{j=1}^{N} J_{ij} S_j(0)\right) \\ &= \operatorname{sgn}\left(\sum_{j=1}^{N} J_{ij} \xi_j^\alpha\right) \\ &= \operatorname{sgn}\left(\frac{1}{N}\sum_{j=1}^{N}\sum_{\beta=1}^{p} \xi_i^\beta \xi_j^\beta \xi_j^\alpha\right) \\ &= \operatorname{sgn}\left(\frac{1}{N}\sum_{j=1}^{N} \xi_i^\alpha \xi_j^\alpha \xi_j^\alpha + \frac{1}{N}\sum_{j=1}^{N}\sum_{\substack{\beta=1 \\ \beta\neq\alpha}}^{p} \xi_i^\beta \xi_j^\beta \xi_j^\alpha\right) \\ &= \operatorname{sgn}\left(\xi_i^\alpha + h_i^\alpha\right) , \end{aligned} \tag{18.11}$$

with the term

$$h_i^\alpha = \frac{1}{N}\sum_{j=1}^{N}\sum_{\substack{\beta=1 \\ \beta\neq\alpha}}^{p} \xi_i^\beta \xi_j^\beta \xi_j^\alpha . \tag{18.12}$$

As ξ_i^α would be the correct output, the term ξ_i^α is called the signal term whereas h_i^α is called the error term [121]. If the patterns are strongly correlated, then $|h_i^\alpha|$ can be of size $p-1$. For uncorrelated patterns, $|h_i^\alpha| \ll 1$, such that $S_i(1) = \xi_i^\alpha$, i. e., the pattern is recognized. Analogously, a slightly changed pattern converges to the corresponding stored pattern. However, this holds true only for a number p of patterns that is small compared to the number N of neurons. If the stored patterns are uncorrelated, then the normalized storage capacity p/N can only reach values up to ≈ 0.14.

Usually, the corrupted pattern does not converge to the original pattern in one step if more than one pattern is stored. Thus, the Hopfield rule has to be applied several times. Furthermore, the pattern might not converge to its original form completely, so that the final pattern $\tilde{\boldsymbol{S}}$ of the Hopfield procedure must be compared with all stored patterns and equated to that one for which the Hamming distance is minimal, i. e., the overlap

$$O(\tilde{\boldsymbol{S}}, \boldsymbol{\xi}^\alpha) = \left|\frac{1}{N}\sum_{i=1}^{N} \tilde{S}_i \xi_i^\alpha\right| \tag{18.13}$$

is maximal. Sometimes the synchronous Hopfield Network, in which all spins are updated in parallel, does not stay constant at some final pattern but finally jumps between two patterns. Then one checks the overlaps between these patterns and the stored patterns in order to find the best solution.

From the point of view of optimization, this NN searches for the ground state of an energy landscape formed by the spin glass Hamiltonian

$$\mathcal{H} = -\frac{1}{2} \sum_{i,j} J_{ij} S_i S_j \,. \tag{18.14}$$

Each neuron can be associated with a spin of the corresponding spin glass problem, and the considered move class of choosing one neuron randomly and applying the Hopfield update rule of Eq. (18.8) can be associated with a single spin flip performed by a greedy algorithm. The parallel update rule of applying Hopfield's rule to all spins simultaneously can be considered as a collective spin flip. The task of the spin glass problem is to find the unknown ground state of a spin glass instance with a given interaction matrix J. On the other hand, the minima of the energy landscape of the NN are known to be the stored patterns (if they are not too correlated and if not too many of them are stored), the task here is to define an appropriate interaction matrix based on the known ground states, so that one speaks of the Hopfield model being an inverse spin glass problem.

Each flip of a spin S_k according to instruction (18.8) leads to a decrease in the energy

$$\mathcal{H} = -\frac{1}{2} \sum_{i,j} J_{ij} S_i S_j = -\frac{1}{2} \sum_{\substack{i,j \\ \neq k}} J_{ij} S_i S_j - S_k \sum_{\substack{i \\ \neq k}} J_{ik} S_i \,, \tag{18.15}$$

as S_k takes the sign of the sum $\sum_i J_{ki} S_i$ if the Hopfield rule is applied. Analogously, if there are additionally local magnetic fields H_i attached to each spin S_i, such that the Hamiltonian is given as

$$\mathcal{H} = -\frac{1}{2} \sum_{i,j} J_{ij} S_i S_j - \sum_i H_i S_i \,, \tag{18.16}$$

then the update rule must be extended to

$$S_i(t+1) = \operatorname{sgn}\left(H_i + \sum_{j=1}^{N} J_{ij} S_j(t) \right) . \tag{18.17}$$

These local magnetic fields H_i can be interpreted in the NN language as the negatives of the thresholds θ_i, which have to be overcome while stimulating a neuron to fire.

Storing a new pattern $\boldsymbol{\xi}^\alpha$ means deforming the energy landscape and introducing two new local minima for $\boldsymbol{\xi}^\alpha$ and $-\boldsymbol{\xi}^\alpha$ due to the Ising degeneracy of the Hamiltonian. However, if this new pattern is correlated with already stored patterns, then further local minima are created that might even become deeper than the minima of the original patterns.

The "Hopfield algorithm" is nothing more than a greedy algorithm that intends to move a slightly changed pattern to the nearest local minimum in which the corresponding unchanged pattern, which serves as an attractor, lies. With an increasing number of stored patterns, the probability increases that this local search process will fail and end up at either another stored pattern or a superposition of stored patterns. Even worse, the stored pattern might not form a local minimum anymore, such that it can no longer be recognized.

The question arises as to how this NN can be used as an optimization algorithm as it proposes that the patterns, i. e., the good solutions, are already known and are used to create the energy landscape in which they form the local minima. However, one can really make use of this model in the following ways:

- If several good solutions for a proposed optimization problem are already known, then an energy landscape can be formed by them in which one searches for the global minimum, which is hopefully a configuration with a smaller energy than the energies of the stored configurations and which is also a solution of the proposed problem. The search process can be performed with the Hopfield approach, but preferably with a nongreedy optimization algorithm, as it is important to get, not to a local minimum near the starting point of the search, but to the global optimum.
- The proposed optimization problem can be formulated in a way like the Hopfield model, i. e., in this description, it consists of neuronlike variables S_i, an interaction matrix J between these variables, and a Hamiltonian of the form of Eq. (18.14). It is not necessary that there be patterns of any meaning; simply a matrix J is needed. Then the Hopfield algorithm can be applied for finding a solution. Furthermore, if a rather good solution is already found but one wishes to get a similar but energetically better solution, then one can make use of this Hopfield model.

18.4 Kohonen Networks

Kohonen invented many NNs. Thus, there are various types of networks nowadays called Kohonen networks, e. g., vector quantization algorithms with unsupervised and supervised learning rules, in which some so-called codebook vectors are used to classify a set of proposed input vectors. Mostly, however, so-called self-organizing maps (SOMs) are meant if the term Kohonen network is used.

These SOMs belong to the class of networks with an unsupervised learning rule. Here the main aspect of the NN lies in the topology of the individual neurons, i. e., in the neighborhood relation between the neurons. The aim is that similar input values are represented by neighboring neurons such that the space of possible input values is mapped topologically correctly onto the neuron space. This idea is biologically inspired e. g., from the so-called sensoric maps in the brain.

Although the cerebral cortex is strongly connected with its 10^4 synapses per neuron, there are bounded areas for certain tasks like the visual cortex and the somatosensoric cortex. Within such an area, there are again subareas that are responsible for some subparts, e. g., the somatosensoric cortex is divided into several subareas, like, e. g., an area for feeling with the right arm that is neighboring the area for feeling with the right hand. As the feeling with the hand is more important, this subarea for the hand is larger than that for the arm. However, the basic topology of the body is completely and correctly mapped into the brain, so that the neighborhoods are preserved. Analogously, neighboring parts of the field of vision are mapped onto neighboring parts of the visual cortex.

A Kohonen network thus consists of N neurons represented by N weight vectors $\boldsymbol{S}_1, \ldots, \boldsymbol{S}_N$ in the input space. Furthermore, these neurons are connected with each other within the set of neurons $\mathcal{N}$ such that there is a topology given. This topology is described via a metric $d_{\mathcal{N}}(i,j)$, where i and j are neurons and $d_{\mathcal{N}}(i,j)$ is the distance between them. A topology has to be chosen appropriate to the topology of the proposed optimization problem, e. g., a line, a circle, or a grid in some higher dimension.

Furthermore, there is a distance metric d defined on the input space, which is often the Euclidean distance. If a stimulus $\boldsymbol{\xi}$ is presented to the Kohonen network, then that neuron i whose weight vector $\boldsymbol{S}_i$ is closest to the stimulus $\boldsymbol{\xi}$ is activated; in mathematical terms

$$i = \mathrm{argmin}_{j \in \mathcal{N}} \{ d(\boldsymbol{S}_j, \boldsymbol{\xi}) \} . \tag{18.18}$$

According to Kohonen, the learning process is not restricted to this activated neuron. Also those neurons j in the neighborhood of neuron i are influenced by this stimulus $\boldsymbol{\xi}$, with the size of the influence being defined by a function $h(d_{\mathcal{N}}(j,i), r_{\mathrm{a}})$, which usually decreases with increasing distance $d_{\mathcal{N}}(j,i)$. This function h is called the activation profile. The parameter r_{a} is called the activation radius and defines the size of the neighborhood that learns (significantly) together with the activated neuron. Function h is normalized in such a way that $h(0, r_{\mathrm{a}}) = 1$ and $h(\infty, r_{\mathrm{a}}) = 0$. Common forms of function h are the Gaussian bell

$$h(d_{\mathcal{N}}(j,i), r_{\mathrm{a}}) = \exp\left(- \frac{d_{\mathcal{N}}(j,i)^2}{2 r_{\mathrm{a}}^2} \right) \tag{18.19}$$

and the 0-1-function

$$h(d_{\mathcal{N}}(j,i), r_{\mathrm{a}}) = \begin{cases} 1 & \text{if } d_{\mathcal{N}}(j,i) \le r_{\mathrm{a}} \, , \\ 0 & \text{otherwise} \, . \end{cases} \tag{18.20}$$

Learning in this network then means that the weight vectors $\boldsymbol{S}_j$ are adapted to the stimulus according to the learning rule

$$\boldsymbol{S}_j^{\text{new}} = \boldsymbol{S}_j^{\text{old}} + \eta \cdot h(d_{\mathcal{N}}(j,i), r_{\text{a}}) \cdot \left(\boldsymbol{\xi} - \boldsymbol{S}_j^{\text{old}}\right) , \tag{18.21}$$

with η being the learning rate, a real number between 0 and 1.

According to [172] and [164], the learning rate η, which is decreased in time, has to fulfill the requirements

$$\int_0^\infty \eta(t)\,\mathrm{d}t = \infty \tag{18.22}$$

such that the learning rate does not decrease too fast, and

$$\lim_{t\to\infty} \eta(t) = 0 \tag{18.23}$$

such that the system arrives at a stationary state. Analogously, r_{a} must be decreased in time: at the beginning, when r_{a} is large, then a large group of neurons learns significantly together with the activated neuron, whereas in the end, the learning neighborhood consists just of the closest neighbors. There are also extensions of the Kohonen network in which the explicit time dependence of r_{a} and η is overcome by a learning-situation-dependent adaptive control by the network itself [80].

19 Genetic Algorithms and Evolution Strategies

19.1 Charles Darwin's Natural Selection

When on board H.M.S. Beagle, *as naturalist, I was much struck with certain facts in the distribution of the inhabitants of South America, and in the geological relations of the present to the past inhabitants of that continent. These facts seemed to me to throw some light on the origin of species—that mystery of mysteries, as it has been called by one of our greatest philosophers.*

With these words the introduction to the famous work by Charles Darwin on the origin of species begins [45]. Darwin was not the first and only one in his time to introduce the idea that species can change during the long time span of evolution. For example, the economist and theologian Malthus derived a mathematically based formulation of the development of populations long before Darwin [131]. He starts out with two postulates: first, that food is necessary for the existence of man. The second postulate is that the passion between the sexes is necessary and remains nearly in its present state. From these two postulates he derives the notion that the size of a population increases at a geometric rate (exponentially in time), as long as there are no restrictions for this growth, like a lack of food supply. On the other hand, subsistence increases only at an arithmetic rate. Therefore, individuals are engaged in a struggle for survival.

Darwin's work is not only based on the work of his predecessors and on theoretical models and philosophical thoughts. His journey to South America enabled him to present some real-world examples that he relied on when formulating his own evolutionary theory. According to Darwin, the discrepancy detected by Malthus leads to a "selection pressure" by which the size of a population decreases again and then stays virtually constant. As nearly all beings produce more offspring than survive, most offspring die before they are themselves able to reproduce.

Furthermore, Darwin states that members of a given species are rather similar to each other but never completely identical. Each individual is thus unique. There are always some small differences among individual members of a species. For some species, these differences might be more pronounced than for other species. Individuals pass on their special properties to their offspring.

Combining these two points, those variations that are seen to work well in the struggle for survival will be found in succeeding generations with a larger probability. In this way the species will change in iterative generations.

In conclusion, certain individuals of a species are better adapted than others to the influences on their lives from their surroundings. These influences are first of all climate, plants and prey for consumption, enemies that see, e. g., them as prey, but also things like a species-specific ideal of beauty, which seems to be important not only for mankind. Incidentally, according to some recent findings, the subjective feeling of beauty is nothing more than some instinct that the beautiful seeming counterpart is healthy and has many variations besides than itself, such that a common offspring would have a good fitness. There are, of course, other concepts as well. Nature can be quite inventive in devising strategems that prove attractive. The males of many species put considerable effort into impressing the females of their species, e. g., some male birds show off while dancing. Apes and humans bring presents such as branches and diamonds, thus signaling to their female counterparts that the males would care for them also around the time of delivery and for raising common offspring. Male spiders of some species perform a drumming solo, although female spiders finding them attractive eat them after mating.

Those individuals that are better adapted survive longer and have a larger probability to mate, thus passing on their variations to the next generation. This insight is summarized in the term natural selection.

19.2 Mutations and Crossovers

In nature, there are two general ways to produce offspring. In the first way, an individual splits into two identical halves and produces a scion. All in all, the individual clones itself such that there are two individuals with the same properties. However, during this type of reproduction, some small errors might occur, such that the two individuals are not entirely identical. The other possibility for producing offspring is that two individuals of one (or two related) species mate with each other and have offspring together. Their common offspring exhibit mostly a mix of their parents' traits. Here, too, some small errors might occur in the process of producing offspring so that the offspring also exhibit traits the parents do not have and do not have stored in their genes, either. Sometimes the set of species is split into two subsets of so-called low-level species such as those that only reproduce by cloning themselves and high-level species that produce offspring by mating. But there are also species like polyps and perhaps even some snakes (this is currently under research) that can reproduce both by cloning and mating. Interestingly, some bacteria species usually reproduce in a sexual way. However, if the environment changes dramatically, then individuals start to prefer cloning instead of mating because the possible number of offspring individuals is much larger in this case such that the probability of survival of at least one individual, and thus of the local colony, increases. But also, the strategy to mate very often and to produce many generations with a large number of offspring in a rather short time can help a population to survive, as demonstrated by the example of the rabbit colonies in Australia that were nearly exterminated by some illness a few years ago. However, some offspring were ultimately immune to this illness so that the rabbit population in Australia did not die out but rebounded to its former population size in a rather short time.

Thus, new traits or new combinations of traits of individuals are brought about by two natural approaches, namely, mutations, the small random changes that occur with a small probability, and crossovers, by which individuals with a completely new mix of traits can be created as a common offspring of two individuals of usually different sexes. Combined with Darwin's natural selection, those mutations and mixings of traits that are advantageous will make their possessors stronger in the struggle for life such that they have a greater probability of passing on their mutations or mixings to the next generation. Iterating this, the whole species will change during the process of evolution and can also split into several species, especially if external influences on various colonies of this species are spatially different. Of course, such natural selection processes can be modified, e. g., by breeding: those animals with traits that humans like get more chances to reproduce than other animals, such that these traits are passed on to many individuals in the next generation.

The Augustinian abbott Gregor Mendel performed many breeding experiments on the color of peas. In his "Untersuchungen über Pflanzenhybride" published in 1865, he published three laws found in his breeding experiments. But he found that his laws dictated that the single traits of two individuals could not be mixed randomly in a crossover. Furthermore, it can be concluded from the probabilities he found in his experiments that the information about each trait is stored twice. There are two possibilities for an individual to exhibit a specific trait:

- Either the information about this trait is stored identically twice in the trait. Then one speaks of homozygous storage of this trait.
- On the other hand, it might be the case that an individual shows a specific trait, although the information about it is stored only once. But the second storage field is filled with different information about a different manifestation of this trait, also known as the phenotype. As the actual phenotype corresponds to, e. g., the first information, the first information is called the dominant one, as it overwrote the second one, which is therefore said to be recessive.
- Such an occurrence of two different pieces of information about the same property can instead lead to an intermediate development of this property that is between those phenotypes that are described by each of the two bits of information. An example are flowers of the same species, one with

the trait to develop red blooms and one with the trait to develop white blooms. If their traits are mixed, the offspring individuals develop pink blooms. Besides the color, the size of the individual can also be determined in this way.

Based on these findings, Mendel stated three laws:

- The uniformity law says that when crossing two homozygous parents that differ in some of their traits, all their children will look the same and show exactly the same traits, and thus the offspring will be uniform.
- The splitting law says that when crossing offspring with each other, their offspring will not be uniform but will show different traits occurring with probabilities that can be easily derived from the fact that the information about a given trait is stored twice and is either working in the dominant–recessive way or in the intermediate way.
- The recombination law states that different properties are transferred to the next generation independently of each other.

As we now know, information about traits is stored in the chromosomes of the cell nucleus. These chromosomes consist of two halves called chromatides. If one cell splits into two, each chromosome breaks into two halves. These two halves move in different directions, so that each of the two cells to be created gets one of the two chromatides of each chromosome. Each of these chromatides is then completed by some biochemical process to a full chromosome. This splitting and completing process might not be error free, so that mutations can occur. On the other hand, in the process of creating new individuals by two members of a species, the chromosomes are again split into two halves, but additionally parts of these chromatides are exchanged in so-called crossover processes so that individual traits are remixed.

A further finding is that the evolution of a particular species cannot be considered independently of the evolution of all other species. Instead, one finds that one has to take coevolution into account. There are examples of individuals from strongly different species that live together and also have developed together during the evolutionary process as this symbiosis is advantageous for both sides. A standard example of coevolution is the development of plants who need insects or other animals for transporting pollen to other individuals of their species. The plants provide nectar for the insects, which get attached to the pollen while drinking the nectar. On the next bloom, they attach the pollen they unwillingly gathered in its stigma, thus initializing the reproduction process. But coevolution is mostly a fight between a prey and its predator. For example, both the cheetah and the gazelle have evolved the ability to run at high speeds, the cheetah to hunt the gazelle, the gazelle to escape the cheetah. However, often a positive symbiosis is also needed for survival. A standard example here are lichen, which consist of algae and fungi. In this symbiosis, the fungi provide water and minerals, which are dissolved from the ground by the fungi, for the algae, which in turn produce oxygen and carbohydrates by photosynthesis. These are in turn needed by the fungi for survival. Thus, lichen can occupy land in which no other species can survive. Another type of connection between individuals of different species is the parasite/host connection. Here at least one species suffers a disadvantage because another species tries to make long-time use of it. Thus the first species develops defense strategies against the parasite and the parasite tries to break this defense.

19.3 Application to Optimization Problems

There are two related types of optimization heuristics that make use of Darwin's principle of natural selection, namely, the German school of evolution strategies (ESs) associated with Rechenberg [169] and the American school of genetic algorithms (GAs) associated with Holland [87]. The various heuristics do not fully imitate the evolutionary process in nature but use three key ingredients based on the biological findings mentioned above, namely, Darwin's principle of natural selection, to establish an artificial population of individuals that can produce offspring by applying the concepts of (a) mutations, (b) crossovers, and (c) selection pressure.

The applications of these two schools differ mainly in the coding of the individuals: the individuals are mostly coded as bitstrings in the GAs, whereas they are mostly coded as a set of real numbers in the ESs. One can refer to these approaches as a wide coding in the case of the GAs, which is even close to the quadruple coding of nature, and as a short coding in the case of the ESs. Both approaches have their advantages and disadvantages: the bitstring approach of the GAs is mostly preferable for combinatorial optimization problems, whereas the real-number approach of the ESs is better for continuous problems. In the bitstring approach, it might often be harder to define appropriate mutation and crossover operators, especially if one has to end up at a feasible configuration. A further disadvantage of this bitstring approach is that often several bits together form a gene, whereas a gene can often be stored as one real number in the ES approach.

The application of mutations differs naturally strongly between these two approaches:

- The bits of a single individual that is coded as a bitstring $\boldsymbol{b} = (b_1, \ldots, b_N)$ in a GA with the bits $b_i = 0, 1$ can be changed in various ways that are often inspired by their natural counterparts:
  - A randomly selected bit b_i can be switched, i. e., $b_i = 1 - b_i$, or set randomly to either 0 or 1.
  - Several randomly chosen bits or a sequence of bits can be switched at the same time.

- A partial sequence of the bitstring can be turned around, so that the original bitstring

$$(b_1, \ldots, b_{i-1}, b_i, b_{i+1}, \ldots, b_{j-1}, b_j, b_{j+1}, \ldots, b_N)$$

is turned into:

$$(b_1, \ldots, b_{i-1}, b_j, b_{j-1}, \ldots, b_{i+1}, b_i, b_{j+1}, \ldots, b_N).$$

- One bit b_i can be moved to another place in the bitstring, so that the bit configuration

$$(b_1, \ldots, b_{i-1}, b_{i+1}, \ldots, b_j, b_i, b_{j+1}, \ldots, b_N)$$

is achieved.
- Several randomly chosen bits or a sequence of bits can be moved to another place in the bitstring.

Thus, there are generally two types of bit mutations, bit-switching and bit-shifting mutations. According to the meaning of the bits and their impacts on the fitness of the individuals, they might be chosen with different probabilities for applying mutation operators to them.

• In the ESs, in which the property i of an individual is coded as a real number p_i, this property is changed by simply adding a random number r_i to it, i. e., $p_i^{\text{new}} = p_i^{\text{old}} + r_i$. As small mutations are preferred, often Gaussian distributed random numbers with zero mean and some variance σ_i^2 are used. σ_i is decreased during the optimization run to put more emphasis on the small mutations at the end and thus to enlarge the probability for an offspring individual to survive. These parameters σ_i need not be identical for all traits. Often the parameters are adaptively changed to keep the average acceptance rates for mutations at a desirable value, typically ≈ 0.2. Rechenberg states that there is an "evolution window" around this value: larger "revolting" mutations would almost never lead to any improvement. Thus, one would lose calculation time if one were always trying these and failing to survive with such a mutation. Small "archconservative" mutations lead to a stagnation or at least to a very slow velocity of the search process, too, so that the optimum width σ_i has to be of some intermediate value.

Also the crossover operators, i. e., the rule sets to create one or two children individuals $\boldsymbol{c}$ and $\boldsymbol{d}$ from two parental individuals $\boldsymbol{p}$ and $\boldsymbol{q}$, differ in these two approaches:

• The crossover operators mostly used in GAs are inspired by the breaking and recombining of chromatides:
  - In the one-point crossover, the bitstrings $\boldsymbol{p}$ and $\boldsymbol{q}$ that are given to their children $\boldsymbol{c}$ and $\boldsymbol{d}$ break after some position i such that $\boldsymbol{c}$ and $\boldsymbol{d}$ get the following bitstrings:

$$\boldsymbol{c} = (p_1, \ldots, p_i, q_{i+1}, \ldots, q_N),$$
$$\boldsymbol{d} = (q_1, \ldots, q_i, p_{i+1}, \ldots, p_N).$$

- In the two-point crossover, the bitstrings break after positions i and j such that the child configurations contain an exchanged subsequence:

$$\boldsymbol{c} = (p_1, \ldots, p_i, q_{i+1}, \ldots, q_j, p_{j+1}, \ldots, p_N),$$

$$\boldsymbol{d} = (q_1, \ldots, q_i, p_{i+1}, \ldots, p_j, q_{j+1}, \ldots, q_N).$$

- Of course, also higher-order point crossover operators can be introduced. Even a random bitstring $\boldsymbol{r}$ can be created such that $c_i = p_i$ and $d_i = q_i$ if $r_i = 1$ and $c_i = q_i$ and $d_i = p_i$ if $r_i = 0$.
- Also, strategies simulating dominant–recessive and intermediate scenarios can be applied. However, the genes stored in the bitstring have to be examined closely and then it has to be determined whether a gene is dominant.

One often has to be especially mindful of the problem that some bits form a gene together such that positions i and j may not be chosen completely at random but only after the ends of some genes.

- For ESs, the crossover operators are similar to the above: children inherit the real numbers of their parents. One can again perform one-point, two-point, and random crossovers with these values. Also, dominant–recessive and intermediate strategies of nature can be simulated easily: if, according to some criterion, a real value for one specific property is dominant over the other value, then both children might inherit the dominant value. In the intermediate approach, the child gets, e. g., the mean value of the real numbers of its parents.

However, often it is necessary to choose a coding other than bitstrings or real numbers for the individuals of the proposed optimization problem. This is mostly done in order to incorporate some constraints of the problem in the individuals. Thus, the mutation and crossover operators must lead to new individuals that do not violate these constraints either. As these mutations correspond to the moves used in the iterative improvement heuristics like greedy and simulated annealing, one can in fact take the move routine from such a program and simply copy it in a GA program. Thus, if one already knows good mutations from the Markovian approach, only new crossover operators have to be developed.

The remaining question is how Darwin's principle of natural selection is applied to the population of individuals:

- Rechenberg and others introduced several types of ESs [169, 192]:
  - The simplest ES is the $(1 + 1)$ ES. Here one starts with a single individual σ, which is then cloned. The clone τ is mutated slightly. Then the fitness values of σ and τ are compared with each other. The individual with the better fitness survives. One iterates this approach of cloning the current individual, mutating the clone, and selecting the better individual until no further improvement can be found.

This approach is exactly the same as the greedy approach in which one also starts with a random configuration on which a series of move trials is applied and then accepted if they lead to an improvement. Thus, the $(1+1)$ ES is a common iterative Markovian improvement heuristic and is identical to the greedy algorithm.

- The $(\mu+\lambda)$-ES is a generalization of the $(1+1)$ ES. Here one starts out with μ individuals at the beginning, which produce λ individuals as offspring. Then the population consists of $\mu+\lambda$ individuals, from which the μ best are able to survive. Again this approach where μ individuals create λ individuals and then the μ best of $\mu+\lambda$ individuals survive is iterated. Mostly, λ is chosen to be larger than μ. An individual can thus produce more than one offspring. Note that the fitness of the best individual in each generation does not decrease. Thus, good individuals have the chance to live over many generations of the evolutionary process. This can be advantageous if an individual represents a very good solution to the considered optimization problem. However, this approach can also lead to the problem that the evolutionary process gets stuck in some high-lying local minimum of the energy landscape, as a long-living individual generates many clones via successive iterations, which are nearly identical to it, such that the search process gets stuck in some local valley.
- In order to overcome this problem, the (μ,λ) ES, which works the same as the $(\mu+\lambda)$ ES, only selects the best μ individuals to survive from the λ offspring individuals. Here λ has to be larger than μ in order to apply a selection pressure. Furthermore, here each individual lives only for one generation, whereas an individual could theoretically live forever in the $(\mu+\lambda)$ ES. The fitness of the best individual can also decrease here such that the best-so-far individual should always be stored and then printed as the final output of the (μ,λ) ES. For this ES, the selection pressure is given by the ratio λ/μ. The larger this ratio, the larger the pressure on individuals.
  As the $(\mu+\lambda)$ ES and the (μ,λ) ES are closely related to each other, one sometimes refers to them both as $(\mu\#\lambda)$ ES. In both scenarios, λ new individuals are created by a cloning and mutation process in each generation. At the end of each generation step, μ individuals are selected to survive. Within this $(\mu\#\lambda)$ ES one can also invent a scenario in which basically the (μ,λ) approach is used but in which also the best parental individuals are able to survive. The numbers μ and λ are usually constant. Sometimes population waves are simulated by changing λ (and sometimes even μ) in time.
- There is no recombination between two or more individuals in the $(\mu\#\lambda)$ ES. These are introduced in the $(\mu/\rho\#\lambda)$ ES. Here again λ new individuals are generated. But these are not mutations of single individuals. Instead, the information stored in ρ individuals is used for creating a new

individual. Usually, ρ is chosen to be 2, in accordance with biological systems. Also in accordance with biological systems, mainly two types of crossovers are used:
  - In the dominant–recessive recombination of genes, the individual traits of the ρ individuals are either randomly chosen to be either dominant or recessive or are already marked for that from the very beginning. Then the offspring individual inherits the dominant traits.
  - If some genes are to be combined in an intermediate way, often the arithmetic mean value of the properties of the ρ parental individuals can be chosen. The notation "$/\rho$" refers exactly to this scenario.
- This Rechenberg notation, which so far refers to the numbers of individuals, can also be extended to systems with several subpopulations: the term $[\alpha/\beta\#\gamma(\mu/\rho\#\lambda)^{/n}]$ ES refers to a system with an overall number of α subpopulations. Individuals of β subpopulations are chosen in order to create γ new subpopulations. Again the α best subpopulations survive. Each of the subpopulations is simulated with a $(\mu/\rho\#\lambda)$ ES over n generations in which there is no interaction between different subpopulations. Thus, the α subpopulations evolve independently of each other for n generation steps. After that a global remixing takes place.
  This approach can also be iterated such that subpopulations of subpopulations are remixed within a $[\ldots[\ldots^{/m}]^{/n}]$ ES. Usually, one does not consider more complicated scenarios [192].
- Scientists from the school of GAs faced the same problems and thus developed analogous approaches:
  - The "general replacement" approach replaces the overall current population completely by their offspring or the best individuals within the offspring. Thus, it is the analogon to the (μ, λ) ES.
  - The "elitism" approach allows the best individuals of the parental generation to survive together with the offspring.
  - The "delete-n" approach replaces n randomly chosen individuals from the current population by n offspring individuals. It can be combined with the elitism approach so that it generally keeps the best individuals of the parental generation.

Of course, more complicated scenarios can be introduced. For example, new individuals might not take part in the mating process as they are in a "kindergarten." These younger individuals are winnowed by the survival rules and must age a bit before becoming parents of the next generation.

Other concepts of nature are also used in these simulation techniques. For example, the above-mentioned coevolution is simulated by optimizing both the proposed optimization problem and the optimization parameters of the GA in order to get even better results. Thus, a coevolution of the configurations of the proposed optimization problem, and of the optimization algorithm itself, takes place.

GAs and ESs can be combined with other algorithms to form hybrid algorithms: for example, the temperature of simulated annealing can be inserted in such an algorithm in order to govern the probability for mutations and the selection pressure. With decreasing temperature, the probability or the size of mutations decreases, thus freezing the genes in the population. Additionally, the individuals selected for survival might be selected according to their Boltzmann weight. Thus, they are selected nearly at random at very high temperatures, whereas only the fittest are able to survive at very low temperatures.

19.4 Parallel Applications

GAs and ESs are very well suited for parallel enablement. The parallelization can be performed in various ways: for some problems, the dominating part of the calculation time might be spent calculating the fitness values of individuals. Then this calculation can be performed in parallel either for several individuals at the same time or sometimes even for a single individual if, e. g., its fitness depends on all other individuals. On the other hand, if the creation of new individuals by mutations and crossovers takes most of the calculation time, then one will perform several creation processes in parallel. In both approaches, one gets a large speedup of the simulation compared to the original sequential approach. Mostly, however, in parallel algorithms, the overall population is split into subpopulations that evolve on one processor each.

In the most trivial approach, the so-called island model, there are no interactions between the subpopulations: the individual subpopulations evolve independently of each other on one processor each. Thus, no communication between the processors during the optimization run is required. At the end, one simply takes the best solution of all subpopulations. The only merit of this approach is that one can watch different evolution processes starting with, e. g., identical initial populations at the same time and thus better understand the outcome of some programmed properties of the evolution process. The main disadvantage of this model is that each subpopulation can only work with the genes of its own initial members. Due to this smaller variety, if compared to the overall population, the evolution process converges much faster to only one or a few related types of individuals. (One can compare this approach with the greedy approach, which also searches only in a small area in the energy landscape for a local minimum.)

In order to overcome this problem, connections between the subpopulations have to be established. Via these connections, some of the individuals move from their original subpopulations to other subpopulations. The various implementations of these network models differ with respect to several properties:

- One of the main properties is the network graph itself. The subpopulations can be fully connected or a connection exists with a predetermined probability such that one gets a random network. However, also certain topologies can be established like a closed ring of subpopulations or the subpopulations might be placed on a regular lattice. Then individuals are, e. g., only exchanged between neighboring subpopulations.
- The links between the subpopulations can be symmetric, such that individuals can be exchanged between the subpopulations, or unidirectional, such that a connection carries only one-way traffic. For example, the pollen flow model, in which the subpopulations are placed somewhere in the 2D space, establishes a computer wind that drives individuals to other subpopulations downwind. The more distant, then, a subpopulation, the fewer individuals will be transferred to it.
- A further important property is the choice of the individuals that are to be transferred to another subpopulation. One can, e. g., select the best or the worst individuals or one can simply choose individuals at random. But one can also select some members of a family of (nearly) identical individuals in order to decrease the dominance of some individual type within a subpopulation.
- A question correlated with this is how many individuals will be transferred and when and how often such a transfer will take place. Here the simulation possibilities are of course limited by the physical computer network.
- Of course, the transferred individuals must be integrated in their new subpopulations in some way. Either they might be immune to the first selection processes, such that they are able to survive and produce offspring at the beginning, or the selection pressure in their new subpopulations might be applied to them already after their arrivals, such that only good new individuals have the chance to survive in their new subpopulations.
- A further general property is whether this network is static or dynamic. In the dynamic case, the number of individuals moving over some connections and also the existing connections themselves change in time, so that migration processes can be rather nicely simulated. The various subpopulations might exhibit different selection pressures such that they more or less attract individuals from other subpopulations, thus altering the selection pressures in the participating subpopulations, which in turn again changes the number of individuals moving between these subpopulations.

There are so many possibilities for parallel genetic scenarios that one can simulate almost all real-life migration processes, such as the above-mentioned pollen distribution by wind or by animals and also the migration of herds or nations. But also artificial civilizations with strange migration processes can be invented: for example, in the commune model, each commune consists of a set of households. The number of households remains constant. The offspring are produced by a couple of individuals in each of these households. When the offspring reach a certain age, they move to some singles bar or some marriage institute where they select another individual in order to form a couple. Here selection processes take place as only the fittest individuals manage to get a partner. In order to reproduce, however, they need to get a household of their own. Here natural selection enters the picture for the second time, so that only the fittest couples get a household for themselves. The other couples go to the bus or train station, move to other towns, and try to get a household there [69]. This model serves as an example, illustrating what scenarios are possible for parallel applications.

20 Optimization Algorithms Inspired by Social Animals

20.1 Inspiration by the Behavior of Animals

Besides the field of genetics and evolution, from which genetic optimization algorithms were developed, other aspects of biology can inspire optimization algorithms.

One of the most interesting fields of biology is behavioral biology. The question is always why a certain animal behaves under specific circumstances in some special way in relation to other members of its own species or also in relation to members of other species. Usually, it is quite obvious that the animal acts in a way that is advantageous for it. However, sometimes it seems that the animal would only have disadvantages from its behavior, and it takes intensive research to find out where the advantages lie. Often a behavior that at first sight seems unselfish gets its reward only after some time has passed. Of course, we humans have to wonder whether this is the origin of our morality.

Therefore, animals living together in complex social structures with other members of their own kind are studied, in part, to answer the question of whether mankind is unique or at least special. Biologists are also interested in the self-organization of much simpler groups of animals like a flock of birds. It is also an interesting subject to study the complex states of insects as these are extremely simple-minded animals. Insights in these areas can lead to new types of optimization heuristics, as the following examples will show.

20.2 Ant Colony Optimization

Ants are very interesting insects: although they have only limited individual capabilities, they behave in a rather complex way when part of a collective. For example, although they are almost blind, they manage to find the shortest distance between their colony and their food. For this purpose, they communicate with each other on a chemical channel by the use of pheromones. Each ant automatically leaves some pheromone on its trail, thus marking its path. While an isolated ant moves essentially at random, a group of ants can make use of these pheromones: when a second ant detects the pheromone trail of the first ant, it follows it with a high probability, thus leaving its own pheromone on the path, such that the path is even more attractive to a third ant. This leads to an autocatalytic behavior: the more ants follow the trail, the stronger and therefore more attractive the trail becomes. If now some obstacle is put asymmetrically on the path of the ants, the first ants arriving at the obstacle will randomly select whether they go to the left or to the right in order to get around this obstacle and to follow the old path afterwards. Therefore, roughly 50% of the ants will turn to the left and 50% to the right, thus distributing their pheromones in equal amounts on both paths. However, those ants that had chosen the shorter way arrive earlier at the other side of the obstacle so that there suddenly is more pheromone on one of the two paths. Therefore, the ants coming from the other side will with a larger probability now follow the shorter path instead of choosing the longer path. This initially somewhat small asymmetry will increase until finally all ants choose the shorter path. Furthermore, the pheromones evaporate over time, so that longer paths lose their attraction to ants also in this way.

This principle is used in the ant colony optimization (ACO) algorithm for finding (quasi) optimum solutions for given optimization problems [40]. To enable the application of an ant algorithm, the optimization problem considered must be split, in either a natural or artificial way, into several objects that interact with each other. Imagine these objects as flowers on a meadow, between which the ants must find the shortest way. The interactions between the objects correspond to the distances between the locations of the flowers on the meadow. But also note that such a set of objects can be very different from a set of locations. Often the optimization problem must be reformulated in a rather artificial way, such that some variables like flowers and the distances between them can be defined. Several ants are placed on each item at the beginning of the optimization run. Then they check the interactions with the other items. After that they select some interaction that must be fulfilled. In their choice, they consider both the strength of the interaction and the amount of pheromones with which the interaction is already marked. While fulfilling the interaction, they create a pheromone trail between the items, which changes the interaction between them. The proposed rules in [40] suggest making a compromise between the "visibility" of the original interaction, which is, e. g., given by its strength, and the trail intensity, which governs the desire of the ants. The first point can be considered as making use of the intelligence of individual ants, which will choose the best locally optimum way if only referring to the "visibility". The second point makes use of the group intelligence of the ants, thus making this approach a global optimization algorithm.

The various ant algorithms differ in the amount of pheromone placed by a single ant on the trail: in the ant-density model, each ant puts an equal amount of pheromone on each interaction, which the optimization process fulfills. In the ant-quantity model, the smaller the amount, the worse the local use of this interaction for the overall system. For these two algorithms, the amount of pheromones is updated after each time step, i. e., after some interaction has been fulfilled. In contrast, the ant-cycle algorithm allows each ant first to complete its generation of a full solution to the proposed optimization problem. Then the various solutions are evaluated. The higher the quality of the overall solution, the more pheromone the ant is allowed to put equally distributed on all fulfilled interactions of its solution. Therefore, at each time step the amount of pheromone on each interaction is determined: it is given by the fraction of the previous amount of pheromone that has not yet evaporated plus the new pheromone of all ants added according to one of the rules mentioned above and in [40].

When the algorithm concludes, all ants should have decided on one common configuration, which should correspond to a (quasi) optimum solution of the optimization problem considered.

20.3 Particle Swarm Optimization

Another approach to optimization is derived from a general view of bird flocking and fish schooling. Obviously, it is advantageous for individuals to belong to a group although they must compete for food items. This disadvantage seems to be outweighed by some profit for the individuals that consists of the previous discoveries of all other members of the group. This suggests that social sharing of information offers an evolutionary advantage. Based on this theory, Kennedy and Eberhart introduced a new optimization method called particle swarm optimization (PSO) [107, 108] and bird flock model: each individual or particle is assumed to move around in a multidimensional space. Each individual i memorizes a certain point $\boldsymbol{P}_i$ in this space, where its fitness is maximal. Sharing the information means now that the other individuals, at least those in the neighborhood, get the information about the optimum point of that individual i. The movement of the individual i is influenced by the following factors at each time step:

- Generally, the previous velocity vector $\boldsymbol{v}_i$ of the individual i is used as an initialization for the new velocity vector. This influence of the previous velocity is sometimes enlarged or shortened by a stiffness factor ω.
- Then each individual i would like to return to its own optimum point, so that there is a trend to change the movement toward this optimum point: let $\boldsymbol{X}_i$ be the current position of the individual i; then a term proportional to $s_{\text{cognition}} \times (\boldsymbol{P}_i - \boldsymbol{X}_i)$ is added to its velocity $\boldsymbol{v}_i$, with $s_{\text{cognition}}$ denoting the strength of this trend.
- Secondly, there is also a trend to follow the best neighboring individual $n(i)$ to its optimum point $\boldsymbol{P}_{n(i)}$, so that an analogous term is added to $\boldsymbol{v}_i$.
- Thirdly, each individual would like to follow the overall best individual g to its optimum point $\boldsymbol{P}_g$, so that a term proportional to $s_{\text{social}} \times (\boldsymbol{P}_g - \boldsymbol{X}_i)$ is added to $\boldsymbol{v}_i$.

- In order to achieve good results for some problems, it is necessary to introduce some randomness, which is called the craziness of the individuals in [107]. A random movement vector $\boldsymbol{R}$ is therefore also added to $\boldsymbol{v}_i$.

The new position of the individual i is then determined by $\boldsymbol{X}_i = \boldsymbol{X}_i + \boldsymbol{v}_i$. The size of each influence is at each move determined by means of a random number generator, which weighs the individual influences with factors in the interval $[0;1]$ or $[0;2]$, the latter in order to simulate the flocking of birds, which circulate around their common target before landing there.

This full model can be reduced by omitting some terms, e. g., usually either the local best or the global best individual is considered. Furthermore, if dropping the social component, one gets the cognition-only model, whereas dropping the cognition component defines the social-only model. Furthermore, there is a selfless model, in which the best individual does not follow itself as a best individual but follows the second best individual of the group or in the neighborhood [106].

In their original publication [107], the inventors of this method showed that the tendency of each individual to approach its own "nostalgia" point $\boldsymbol{P}_i$ and the tendency to approach the best-so-far point $\boldsymbol{P}_g$ should be of roughly the same size: a dominating tendency toward $\boldsymbol{P}_i$ ($s_{\text{cognition}} \gg s_{\text{social}}$) leads to excessive wandering by isolated individuals, while the reverse results in the flock rushing fast toward a bad local minimum.

This PSO approach has till now mainly been used for finding the global optimum of nonlinear functions and for optimizing the weights of neural networks. The birds and fish are placed on random positions in the energy landscape of the proposed optimization problem and move to other locations according to the rules above.

20.4 Fighting and Ranking

The struggle for survival, which is fundamental to Darwin's natural selection and which is used in genetic algorithms (GAs), can result in individuals of a species forming a group if it is advantageous for them to do so. The advantage can lie in hunting together and thus getting more food or getting food more easily than by hunting alone. But then the question arises of how to distribute the prey. Of course, each individual knows that the more it gets, the larger its probability is of surviving when hunting conditions deteriorate. Furthermore, if there are two genders, there is the wish to have as many offspring as possible with as many individuals of the other gender as possible. These two factors lead to fights within the group. However, these fights cost energy, can lead to injuries, and might also hurt or kill one's own offspring, such that the probability of surviving decreases with an increasing number of fights.

In order not to descend into constant fighting, all individuals of a group must respect some sort of hierarchy, according to which the food is distributed in shares of different sizes and quality and according to which the mating chances are regulated. Such a "contract" within a group might even be advantageous for those individuals that get a smaller share, as, e. g., stronger individuals that get a larger share have more energy to protect the whole group if it is attacked. Of course, the smaller share has to be sufficient for survival. If one or more individuals of a group becomes dissatisfied with the hierarchy, new fights break out and a new hierarchy is established.

Due to this hierarchy, a ranking between the individuals of the group is established. Such a ranking often remains constant for a long period of time, so that social relationships of a complex kind can develop within the constraints of the overlying ranking. This ranking is often reflected in the seating arrangements of a group: usually, the β-male will not be found at the side of the α-male. Instead, both are surrounded by their supporters, with the club of the β-male usually much smaller than that of the α-male. It even happens that the β-male and the γ-males form a coalition against the α-male and its supporters, with the intention to take over.

The general development and features of ranking can be very well simulated on a computer. However, this behavior can also be used for an optimization algorithm: each individual again represents a configuration of the proposed optimization algorithm. In a fight, the individual who exhibits greater fitness, i. e., the lower cost function value, wins with a higher probability. This probability difference should be proportional to the difference in these values between the two fighters. Then there are several possibilities:

- Each win is recognized by the right to produce one offspring, which is either nearly identical to the winner except for one or more mutations or which is produced by performing a crossover operation with a mate. In this case, the algorithm becomes a hybrid algorithm with elements of GAs, so that Darwin's principle of natural selection must be introduced in order to impose some limits on the size of the society.
- The winner could also adopt good local parts of the loser's configuration. This approach would correspond to what happened in ancient times, as one militarily strong civilization captured a militarily weaker but culturally more progressive nation. In this way, Egypt was able to retain its identity over thousands of years, although it was captured several times. The conquerors adopted Egyptian culture. Analogously, the Romans adopted Greek philosophy after capturing Greece. Furthermore, Greek became the *lingua franca* in the eastern part of the Roman empire, and knowledge of Greek became a symbol of higher education in ancient Rome. Although the Greeks and the Egyptians lost their respective wars and were no longer free when first occupied, their cultures were able to survive.

21 Optimization Algorithms Based on Multiagent Systems

21.1 Motivation

In the last chapter, we introduced algorithms that are based on insights into the complex behaviors of animals living in large communities. Among the genetic algorithms (GAs), populations of individuals are considered on which an evolution process is started. These and other algorithms have in common that there are some "beings" that are mainly given as configurations of the considered optimization problems. Furthermore, some optimization problems can be split in a natural way, for example, the vehicle routing problem (VRP), in which several truck drivers perform a closed tour and serve their customers can be considered as "beings", each representing a part of the whole configuration. In computer science, where researchers have found related similarities between different "beings" of various problems, one uses the term "agents" for these beings. Thus, an agent can be everything, an ant in the ant colony optimization problem, an individual in a GA, a single part of a configuration, etc. This agent concept or this language of speaking about agents is very general.

Thus, first of all, one has an agent with more or less elaborate individual properties and capabilities. But furthermore, there are interactions between the individual agents. For example, an ant uses pheromones to disseminate information to the whole group of ants, an individual in a GA selects another individual for generating child configurations, and so on. Generalizing these concepts, we have multiagent systems consisting of a number of agents with individual properties and capabilities and of interactions between these agents.

The most prominent examples of multiagent systems are financial markets: Every day, new exchange rates between various currencies, stock prices, and oil and gold prices, among prices for many other goods, are determined at markets by means of trading: one person offers something to sell at a certain price, and another person offers to buy something at some price. In the end, a price for a good or some exchange rate is determined. The exchange of the currencies is then carried out according to the determined exchange rate; analogously, goods are sold according to set prices.

This trading scenario has been simulated in various ways in order to understand the development of the prices and mechanisms of markets. Of course, the final aim of these simulations is usually to anticipate exchange rates and prices. However, one can also make use of a simulated trading ansatz for optimization as Bachem et al. showed in their approach, which was originally developed for VRPs [13, 14].

21.2 Simulated Trading

A system must be divided into various parts in some natural way corresponding to the properties of the system. These parts of the system are considered as agents who try to get rid of items that are "expensive" to handle and to get "cheaper" ones instead so as to increase profits, by which the system is indirectly optimized.

The algorithm consists of two phases: in the sell-and-buy phase, each agent checks for the best trading possibilities, also regarding previous actions. First, all the put-orders are given to the central stock market with offers stating the price at which the items are to be sold. These prices can be given, e. g., by the improvements the agents make by removing these items from their parts of the system. The central stock market provides a public selling list consisting of all these sell orders. The individual agents can then decide which items on the selling list they want to provide buy orders for. Again they determine prices by estimating the benefits or costs for introducing an item in their local part of the system. Based on these orders, the stock market constructs a so-called trading graph consisting of feasible orders.

In the second phase of the algorithm, the trading-matching-search phase, the trading graph is searched for a maximum weighted trading matching between pairs of sell and buy orders. According to the matching found, items are shifted to the new owners, i. e., they are removed from their current system section and transferred into another system section.

This algorithm can be implemented in various ways: a simple approach would be for each agent to give only one order, that is, half of the agents give a sell order, the other half a buy order. Whether an agent buys or sells can be determined either randomly or by assigning desired values for buying or selling to individual agents. These desired values can depend on the number of items an agent serves or the maximum values for the savings by removing an item or the costs for inserting one. More difficult approaches would allow for selling and buying a sequence or a bunch of items or even for providing several simultaneous sell and buy orders for each agent.

Individual agents can follow various strategies: they might want to get a maximum profit as fast as possible. In this case, they will select the optimum order, i. e., they will offer that item for sale for which they obtain the largest savings for removing it and offer to buy that item for which the costs of insertion are smallest. However, this greedy behavior leads to worse results than in a probabilistic strategy in which probability values are assigned to all possible orders one agent can make according to their quality. The quality of a sell order is usually given by the size of the savings for removing the corresponding item, i. e., the larger the savings, the better the order. The actual sell order the agent gives is selected in a randomized way according to these probabilities. Analogously, the buy order a buying agent gives should not be the best possible one, i.e, the one with minimum costs for inserting an item. It should also be chosen in a probabilistic way considering the costs of the insertions. The smaller the costs are, the larger the probability for the buy order should be.

Therefore, promising actions are taken more often and getting stuck in a static situation is prevented. These probabilities can be changed during the optimization run, e. g., by introducing a temperaturelike control parameter, which leads to random orders at the beginning and a greedy behavior at the end.

But also the behavior of the stock market can be changed: instead of finding the best matching between buy and sell orders, it is sufficient to work with any matching that improves the current solution. Furthermore, one can also allow deteriorations during the simulation process, with the size or probability of the deteriorations determined by some control parameter as in simulated annealing (SA) or threshold accepting.

Furthermore, one can weaken the exactness of the stock market: for example, some item could be sold to several agents such that it is inserted into various parts of the system several times, or it could be removed from the system completely if, e. g., one agent sells it but no other agent buys it. Of course, one must end up at a feasible solution at the end of the optimization run, so that one must add a penalty for such items left at the stock market, thus decreasing the costs for reinserting them into the system. Analogously, one must make agents get rid of items inserted several times into the system.

As each agent can develop buy and sell orders independently of all other agents, this algorithm is well suited for parallel enablement. An obvious parallel approach would be that each agent is run on one processor. Furthermore, one processor is dedicated to the simulation of the stock market. However, this type of parallelization is very poor, as all agent processes have to wait until the stock market process has gathered its selling list and even more until it has determined the matching of the orders. If there are many agents, this matching might take even more time than the determination of the orders by the agents.

Therefore, it is advantageous to parallelize the stock market itself: each part of the stock market that is now run on a single processor serves some partition of agents that are in some sense related to each other. This relationship can, e. g., be given by the neighborhood of the system parts they represent. Thus, each processor handles some related agents that (partially) give their sell orders and handles a partial stock market that summarizes the orders of these agents in a local selling list. Then those agents that are located on this processor can give their buy orders, if they want to give any, due to this local selling list, so that a local trading graph can be built. In conclusion, trading is done only between agents located on the same processor.

However, this partitioning of the agents onto the stock market parts must be changed dynamically; otherwise some useful trading possibilities might be prohibited.

21.3 Selfish vs. Global Optimization

The simulated trading algorithm as described above is only one example of the various possibilities for optimizing a proposed optimization problem by using a multiagent approach. Generally, several agents are introduced. Often, each of these agents is supposed to be responsible for a part of the system and to try to optimize it in the best possible way. Thus, each agent can be equipped with a selfish behavior: he/she tries to optimize his/her local part, regardless of whether the outcome of his/her moves improves or worsens the local parts of the other agents.

This selfish approach has some advantages over optimization methods like the standard greedy algorithm, which tries to arrive at an overall good solution by accepting only changes that lead to an improvement of the whole system:

- First, each agent can find out more easily how to improve his local part as he/she does not need to consider the whole system.
- The time for calculating such a local energy difference $\Delta\mathcal{H}_{\text{local}}$ is much shorter than the calculation time for the energy difference $\Delta\mathcal{H}$ of the overall system. Thus, the decision of whether to accept or reject such a change can be made much faster [188].
- From the outside, one obtains a deeper insight into how the system is optimized as one can have a better view at the moves of the agents as these are local when compared to the global system.

All in all, this selfish optimization approach by using a multiagent system saves calculation time and provides deeper insight into how the system is optimized. If none of the agents and none of the pairs of groups of interacting agents can find any improvements, the optimization run has ended at a configuration known as a Nash equilibrium.

However, this selfish approach also has some disadvantages when compared with global optimization approaches like the greedy algorithm or more elaborate techniques.

- As already mentioned, an agent that just now improves his/her part of the system might worsen the parts of other agents. When it is their turn to optimize their parts, they might in turn worsen the part of the first agent. Thus, this selfish optimization approach might not lead to a Nash equilibrium in the end but might cycle between various configurations and might never lead to really good configurations.

- Furthermore, as each agent only considers his/her local part, long-range interactions with other parts may be suppressed or even completely neglected. Even if the agents do a good job on their parts of the system, the optimization process will often not end up at a good local or even the global minimum as the global minimum is often not the sum of local minima if considering complex problems due to the existence of long-range interactions between the parts of a complex problem.

Therefore, this selfish approach might lead to worse results than the standard global approach and might not even be applicable.

21.4 Introduction of a Social Temperature

In the last section, agents were described as selfish beings. In real life, when a person is very selfish, we usually say that this person is cold and egotistic, whereas if we find just the contrary, namely an altruistic behavior, we speak of a warm person.

It was already shown that if the agents of a multiagent system do not always act in a completely selfish way, then one arrives at better solutions. The simulated trading algorithm makes use of this insight by assigning probabilities to the individual orders an agent can make: of course, the larger the gain for a particular order, the larger the probability that this order will then be used. However, other orders, which are worse from the point of view of the individual agent, can be chosen with specific probabilities.

This probability concept can be altered by introducing a social temperature: we can assign a social temperature to a multiagent system that then governs the behaviors of the individual agents: high temperatures are associated with heaven and cold temperatures with hell. We can even assign social temperatures T_i to each agent i: if T_i is large, agent i can be considered an angel; if T_i is small, then agent i behaves like a small devil, agents with intermediate T_i act like normal people: they have their interests in mind but also know that they should not destroy the work of the other agents completely.

There are various ways to use this social temperature concept in order to make the behavior of an agent less selfish: as stated above, one could govern with this temperature the extent to which an agent considers, e. g., neighboring agents when making its decision. Here the cost functions of these neighboring agents are added to the cost function of the agent, but only in a reweighted form. The weight of these cost functions is determined by the social temperature. Another approach is that the individual agents retain their local cost functions. When they want to change their local parts, they do this in a less selfish way in the sense that they not only accept improvements but also deteriorations with some probability. For example, the Metropolis criterion $\min\{1, \exp(-\Delta\mathcal{H}_{\text{local}}/(k_B T_i))\}$ can be used as an acceptance rule for the moves the agents make.

Often, only one social temperature T is assigned to the overall system. Each agent of the multiagent system is thus at the same temperature. Furthermore, this temperature is often decreased during the optimization run, such that the individual agents become more and more selfish. This approach recalls the SA algorithm, which analogously ends up with the greedy algorithm at the end. Just as SA led to better results than the greedy algorithm as it can climb over barriers in the energy landscape, this social temperature approach is superior to the selfish approach, as barriers due to the selfish decisions of the individual agents can be overcome. Often the same approach as within SA is used: one starts with a high social temperature and decreases this temperature gradually until all agents behave in a completely selfish way at the end. One finds that one gets better results with this approach as with an overall selfish approach. This finding is the analog to the finding that one gets better results using SA than with the greedy algorithm.

22 Tabu Search

As there is already an excellent book about tabu search and the philosophy behind it by its inventor Fred Glover and Manuel Laguna [67], we only sketch here the general ideas and the important ingredients of tabu search.

22.1 Tabu

The Encyclopaedia Britannica [209] defines the word *tabu* as follows: *taboo, also spelled tabu, Tongan tabu, Maori tapu, the prohibition of an action or the use of an object based on ritualistic distinctions of them either as being sacred and consecrated or as being dangerous, unclean, and accursed. The term taboo is of Polynesian origin and was first noted by Captain James Cook during his visit to Tonga in 1771; he introduced the term into the English language, ...*.

Based on this tabu paradigm, Fred Glover introduced a field of optimization algorithms called tabu search: starting out at some initial configuration, which is either random or preoptimized according to a specific rule set or proposed in some other way, these algorithms apply a sequence of moves to the system being optimized. Configurations that are either declared as tabu on the whole or that contain properties declared as tabu are forbidden.

All these forbidden configurations or properties have to be stored in some tabu list. The number of the items stored in the tabu list is usually called the tabu list size (TLS). Initially, the tabu list is usually empty. Then the tabu search algorithm searches in the neighborhood of the current configuration for the best neighboring configuration, much like a steepest descent algorithm. But in contrast to the steepest descent algorithm, the tabu search algorithm does not get stuck in local minima as it accepts the minimum deterioration if the system is currently in a local minimum. Here a problem can arise: the system might then jump back and forth all the time between the local minimum configuration and its best neighboring configuration. Thus, after perhaps several such 2-cycles, the local minimum configuration must be declared as tabu: then the move leading back to it is forbidden and the algorithm must then choose the best move among all possible allowed moves leading to configurations not marked as tabu. Then the system can get to other local minima in the energy landscape. But it could be that the local minimum now declared as tabu was already the global minimum. Therefore, one must always store the best configuration that has been found so far and return this best configuration as the result of the optimization run.

At every subsequent move, the tabu search algorithm goes through all configurations in the neighborhood of the current configuration, checks whether they are tabu, and then selects the best neighboring configuration that is not marked as tabu as the new configuration. The number of configurations that are declared as tabu increases in time. It takes more and more time to evaluate whether any of the neighboring configurations are tabu, as each of these has to be compared with every configuration stored in the tabu list for identity. Thus, the requirements for both memory size and calculation time increase with increasing TLS. Therefore, instead of using this explicit memory type in which complete configurations that have been declared as tabu are stored, for many problems the attributive memory is used in which properties common to many visited configurations are marked as tabu. Avoiding putting complete configurations on the tabu list, but extracting the essence of what has now been explored, so that one can go and find fresh things to try, saves both calculation time and computer memory. The tabu search algorithm then moves on in its search for the optimum configuration in other parts of the configuration space in which the configurations do not contain these tabu properties.

22.2 Use of Memory

For some problems, this type of memory, in which more items are gradually added to the tabu list, might be sufficient. However, for many other optimization problems, one faces the problem that the number of properties stored in the tabu list becomes so large that the tabu search algorithm is unable to get to further good configurations, as the ways to them lead via configurations that are forbidden due to tabu constraints. The system might even freeze at some configuration if no further move is allowed.

In this case, usually a tabu tenure is introduced: items marked as tabu lose this tabu stamp after some time; they become tabu-inactive and can thus be again part of new configurations. Here the long-term memory is replaced by a short-term memory. Very complex tabu search algorithms use both long-term memory and short-term memories with various tabu tenures. Here not only complete configurations or their properties are stored because they were found either to be good or bad, but also information about choices made in the past that have been proved to lead to good or bad results is stored as well. Thus, these tabu search algorithms also incorporate some learning methods by which the way through the configuration space can be directed. One may therefore consider the tabu search algorithm an artificial intelligence algorithm, as there is not only a memory but also a mechanism for an excellent exploitation of the contents stored in it.

22.3 Aspiration

A further concept of tabu search is aspiration. Sometimes one simply wants to accept a configuration that is forbidden due to tabu constraints. Then these tabu constraints can be overridden by some aspiration criterion: usually, aspiration is used if a move could lead to a configuration of very good quality or with some desired properties if it did not contain some properties marked as tabu. Then these tabu constraints are overridden by the aspiration criterion.

For this purpose, usually some elite configurations are stored during the optimization run. If the tentative new configuration is, e. g., better than any of the stored elite configurations, then it should of course be accepted. Furthermore, it might also be worth accepting if it contains several properties that are part of various elite configurations, in the hope that by including these properties one might get to even better configurations, as these properties are part of different elite configurations already.

22.4 Intensification and Diversification

Two key ingredients of tabu search are intensification and diversification. The concept of intensification modifies choice rules for moves in order to encourage move combinations and solution properties historically found to be good. It may also initiate a return to regions in the configuration space in which some stored elite solutions lie. These regions can then be searched more thoroughly. But intensification also allows for direct jumps to these elite configurations, such that the size of the neighborhood is enhanced by the intensification concept. The concept of intensification is usually applied when the tabu search optimization run has not found for some time any configuration that is quite as good as one of the stored elite configurations.

On the other hand, diversification encourages the system to examine unvisited regions of the configuration space and thus to visit configurations that might differ strongly from all configurations touched before. Intensification and diversification are often used together: instead of jumping to one of the stored elite configurations, the system jumps to a configuration that has been created by changing one of the elite configurations in some significant way, i. e., slightly enough to search the neighborhood of the elite configuration and strongly enough so that the new configuration contains properties that are not part of the elite configuration from which it was constructed.

Summarizing, Glover states that the use of memory within tabu search is coupled to four principal dimensions—recency, frequency, quality, and influence—that partially contradict each other [67]. Recency is governed by the short-term memory, frequency by the intensification strategies, quality by the general approach to searching for the best configuration in the neighborhood and also by the intensification strategies, and influence by aspiration, intensification, and diversification strategies. One finds that all these strategies that partially lead to contrary effects (like the pair tabu list–aspiration and the pair intensification–diversification) must be outbalanced in an optimum way in order to achieve optimum results.

23 Histogram Algorithms

Many algorithms make use of a histogram as an adaptive memory. Thus, they count how often configurations or properties of configurations occurred during an optimization run and store this information in some histogram in order to make use of this information later on. These algorithms can therefore be classified as algorithms that are closely related to tabu search or even as tabu search implementations if defining tabu search in a wider scheme. However, the background of these algorithms is mostly not in the tabu search field but in statistical physics or smoothing techniques. In this chapter, we will introduce a few of these algorithms.

23.1 Guided Local Search

Guided local search (GLS), which was invented by Voudouris and Tsang [212, 213, 214], is a method that can be classified as changing the energy landscape. It is indeed closely related to methods like weight annealing, which was introduced in Sect. 13.4. On the other hand, GLS uses an adaptive memory structure like that used by tabu search.

GLS focuses on features or properties f_i a solution σ does or doesn't have. For this purpose, indicator functions

$$I_i(\sigma) = \begin{cases} 1 & \text{if } \sigma \text{ has the feature } f_i \\ 0 & \text{otherwise} \end{cases} \tag{23.1}$$

are introduced. Constraints on features are introduced by extending the original cost function $\mathcal{H}$ with a set of penalty terms to

$$\tilde{\mathcal{H}}(\sigma) = \mathcal{H}(\sigma) + \lambda \times \sum_i p_i I_i(\sigma) \tag{23.2}$$

with a Lagrange multiplier λ, with which the size of the penalties compared to the original cost function is controlled and with the penalties $p_i \geq 0$. The penalties are initialized with zeroes at the beginning and are incremented based on information collected during the GLS iterations.

Let us consider problems in which these features have assigned some costs c_i. These costs raise the value of the energy the configuration σ has if it contains the feature f_i. For example, the Hamiltonian can be written as

$$\mathcal{H}(\sigma) = \sum_i c_i I_i(\sigma) \,. \tag{23.3}$$

Then the GLS algorithm proceeds as follows:

1. First, the penalty values are initialized with 0 such that $\tilde{\mathcal{H}} = \mathcal{H}$. An initial configuration called σ_0 is generated and an iteration counter is introduced with $k = 0$.
2. A conventional greedy or steepest descent algorithm starting from the configuration σ_k and using the Hamiltonian $\tilde{\mathcal{H}}$ is performed in order to get into a local minimum in the energy landscape. Let the final configuration of this run be σ_{k+1}.
3. A utility value u_i is calculated for every feature:

$$u_i = I_i(\sigma_{k+1}) \times c_i/(1 + p_i) \,. \tag{23.4}$$

4. The penalty values p_i for those features f_i for which u_i is maximum are increased.
5. k is incremented by 1.
6. If the maximum number of iterations is not exceeded, i. e., if $k < k_{\text{max}}$, then the algorithm returns to step 2.

Now, initially the algorithm starts off with $p_i = 0$ for all features. Thus, in the first iteration, the algorithm works with the original Hamiltonian in the original energy landscape. Then, after the first local optimization run, the worst features that are part of the final configuration are punished, and λ or a few λs are added to the Hamiltonian. Of course, the next local optimization run, which starts from the final configuration of the previous optimization run, will try to reduce the energy again so that these already punished features might be removed from the solution.

Note that the construction of the utility values u_i is rather elaborate: how bad a feature is with respect to its costs c_i, as well as how often it was already punished with respect to penalties p_i, is taken into account. Sometimes a bad feature must be part of a solution if all alternatives are worse.

23.2 Multicanonical Algorithm

When introducing the multicanonical algorithm (MUCA) [19, 20], we must first reconsider the physical formulas that led to the derivation of simulated annealing (SA): we consider the problem to be optimized a classical physical system, in which every state σ occurs in equilibrium with the probability $\pi_{\text{equ}}(\sigma) = \exp(-\mathcal{H}(\sigma)/(k_B T))/Z$ according to the Boltzmann equilibrium distribution (Chap. 11). In a SA run, usually the Metropolis criterion for accepting a move is used in order to generate this Boltzmann distribution.

In contrast, the MUCA is designed to generate a probability distribution $\pi_{\text{mu}}(\sigma)$ in such a way that

$$\pi_{\text{mu}}(\sigma) \times n(\mathcal{H}(\sigma)) = P(\mathcal{H}(\sigma)) \equiv \text{const} \qquad (23.5)$$

for all temperatures, with $n(\mathcal{H})$ being the density of states with the energy $\mathcal{H}$ and $P(\mathcal{H})$ being the probability of the energy $\mathcal{H}$, which will be a constant. Thus, the probability of a state is inversely proportional to the number of states having the same energy:

$$\pi_{\text{mu}}(\sigma) \propto \left(n(\mathcal{H}(\sigma))\right)^{-1} . \qquad (23.6)$$

As in SA, the detailed balance criterion should be fulfilled:

$$\pi_{\text{mu}}(\sigma) \times p(\sigma \to \tau) = \pi_{\text{mu}}(\tau) \times p(\tau \to \sigma) . \qquad (23.7)$$

Combining these two criteria, one gets

$$\frac{p(\sigma \to \tau)}{p(\tau \to \sigma)} = \frac{\pi_{\text{mu}}(\tau)}{\pi_{\text{mu}}(\sigma)} = \frac{n(\mathcal{H}(\sigma))}{n(\mathcal{H}(\tau))} . \qquad (23.8)$$

As in SA, some arbitrariness in the explicit choice of the transition probability p remains. One choice would be a Metropolis-like criterion:

$$p(\sigma \to \tau) = \begin{cases} 1 & \text{if } n(\mathcal{H}(\sigma)) \geq n(\mathcal{H}(\tau)) , \\ \dfrac{n(\mathcal{H}(\sigma))}{n(\mathcal{H}(\tau))} & \text{otherwise} . \end{cases} \qquad (23.9)$$

If this or another criterion based on the multicanonical distribution and on detailed balance is applied to the system, then the system performs a random walk (RW) in the energy space, as the probabilities for all energies are identical. Note that this does not mean that the algorithm performs a RW in the search space, as such a RW in the search space or energy landscape would have to accept every tentative new configuration. Furthermore note that, in contrast to SA, there is no control parameter like the temperature that is an external variable applied to the system. Instead, some system-intrinsic variable, namely the energy density of the states, is used.

However, this is the main problem of this algorithm: the energy density of the states is often unknown a priori and thus must be measured either before performing the multicanoncial optimization run or during the multicanonical optimization run. Mostly, this is done using a histogram technique: one splits the interval of all possible energies the configurations can have mostly into parts of equal length either on a linear or on a logarithmic scale. Then one initializes the counters H_i for each bin with some initial value. This can be a value of 0 or 1 for all bins. But one can also start with other values if one has a rough estimate of the shape of the energy density of the states.

Starting from these initial values of the counters, one performs a simulation. During this simulation, one increments the counter H_i for the energy interval $[\mathcal{H}_i; \mathcal{H}_{i+1}]$ if the energy $\mathcal{H}(\tau)$ of the accepted new configuration τ lies between these marginal values.

If one starts out by setting the initial values for the counters to 1 and starting the multicanonical simulation from a random configuration, then one replaces the energy density values $n(\mathcal{H}(\sigma))$ and $n(\mathcal{H}(\tau))$ by the corresponding current counter values $H(\sigma)$ and $H(\tau)$. $H(\sigma)$ denotes the counter for the energy interval in which $\mathcal{H}(\sigma)$ lies. One might first increase the counter for some interval, but if a move then leads to another energy interval, the system jumps to this other energy interval and prefers it as long as its counter value is smaller than the counter value for the first interval. Thus, some pressure is performed on the system to gradually leave the range of the random configurations and to go either downwards to the optimum solution or upwards to the worst solution. Of course, the widths of the bins influence this behavior, the following situations arise:

- If the bins are not wide, then many bins are needed. In order to get a proper distribution, one must take a number of measurements that is at least proportional to the number of bins. Therefore, the more bins there are, the more calculation time required by the simulation.
- However, all energy values within an energy interval are considered equal, as there is only one counter for all of them. If the simulation reaches the energy interval containing the unknown global optimum value, and if this is very wide, then there is no pressure on the system to perform the last necessary improvements in order to get into the global optimum. Instead, within these intervals, the system performs a restricted RW.

Often this approach is rewritten in a way that resembles the notations of SA and simulated tempering (ST). The multicanonical probability π_{mu} is written as

$$\pi_{\mathrm{mu}}(\sigma) = \frac{1}{Z_{\mathrm{mu}}} \exp\left(-S(\mathcal{H}(\sigma))\right) = \frac{1}{Z_{\mathrm{mu}}} \exp\left(\alpha(\mathcal{H}(\sigma)) - \beta(\mathcal{H}(\sigma)) \times \mathcal{H}(\sigma)\right) , \tag{23.10}$$

with Z_{mu} being the multicanonical partition sum. $S(\mathcal{H})$ is the so-called microcanonical entropy. Note that there is this parameter $\alpha(\mathcal{H})$, called here the fugacity, similar to ST, but in contrast to ST it does not depend on T but on the energy. Furthermore, in contrast to SA and ST, the inverse temperature β is not a constant but a function of the energy and is therefore called a microcanonical inverse temperature as one could thus say that there is a temperature for each energy. The Metropolis-like acceptance criterion is then written as

$$p(\sigma \to \tau) = \begin{cases} 1 & \text{if } \Delta \leq 0\,, \\ \exp(-\Delta) & \text{otherwise}\,, \end{cases} \tag{23.11}$$

with

$$\Delta = \alpha(\mathcal{H}(\tau)) - \alpha(\mathcal{H}(\sigma)) + \beta(\mathcal{H}(\tau)) \times \mathcal{H}(\tau) - \beta(\mathcal{H}(\sigma)) \times \mathcal{H}(\sigma) . \quad (23.12)$$

Some researchers prefer to work with these parameters, which have to be derived from an estimate for the energy density n, which in turn has to be measured first. Of course, there is the additional difficulty of deriving either values α and β for all energy intervals or even better real continuous functions $\alpha(\mathcal{H})$ and $\beta(\mathcal{H})$. But if one can actually write down such functions of α and β, then this approach is superior, as then one has continuously changing functions instead of the intervals and thus generally a pressure on the system to move to the extremes, as there the density of the states is smaller.

There are two possible approaches when working with the multicanonical algorithm: as one continuously samples the histogram for the approximation of the density $n(\mathcal{H})$, one can of course use the current values of the histogram as mentioned above. Mostly, however, one starts off with some initial values for the weights π^0_{mu}, which can, e. g., be set to 1 initially. In this first iteration, one works with these weights while getting histogram values $H(\mathcal{H})$. Then one obtains an estimate for the weight values for the next iteration $i+1$ by

$$\tilde{\pi}^{i+1}_{\text{mu}}(\mathcal{H}) = c \times \frac{\pi^i_{\text{mu}}(\mathcal{H})}{\hat{H}^i(\mathcal{H})} \quad (23.13)$$

with some constant c. As one wants to avoid a division through zero, one uses slightly changed histogram values [18]

$$\hat{H}(\mathcal{H}) = \max\{\epsilon, H(\mathcal{H})\} \quad (23.14)$$

with $\epsilon > 0$ and $\epsilon \ll 1$. When using this approach, one usually derives formulas for how to change the values of $\alpha(\mathcal{H})$ and $\beta(\mathcal{H})$ between successive iterations. One way to derive a multicanonical recursion for these parameters is described in [18]: the microcanonical entropy $S(\mathcal{H})$ is given by definition as

$$\beta(\mathcal{H}) = \frac{\partial S(\mathcal{H})}{\partial \mathcal{H}} . \quad (23.15)$$

If the width of a bin is given by $\Delta\mathcal{H}$, this derivation is numerically given by

$$\beta(\mathcal{H}) = \frac{S(\mathcal{H} + \Delta\mathcal{H}) - S(\mathcal{H})}{\Delta\mathcal{H}} . \quad (23.16)$$

On the other hand, $S(\mathcal{H}) = \beta(\mathcal{H}) \times \mathcal{H} - \alpha(\mathcal{H})$ holds such that

$$\begin{aligned} S(\mathcal{H}) - S(\mathcal{H} - \Delta\mathcal{H}) = \beta(\mathcal{H}) \times \mathcal{H} - \beta(\mathcal{H} - \Delta\mathcal{H}) \times (\mathcal{H} - \Delta\mathcal{H}) \\ -\alpha(\mathcal{H}) + \alpha(\mathcal{H} - \Delta\mathcal{H}) . \end{aligned} \quad (23.17)$$

Inserting $\Delta\mathcal{H}\beta(\mathcal{H} - \Delta\mathcal{H}) = S(\mathcal{H}) - S(\mathcal{H} - \Delta\mathcal{H})$ yields

$$\alpha(\mathcal{H} - \Delta\mathcal{H}) = \alpha(\mathcal{H}) + (\beta(\mathcal{H} - \Delta\mathcal{H}) - \beta(\mathcal{H})) \times \mathcal{H} . \quad (23.18)$$

Thus, if fixing $\alpha(\mathcal{H}_{\max}) = 0$, $\alpha(\mathcal{H})$ is determined once $\beta(\mathcal{H})$ is given.

The values for $\beta(\mathcal{H})$ have to be determined in an iterative way: the values $\beta^{i+1}(\mathcal{H})$ for iteration $i+1$ have to be derived from the values $\beta^i(\mathcal{H})$ for the previous iteration and the resulting histogram $\hat{H}^i(\mathcal{H})$ from the previous iteration. From

$$\tilde{\pi}^{i+1}_{\text{mu}}(\mathcal{H}) = \exp\left(-\tilde{S}^{i+1}(\mathcal{H})\right) = c \times \frac{\pi^i_{\text{mu}}(\mathcal{H})}{\hat{H}^i(\mathcal{H})} \tag{23.19}$$

(note that variables with a $.\tilde{.}.$ above denote that these variables are not the final values but only first estimates) one gets

$$\tilde{S}^{i+1}(\mathcal{H}) = -\log(c) + S^i(\mathcal{H}) + \log\left(\hat{H}^i(\mathcal{H})\right) . \tag{23.20}$$

Combining this with $\beta(\mathcal{H}) = (S(\mathcal{H}+\Delta\mathcal{H}) - S(\mathcal{H}))/\Delta\mathcal{H}$ [see Eq. (23.16)], one gets

$$\tilde{\beta}^{i+1}(\mathcal{H}) = \beta^i(\mathcal{H}) + \frac{\log\left(\hat{H}^i(\mathcal{H}+\Delta\mathcal{H})\right) - \log\left(\hat{H}^i(\mathcal{H})\right)}{\Delta\mathcal{H}} . \tag{23.21}$$

The estimator of the variance of $\tilde{\beta}^{i+1}(\mathcal{H})$ is obtained from

$$\begin{aligned} \operatorname{Var}\left(\tilde{\beta}^{i+1}(\mathcal{H})\right) &= \operatorname{Var}\left(\beta^i(\mathcal{H})\right) \\ &+ \frac{\operatorname{Var}\left(\log\left(\hat{H}^i(\mathcal{H}+\Delta\mathcal{H})\right)\right)}{\Delta\mathcal{H}} + \frac{\operatorname{Var}\left(\log\left(\hat{H}^i(\mathcal{H})\right)\right)}{\Delta\mathcal{H}} . \end{aligned} \tag{23.22}$$

As $\beta^i(\mathcal{H})$ is a fixed function used in the ith simulation, its variance is zero. The variance of the histogram can be rewritten as

$$\operatorname{Var}\left(\log\left(\hat{H}^i\right)\right) = \left(\log\left(\hat{H}^i + \Delta\hat{H}^i\right) - \log\left(\hat{H}^i\right)\right)^2 , \tag{23.23}$$

where $\Delta\hat{H}^i$ is the fluctuation of the histogram, which is known to grow with the square root of the number of entries [18],

$$\Delta\hat{H}^i \propto \sqrt{\hat{H}^i} , \tag{23.24}$$

such that

$$\begin{aligned} \operatorname{Var}\left(\log\left(\hat{H}^i\right)\right) &= \left(\log\left(1 + \frac{\Delta\hat{H}^i}{\hat{H}^i}\right)\right)^2 = \left(\log\left(1 + \frac{c'\sqrt{\hat{H}^i}}{\hat{H}^i}\right)\right)^2 \\ &= \left(\log\left(1 + \frac{c'}{\sqrt{\hat{H}^i}}\right)\right)^2 \approx \left(\frac{c'}{\sqrt{\hat{H}^i}}\right)^2 = \frac{c'^2}{\hat{H}^i} . \end{aligned} \tag{23.25}$$

Hence

$$\begin{aligned}\mathrm{Var}\left(\tilde{\beta}^{i+1}(\mathcal{H})\right) &= \frac{c'^2}{\hat{H}^i(\mathcal{H}+\Delta\mathcal{H})} + \frac{c'^2}{\hat{H}^i(\mathcal{H})} \\ &= c'^2 \frac{\hat{H}^i(\mathcal{H}+\Delta\mathcal{H}) + \hat{H}^i(\mathcal{H})}{\hat{H}^i(\mathcal{H}+\Delta\mathcal{H}) \times \hat{H}^i(\mathcal{H})}\end{aligned} \tag{23.26}$$

holds, with c' an unknown constant. Of course, the variance explodes when there is zero statistics, i. e., $H^i(\mathcal{H}) = 0$ or $H^i(\mathcal{H}+\Delta\mathcal{H}) = 0$. The statistical weight for $\tilde{\beta}^{i+1}(\mathcal{H})$ is inversely proportional to its variance and the overall constant is irrelevant [18], so that one can choose the convenient weight factor

$$\tilde{g}^i(\mathcal{H}) = \begin{cases} 0 & \text{if } H^i(\mathcal{H}) = 0 \\ & \text{or } H^i(\mathcal{H}+\Delta\mathcal{H}) = 0 \\ \dfrac{\hat{H}^i(\mathcal{H}+\Delta\mathcal{H}) \times \hat{H}^i(\mathcal{H})}{\hat{H}^i(\mathcal{H}+\Delta\mathcal{H}) + \hat{H}^i(\mathcal{H})} & \text{otherwise} \end{cases} \tag{23.27}$$

for the impact of $\tilde{\beta}^{i+1}(\mathcal{H})$ on the utlimately chosen value of $\beta^{i+1}(\mathcal{H})$. Furthermore, it must be borne in mind that the ith iteration was carried out using $\beta^i(\mathcal{H})$. According to [18], it is now straightforward to combine $\tilde{\beta}^{i+1}(\mathcal{H})$ and $\beta^i(\mathcal{H})$ by their respective statistical weights into the desired estimator

$$\beta^{i+1} = G^i(\mathcal{H})\beta^i(\mathcal{H}) + \tilde{G}^i(\mathcal{H})\tilde{\beta}^{i+1}(\mathcal{H})\,, \tag{23.28}$$

where the normalized weights

$$\tilde{G}^i(\mathcal{H}) = \frac{\tilde{g}^i(\mathcal{H})}{g^i(\mathcal{H}) + \tilde{g}^i(\mathcal{H})} \tag{23.29}$$

and

$$G^i(\mathcal{H}) = 1 - \tilde{G}^i(\mathcal{H}) \tag{23.30}$$

are determined by the recursion

$$g^{i+1}(\mathcal{H}) = g^i(\mathcal{H}) + \tilde{g}^i(\mathcal{H}) \tag{23.31}$$

with the starting value

$$g^0(\mathcal{H}) = 0\,. \tag{23.32}$$

As in the other algorithm inspired by statistical physics, SA, one can show that this algorithm leads to the global optimum. However, again the available calculation time is finite. Unlike SA, in which the system freezes in a local or the global minimum at the end, the MUCA never comes to such an end: if the area around a local minimum has been explored to some extent, then the algorithm tries to leave this valley again, as it is more desirable to go to the next higher energy interval if this one has been less explored so far. This method is of course very nice when it comes to overcoming barriers in the energy landscape, but as the algorithm does not remain in the global optimum, one must always store the "best so far" solution.

23.3 MUCAREM and REMUCA

The MUCA can also be combined with other algorithms in order to get even better simulation or optimization results. An example of such a combination is the combination of the MUCA, which is a serial algorithm, with the parallel algorithm replica exchange method (REM), which is also called parallel tempering (PT), to produce either REMUCA or MUCAREM [143, 144].

In the replica exchange multicanonical algorithm (REMUCA), the multicanonical weight factor is determined from a short PT simulation with the multihistogram reweighting techniques. This weight factor is then used in a long multicanonical production run with high statistics. In the MUCAREM, the REMUCA algorithm is performed with a small number of replicas; thus the process of determining the multicanonical weight factor is faster and simpler than in the usual iterative determination.

23.4 Multicanonical Annealing

Based on [19, 20], Lee developed a method called multicanonical annealing [127]. MUCA allows one to sample the density of the states and thus the entropy $S(\mathcal{H})$ directly by imposing the appropriate detailed balance condition. The framework of the multicanonical annealing algorithm is based on the feedback of information about the local entropy. After letting the system run into a local minimum, the multicanonical annealing algorithm tries to obtain some new information about the local entropy S near the local minimum via a short sampling of the surrounding area, and the old value of S is updated accordingly. Then the transition probability for a move is modified according to the new value of S, such that the system is driven out of the local minimum and is able to get to a new local minimum. This approach is iterated until some convergence criterion is met.

24 Searching for Backbones

24.1 Comparing Different Good Solutions

If a stochastic optimization algorithm is used for finding a good or quasioptimum solution for a proposed problem instance, then one usually receives different solutions depending on the initial conditions like the seed for the random number generator. These different solutions sometimes differ rather strongly in their quality. However, if one has a closer look at many different solutions, one finds that they usually exhibit many common structures [181, 187, 182, 183, 184].

Obviously, these structures seem to be optimally solved, as each of the optimization runs has independently led to these structures. Note that the emphasis lies on the word "independently". Let us draw a comparison with a situation in everyday life: if a single person argues or claims something, we will not believe him/her in all cases. However, if various sources deliver the same information, then we are more likely to believe this information; the larger the number of sources, the greater the likelihood of our believing it. However, the insight of Albert Einstein that "common sense is the collection of prejudices acquired by age eighteen" does not apply to our computational results. In the real world, there are no real independencies; everybody is in contact with other persons such that prejudices can spread and are even passed on from one generation to the next. Furthermore, the human mind is selective and weighs those experiences stronger that fit into already existing schemes, such that prejudices remain rather stable. Here, however, we have no such preexisting schemes and we have no information from external sources, so that we can really trust that these structures are special to the problem instance if no construction heuristics or other intelligence was used.

As we are generally trying to find a quasioptimum solution to a given problem instance, these structures seem to be optimally solved, at least in a local sense. If the number of compared solutions is large enough, one can assume that these structures are part of any good solution and therefore also part of the optimum solution to the problem.

If one performs further optimization runs, one will obtain further solutions, again containing these structures. Therefore, some of the calculation time seems them to be wasted when we already know these structures. The system should therefore not spend any more calculation time for finding these optimum parts or trying to optimize parts that are already optimally solved. This time would be better spent on parts for which it is obviously not so easy to find out what the optimum solution does look like such that multiple possibilities are provided for these parts in the various solutions. These possibilities must be considered in the further progress, such that they all must be kept at least at first.

Therefore, a parallel algorithm should be implemented making use of these common parts, which will be called the "backbones" of the system. The searching for backbones (SfB) algorithm tries first to find such common structures and then to use this knowledge to obtain even better solutions.

24.2 Determining the Backbone

Usually, it is rather easy to see the common structures in different solutions. However, sometimes there is the problem that one finds various types of common structures. For such a problem, one must find out whether all of these types can or must be used for defining a backbone or which of these types will be preferred. This selection can be performed by considering the type of the problem: when facing, e. g., a sequencing problem, the type of useful structures will usually be sequences; when facing a distribution problem, the distributions will have to be detected.

On the other hand, there might also be problems in which no common structures are apparent if comparing different solutions. However, a detailed investigation of the problem and the various solutions often still leads to the detection of common parts. When this is not the case, it is often because one is dealing with a problem instance that is highly degenerate both in the global optimum and in the local minima. In this case, the SfB algorithm is hard to apply to the considered problem instance, because the degeneracy keeps one from identifying the backbone by the superposition of multiple solutions. However, such a guage ambiguity can be frozen out by picking some initial conditions.

If one has determined common structures by eye or by an elaborate investigation of the solutions, the question arises as to how these structures can be determined by a computer, which does not see this a priori. There are several ways to do that. The method that can likely be used for any optimization problem is to define an overlap matrix η in the following way [182, 183]: The overlap $\eta^\sigma(i,j)$ between two system parts i and j according to the solution σ is given by

$$\eta^\sigma(i,j) = \begin{cases} 1 & \text{if } i \text{ and } j \text{ are related in the same way} \\ & \text{to each other in all solutions}\,, \\ 0 & \text{otherwise}\,. \end{cases} \tag{24.1}$$

One then sums up over the set (σ^k) of all p considered solutions and defines the entry $\eta(i,j)$ of the overlap matrix as

$$\eta(i,j) = \left[\frac{1}{p} \sum_{k=1}^{p} \eta^{\sigma_k}(i,j) \right] , \tag{24.2}$$

with the Gaussian brackets $[x]$ denoting the integer part of x.

The considered type of relation can be kept rather simple. The relation does not necessarily have to be symmetric, i. e., there could be the case where $\eta(i,j) \neq \eta(j,i)$. Based on a simple relation that leads to this overlap matrix η, one can define a more complicated relation by making the basic relation transitive: if i is related to j and j is related to k, then i is related to k in this complicated relation even if $\eta(i,k) = 0$. Using this complicated relation, one can define the backbones of a system according to the considered solutions of the problem: parts that are related to each other according to the transitive relation form a backbone. If there are other parts that are also related to each other according to this transitive relation but not related to the latter parts, then they form another backbone, and so on.

Sometimes it might be the case that one has to work with more entries in the overlap matrix or with several overlap matrices. The basic principle, however, stays the same.

24.3 Outline of the SFB Algorithm

Thus the algorithm starts with the production of p solutions, the comparison of these solutions, and the determination of the backbones. But now the algorithm must make use of the existence of these backbones. Therefore, further optimization runs are started in which these backbones must not be destroyed as they are already optimally solved.

These new optimization runs, which are again performed independently of each other, lead to further solutions. As the backbone parts were not allowed to be changed, each new solution again contains the backbones already found. These new solutions are now assumed to be of a higher quality than the old solutions as the optimization processes were able to concentrate on parts that are more difficult to solve as the easy-to-find backbones were held constant. Therefore, the old solutions are discarded. However, their inheritance exists in the set of backbones.

Then the new solutions are compared with each other. One will again find common structures. Among these, the known backbones will be found. However, one can now expect that even more common structures will be found as the solutions are better. Therefore, a new set of backbones can be built containing the former backbones but also new backbones. It might also be that two or more backbones can be united to one larger backbone.

Then, this approach is repeated: again some optimization runs are performed in which these new backbones are held constant. They lead in turn to new solutions that are again better and again contain more or larger common structures than the solutions of the two previous iterations. This iteration is performed until finally all parts of the system are in one backbone, i. e., until all solutions are identical to each other.

Thus the outline of the SfB algorithm is as follows. As the SfB algorithm is ideally suited for parallel enablement on workstation clusters and parallel computers but can also be performed on a single computer, let us formulate the outline of the SfB algorithm in the master-slave model.

1. The optimization program starts with the master sending the data of the problem instance to the p slaves and the slaves performing their optimization runs independently, each of which usually leads to a different solution.
2. The slaves send their solutions to the master.
3. The master compares the p solutions and determines the backbones.
4. The master sends the information about the backbones back to the slaves.
5. If there is only one backbone containing the whole system, the algorithm stops.
6. The slaves perform new optimization runs in which these backbones must not be destroyed.
7. The algorithm jumps back to step 2.

Sometimes it is necessary to weaken the stop criterion in step 5: if the considered problem instance has a degenerate ground state, it is rather improbable that all slaves will produce the same solution. In this case, it is sufficient to demand that all solutions be of the same quality. However, the convergence of the algorithm in building backbones can also be so slow that one requires the algorithm to stop after a maximum number of iterations.

24.4 Discussion of the Algorithm

The main parameter of this algorithm is the number p of solutions used for comparison. The number p is rather crucial as the following considerations for extreme values show:

- If p is very small, then only a small number of solutions is compared in order to determine common structures. In this case, one will find more common structures than for a larger p. This small statistic leads to false assumptions: one identifies some found structures as optimally solved backbones, although they are not optimally solved. As these backbones must in turn not be destroyed in the next iterations of the algorithm, there is no "undo" for this error. One will therefore end up at worse solutions than if working with a larger p. p must therefore be large enough to provide a good statistical significance.
- On the other hand, if working with a very large p, there are so many solutions to compare that they exhibit so many differences that one will some-

times find hardly any backbones. Only one or a small number of worse solutions with fewer parts optimally solved prohibits the detection of further backbones. Therefore, the second iteration starts with no or only a small number of backbones. This convergence problem will then usually again occur in the following iterations. In this case, it takes such a large number of iterations for some problems that it is impossible to wait for the end of the algorithm. One introduces, therefore, a maximum number of iterations as an additional stopping criterion.

Therefore, one must make a compromise between a number of processors that is on the one hand large enough for statistical evidence and on the other hand small enough not to prevent convergence. Therefore, there is a medium-sized optimum value for p.

Of course, there are ways out of this dilemma: for example, one can adaptively determine a good or even the optimum number p of processors, which can also change for each iteration. This value can be determined according to the convergence speed of the last few iterations. For some problems, it is also possible to determine whether the introduction of a backbone was bad. In this case, the backbone must be split into its individual components. Furthermore, one knows that one has to increase p. A further way is only to use a number p' out of the p solutions for comparison and for determining the backbones. The obvious approach to this is to simply use the best p' out of the p solutions.

Of further importance for the algorithm is the type of solutions that are used as an input for the comparison. As already mentioned above, the underlying local optimization procedures must guarantee that the whole configuration space is sampled for good solutions. Therefore, one cannot use a construction heuristic that leads either always to the same solution or only to a small set of possible solutions or only to solutions of a certain type. Analogously, one must be careful when working with the large ruin & recreate moves. Also, any other intelligence inside the optimization process must be tested for whether it guides the search for a good solution only in a certain direction. One should also not only compare solutions that have been generated over several bouncing iterations as the overlap between these solutions is usually larger than that between independently generated solutions as the bouncing process depends on staying in a particular valley in the energy landscape. Generally, small local moves are to be preferred, the basic serial optimization algorithm should not contain too much intelligence (better no intelligence at all), and the results of independent optimization runs should be used for comparison.

Summarizing, SfB can be considered an algorithm that gradually reduces the complexity of a problem: by finding common structures that are held constant in subsequent iterations, first, the system size is reduced. Only those parts that are more difficult to solve remain. The basic serial optimization runs then work on problems that are smaller and that are often disconnected from each other by the extended backbones that have already been found. Therefore, the global optimization task is partially reduced to several local optimization problems.

One can also interpret the SfB algorithm in “other languages”: one can first of all speak of an inverse tabu search as tabu search forbids structures commonly found in different solutions whereas SfB wants to keep them. Furthermore, one can give a genetic interpretation of the algorithm: in a crossover operation of a genetic algorithm, two parental solutions produce some child solutions containing properties of their parents. SfB marries not only two but p solutions with each other. Subsequent optimization runs serve as crossover operators to produce children that contain the properties of all of their parents.

Part II

Applications

0 General Remarks

0.1 Dealing with a Proposed Optimization Problem

After the first theoretical part, in which an overview of optimization algorithms was given, we now want to show how these algorithms can be applied to a given problem. We will exemplify this with two principal examples, with which we can clarify easily why we have done the application in this and not in another way. In this way, the reader should get some feeling about how to apply an algorithm to proposed problems. However, in practice, the work starts already some levels deeper.

The first step in solving an optimization problem in practice is to find out what exactly must be optimized. Usually, there are only a few people in a company who can say something more about this problem, and often the head of the company or one of the companys departments is not one of them. In the case of larger companies, often several departments of the company are part of the business processes to be optimized. Thus, one must find the "experts" in these departments in order to get an overview of the whole problem. Secondly, one must find out as much as possible about the real costs occurring in the business processes. Often the company management is not fully aware of those processes; sometimes, it does not know about them at all. Furthermore, in the case of large companies, the existing knowledge is spread over various departments that must compete with each other in the way that each department manager tries to minimize his/her local costs, regardless of whether a globally optimum solution is thereby achieved or missed. Therefore, as an internal or external optimization consultant, one can easily get caught in the middle of political considerations and rivalries among individual managers.

Having resolved these questions, one is now able to model the problem on a computer and to write down a cost function $\mathcal{H}$ for the problem in this representation. This is the most important part of the work—how to represent the problem on a computer. Usually, one should follow the advice "Keep it simple!" Closely related to this is the notion that one should also find an appropriate cost function representing all (important) cost factors within the business processes and adding some penalty terms for constraints that should not be violated. Again, the cost function should not be too elaborate, as then the optimization process might lead to unexpected and undesired solutions.

Roughly speaking, frustration within the system increases with the number of addends in each cost function and with the complexity of the individual addends. There is even a third part of the implementation that is also closely related to these two parts, namely, the definition of moves for jumping from one configuration to another. Initially, one should start with small, simple moves that guarantee ergodicity in the random walk mode. Later on, one can switch to more complicated moves of, e. g., the ruin & recreate (R & R) type. Sometimes, one must even do this if one finds out in tests that it is impossible to reach good solutions with these small moves. Therefore, after writing a basic program with one or two small move routines, one usually performs some test runs with the greedy algorithm to see if the program works correctly and to try to obtain good results. Then one switches to more elaborate algorithms and tries to tune these algorithms, using some of the improvement tricks described in the first part, in order to get to (quasi) optimum solutions and in order to achieve such solutions rather fast.

0.2 Programming Languages and Parallelization Libraries

As stated above, once the problem is understood, it must be modeled on a computer. This representation strongly depends on the programming language used. Each language has its advantages and its disadvantages.

- The use of Fortran 77 and earlier versions of Fortran, like Fortran 66 and Fortran IV, aims at really utilizing the full CPU power in order to achieve solutions as fast as possible. Fortran compilers often produce the fastest codes on various machines. Fortran is a programming language that is available on most modern supercomputers. Often it is the standard language on such computers; other programming languages are still sometimes less supported. A further advantage is that the standard numerical precision is rather high if using double precision variables. Furthermore, when using the IEEE standard, one obtains the same result on every machine as the numerical precision of a real number is defined and the rounding is done in the same way.
  On the other hand, Fortran 77 has the disadvantage that the programmer might get bogged down in a large number of variables and lose sight of the big picture if the optimization problem is very complex as it contains no opportunities to define structures or methods of object-orientated programming (OOP). However, after some practice solving complex problems with Fortran 77 programs, this no longer presents a problem, so that the authors of this book prefer to use Fortran 77.
- Another widely used language in scientific computing is C. The code produced by C compilers is usually nearly as fast as that of Fortran compilers, sometimes even faster. An advantage, compared to Fortran 77, is the pos-

sibility of defining structures, i. e., combining several data into one block of data. A further advantage of C is that it allows for hardware-near programming, e. g., it presents opportunities to obtain further memory and to exchange messages with other processes or even to start further processes. However, C has the disadvantage that variable types are not uniquely defined, e. g., one must find out for every new compiler and every new machine whether a double variable contains four or eight bytes. Therefore, the results of C programs depend on the underlying computer.

- C++ and Fortran 90 (and newer Fortran dialects like Fortran 95) are OOP languages based on C and Fortran 77, respectively. They therefore allow for defining objects that represent objects of the real problem. These objects are not only simple structures as in C, they are also able to exchange messages with other objects. A further advantage of these languages is that, due to the encapsulation of the information in the objects, it is more easily possible to split the development of an optimization program between a group of people. Each person only programs certain objects and can rely on the fact that nobody else can do anything unexpected with these objects. The interactions between the objects are strictly defined by interfaces. Furthermore, it is claimed that old program code can be reused as more elaborate objects may be derived from some basic objects.
  However, according to the experience of this book's authors, there are often interaction problems within a group of programmers, so that the interfaces between the individual objects must be altered repeatedly. Furthermore, it is nearly impossible to determine a best OOP approach to a complex problem a priori; this is often done by trial and error. Furthermore, it is only seldom the case that old code from a project can be reused in the next project in an unaltered form. Usually, one must change at least some small sections of the code. Thus the degree to which the philosophy of OOP gets fulfilled depends on the programming discipline of all members of the group. Furthermore, compilers for these languages produce a slower code than C and Fortran 77 compilers. One often finds that the program becomes slower depending on the extent to which OOP methods have been used: the more they are used, the slower the program becomes.
- Java is currently the standard language for OOP and contains standard libraries for creating a graphical user interface (GUI) or for accessing databases. Furthermore, it is the standard language for creating internet applications, so-called applets, which run in any modern internet browser. A test performed by one of the authors in 2001 showed that an optimization program written in Java was more than 100 times slower than the corresponding Fortran 77 program. Although nowadays Java has become faster than before, we still cannot recommend the use of Jave for large optimization problems in practice. On the other hand, one can of course run applets or applications for very small instances of the proposed optimization problem and try to gain more insight into the behavior of the optimization algorithm and the interaction between it and the problem.

Java allows for easy creation of a visually appealing GUI and for watching intermediate results and the change of the current configuration during the optimization run. (Of course, this can also be done using other programming languages and linking external graphical libraries like Xlib, Xt, Motif, pgplot, etc., or by combining it with a graphical interface created with, e. g., Tcl/Tk.)

If a workstation cluster or even a parallel computer is available, then there is also the question of which parallelization library to use. Currently there is the standard library Message Passing Interface (MPI), which is available for most parallel computers and also in a free version for workstation clusters. Furthermore, there is a freeware library called Parallel Virtual Machine (PVM), which has been installed on a large number of different systems. Besides these standard libraries, each vendor of parallel computers delivers his own parallel library. Although this parallel library might make the best use of the parallel computer, we strongly recommend using MPI or PVM to keep the program portable. Furthermore, most vendors these days deliver an optimized version of MPI for their parallel computers.

0.3 Optimization Libraries

Of course, you, the reader of this book, might think of saving time and not implementing the optimization algorithm yourself but downloading an optimization library from the internet to do the work. There already exists a small variety of free optimization libraries for download on the internet. There are libraries for exact algorithms, like the simplex algorithm, in which sometimes even results from current research are included. Each such library requires you to represent the optimization problem in a specific way, as the developers of the library had in mind. This might be the optimal way to solve the problems that they worked on or some class of problems they wanted to consider. However, this might not be the optimal way to solve your particular problem.

Even worse is the situation with optimization libraries that use heuristic algorithms. Sometimes, such libraries simply contain basic objects, which contain hardly anything. Such a library can only serve as a starting point for one's own implementation; therefore, it is virtually useless. Secondly, there are optimization libraries that were developed and tailored for a special problem or for a class of problems. In this case, often the authors of the optimization library claim that it can be used for any problem and will lead to quasioptimum results. If one must solve a problem of the type for which the library was written, it might lead to very good results. Otherwise, the situation might be much worse. In any case, one should look through the source code of the library in order to understand what it does exactly. For example, consider a simulated annealing (SA) optimization library in which the authors fixed the values of the initial and final temperatures. These values might either be much too large for your problem, so that the optimization process spends most of the calculation time in the quasi-random-walk phase and does not freeze at the end in a local minimum, or it might be too small, so that the system is quenched down rapidly, as if with the greedy algorithm.

Thirdly, there are very elaborate libraries that were developed over the course of several years for the optimization of several types of problems. Such libraries, like the elaborate TopC library of the IBM Scientific Center Heidelberg, contain many tricks to automatically find good parameters for the problem instance to be solved. But they also contain many possibilities for setting flags and other parameters, according to the wishes of the people who had already used this library on their projects. (These wishes can even go so far as to ask the developer of the library to make it possible to set an additional flag such that the library beeps three times at the end of the optimization run.) In this case, one first must read a long description of how to make use of the library. Of course, only a few switches must be known in order to be able to work with the library. But it is also good to learn about the switches for experts in order to achieve quasioptimum results in a small amount of time.

However, even when using such an optimization library much programming work is left to be done: one must program the appearance of a configuration of the problem, write down functions like the cost function or an energy difference function, and implement various move routines. If the used optimization library—again assume a SA library—only contains a small main program consisting of a call for an initialization routine, a loop over several temperature steps in which there is a loop calling a move routine several times, and, finally, a routine for showing the end result, it is not worth using, as these take up only a few programming lines.

Generally, we strongly recommend not using any optimization library at first, as one should learn how to apply an optimization algorithm to a proposed problem and get a feeling for how to improve results. Later on, one can either write one's own optimization library, collecting all of the tricks one has found and does not want to implement over and over, or one can make use of an elaborate optimization library, as one already has a feeling as to which effects it might lead to if one of the library's parameters is changed.

0.4 Difficulty of Comparing Various Algorithms

Finally, we want to issue a warning regarding the literature in the field of optimization: one often reads about comparisons of different algorithms. Such a comparison is often performed between algorithms of different optimization schools, like the genetic algorithm (GA) school and the SA school. One usually finds, for example, that the author of a paper entitled "Genetic algorithms are better than simulated annealing" belongs to the GA school. Thus, he/she has much more experience in implementing and tuning GAs than with SA. Therefore, his/her results and conclusions are not surprising.

Secondly, authors who want to introduce a new optimization algorithm must usually prove that their algorithm is doing a good job. This proof can be most easily done by comparing the achieved results with results from existing standard optimization algorithms.

In both cases, often political considerations play a role: scientists must write proposals in order to get money and therefore must prove that they are doing a good job. In these times of increasingly intense competition for funding, one must even prove that one is doing a better job than others. Therefore, some scientists yield to the temptation to compare their tuned algorithm to an untuned algorithm or by only considering problems for which they know their algorithm is superior. In such comparisons, the authors also take advantage of some specific features of their algorithms: for example, some authors introduce a time limit by which the algorithms being compared should return their final solutions. Of course, often this time limit is set in such a way that it is advantageous to their algorithms. But if one allows, for example, only a minimal amount of time of computing time for a very large problem, then typically the greedy algorithm will turn out to be superior to SA, as the greedy algorithm only allows for improvements whereas SA must work its way down from high to low temperatures, thus losing time initially. In some papers, the results of some specific computing times are also compared to results from other papers for the same computing times, although the computers were much slower formerly. Some authors try to find out what can be generally achieved with their algorithm, investing a huge amount of computing time, taking the best solution, and comparing it to published results of other algorithms, for which much less computing time was invested.

There is simply no one best optimization algorithm. If one wants to obtain a solution as fast as possible, one must work with the fastest construction heuristics, which try to meet the constraints but do not care if a good solution is achieved. If one must obtain the optimum solution, regardless of the time required for achieving it, then one uses an exact method, but of course only if the system size is small enough such that one can obtain the solution in one's own lifetime. All types of heuristics mentioned in the first part of this book lie between these two extremes.

In subsequent chapters, we will present the application of some algorithms to specific problems and provide results, so that the reader may see what can be obtained with these algorithms. We are both from the school of physical optimization and therefore are more used to tuning algorithms related to SA than, e. g., those related to GAs or tabu search. We are furthermore fully aware of the fact that many algorithms have been slightly altered and improved for some specific problems by a series of authors. However, this book will only give a general overview of various stochastic optimization algorithms and can therefore only serve as an introduction to the implementation of these algorithms for a given problem. Therefore, we will restrict ourselves to the original flavors of some optimization methods or some special implementations we have heard and read about.

For some methods, we will show the extent to which the results can be improved by tuning the algorithm. However, this book must also be kept finite. Of course, one might consider, e. g., showing the results for SA combined with a sophisticated bouncing scheme and with special R & R moves, etc. This might not be helpful, as the question remains as to whether the improvement was due to R & R or bouncing. Therefore, we will repeatedly start from a basic optimization system and change this system only in one step, like using R & R moves instead of small moves in order to learn about the effects of using R & R moves. We will also introduce some variations of existing algorithms. This is because, in part, these variations can lead to a better understanding of the algorithm itself. Mostly, however, we want to give the reader some hints of what ideas one can come up with while reading other people's descriptions of optimization algorithms. Sometimes these variations lead to improvements; sometimes they unexpectedly worsen the results, so that a closer look at the algorithm is necessary in order to explain the new results.

Thus the first part of the book presented the philosophy behind various optimization algorithms. This second part of the book is dedicated to the implementation and adaption of these algorithms and the effects of clever tricks one might think of.

Applications A
The Traveling Salesman Problem

1 The Traveling Salesman Problem

1.1 The Task of the Traveling Salesman

The first problem to which we apply the algorithms described in the chapters above is the traveling salesman problem (TSP). Nowadays it is sometimes, in a politically correct way, called traveling salesperson problem in order to emphasize that there are also traveling salesladies.

A set of N nodes that the traveling salesman must visit is given. These nodes can be cities or, more exactly, the locations of customers. Generally, the distances $D(i, j)$ between the various pairs (i, j) of nodes are known to the traveling salesman. Mostly, the distances are measured in units of length or of time or most generally of money. Usually, the symmetric TSP is considered, i. e., the distance from i to j is identical to the distance from j to i for all pairs (i, j) of nodes.

The task of the traveling salesman is now to find the shortest tour through the given set of nodes, touching each node exactly once and returning at the end to the first node in the tour.

There are also other problems that can easily be mapped on the TSP such that a node might have another meaning than spatial location or that the meaning of a node is not important at all. We will discuss one such problem in the next chapter.

1.2 Distance Metrics

Sometimes the values of the distances $D(i, j)$ are given explicitly. However, often the distances between nodes are calculated according to an L_p-metric, i. e., the the distance between nodes i and j is given as

$$D(i, j) = \left(|x_i - x_j|^p + |y_i - y_j|^p\right)^{1/p} . \tag{1.1}$$

The cases $p = 2$, $p = 1$, and sometimes $p = \infty$ are the ones most often considered:

- Usually, the Euclidean metric or L_2-metric is used: the distance between nodes i and j is given as

$$D(i,j) = \sqrt{(x_i - x_j)^2 + (y_i - y_j)^2} \, . \tag{1.2}$$

In some applications in which the heights of the nodes play a role, like in Switzerland, also the third dimension is considered:

$$D(i,j) = \sqrt{(x_i - x_j)^2 + (y_i - y_j)^2 + (z_i - z_j)^2} \, . \tag{1.3}$$

- In some cities, like in New York or in Mannheim, the Manhattan or L_1-metric is preferable:

$$D(i,j) = |x_i - x_j| + |y_i - y_j| \, . \tag{1.4}$$

- Sometimes the Tschebyscheff or L_∞-metric is used:

$$D(i,j) = \max\{|x_i - x_j|, |y_i - y_j|\} \, . \tag{1.5}$$

1.3 The Dijkstra Algorithm

In practical applications of the TSP and derived problems, the traveling salesman has a more complex task than the abstract task defined above because he can neither know all distances exactly a priori nor approximate them by using the straight lines between pairs of nodes. Of course, there are approximations like multiplying the Euclidean distance between the nodes by a constant factor. Such a factor can be derived from fractal considerations, but we want to calculate the distances more exactly.

A digital map of the area in which the nodes of the TSP instance lie must be used. A digital map contains a large set of nodes and the lengths of the edges between them. The quality of the digital map is therefore given by the exactness of the lengths and by the number of the nodes. Sometimes, only nodes representing a splitting of important streets or the end of a street are in such a digital map. Better maps also contain the small roads and nodes denoting turns in these roads.

The real nodes of the TSP instance are then introduced in the digital map used, either in a simple way such that a TSP node is mapped onto the nearest node on the digital map or in a more elaborate way where it is mapped on its real position, thus often dividing one edge into two edges. For calculating the distances between these TSP nodes, the Dijkstra algorithm is used.

The Dijkstra algorithm [155] for calculating a distance matrix requires only that the distances not be negative, which is clearly fulfilled due to the positive street lengths. The Dijkstra algorithm is unable to detect negative cycles, so, for example, the distance from A to D in Fig. 1.1 becomes smaller the more often the triangle Δ(BCE) is traversed. Other algorithms like the Ford–Bellmann algorithm are able to detect such negative cycles, but they generally require more calculation time than the Dijkstra algorithm.

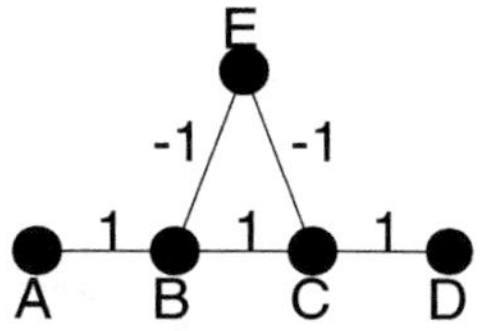

Fig. 1.1. Negative cycle in a graph: the distance of A to D is, if the way leads only over B and C, the length 3; the more often the triangle Δ(BCE) is traversed on the way from A to D, the more often the distance is decremented by 1

Most algorithms for the calculation of distances use the Bellmann optimality principle, according to which the subparts of shortest ways must be minimal. This becomes clear with the following example: let us assume that on the way from A to D via B and C one has two connections between B and C, one via E and one via F. Of course, the connection between A and D becomes shorter if the shorter connection between B and C is chosen. The Bellmann optimality criterion forms from this apparent comparative sentence the proof for the superlative sentence. It is also of crucial importance for the Dijkstra algorithm.

Dijkstra's distance matrix algorithm, which also works if one-way streets exist, can be used efficiently both for calculating the distance of one node to another node and for calculating a whole distance matrix. The latter case is done by a loop over all nodes in which all distances from a chosen start node to all other nodes are calculated. In this loop, the outline of the Dijkstra algorithm is as follows:

First, the nodes are distributed among three sets. Set 1 contains all nodes for which the distance from the initial node r is already known. Some upper bound for the distance can be given for all nodes in set 2. Set 3 contains the remaining nodes. In the initialization part of the algorithm, only the starting point r is put in set 1 because the distance from it to itself is known to be 0. All nodes that are directly connected to r by an edge, the neighbors of r, are put in set 2. Set 3 contains the remaining nodes, as shown in Fig. 1.2.

After this initialization procedure, the algorithm starts by searching the node s in set 2 that has the smallest preliminary value for the distance $D(r, s)$. Therefore, s cannot be reached faster from r via a detour over nodes in set 2 or 3, as all distances are nonnegative. Therefore, the distance $D(r, s)$ from r to s is known already; it is equal to the length of the edge. Therefore, s is moved from set 2 to set 1. However, one must consider not only s but also the neighbors t_i of s that are connected via edges with s.

One must distinguish the following cases for these neighbors t_i:

- t_i is in set 3:
  Then a preliminary upper bound for the distance from r to t_i can be calculated, namely, $D(r, t_i) \leq D(r, s) + D(s, t_i)$. This distance is inscribed

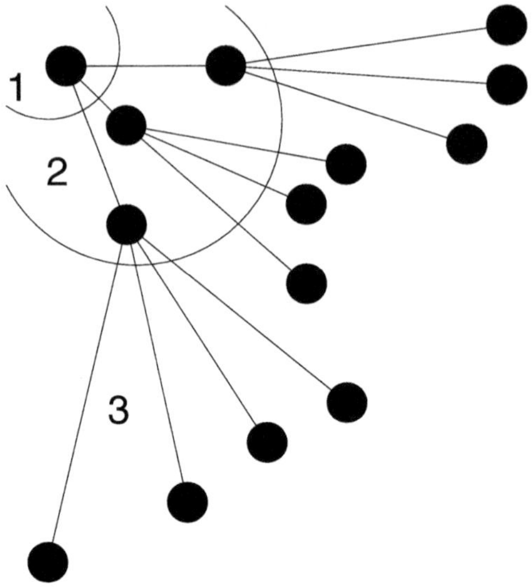

Fig. 1.2. Initialization of the Dijkstra algorithm: the points are distributed in three sets. Only the starting point, from which one wants to know the distances to all other nodes, is in set 1. All nodes with a direct connection to this starting point are in set 2. For these nodes, one already knows a preliminary distance value, namely, the length of the edge. The remaining nodes are in set 3. No estimation for the distance to the initial node is given for them

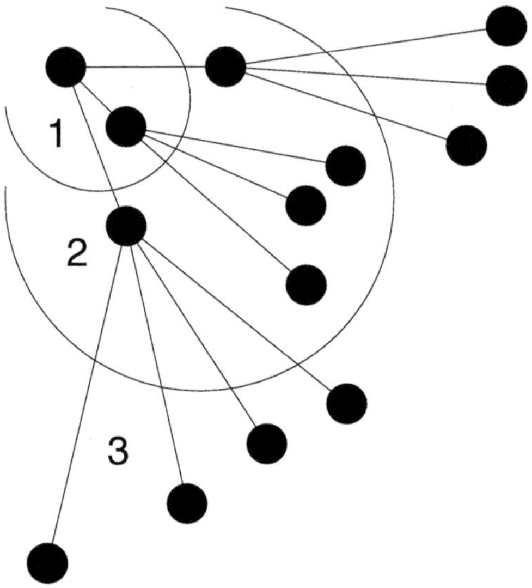

Fig. 1.3. First step of the Dijkstra algorithm: here the nearest neighbor of the starting point is already moved from set 2 to set 1. The neighbors of this nearest neighbor that had been in set 3 are moved to set 2

as a preliminary distance from r to t_i. t_i is transferred from set 3 to set 2, as shown in Fig. 1.3.

- t_i is in set 2:
  Then there already exists another connection from r to t_i. However, the distance from r via s to t_i might be smaller than the previous distance bound. Therefore one must check whether the old inscription for $D(r, t_i)$ is larger than $D(r, s) + D(s, t_i)$ and then must be replaced by this smaller value.
- t_i is already in set 1:
  Then nothing need be done, as the distance from r to t_i would become larger when driving via the intermediate node s.

This step is repeated for all further nodes. The breaking condition of the algorithm is that there is no longer any node in set 2. If set 3 contains at least one node in this case, then there are nodes that cannot be reached from r via the known edges. All distances to the reachable nodes have been calculated. If the task only consists in determining the distance between two nodes instead of calculating the full distance matrix, then there is a further break condition that is usually fulfilled earlier, namely, that the destination node is moved to set 1.

Mostly, one wants not only to know the distances between the nodes but also the sequence of nodes via which one must pass in order to use the shortest path from r to a node t. Conveniently, the sequence of these nodes is not stored directly. Instead, for all nodes t a node s is stored that is the last intermediate node on the way from r to t. By that approach, the way from r to t can be reconstructed by going backward from t via s and further intermediate nodes to r. This approach is also used for graphical output.

If one only wants to know the distance between two nodes, then the algorithm is of the order $\mathcal{O}(N^2)$, with N denoting the number of nodes, as in the worst case $N - 1$ nodes must be transferred from set 2 to set 1 and as in this case all neighbors t_i of s must be updated that might also be $N - 1$ in the worst case. If a full distance matrix must be calculated, then one must iterate over all starting nodes r. The calculation time is then proportional to $\mathcal{O}(N^3)$. Therefore, in this case, the faster algorithm by Dantzig is used.

The Dijkstra algorithm can be simplified further if one has integer street lengths and an upper bound for the distances. The time for calculating a distance from r to t is then only of order $\mathcal{O}(M)$, with M denoting the number of edges. As these are again of order $\mathcal{O}(N)$ in the case of real maps (mostly there are not more than, say, ten streets originating at a node, such that one has $10 \times N$ streets in the system), the algorithm is linear with the number of nodes, i. e., of order $\mathcal{O}(N)$ [202].

1.4 Various Possible Codings

The simplest way to describe a configuration of an instance of the TSP is as a permutation σ of the numbers $1, \ldots, N$. The meaning of, e. g., $\sigma(3) = 7$ is that the third node in the tour is the node with the number 7. The energy function of this problem is simply the length of the configuration and in this coding is given as

$$\mathcal{H}(\sigma) = D(\sigma(N), \sigma(1)) + \sum_{i=1}^{N-1} D(\sigma(i), \sigma(i+1)) \,. \tag{1.6}$$

The first addend is needed to close the tour; it denotes the length between the last node in the tour and the first node in the tour. The following addends denote the lengths between all pairs of successive nodes in the tour. If the distance matrix D is symmetric, one gets an at least two-fold degeneracy of each energy level, as a tour is the same length after being turned around, i. e., whether it is driven through clockwise or counterclockwise.

This permutation σ can be implemented in various ways. The standard way of implementing it is by working with an array σ of N integers, with $\sigma(i) = j$ denoting that the node with number j is the ith node in the tour. This is an implementation following strictly the definition. However, the permutation of the nodes can also be implemented as a connected list of some structure. Each structure contains at least a minimum information about the node, i. e., the number of the node or its location, and additionally a pointer to a further structure that is the successor of the current structure in the connected list. In this way, a closed ring of such structures, each representing a node of the TSP instance, results in a tour of the traveling salesman. In contrast to the implementation as an array, this method has the disadvantage that the individual structures might be distributed over the whole memory such that the cache and its accelerator effects cannot be used. However, this problem can also be overcome by simulating a connected list with an array l of integers: $l(i) = j$ denotes that the successor of node i is node j. If desired, one can store more information about the individual nodes in a second array containing the structures described above. Besides this single connected list, one could also work with a double connected list, in which each structure contains both a pointer to its successor and a pointer to its predecessor. The optimal implementation must be decided for each optimization method. Changing some parameters of the optimization method could result in another implementation becoming optimal.

Secondly, a configuation can be coded as an edge matrix η, which is also called the adjacency matrix:

$$\eta(i,j) = \begin{cases} 1 & \text{if } j \text{ is the successor or the predecessor of } i\,, \\ 0 & \text{otherwise}\,. \end{cases} \tag{1.7}$$

Then the Hamiltonian is written as

$$\mathcal{H}(\eta) = \sum_{i}^{N} \sum_{j=i+1}^{N} D(i,j)\eta(i,j) \,. \tag{1.8}$$

However, not all possible edge matrices correspond to a feasible configuration of a TSP. The constraints for a feasible edge matrix η can be written as

$$\sum_{i=1}^{N} \eta(i,j) = 2 \quad \forall j \qquad \wedge \qquad \sum_{j=1}^{N} \eta(i,j) = 2 \quad \forall i \,, \tag{1.9}$$

plus the constraint that all nodes must be in one tour. This prohibition of closed subtours can also be expressed with the edge matrix. Let $V = \{1, \ldots, N\}$; then

$$\sum_{i,j \in U} \eta(i,j) < 2|U| \quad \forall U \subset V \quad \text{with} \quad U \neq \emptyset \wedge U \neq V \,. \tag{1.10}$$

This representation of the TSP is used if solving it with an exact algorithm based on integer programming methods. The main problem when working with such an edge matrix is the prohibition of subtours, as the number of these subtour constraints increases exponentially with the number of nodes in the tour. However, there is another possibility of an edge matrix for which this problem is strongly simplified [58]: let

$$\eta_{i,a} = \begin{cases} 1 & \text{if node No. } i \text{ is the } a\text{th} \\ & \text{node in the tour}\,, \\ 0 & \text{otherwise}\,. \end{cases} \tag{1.11}$$

This edge matrix corresponds to an incidence matrix. Note that we use the letters i, j, ... for the "name numbers" of the nodes, which are given to them at the beginning and which stay constant, and the letters a, b, ... for the tour positions of the nodes, which can change when a move is performed. Then the length of the tour is given by

$$\mathcal{H}_0(\eta) = \sum_{ija} D(i,j)\eta_{i,a}\eta_{j,a+1} \tag{1.12}$$

with $\eta_{i,N+1} \equiv \eta_{i,1} \, \forall \, i$. The constraints that each node is visited exactly once, i. e.,

$$\sum_{a} \eta_{i,a} = 1 \; \forall i \,, \tag{1.13}$$

and that each stop in the tour contains exactly one node, i. e.,

$$\sum_{i} \eta_{i,a} = 1 \; \forall a \,, \tag{1.14}$$

are considered by adding penalty functions. The system exhibits therefore the Hamiltonian

$$\mathcal{H}(\eta) = \sum_{ija} D(i,j)\eta_{i,a}\eta_{j,a+1} + \lambda_1 \sum_i \left(\sum_a \eta_{i,a} - 1 \right)^2 + \lambda_2 \sum_a \left(\sum_i \eta_{i,a} - 1 \right)^2 \tag{1.15}$$

containing the Lagrange multipliers λ_1 and λ_2. Removing the constant terms from this formula, one gets the effective Hamiltonian

$$\mathcal{H}_{\text{eff}}(\eta) = \sum_{ija} D(i,j)\eta_{i,a}\eta_{j,a+1} + \lambda \left(\sum_{ija} \eta_{i,a}\eta_{j,a} + \sum_{iab} \eta_{i,a}\eta_{i,b} - 4 \sum_{ia} \eta_{i,a} \right) . \tag{1.16}$$

In this way, the TSP is mapped on a spin system in which the single spins $\eta_{i,a}$ are on a 2D lattice with periodic boundaries, can take the values 0 and 1, and interact with all other spins in the same row, the same column, and the next column.

There are of course still other possibilities for coding a configuration of the TSP. Summarizing, usually the first coding mentioned is used due to its simplicity, as it already contains all constraints. When working with this coding, mostly a 1D array is used, containing the numbers $1, \ldots, N$ in that order in which the corresponding nodes appear in the tour. Contrarily, sometimes a single or a double connected list is used for storing the sequence of nodes.

1.5 Four Approaches to the TSP

Usually, one of the following four approaches is used for investigating the quality of an algorithm for the TSP.

- Often TSP instances are used consisting of a fixed number of N randomly placed nodes. Always several such instances are created and the average of the results for the individual instances is taken. From these results the behavior of the algorithm used is described for a TSP of size N. Then N is increased step by step, and for each N a new average is taken, so that the quality of the algorithm is investigated as a function of the system size. Furthermore, this curve can be extrapolated to a TSP with an infinite number of nodes.

- The second method consists of putting the nodes on a square lattice with n columns such that there are always $N = n^2$ nodes. Here the quality of the algorithm is also described for increasing n. In contrast to the first method, here it is rather easy to determine the optimum length, which is given by

$$\mathcal{H}_{\text{opt}}(N) = \begin{cases} d \cdot N & \text{if } n \text{ even} \\ d \cdot (N - 1 + \sqrt{2}) & \text{if } n \text{ odd} \end{cases}, \tag{1.17}$$

  with d being the lattice constant, i. e., the nearest neighbor distance on the lattice. Furthermore, in contrast to the first method, problems with a highly degenerate ground state are investigated.
- The third approach considers a TSP with randomly chosen values for the entries in the distance matrix. This so-called random link TSP [31] is of a more complex randomness than the TSP with randomly placed nodes, as, e. g., the triangle inequality does not hold. For this type of TSP, only the entries in the distance matrix, but not the locations of the nodes, are given. Of course, the locations could be reconstructed in an $N-1$-dimensional space if D is symmetric. Furthermore, the aim is usually not to solve a particular instance but to derive analytically some estimates for, e. g., the optimum tour length, which is for higher dimensions except small deviations identical to the optimum tour length of a Euclidean TSP with randomly placed nodes.
- The first three approaches use only artificial instances, as nodes are never purely randomly distributed or on a square lattice in practical applications and as edge lengths are not purely random, either. Therefore, several groups have published instances from practical applications. These instances have been collected in some libraries, among them the TSPLIB95 by Gerhard Reinelt [90]. Besides the data of the instances, also the optimum values are given, such that these benchmark instances are a useful tool for investigating the quality of various algorithms. We will use this approach in this book.

1.6 Benchmark Instances

The benchmark instances in Reinelt's TSPLIB95 cover different TSP sizes: the smallest instances only have a few points, like the 16 locations of the odyssey [74, 75] or the 127 beer gardens near Augsburg, a city in southern Germany, which are useful for finding out whether a program works correctly. Then there are medium-sized instances like 442 drilling holes on a computer circuit, the 532 AT&T switch locations in the USA, the 535 airports throughout the world, or 1379 nodes in the German state of Northrhine-Westfalia, which are used for tuning algorithms, as it is not easy to find the optimum solution for one of these problems. Finally, the library is extended step by step with larger and larger instances, like the locations of 4461 nodes in the former German Democratic Republic. A large instance consisting of the

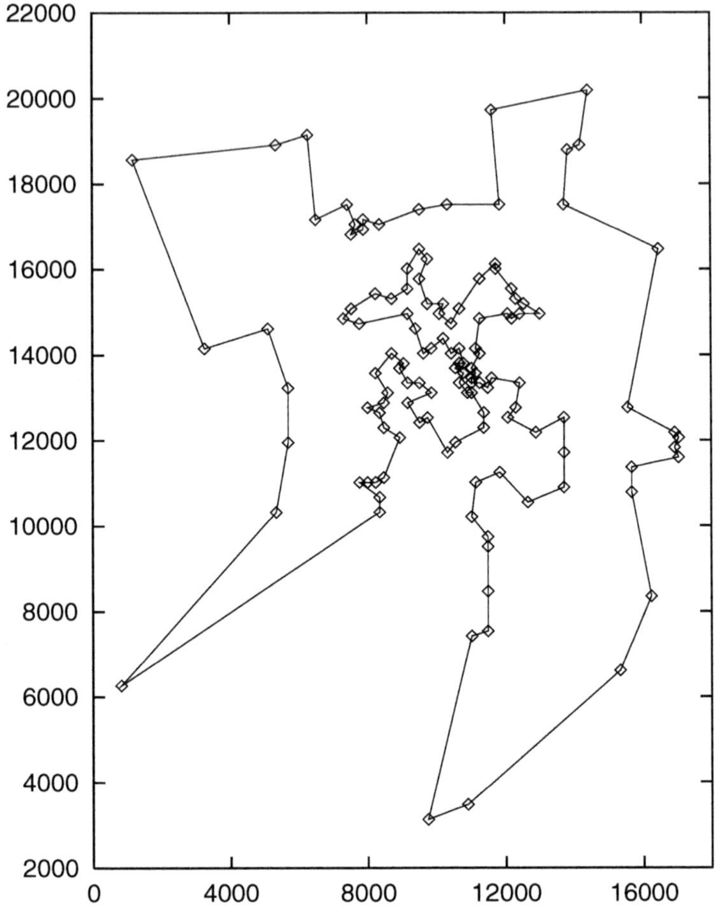

Fig. 1.4. Optimum solution to the BEER127 problem: the BEER127 problem consists of the locations of 127 beer gardens near Augsburg, a city in southern Germany. The optimum tour length is 118,293.52... in double precision

13,509 towns in the USA with at least 500 inhabitants was solved several years ago [74, 75]. The largest instance solved exactly up to the end of the year 2005 consists of 24,978 cities in Sweden [96].

We concentrate on medium-sized TSP instances as they predominate in practice, whereas the large instances are simply enjoyable challenging problems and seldom occur in practical applications, and the small instances can usually be solved manually. We want to provide results mainly for the famous PCB442 problem, which is a very interesting problem, as it has a highly degenerate ground state. An optimum solution to this problem is shown in Fig. 1.6. The PCB442 problem was introduced by Grötschel in 1984. It was first solved in 1987 by his student Holland [73]. The length of its ground state

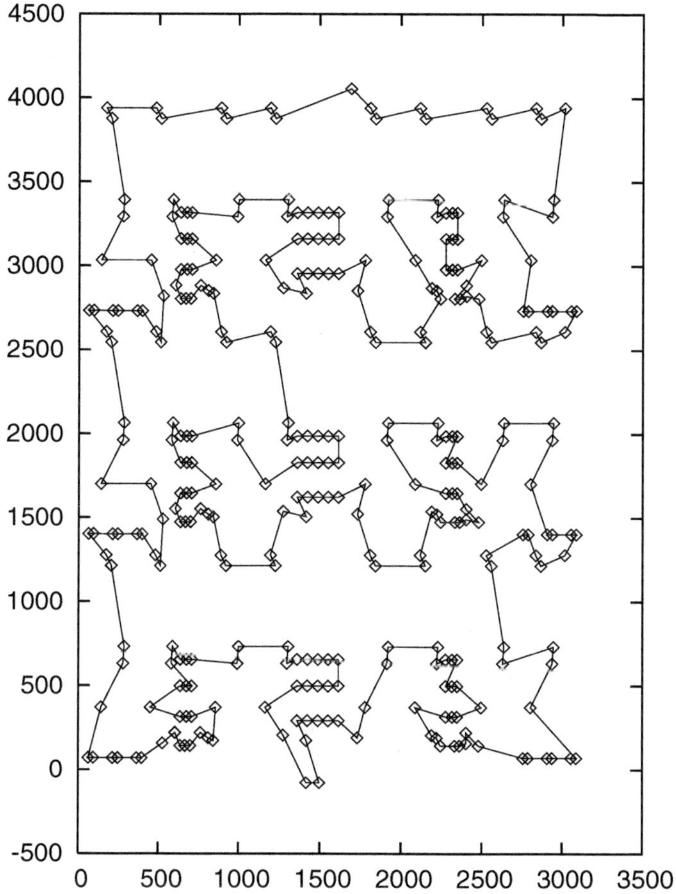

Fig. 1.5. Optimum solution to the LIN318 problem: the LIN318 problem consists of 318 nodes and is also part of the TSPLIB95 [90]. Its optimum tour length is 42,042.535... in double precision. Note that it is not exactly the same problem as the 318-node problem of Lin and Kernighan

is 50,783.547513735... milliinch, if a Euclidean real metric with 8-byte reals is used. If all distances are rounded to integers, then the length of the ground state is 50,778.

Furthermore, we will present results for the BEER127 problem, which was introduced by Jünger and Reinelt and which is famous because most of the 127 beer gardens are located in the city of Augsburg and that there are only a few beer gardens in the suburbs of Augsburg. Usually, it is possible to find the optimum solution, which is shown in Fig. 1.4 and which has a length of 118,293.52... in double precision. For testing the effects of larger moves, we like to use the LIN318 instance (Fig. 1.5). A further well-known test instance are the 532 largest cities of the USA (Fig. 1.7). This instance was introduced

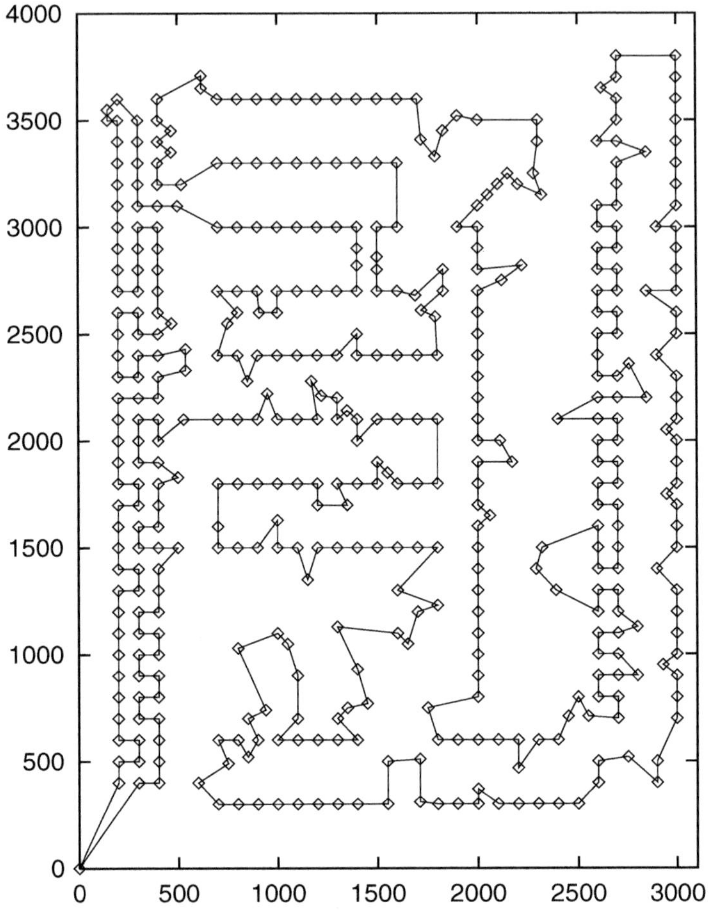

Fig. 1.6. An optimum solution of the PCB442 problem: the PCB442 problem consists of drilling 442 holes on a printed circuit board. The ground state of this problem is highly degenerate. The optimum length is 50,783.5475... milliinch in double precision

by Padberg and Rinaldi [159]. For this problem, a special pseudo-Euclidean distance metric is used:

```
xd = x(i) - x(j)
yd = y(i) - y(j)
rij = sqrt( double( xd*xd + yd*yd )/10.0 )
tij = nint( rij )
if ( double( tij ) < rij ) then
  D(i,j) = double( tij ) + 1
else
  D(i,j) = double( tij )
endif
```

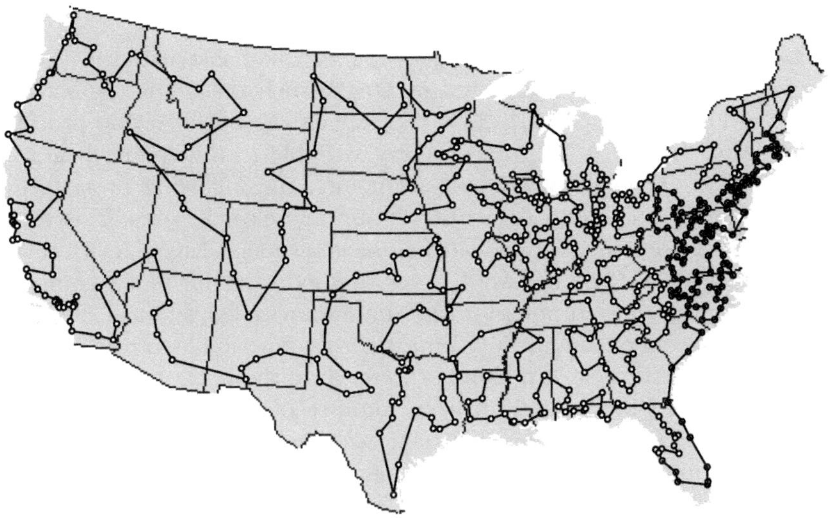

Fig. 1.7. Optimum solution to the ATT532 problem: the ATT532 problem consists of the 532 AT&T switch locations in the USA. For this problem, the special integer ATT metric is used. The length of this configuration is given as 27,686 with this pseudo-Euclidean metric

According to this metric, its optimum has a length of 27,686. In contrast to the PCB442 problem, in which the distance to the nearest neighbors is the same for most nodes, this instance exhibits large differences between these nearest neighbor distances: on the east coast of the USA, these distances are very short, in other areas they are comparatively very long. This results in this TSP instance being split into subproblems with different length scales.

1.7 Bounds for the Optimum Solution

While seeking a TSP solution, there are some other, simpler problems that we may wish to solve along the way. Some of these constructions have fast, exact algorithms, and are useful in determining upper and lower bounds to the length of the optimal TSP tour.

Let us consider a set of nodes. A spanning tree on these nodes is a set of edges by which all nodes are thus connected so that one can get from every node to any other node without a redundant edge. Therefore, each spanning tree on N nodes contains $N-1$ edges. If one edge of a configuration of a TSP instance is removed, then one gets a spanning tree. The number of edges that are adjacent to some nodes is called the degree of the node. It can be shown that the number of nodes with an odd degree is even. The minimum spanning tree (MST) is the spanning tree of shortest length.

Using the algorithm of Kruskal, the MST can be constructed in the following way: first all edges are sorted by length such that e_1 is the shortest and e_M the longest length with $M = \binom{N}{2}$. Then the initialization proceeds with an empty tree $T = \{\}$. An auxiliary variable i is introduced and set to $i = 1$. While T does not connect all points, the following procedure is iterated: as long as $T \cup \{e_i\}$ does not contain any closed cycles, T is set to $T \cup \{e_i\}$ and i is incremented by 1. It can be shown that this algorithm is of order $\mathcal{O}(N^2 \log(N))$ [177]. The MST plays an important role in the computer network field when a net topology with minimum wire length must be found. The MST is both an optimum solution if the cost function is to minimize the wire length and if the cost function is to minimize the delay time [177, 74].

The problem of organizing an even number of points in pairs so that the sum of all connecting edges between each pair of nodes becomes minimal is called minimum weighted matching (MWM). This algorithm can be mapped onto a linear optimization problem that can be solved by the simplex algorithm. However, various implementations have an order $\mathcal{O}\left(\left(\frac{N}{2}\right)^{5/2}\left(\log\left(\frac{N}{2}\right)\right)^4\right)$ or an order $\mathcal{O}\left(\left(\frac{N}{2}\right)^3\right)$ of calculation time, such that often heuristics are used for larger instances to find a rather good MWM in a short time [104].

The length of a MST is a lower bound for the optimum length of a TSP instance, because if one edge is removed from an arbitrary solution of this instance, a spanning tree without branching remains. This spanning tree is at least as long as the MST. Similarly, the MWM can provide a lower bound: each configuration, including the optimum one, of a TSP instance with an even number of nodes can be split into two sets of edges by alternatively putting the succeeding edges into these two sets. Therefore, each of these sets contains a different matching of the nodes, both of which must be at least as long as the MWM. Therefore, each solution for a TSP must be at least twice as long as the MWM.

When doubling the MST, i. e., traversing each link in the MST twice, one gets a roundtrip in which each node is visited at least once. Thus, two times the length of the MST forms an upper bound for the length of an optimal tour of the TSP. This bound can be tightened by removing from the doubled MST those sites which are visited more than once until a tour remains, which is a correct solution to the TSP. There are various heuristics for this shortcutting. The Christofides heuristic guarantees a solution that is not more than 50% longer than the optimum because the MWM is used: the system is initialized by a simple MST, which is unified with a MWM between the nodes of odd degree. (The degree of a node is defined as the number of edges adjacent to this node.) In this way, every node has an even degree, so that one ends again up with a TSP with some nodes visited more than once. These can again be removed by shortcutting, such that a feasible solution for a TSP is achieved. The length of a configuration constructed in this way would not be larger than the sum of the lengths of the MST and of the MWM. This is again smaller than $\frac{3}{2}$ times the optimum length of the TSP.

There are also more elaborate bounds on the optimum length of a TSP instance. One widely recognized bound, even for large TSP instances, is the Held–Karp bound [84, 85, 103]: for the calculation of the Held–Karp bound, the second coding of the TSP mentioned in Sect. 1.3 is used; however, the definition of the edge matrix (1.7), in which all entries $\eta(i,j)$ can only take the values 0 and 1, is replaced by the condition $0 \leq \eta(i,j) \leq 1$. Thus, a linear problem is created whose solution is the so-called Held–Karp bound, a lower bound on the optimum solution of the TSP. This linear problem can be solved in polynomial time with an altered simplex algorithm, despite the fact that the number of subtour constraints increases exponentially with the system size because there exists a polynomial-time "separation oracle" for the subtour constraints [103]. The lower bounds are usually roughly 0.5 to 2% smaller than the optimum tour length of the corresponding TSP, even for large TSP instances, and so it is widely used.

Besides knowing bounds for the optimum solution of a TSP, this knowledge is also useful if working with, e. g., the great deluge algorithm, because if one knows the depth at which the optimum configuration lies, one can tune the algorithm and steer it.

1.8 The Misfit: A Frustration Measure

The misfit parameter was originally introduced by Kobe (see [118] and references therein) for measuring the frustration in a spin glass system. The misfit parameter compares the ground state energy $\mathcal{H}_0$ of the given instance of a problem with the ground state energy $\mathcal{H}^{\text{id}}$ of an idealized instance. The idealized instance is achieved by changing the original instance in such a way that the local energies of the various parts of the system are minimized independently of each other. $\mathcal{H}^{\text{id}}$ is simply the sum of these local energies. Due to competing effects between the various parts of the system, $\mathcal{H}_0$ is usually larger than $\mathcal{H}^{\text{id}}$ because not all local "wishes" can be fulfilled. Thus the misfit is defined as

$$m = \frac{\mathcal{H}_0 - \mathcal{H}^{\text{id}}}{\mathcal{H}^{\text{id}}} . \tag{1.18}$$

For a specific TSP instance, the local energy at each node is minimal if each node i is connected with its two nearest neighbors $n_1(i)$ and $n_2(i)$. Thus, the ground state energy $\mathcal{H}^{\text{id}}$ is given as

$$\mathcal{H}^{\text{id}} = \frac{1}{2} \sum_{i=1}^{N} D(n_1(i), i) + D(i, n_2(i)) . \tag{1.19}$$

The factor $\frac{1}{2}$ is needed as for each node two edges are counted instead of one. This formula holds also for the asymmetric TSP if $n_1(i)$ and $n_2(i)$ are chosen appropriately.

Thus, the misfit measures how complicated the given instance is to solve compared to an idealized instance. For example, the misfit vanishes if all nodes are placed on a circle, which is surely a trivial instance. One can also normalizethis misfit by introducing a second idealized scenario that represents the worst case [118]. The formula for this normalized misfit parameter μ is given by

$$\mu = \frac{\mathcal{H}_0 - \mathcal{H}^{\mathrm{id}}_{\mathrm{best}}}{\mathcal{H}^{\mathrm{id}}_{\mathrm{worst}} - \mathcal{H}^{\mathrm{id}}_{\mathrm{best}}} . \tag{1.20}$$

For the TSP, the best-case scenario is given as above. Now a suitable worst-case scenario must be determined. As a first approach, as was done above, one can connect each node with the two furthest nodes. This leads to a comparison with the worst possible solution. However, this approach contradicts the aim of finding the best possible solution. Thus a comparison with another idealized scenario might be more useful: in this scenario, all distances are equal and are thus given by the mean distance $\bar{D}$ of the instance. $\bar{D}$ is the average over the lengths of all edges that can occur in a TSP configuration:

$$\bar{D} = \frac{1}{N(N-1)} \sum_{\substack{i,j=1 \\ i \neq j}}^{N} D(i,j) . \tag{1.21}$$

Note that the diagonal elements of the distance matrix never occur as edge lengths as there is no edge from a node to itself in the tour. This type of problem is also trivial as all possible configurations have the same length, namely,

$$\mathcal{H}^{\mathrm{id}}_{\mathrm{worst}} = \frac{1}{2} \sum_{i=1}^{N} (\bar{D} + \bar{D}) = N \times \bar{D} . \tag{1.22}$$

$\bar{D}$ is calculated from the original instance. There might actually be an instance in which all distances are equal. In this case, $\mathcal{H}^{\mathrm{id}}_{\mathrm{worst}}$ and $\mathcal{H}^{\mathrm{id}}_{\mathrm{best}}$ coincide. As this problem is trivial, the misfit parameter μ can be set to 0.

Summarizing, it is not the size of an instance, i. e., the number of its nodes, that determines how difficult the instance is to solve. A relatively small misfit value μ for a given instance is a hint that the instance is a relatively easy one compared to other instances.

1.9 Order Parameters for the TSP

The next question is how to define an order parameter ξ for the TSP. Naturally, this order parameter ξ should refer to how well the problem instance is already ordered during an ongoing optimization process. Furthermore, it should not be identical to the energy or depend directly on it.

A first approach for such an order parameter could be to consider the area $\mathcal{A}$ that is included in the closed polygon of the current configuration. This area can be easily calculated by

$$\mathcal{A}(\sigma) = \frac{1}{2}\left|\sum_{i=1}^{N}(x_{\sigma(i)} \cdot y_{\sigma(i+1)} - x_{\sigma(i+1)} \cdot y_{\sigma(i)})\right| , \tag{1.23}$$

with $\sigma(N+1) \equiv \sigma(1)$ [177]. At first glance, this seems to be a reasonable approach: if minimizing the number of intersections in the tour, both the length $\mathcal{H}$ is minimized and the absolute value of the signed surrounded area $\mathcal{A}$ is maximized. Thus, the ratio

$$\mathcal{V}(\sigma) = \frac{\mathcal{A}(\sigma)}{\mathcal{A}_0} , \tag{1.24}$$

with $\mathcal{A}_0$ being the included area in the ground state configuration, seems to be a good definition of a normalized order parameter.

However, an order parameter should be maximal if the ground state is reached. This constraint is violated for the included area, as the counterexample in Fig. 1.8 shows. Thus, the size of the included area is not a good order parameter for the TSP.

A further property by which a configuration can be compared with the ground state or even projected on the ground state is the antisymmetric edge matrix η_σ with

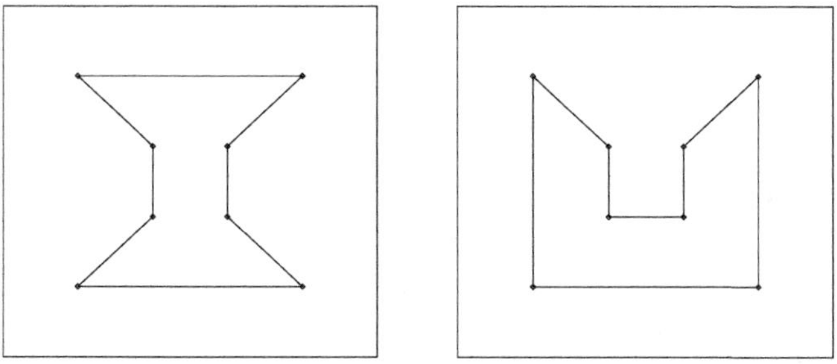

Fig. 1.8. This TSP instance consists of eight nodes lying at the edge points of two concentric squares. The interior square has an edge length of 1, the outer square an edge length of 3. In the graphic on the *left*, the ground state with a length of $\mathcal{H}_0 = 8 + 4\sqrt{2} = 13.6\ldots$ and an included area of $\mathcal{A}_0 = 5$ is shown. The graphic on the *right* shows a quasioptimum configuration with $\mathcal{H}_1 = 12 + 2\sqrt{2} = 14.8\ldots$ and $\mathcal{A}_1 = 6$. Comparing these two configurations, the longer solution has the larger included area

$$\eta_\sigma(i,j) = \begin{cases} +1 & \text{if } j \text{ is successor of } i \text{ in configuration } \sigma\,, \\ -1 & \text{if } j \text{ is predecessor of } i \text{ in configuration } \sigma\,, \\ 0 & \text{otherwise}\,. \end{cases} \tag{1.25}$$

Let η_σ be the antisymmetric edge matrix corresponding to the configuration σ, and let η_0 analogously be the edge matrix of the ground state configuration. Then the edge projection order parameter ξ can be defined by

$$\xi(\eta_\sigma) = \frac{1}{N}\left|\sum_{i=1}^{N}\sum_{j=1}^{N}\eta_\sigma(i,j)\eta_0(i,j)\right| \tag{1.26}$$

or more simply by

$$\xi(\sigma) = \frac{1}{N}\left|\sum_{i=1}^{N}\eta_0(\sigma(i),\sigma(i+1))\right|\,, \tag{1.27}$$

with $\sigma(N+1) \equiv \sigma(1)$.

If this order parameter ξ were not normalized it could take values between -1 and $+1$, as is the included area parameter. One must define order parameters of this kind in such a way as to take the absolute value of some sum over entries of an antisymmetric matrix. Furthermore, note that for both order parameters knowledge about the ground state is needed.

When working with improvement heuristics, if the ground state of the problem is not known, one studies "freezing order parameters" like the Edwards–Anderson order parameter q_{EA} for spin glasses, which measures the mobility of a system. One could thus analogously derive such an order parameter using the antisymmetric edge matrix η_σ: let $\langle\eta(i,j)\rangle$ be

$$\langle\eta(i,j)\rangle = \sum_\sigma \eta_\sigma(i,j)\pi(\sigma)\,, \tag{1.28}$$

with $\pi(\sigma)$ the probability of the configuration σ, e. g., the Boltzmann weight $\exp(-\mathcal{H}(\sigma)/(k_{\mathrm{B}}T))/Z$ when working with simulated annealing. Then q_{EA} is simply given as

$$q_{\mathrm{EA}} = \frac{1}{2N}\sum_{i,j=1}^{N}\langle\eta(i,j)\rangle^2\,, \tag{1.29}$$

as $2N$ entries of $\langle\eta(i,j)\rangle$ are ±1 if the system is frozen and all other entries of the matrix vanish.

But for this type of order parameter again one must have some basic a priori knowledge about the system. A general way of studying the freezing behavior without additional knowledge of the system is to investigate the acceptance rates of the individual moves and the total acceptance rate, which is a measure, applicable to every system, of whether the system is already frozen in some local minimum. Note that these freezing order parameters always tend to 1 if no further improvement can be found. The final value does not depend on whether or not the ground state is reached in the end.

1.10 Short History of TSP

The TSP was, as far as is known, first defined in an 1832 manual for the successful traveling salesman [154, 193], *Der Handlungsreisende – wie er sein soll und was er zu thun hat, um Aufträge zu erhalten und eines glücklichen Erfolgs in seinen Geschäften gewiß zu sein – von einem alten Commis-Voyageur* (The Traveling Salesman – How He Should Be and What He Must Do to Obtain Orders and to Be Sure of Good Success in His Business – by an Old Traveling Salesman).

In this handbook the problem is stated but not formulated in a mathematical way: *Die Geschäfte führen die Handlungsreisenden bald hier, bald dort hin, und es lassen sich nicht füglich Reisetouren angeben, die für alle vorkommende Fälle passend sind; aber es kann durch eine zweckmäßige Wahl und Eintheilung der Tour, manchmal so viel Zeit gewonnen werden, daß wir es nicht glauben umgehen zu dürfen, auch hierüber einige Vorschriften zu geben. Ein Jeder möge so viel davon benutzen, als er es seinem Zwecke für dienlich hält; so viel glauben wir aber versichern zu dürfen, daß es nicht wohl thunlich sein wird, die Touren durch Deutschland in Absicht der Entfernungen und, worauf der Reisende hauptsächlich zu sehen hat, des Hin- und Herreisens, mit mehr Oekonomie einzurichten. Die Hauptsache besteht immer darin: so viele Orte wie möglich mitzunehmen, ohne den nämlichen Ort zweimal berühren zu müssen.* [Business brings the traveling salesman now here, then there, and no travel routes can be properly indicated that are suitable for all cases occurring; but sometimes, by an appropriate choice and arrangement of the tour, so much time can be gained, that we don't think we may avoid giving some rules also on this. Everybody may use that much of it, as he takes it as useful for his goal; so much of it however we think we may assure, that it will not be well feasible to arrange the tours through Germany with more economy in view of the distances and, which the traveler mainly must consider, of the trip back and forth. The main point always consists of visiting as many places as possible, without having to touch the same place twice (translation taken from [193]).]

The manual suggests five tours, four in Germany only, the fifth in Germany and part of Switzerland, which might even be optimal considering street layout in early-19th-century Europe. Whereas nowadays it is politically correct to speak of the traveling salesperson problem, the manual considers only sales*men* and warns about the risks of women in this business.

More books on traveling salesmen were published later, e. g., *The Commercial Traveller's Guide Book*, 1871, by Linus Pierpont Brockett, *How to Become a Commercial Traveller*, 1893, *The Tales of a Traveller*, 1916, by E. Cadwell, Simon S. Skidelsky, and *Tips for Traveling Salesmen*, 1927, by Herbert N. Casson [93]. The TSP was reintroduced as the *Botenproblem* (messenger problem) by Karl Menger in 1930 [193], in which the problem is coded as a permutation of the cities. Besides a complete search through all configurations, a heuristic for solving this problem is mentioned by Menger with the remark that in general it does not lead to the shortest roundtrip.

Furthermore, in the 1930s, the problem reappeared in mathematical circles at Princeton University [93] and was posed by Hassler Whitney [193].

A breakthrough in the mathematical formulation of the TSP came in 1954, when Dantzig, Fulkerson, and Johnson coded the TSP with an edge matrix and with linear constraints for the column and row sums of this matrix (the so-called convex hull of the permutation matrix) and for the subtour elimination. There, for the first time, it was mathematically proven that a given tour for a problem instance was optimal.

In subsequent decades, ever larger instances were solved, as shown in Fig. 1.9. Not only did computers become faster, but mathematicians were also able to develop new algorithms and improve them. The best algorithm so far for solving the TSP exactly is branch & cut. TSP instances with fewer than 1,000 nodes can be solved exactly within a couple of seconds or minutes. The solution of the instance with 15,112 nodes, however, still took 22.6 CPU years. The calculation was performed on 110 distributed processors. The benchmark instances that have been introduced in the last few years as challenges contain up to 10,000,000 nodes. So far, exact methods have not been able to provide optimum solutions for these huge instances. However, some maximum gap to the real optimum can be determined: for some problems, it is known that the best solution found so far is at maximum less than 0.1% worse than the optimum configuration, which is mostly good enough for practical applications.

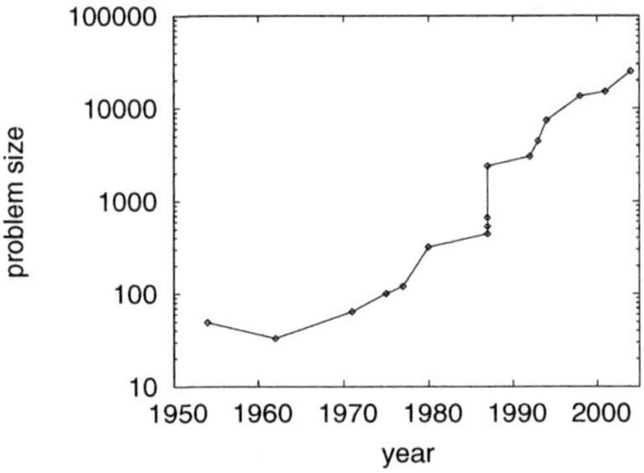

Fig. 1.9. Increase of the size of the TSP instances solved optimally in recent decades: the first reported optimally solved problem was an instance containing 49 nodes, namely, cities chosen in the 48 coterminous states of the United States and Washington, D.C., in 1954. The largest instances solved so far contain 15,112 nodes (solved in 2001) and 24,978 nodes (solved in 2004). (data from [93])

For the problem of the real-life traveling salesman, this research is of course no longer of any importance, e. g., no traveling salesman would be able to perform a roundtrip through the World TSP, containing all 1,904,711 populated cities or towns in the GEOnet Names Server (Fig. 1.10), within his

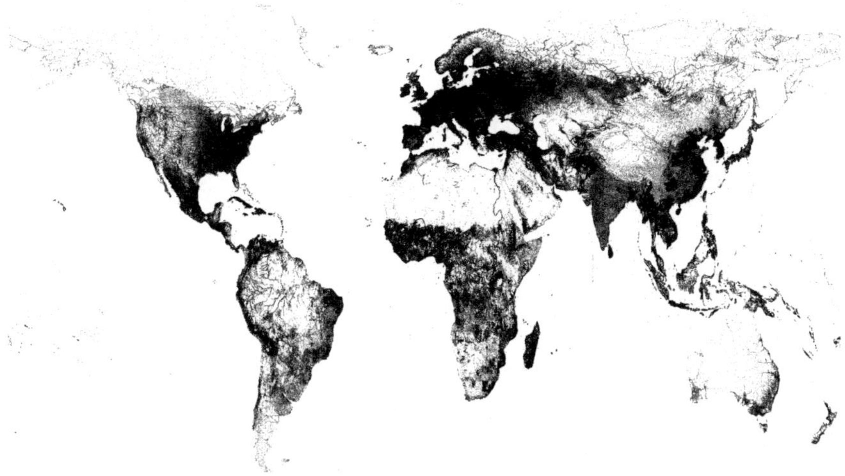

Fig. 1.10. "World TSP" containing all 1,904,711 populated cities or towns in the GEOnet Names Server

lifetime. However, there are applications of the TSP in circuit boards, VLSI, genome sequencing, and genetic engineering, for which one must deal with thousands or even millions of nodes.

The question arises now as to why one should care about stochastic optimization algorithms if the exact algorithms are able to deal with such large problem instances. The reason for this is that, first of all, to perform their work (faster), exact algorithms need the output of heuristic methods as an input. Further, additional constraints are often introduced in practical applications, and these require developing new exact algorithms from scratch, while we shall find that heuristics can often be extended simply. Furthermore, the exact methods developed for extensions of the TSP (see next chapter) are only able to deal with very small system sizes due to finite calculation time and often then fail to solve the problems actually occurring in practice.

2 Extensions of Traveling Salesman Problem

2.1 Temporal Constraints

Although the traveling salesman problem (TSP) was introduced in the last chapter in an idealized form, it has practical import. There are related problems of which the basic problem is a TSP, but that are extended with additional constraints. As these problems occur even more often in practice, we will discuss them here.

The greatest simplification of the classic TSP is the distance matrix, which is usually considered to be symmetric and time-independent. However, the distance matrix is asymmetric in most real-life applications due to the existence of one-way streets. Thus, onc has to deal with the asymmetric traveling salesman problem (ATSP). This ATSP can easily be mapped on a symmetric TSP [170]: for an instance of the ATSP with N points and the asymmetric distance matrix D_{A}, one defines an instance of a TSP with twice as many points. Each point is simply doubled, such that there are instead of one point with the index i now one point with the index i and a further point with the index $N+i$. The entries of the symmetric distance matrix D_{S} of the corresponding symmetric TSP are given as follows:

$$\begin{aligned}
&D_{\mathrm{S}}(i, N+j) = D_{\mathrm{S}}(N+j, i) = D_{\mathrm{A}}(i,j) \text{ for all } 1 \le i,j \le N \text{ and } i \ne j\,, \\
&D_{\mathrm{S}}(j, N+i) = D_{\mathrm{S}}(N+i, j) = D_{\mathrm{A}}(j,i) \text{ for all } 1 \le i,j \le N \text{ and } i \ne j\,, \\
&D_{\mathrm{S}}(i,i) = 0 \text{ for all } 1 \le i \le 2N\,, \\
&D_{\mathrm{S}}(i, N+i) = D_{\mathrm{S}}(N+i, i) = -M \text{ for all } 1 \le i \le N\,.
\end{aligned} \tag{2.1}$$

M must be some sufficiently large number, e. g., $M = \sum_{i,j} D_{\mathrm{A}}(i,j)$. Every configuration of the ATSP with a length l can be transformed into a configuration of the symmetric TSP with the length $l - NM$. This is simply done by replacing the edge (i,j) of the ATSP configuration by the edge $(i, N+j)$ and adding the edge $(N+j, j)$ right after that. Furthermore, an optimum configuration of the symmetric TSP contains all N edges with length $-M$ (this is of course only the case if M is large enough) such that the ATSP can be solved by first finding the optimum configuration of the symmetric TSP and then removing the edges with length $-M$. At first sight this transformation of an ATSP into a symmetric TSP looks rather elegant. However, it is usually not used for practical applications as the system size is doubled.

Instead, one alters the algorithms developed for the symmetric case in order to deal with the asymmetries.

Even worse than the asymmetry problem, the entries in the distance matrix are usually not constant in time due to varying traffic densities. The traveling salesman can drive at the maximum velocity allowed if the traffic density is light; however, he mostly has to drive at slower velocities. This effect is most dramatic at rush hours in which all traffic might be stop and go. However, rush hours are not noticed on all streets at the same intensity; it usually affects traffic on streets in downtowns and on highways but not on side streets in the suburbs. Therefore, the entries in the distance matrix vary differently in time, so that one must deal with the time-dependent traveling salesman problem (TdTSP), in which the traveling salesman is forced to find a compromise between taking detours and waiting in traffic jams [17].

The Hamiltonian (1.6) of the original TSP can still be used if one considers the entries of the distance matrix no longer as constants. However, there are additional constraints: those that probably occur most often to a real traveling salesman are time windows: he must arrive at the home of a customer i within a time interval $[t_{\mathrm{S}}(i); t_{\mathrm{E}}(i)]$. Thus, with such problems the entries in the distance matrix are measured in time units. Furthermore, a specific node is fixed at which the traveling salesman starts his tour. Additionally, service times $\tau_s(i)$ must be considered for each customer i. If the traveling salesman arrives too late at a customer's home, then the solution is, of course, not feasible if the time window is a hard constraint. Also, if he arrives too early, he might be asked to wait. However, he should not wait too long or too often as then he will miss time windows of later customers and return too late to the starting node. This extension of the TSP with time intervals for the arrival time and for the service time is called the traveling salesman problem with time windows (TSPTW), which was studied, e. g., in the asymmetric ATSPTW case [11]. If additionally the entries in the distance matrix depend on the time of day, then the problem is a TdTSPTW.

2.2 Vehicle Routing Problems

There are also other problems related to the TSP. Let us again start from the standard TSP in which a lone traveling salesman performs a closed roundtrip through a given set of nodes. If there is not just one but several traveling salesmen starting from a central depot performing roundtrips, such that every node is visited once by one of the traveling salesmen, and returning to the depot, then this is a multiple traveling salesman problem (MTSP). If there are no further constraints, this MTSP can be mapped on a standard TSP by simply merging the tours of all traveling salesmen into one tour. If omitting the intermediate visits at the depot in this tour, the length of this solution can be reduced. Of course, the optimum solution for this TSP would usually not be this merged roundtrip but an even shorter tour. Thus, if there are no constraints, it is optimal to work with one traveling salesman only.

However, usually there are additional constraints.

- Again, individual customers might demand time windows, such that it becomes impossible for one traveling salesman to visit all of the customers as different time windows at the same time of day cannot be fulfilled by one person. Therefore, the individual tasks must be distributed among several traveling salesmen, so that the time windows are fulfilled and the overall tour length is as small as possible. Thus, a configuration is here given not as a permutation σ of the numbers $1, \ldots, N$ but as a tour plan $\sigma(i,j)$. Every row of this plan contains a tour of one of the M traveling salesmen, starting and ending at the depot. The depot is usually denoted as node No. 1. Let there be N_j entries for the tour of salesman j. Thus, $\sigma(1,j) \equiv \sigma(N_j,j) \equiv 1$; furthermore, the salesman serves $N_j - 2$ customers, so that $\sum_{j=1}^{M} N_j = N - 1 + 2M$. This problem is, e. g., called a MTSPTW.
- More often, the constraint occurs that the customers want to get goods of a certain amount of size or weight delivered or fetched. Thus, the traveling salesman becomes a trucker driving a truck with a finite capacity. Due to this restriction, it is usually again impossible for one traveling salesman to fulfill all these demands. Therefore, there must again be several trucks, and a tour plan for all these trucks has to be created.
  In scientific investigations, usually a simplified version of this problem is used called the (capacitated) vehicle routing problem, or (C)VRP. In this problem type, all trucks start and end at the same depot. One considers the fleet to be homogeneous, i. e., all trucks have the same capacity and all other properties of the trucks are also identical. Additionally, there are either only demands to deliver goods or only demands to pick up goods. The goods are considered to be nonsplittable, i. e., it is impossible that one truck takes half of the goods for a certain customer and another truck takes the other half. These goods are of some abstract amount, and the capacity of the trucks is also measured in this unit. For example, a problem is given in such a way that a customer orders 25 units of a good and another customer orders 10 units, and so on, and each truck has a capacity of 100 units.
  The Hamiltonian of this problem is given as follows: first, the overall tour length shall be as small as possible. If we denote the depot again as node No. 1 and if the array vectors of the tour plan $\sigma(i,j)$ contain as above the tours of M trucks (starting and ending at the depot and containing $N_j - 2$ customers), then the basic Hamiltonian measuring the overall tour length is given as
  $$\mathcal{H}_0(\sigma) = \sum_{j=1}^{M} \sum_{i=1}^{N_j-1} D(\sigma(i,j), \sigma(i+1,j)) \,. \tag{2.2}$$
  Additionally, we constrain the trucks not to be overloaded. Let $m(i)$ be the amount to be delivered to or picked up from customer i and κ the capacity

of a truck. Usually, a linear penalty function for trucks being overloaded is sufficient:

$$\mathcal{H}_1(\sigma) = \sum_{j=1}^{M} \left(\left(\sum_{i=2}^{N_j-1} m(\sigma(i,j)) \right) - \kappa + \gamma \right) \times \Theta \left(\left(\sum_{i=2}^{N_j-1} m(\sigma(i,j)) \right) - \kappa \right), \tag{2.3}$$

with the Heaviside function

$$\Theta(x) = \begin{cases} 1 & \text{if } x > 0\,, \\ 0 & \text{otherwise}\,. \end{cases} \tag{2.4}$$

Often it is useful to introduce an additional small offset $\gamma > 0$ such that no small overloadings remain at the end of the optimization run. Thus, the complete Hamiltonian is given as

$$\mathcal{H}(\sigma) = \mathcal{H}_0(\sigma) + \lambda \mathcal{H}_1(\sigma)\,, \tag{2.5}$$

with $\lambda > 0$ denoting the Lagrange multiplier. λ determines how strongly the penalty function $\mathcal{H}_1$ is weighted in relation to the basic Hamiltonian $\mathcal{H}_0$. If λ is too small, then there might remain some overloadings in the resulting configuration of the optimization run. The penalty function $\mathcal{H}_1$ must vanish at the end of the optimization run. On the other hand, if λ is increased too much, then the probability of coming up with a feasible tour plan with minimum length decreases again.
Several investigations were done on this VRP. As in the TSP, collections of benchmark libraries with VRP instances exist; see, e. g., the collection of Augerat [97].

- An additional problem often considered in science is the vehicle routing problem with time windows (VRPTW), which is given as a combination of the problems discussed above. It is based on the VRP, but additionally individual customers want to be served in a given time interval.
If the time intervals for the arrival times are rather narrow, then they can be treated as punctual time intervals. Let $\delta(i,k)$ be the temporal distance between the arrival times of customers i and k, $\tau_s(i)$ the service time for customer i, and $D(i,k)$ the time for driving from customer i to customer k. Then $\delta(i,k)$ should be equal to $\tau_s(i) + D(i,k)$. Thus, one could introduce penalty terms for each customer $\sigma(i,j)$ proportional to

$$\begin{aligned} &\left(\delta(\sigma(i-1,j),\sigma(i,j)) - \tau_s(\sigma(i-1,j)) - D(\sigma(i-1,j),\sigma(i,j)) \right)^2 \\ &+ \left(\delta(\sigma(i,j),\sigma(i+1,j)) - \tau_s(\sigma(i,j)) - D(\sigma(i,j),\sigma(i+1,j)) \right)^2 . \end{aligned} \tag{2.6}$$

Another approach must be chosen if the time windows are more extended. The arrival time t_A of the traveling salesman has to be between the start time t_S and the end time t_E of a customers time window. Thus, the arrival times at the customers could be calculated according to

$$t_{\mathrm{A}}(\sigma(i,j)) = \sum_{k=2}^{i} D(\sigma(k-1,j),\sigma(k,j)) + \sum_{k=3}^{i} \tau_s(\sigma(k-1,j)) \tag{2.7}$$

for $2 \leq i \leq N_j$. Following this definition, one would have to ensure that the traveling salesmen arrive neither too early nor too late at the customers' homes and to define penalty functions for this constraint. However, then there arises the problem that a salesman may detour to other customers in order not to wait at a customer where he arrives early.
Therefore, it is advantageous to let the traveling salesman wait if he arrives too early at a customer and only to punish him if he is too late [49]. This asymmetric approach results in another rule for calculating the arrival times that has to be formulated in a recursive way, i. e.,

$$t_{\mathrm{A}}(\sigma(i,j)) = \max\left\{t_{\mathrm{S}}(\sigma(i,j)), \tilde{t}_{\mathrm{A}}(\sigma(i,j))\right\} \tag{2.8}$$

for $i \geq 3$ with

$$\tilde{t}_{\mathrm{A}}(\sigma(i,j)) = t_{\mathrm{A}}(\sigma(i-1,j)) + \tau_s(\sigma(i-1,j)) + D(\sigma(i-1,j),\sigma(i,j)) \tag{2.9}$$

and the starting condition $t_{\mathrm{A}}(\sigma(2,j)) = \max\{t_S(\sigma(2,j)), D(1,\sigma(2,j))\}$. Thus, long detours are avoided, but a traveling salesman might wait a rather long time for the time window of a particular customer. This waiting time decreases automatically during the optimization process as it is advantageous to drive to another customer instead of waiting all the time. The penalty function for arriving too late can be coded in, e. g., a linear way as follows:

$$\mathcal{H}_2(\sigma) = \sum_{j=1}^{M}\sum_{i=2}^{N_j} \left(t_{\mathrm{A}}(\sigma(i,j)) - t_{\mathrm{E}}(\sigma(i,j))\right) \times \Theta\left(t_{\mathrm{A}}(\sigma(i,j)) - t_{\mathrm{E}}(\sigma(i,j))\right) . \tag{2.10}$$

Due to the Heaviside function (2.4) there is a penalty only if a traveling salesman arrives too late at a customer's home. Note that this formula also considers a time window for the depot within which each truck has to return. The complete Hamiltonian of the VRPTW consists of

$$\mathcal{H}(\sigma) = \mathcal{H}_0(\sigma) + \lambda\mathcal{H}_1(\sigma) + \mu\mathcal{H}_2(\sigma) . \tag{2.11}$$

The Lagrange parameters λ and μ must be chosen appropriately so that both $\mathcal{H}_1$ and $\mathcal{H}_2$ vanish at the end of the optimization run and that there is a good compromise between these competing objective functions during the whole optimization run. Of course, one could introduce an additional offset in Eq. (2.10) as in the penalty function for overloadings.

A library of VRPTW instances as described above was created by Solomon [89], who distinguishes individual instances based on whether customers are placed randomly (R-problems), are composed in clusters (C-problems), or consist of both random and clustered customers (RC-problems).

Although these scientific problems inherit the most important aspects of the corresponding real-life optimization problems, they are still rather abstract: usually, a fleet of trucks is inhomogeneous. This inhomogeneity leads not only to different capacities one must consider but also to different maximum velocities, such that the temporal distance $D(i, j)$ between two customers i and j also depends on the truck that wants to serve these two customers. Furthermore, there is also often the restriction that not every truck can serve every customer. Think, e. g., of a dairy problem in which the trucks have to collect the milk from the farmers. Some trucks might not be able to reach a farm due to the low quality of the roads or their tubes might be too short for pumping the milk into the truck or they are otherwise inappropriate. Trucks may have to be assigned to handle specific nodes. An additional problem occurring in this type of problem is that some trucks have trailers with a second tank. The capacity of this trailor is usually larger than the capacity of the truck itself. The trailer is pulled by the truck from the depot to the first customers in the tour, but usually not to all customers. Instead, it is parked in a special parking place while the truck goes to a farm that it cannot reach with the trailer. The truck drives a few loops around the trailer's parking place and pumps the milk from the truck into the trailer every time it returns to the parking place. It might even be advantageous for a second truck to pump the milk into the trailer of the first truck before continuing its tour. Thus, the used part of the capacity of the trucks can change; furthermore, if a truck goes with its trailer to a farm, then it could also pump the milk directly into the trailer so that the capacities of the truck, the trailer, and the whole "truck and trailer" system must be considered.

This capacity problem can be even worse if more than one good is transported. In some countries, one differentiates between A-milk and B-milk, which have to be transported in different tanks such that the quality of the A-milk is not affected. Analogously, oil companies also have trucks with several chambers for, e. g., heating oil, gas, and diesel fuel. The capacities of these chambers can usually be considered separately. But there might also be a bridge that can only be crossed if the total weight of the truck does not exceed a certain value. Furthermore, sometimes one must deal with a pickup and delivery problem, i. e., some customers want to get goods delivered, other want to provide goods; additionally, some customers want to get goods from other customers [124]. Then the problem of overloading must be checked at every single customer.

There is also the problem that there is often more than one depot from which goods can be picked up or to where they can be delivered. Therefore, a decision must be made as to which of the depots will be the traveling salesmen's starting-out point and which depots they will return to. This problem also occurs for the parking places for the trailers if there are more parking places available. The problem can be extended even more if not only a list of possible depots is given but also if there is a demand for reducing the number of depots used. Then one must deal with a combination of a vehicle routing problem and a localization problem.

The VRP is often also combined with a truck packing problem: individual goods must be packed into the truck in such a way that the packing and unpacking can be done rather fast and that no goods are in danger of being damaged. (For example, a hat box below a cupboard is a bad idea.) Of course, the goods must be packed closely in so that the spatial capacity of the truck is used efficiently.

2.3 Probabilistic Models and Online Optimization

So far we have only considered the case where all customers are exactly known in advance, together with the exact data for the amount of goods they are receiving or shipping and their time windows. However, this is sometimes not the case. The traveling salesman only has a list of possible customers and knows from past experience that the probability that a given customer will call to get goods picked up or delivered or for a traveling salesman to pay him a visit. Potential applications include school-bus routing and waste collection where the nodes are known but the amounts are unknown. Another application are taxi companies for which there is a small probability for each node being part of the tour.

This problem type has been investigated as an extension of the TSP called the probabilistic traveling salesman problem (PTSP) [101], in which each node i gets a probability $p(i)$ with which it will actually be a part of a tour. The individual probabilities are usually independent of each other. But there is also the case that the probabilities are correlated between pairs of customers. Furthermore, such extensions of the VRP resulting in a probabilistic vehicle routing problem (PVRP) have been studied [102, 21]. There are two different strategies for solving such problems:

- One starts with the problem that all customers who might call actually want to be served, so that the traveling salesman or one of the trucks must visit each of them. In the case of the PTSP, one creates a solution for this TSP instance. In the case of the PVRP, one produces a solution for this problem while assuming that the customers will ask for slightly more goods than usual so as to be on the "safe side".Then, according to the application, the traveling salesman either drives the whole tour, asking customers whether they want to be served, or skips some nodes from the tour plan if it is known that a particular customer does not want to get served. In the case of the PVRP, sometimes there might occur the problem that one must perform an additional intermediate stop at the depot in order to load or unload goods or to ask another truck to take some of the orders. However, this should be the exception and not the rule.

- The other strategy considers all possible instances of the given problem together with their corresponding probabilities, which are given by the probabilities for serving the customers. Thus, if these probabilities are independent of each other, one easily multiplies the probabilities $p(i)$ for the customers being served by the probabilities $1 - p(i)$ for the customers not being served; the total tour length of an instance [124] and the mean tour length over all instances can be calculated analogously. Furthermore, one can study several instances, chosen either randomly or because of their relatively large probability, and determine by comparing them what the tour plan will look like in order to easily incorporate additional customers or substitute assumed ones if necessary. An application of this problem is a taxi company that needs to distribute its taxis over some area in order to reach potential customers as fast as possible.

2.4 Supply Chain Management

These problems may arise in isolation or be embedded in much larger problems. These large problems should be optimized globally as otherwise synergy effects might be lost.

In the field of supply chain management (SCM), many interlocking problems must be considered: first, one must determine what types of goods will be produced, how many of them can be sold on the markets, what prices for them can be demanded, how large the investment costs are, and whether parts of the production will be outsourced to subcontractors from whom some preproducts or intermediate products are bought. Secondly, one must solve the location problem and optimize the factory for the production processes. Then one must deal with stock-keeping problems, with the VRP for delivering the preproducts from the component suppliers to the factory, and with the distribution problem of the preproducts inside the factory. Then come the main production processes themselves, which must be supervised and controlled. After that the products must be delivered to the customers, which is again a VRP. Finally, one must determine what to do with the income, whether the money should be reinvested or given partially as dividends and employee awards [185].

All these problems put together result in the problem of SCM or advanced planning and scheduling (APS). However, there are additional processes besides vehicle routing that are related to the TSP. For example, the optimum sequence in which the products are produced on the assembly line must be found. Assembly line problems can be mapped on the TSP with additional constraints if an array like a distance matrix can be defined [185, 189]. But there are also many other applications that can be rewritten in a TSP-like way. Some archeologists even mapped once the task to determine the chronological sequence of various gravesites in a graveyard onto a TSP with open end points. The "distance" between two gravesites was given by the diversity between their respective contents.

Thus, we want to concentrate on the TSP in the following chapters as it is the basis for many other problems. We will show how the algorithms that were described in a general way in the first part of the book can be applied to the TSP and we will sometimes give hints about how to incorporate additional constraints.

3 Application of Construction Heuristics to TSP

Many construction heuristics have been developed by several authors and been altered and partially improved by other groups. It is impossible to mention all of them in this book. We will only provide some examples here, each of them standing for a certain type of construction heuristic.

3.1 Nearest Neighbor Heuristic

Probably the most natural way of constructing a solution for the TSP is the nearest neighbor heuristic: it starts at an arbitrary node i, which becomes the first node in the tour: $\sigma(1) = i$. Then that node j that is closest to node i is selected to be the second node in the tour: $\sigma(2) = j$. After that the neighborhood of j is considered: that node k that is closest to j and is not identical to i becomes the third node in the tour: $\sigma(3) = k$. This approach is repeated again: one checks the distances of all nodes except i and j, which are already part of the tour, to node k and chooses the node l that is closest to k, which becomes the fourth node of the tour: $\sigma(4) = l$. This algorithm is iterated until the tour contains all nodes.

Therefore, the outline of this algorithm is as follows:

- Choose an arbitrary node i as the starting point of the tour—$\sigma(1) = i$—and set a counter c denoting the number of nodes included in the tour to 1.
- While not all points are included in the tour, i. e., while $c < N$, select

  $$j \in \{1, \ldots, N\} \backslash \{\sigma(1), \ldots, \sigma(c)\},$$

  with $D(j, \sigma(c))$ minimal,
  set $c = c + 1$ and $\sigma(c) = j$.

At first glance, this seems to be a rather nice construction heuristic. In particular, one can expect that at the very beginning (nearly) always the shortest possible distances are chosen. Looking at Figs. 3.1 and 3.2, one finds that this expectation is fulfilled. However, in the intermediate part of the construction process, it is sometimes impossible to find nodes in the neighborhood that have not yet been inserted, so that some longer edges must be

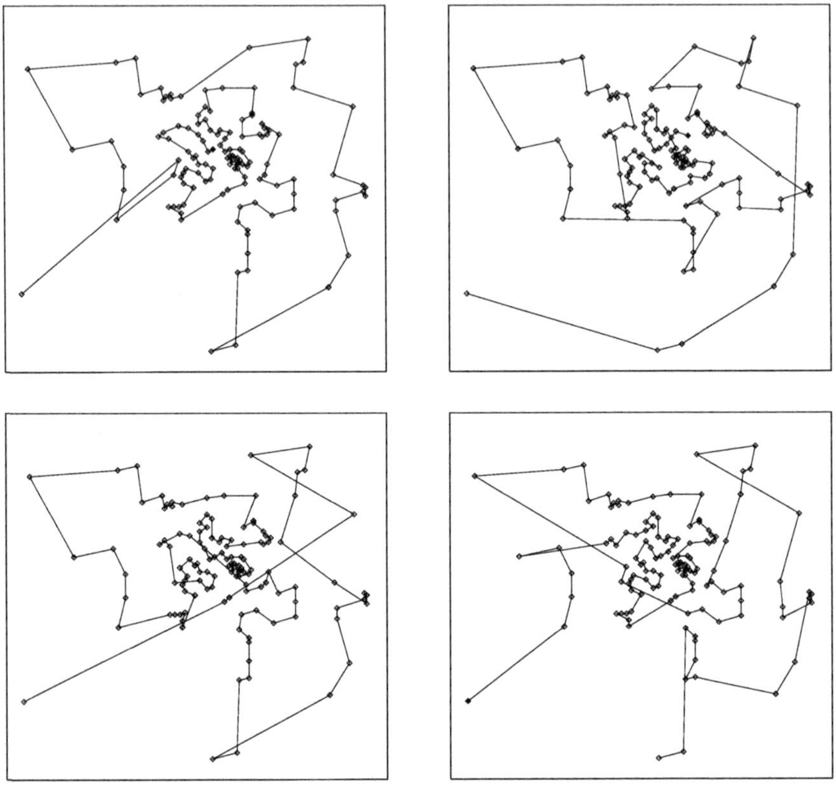

Fig. 3.1. Solutions of the BEER127 instance created by the nearest neighbor heuristic: a node in the center was used as the initial node for the heuristic in the first three solutions. For the *bottom right* solution, the common final node of the previous solutions was chosen as initial node. Note that the last line connecting the last node with the first node is omitted to clarify the direction of the appending steps of the algorithm

used. Even worse, the last remaining nodes can only be appended via rather long edges.

Table 3.1 shows the results for this heuristic for five benchmark instances. The results clearly show that this nearest neighbor heuristic has the disadvantage that the mean results are rather bad—20 to 30% worse than the optimum configuration. Furthermore, whether the result is, e. g., "only" 15% worse or even 35% worse depends on the starting node. As the graphics and also further investigations by us show, there is no general criterion for determining which node should be chosen as the starting node for the heuristic, e. g., there is no advantage from using a starting node in the center of the area in which the nodes lie.

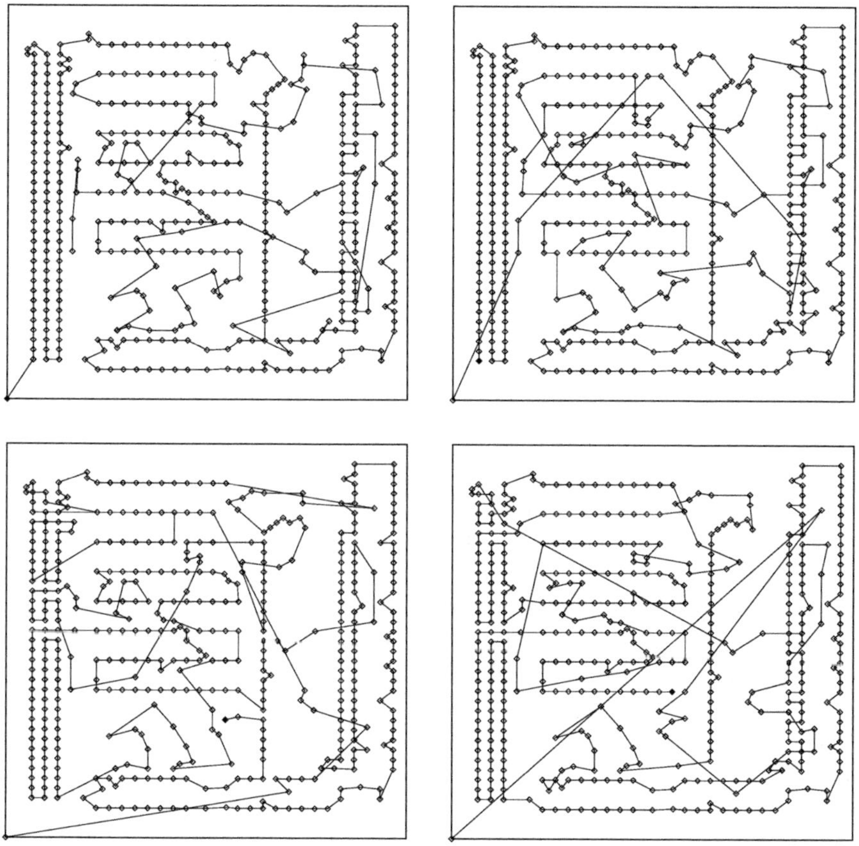

Fig. 3.2. Solutions of the PCB442 instance created by the nearest neighbor heuristic: a node at the *bottom left edge* serves as initial node for the solutions on *top*, whereas a node in the *lower middle part* was chosen for the solutions at *bottom*. As in Fig. 3.1, the last connection is omitted

Table 3.1. Results for the nearest neighbor heuristic: for each instance, each of the nodes was once chosen as starting node of the heuristic. The mean value denotes the average over N results

Instance	Minimum	Maximum	Mean value ± error
BEER127	133,970.646	158,890.152	146,597.222 ± 397
LIN318	49,215.6125	58,862.7628	52,599.1611 ± 70.3
PCB442	58,124.402	69,843.9295	62,703.2979 ± 90.1
ATT532	33,151	37,928	34,982.7105 ± 37.2
NRW1379	68,326.8723	72,379.6559	70,342.8998 ± 21.0

3.2 Insertion Heuristics

The nearest neighbor heuristic appends a new node at the end of the already existing sequence of nodes in each step of the algorithm. However, one can easily think of a variation improving this general approach to constructing a configuration. The improvement would be that a new node might not only be appended at the end but inserted somewhere inside the sequence. This ansatz opens many possibilities for insertion heuristics.

Again one starts with a single node as the starting point of the roundtrip. Of course, the individual nodes to be inserted will be inserted in a somehow optimal way. But two questions arise at this point:

- In which order should the individual nodes that are not yet part of the tour be inserted?
- Where is the optimum place to insert them?

Insertion heuristics differ in their answers to these questions.

A widely used insertion heuristic is the bestinsertion heuristic. Initially, it selects three nodes randomly. The triangle through these nodes serves as a starting tour of the algorithm. The other $N - 3$ nodes are put in a bag. A counter c is set to 3. Then some node o inside the bag is chosen randomly to be inserted into the tour. As the name of the heuristic implies, this insertion will be done in the best possible way. Thus, one calculates for each edge of the tour what it would cost to cut the edge and to insert the new node o between the end points of the edge:

$$\Delta_{i,o} = \begin{cases} D(\sigma(i), o) + D(o, \sigma(i+1)) - D(\sigma(i), \sigma(i+1)) & \text{for } 1 \leq i \leq c-1\,, \\ D(\sigma(c), o) + D(o, \sigma(1)) - D(\sigma(c), \sigma(1)) & \text{for } i = c\,. \end{cases} \tag{3.1}$$

The new node is then inserted where the insertion costs $\Delta_{i,o}$ are minimal. Thus let $\Delta_{j,o} = \min_i \Delta_{i,o}$; then o is inserted between $\sigma(j)$ and $\sigma(j+1)$ in the tour. The counter c is incremented by 1. This procedure is repeated until all nodes are inserted in the tour.

Figure 3.3 shows how a solution is built using the bestinsertion heuristic. One finds here, too, that the start and the intermediate part of the heuristic seem to perform rather well. However, looking at the final result one finds first of all that a solution could easily be improved locally in some areas. Furthermore, the random choice of the nodes to be inserted seems sometimes to guide the heuristic in a wrong way. The results for applying this heuristic to various TSP instances are composed in Table 3.2. Clearly, the bestinsertion heuristic leads to much better results than the nearest neighbor heuristic, which was only able to append a new node at the end of the existing sequence of nodes.

However, the question arises as to whether one could do it even better with this insertion type of heuristic. A point of criticism of the bestinsertion heuristic could be that the nodes to be inserted are selected in a random order.

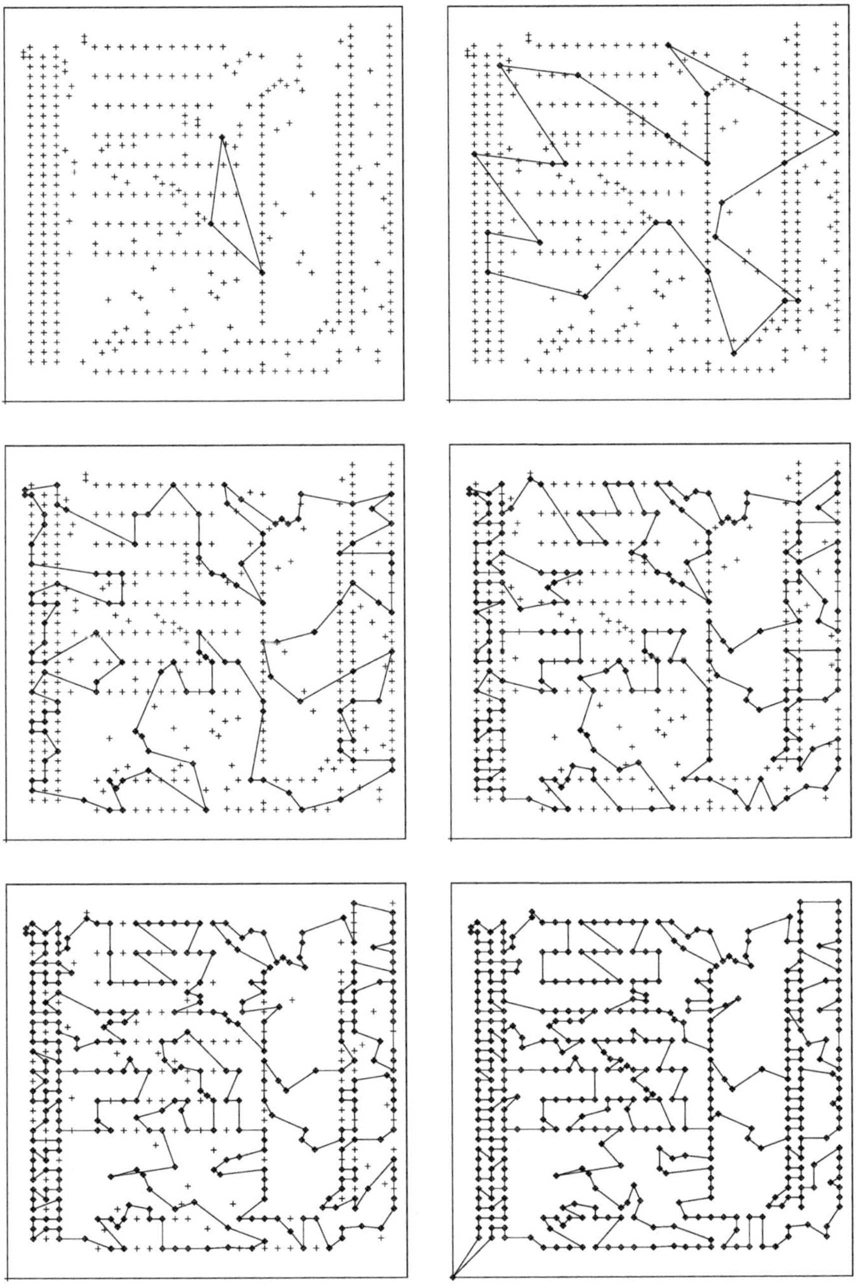

Fig. 3.3. Constructing a solution of the PCB442 instance using the bestinsertion heuristic: the six pictures show various stages of one application of this algorithm, from the initialization via intermediate stages to the final solution

Table 3.2. Results for the bestinsertion heuristic: for each instance, the bestinsertion heuristic was performed 100,000 times

Instance	Minimum	Maximum	Mean value ± error
BEER127	121,062.674	146,446.669	131,970.892 ± 10.2
LIN318	43,807.0156	49,595.9488	46,569.0740 ± 2.1
PCB442	54,796.1778	61,561.2966	58,153.4797 ± 2.5
ATT532	29,206	32,281	30,627.9032 ± 1.1
NRW1379	61,913.7975	65,399.0571	63,638.0352 ± 1.3

One might wonder whether it is not better to derandomize this probabilistic algorithm and to use an order constructed according to special criteria. A first thought would be that that node o should be inserted next in the tour for which the insertion costs $\Delta_{j,o}$ are minimal. Thus, the algorithm above is altered in the following way, leading to a new "best-bestinsertion heuristic" called the cheapest insertion heuristic [170]:

- Start with a randomly selected node $\tilde{o}$ as the first node of the tour, $\sigma(1) = \tilde{o}$. Put all nodes except $\tilde{o}$ in a bag.
- Select that node $\hat{o}$ as the second node in the tour for which the distance to $\tilde{o}$ is minimum, i. e.,

$$D(\hat{o}, \tilde{o}) = \min_{o} D(o, \tilde{o}) .$$

  Set $\sigma(2) = \hat{o}$, remove $\hat{o}$ from the bag, and set a counter $c = 2$.
- As long as $c < N$ do:
  - Determine the minimum insertion costs $\Delta_{j,o} = \min_i \Delta_{i,o}$ for each node o in the bag and memorize the edge position j where the best place to insert node o is.
  - Select that node $\hat{o}$ for which $\Delta_{j,\hat{o}}$ is minimal, insert it in the partial tour between $\sigma(j)$ and $\sigma(j+1)$, remove it from the bag, and increment the counter, $c = c + 1$.

One might guess that this cheapest insertion heuristic would lead to better results than the bestinsertion heuristic as the minimum of all minimum costs for inserting a further node is used. However, a comparison of Table 3.3 with

Table 3.3. Results for the cheapest insertion heuristic: for each instance, each of the nodes was once chosen as starting node of the heuristic. The mean value denotes the average over N results

Instance	Minimum	Maximum	Mean value ± error
BEER127	135,496.102	141,685.394	139,770.738 ± 93.7
LIN318	49,393.1998	50,808.5208	50,035.6234 ± 17.0
PCB442	57,415.3716	61,094.7875	59,353.3754 ± 55.3
ATT532	32,066	32,840	32,491.2481 ± 4.4
NRW1379	65,396.0661	66,478.0767	66,100.9075 ± 5.0

Table 3.2 shows just the opposite. The bestinsertion heuristic leads to clearly better results than the cheapest insertion heuristic. Obviously, this is the case not *although* but *because* the bestinsertion heuristic allows for random choices during its application.

Thus, it seems generally a good idea to insert new pieces into the system in a random and not in a predefined order. However, one can also think of many other rules defining an order for inserting the nodes (for a small overview see, e. g., [170]). All of these rules have in common that they determine the insertion order of the nodes according to some extremality conditions. One might select that node from the bag for which the minimum insertion costs are minimal or maximal, for which the maximum insertion costs are minimal or maximal, for which the minimum distance to one of the nodes in the partial tour is minimal or maximal, or for which the maximum distance to one of the nodes in the partial tour is minimal or maximal, and so on.

Table 3.4. Results for the farthest insertion heuristic: for each instance, each node was once chosen as starting node of the heuristic. The mean value denotes the average over N results

Instance	Minimum	Maximum	Mean value ± error
BEER127	124,000.033	140,965.911	129,789.501 ± 290
LIN318	45,065.7241	49,331.1769	46,633.8460 ± 38.1
PCB442	56,646.3008	61,013.0087	58,759.8852 ± 34.4
ATT532	29,918	32,024	30,986.2782 ± 17.0
NRW1379	62,862.6274	65,957.9057	64,214.9173 ± 11.9

Table 3.4 shows the results for the so-called farthest insertion heuristic. Here that node is selected among all nodes whose minimum insertion costs are maximal [170], i. e., one again determines for all nodes o and all possible insertion points i the insertion costs $\Delta_{i,o}$. Then one asks for $\Delta_{j,\hat{o}} = \max_o \min_i \Delta_{i,o}$ and inserts the node $\hat{o}$ by removing the jth edge. Comparing Table 3.4 with the results in Table 3.2, one finds that this heuristic leads for one of the considered instances to better results, but mostly to worse results than the bestinsertion heuristic.

Then we want to have a look at the type of solutions that are created by these heuristics. Figure 3.4 shows typical solutions of the BEER127 and the PCB442 instances produced by the bestinsertion, cheapest insertion, and farthest insertion heuristics. The starting node of the cheapest insertion and of the farthest insertion heuristics are specially marked with a filled dot. We find that the solutions of the bestinsertion heuristic and the farthest insertion heuristic look rather similar to each other. Furthermore, the solutions of the cheapest insertion heuristic look somewhat strange, but one can clearly see what solutions such a selection rule for the node to be inserted leads to. For all solutions, one finds that there are still crossings and other local things that could be improved.

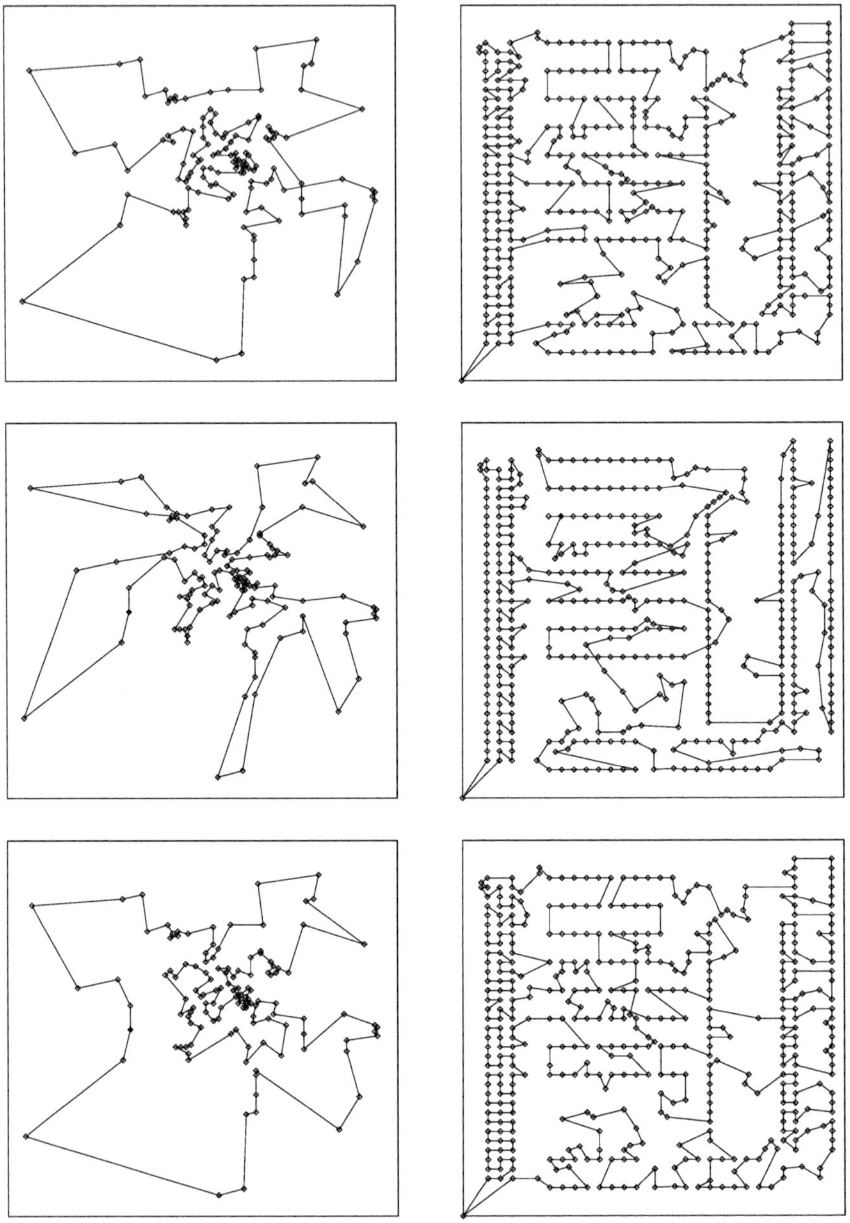

Fig. 3.4. Comparison of solutions produced with the insertion heuristics mentioned in the text. *Left column*: solutions for the BEER127 instance; *right column*: corresponding solutions for the PCB442 instance. The solutions in the *top row* were created with the bestinsertion heuristic, those in the *middle* with the cheapest insertion heuristic, and those at the *bottom row* with the farthest insertion heuristic

Summarizing, insertion heuristics lead very quickly to good, but not very good, solutions. Generally, working with this type of heuristic seems to be a very good way to randomly select the piece to be inserted in the system and to insert it in the best possible way. If one really wants or has to define a rule set governing the order in which individual pieces will be inserted in the system, then one should work with minimax conditions, like taking "the best of the worst" or "the worst of the best" item. Inserting "the worst of the worst" naturally cannot lead to good solutions; furthermore, inserting "the best of the best" has been shown not to be optimal in general. For more complex problems, one should not develop a stiff rule set for which item should be inserted and when and where to insert it, and one should not try to perform insertions purely at random. Instead, one must find some compromise between the paradigms of randomness and stiffness.

3.3 Using Deeper Insight into the Problem

The question arises as to whether one could make even better use of the insertion approach to heuristics. One might have additional knowledge of the problem one wants to use for such a construction heuristic in order to achieve better results.

We will now give an example of how such deeper insight into a given problem can actually lead to better results. There is considerable additional knowledge about the properties of the TSP as it has been studied exhaustively. For example, one might consider the convex hull (CH) of all nodes. The CH is the smallest convex area in which all nodes lie. It is limited by some outer nodes and the connecting lines between them. The order in which these nodes of the CH are arranged is the same order in which they lie in the optimum solution. As the insertion heuristics described above only insert additional nodes between other nodes in the tour, they do not change the order of these nodes. Thus, the CH can be considered a good starting point for insertion heuristics.

There are several ways to construct a CH [177]. An easy way to do this for a set of nodes in two dimensions is as follows:

- First determine the extreme values of the x- and y-coordinates of the nodes. Let $x_{\min}$, $x_{\max}$, $y_{\min}$, and $y_{\max}$ be these extreme values. The distances $D(i,j)$ between the single nodes shall be calculated according to the Euclidean metric for this algorithm here.
- Then select among the nodes with $y = y_{\min}$ the one with the largest x-coordinate. This node is part of the CH and is set as $\sigma(1)$. Set a counter $c = 1$.
- While $x(\sigma(c)) < x_{\max}$, calculate for each node o with an x-coordinate larger than $x(\sigma(c))$, i. e., $x(o) > x(\sigma(c))$, the angle between the connection line from $\sigma(c)$ to o and the x-axis. The node $\hat{o}$ with the smallest angle is

the next node in the convex hull. One need not calculate the angle $\alpha(o)$ explicitly but only its cosine. This cosine is given by

$$\cos(\alpha(o)) = \frac{\begin{pmatrix} x(o) - x(\sigma(c)) \\ y(o) - y(\sigma(c)) \end{pmatrix} \circ \begin{pmatrix} 1 \\ 0 \end{pmatrix}}{\left|\begin{pmatrix} x(o) - x(\sigma(c)) \\ y(o) - y(\sigma(c)) \end{pmatrix}\right| \times \left|\begin{pmatrix} 1 \\ 0 \end{pmatrix}\right|} = \frac{x(o) - x(\sigma(c))}{D(\sigma(c), o) \times 1} . \tag{3.2}$$

Then that $\hat{o}$ is selected for which $\cos(\alpha(o))$ is maximal. $\hat{o}$ becomes a part of the convex hull, and the counter c is thus incremented by 1 and $\sigma(c) = \hat{o}$. This step of the algorithm is repeated if $x(\sigma(c)) < x_{\max}$.

- Now the algorithm has reached the rightmost region of the nodes. One selects then from all nodes with $x(o) = x_{\max}$ the one with the largest y-coordinate. If this node is not identical to $\sigma(c)$, then the counter must be incremented again and the node inserted at the end of the convex hull.
- While $y(\sigma(c)) < y_{\max}$, calculate for each node o the angle $\alpha(o)$ to the y-axis, as above:

$$\cos(\alpha(o)) = \frac{\begin{pmatrix} x(o) - x(\sigma(c)) \\ y(o) - y(\sigma(c)) \end{pmatrix} \circ \begin{pmatrix} 0 \\ 1 \end{pmatrix}}{\left|\begin{pmatrix} x(o) - x(\sigma(c)) \\ y(o) - y(\sigma(c)) \end{pmatrix}\right| \times 1} = \frac{y(o) - y(\sigma(c))}{D(\sigma(c), o)} . \tag{3.3}$$

That node $\hat{o}$ for which this cosine is maximal is then appended to the convex hull; thus $c = c + 1$ and $\sigma(c) = \hat{o}$. This step of the algorithm must be repeated if $y(\sigma(c)) < y_{\max}$.

- As the y-coordinate of the last node of the CH is maximal, the algorithm has reached the topmost region of the problem. One searches through all nodes with $y(o) = y_{\max}$ for the one with the minimum x-coordinate. If this node $\hat{o}$ is not identical to $\sigma(c)$, then it must be appended to the CH in the described way.
- Now one must search for the node with the smallest angle to the $-x$-axis. Thus let $\hat{o}$ be the node with the maximum value for

$$\cos(\alpha(o)) = \frac{\begin{pmatrix} x(o) - x(\sigma(c)) \\ y(o) - y(\sigma(c)) \end{pmatrix} \circ \begin{pmatrix} -1 \\ 0 \end{pmatrix}}{\left|\begin{pmatrix} x(o) - x(\sigma(c)) \\ y(o) - y(\sigma(c)) \end{pmatrix}\right| \times 1} = \frac{x(\sigma(c)) - x(o)}{D(\sigma(c), o)} . \tag{3.4}$$

This node $\hat{o}$ is again appended to the CH. If its x-coordinate of the new $\sigma(c)$ is larger than $x_{\min}$, then this step of the algorithm must be repeated.

- Now the algorithm has reached the leftmost area. Among all nodes with $x(o) = x_{\min}$ the one with the minimum y-coordinate is selected. If this node $\hat{o}$ is not identical to $\sigma(c)$, then it must be added to the CH.

- As long as $y(\sigma(c)) > y_{\min}$, calculate for each node

$$\cos(\alpha(o)) = \frac{\begin{pmatrix} x(o) - x(\sigma(c)) \\ y(o) - y(\sigma(c)) \end{pmatrix} \circ \begin{pmatrix} 0 \\ -1 \end{pmatrix}}{\left| \begin{pmatrix} x(o) - x(\sigma(c)) \\ y(o) - y(\sigma(c)) \end{pmatrix} \right| \times 1} = \frac{y(\sigma(c)) - y(o)}{D(\sigma(c), o)} \tag{3.5}$$

  and select that node $\hat{o}$ for which this cosine is maximal. Then $\hat{o}$ is appended to the CH. Again, this step is repeated until $y(\sigma(c)) = y_{\min}$.
- Finally, the bottom region is reached again. If $\sigma(c) = \sigma(1)$, then the last node must be removed, as it appears both at the beginning and at the end of the CH. This can simply be done by decrementing c by 1.

Using this approach, there are sometimes too many nodes in the CH due to collinearities that occur when two or more nodes have the same cosine value. However, the CH is supposed to contain the minimum number of nodes necessary. Thus, one must remove nodes that are not needed.

Of course, there are also other and partially more elegant ways to construct the CH of a set of nodes in two dimensions. Some of them work with angular approaches like the algorithm above, others start with nonoverlapping CHs for two and three nodes and unite these CHs successively into one CH for all nodes [177].

The CH of the nodes is then used as the starting point of an insertion heuristic. Figure 3.5 shows an application of the bestinsertion heuristic starting with the CH as initialization. The pictures denoting various stages of the algorithm are really typical: one often finds that an insertion heuristic starting with the CH first inserts many inner nodes only between two neighboring nodes of the CH before further connections of the CH are destroyed. Table 3.5 presents the results of this approach. One finds that the results are mostly slightly better than using the pure bestinsertion heuristic, which starts from three randomly chosen nodes.

Of course, one can also start the cheapest insertion and the farthest insertion heuristics from the CH. Due to the deterministic order of the insertions of the nodes, the algorithm will end up at a unique solution. The results for these heuristics are presented in Table 3.6. Comparing the results for the cheapest insertion with the results for the pure cheapest insertion heuristic in Table 3.3, one finds that the results here are much better than the minimum values given in Table 3.3, such that the CH is a very good starting point for an altered cheapest insertion heuristic. However, this trend is not true for the results for farthest insertion, as a comparison with the results for the pure farthest insertion heuristic in Table 3.4 shows. Here the results are sometimes better and sometimes worse than the mean values in Table 3.4. One can also compare the results in Table 3.6 with those in Table 3.5 and finds that the cheapest and the farthest insertion heuristics lead to sometimes better and sometimes worse results than the corresponding mean value of the bestinsertion heuristic.

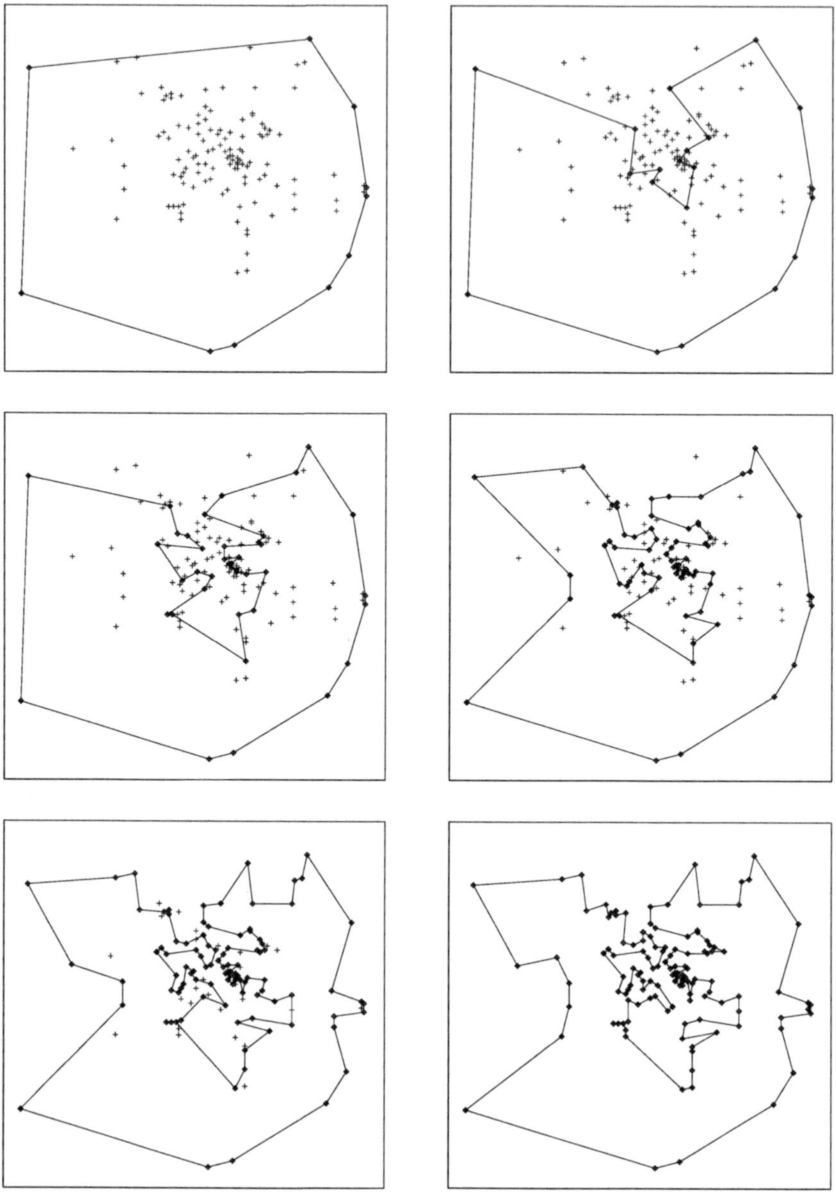

Fig. 3.5. Constructing a solution of the BEER127 instance, starting from the CH of the nodes and using the bestinsertion heuristic: the six pictures show various stages of one application of this algorithm, from the initialization via intermediate stages to the final solution

Table 3.5. Results for the bestinsertion heuristic starting with the CH of the nodes as initial tour: for each instance, the heuristic was performed 100,000 times

Instance	Minimum	Maximum	Mean value ± error
BEER127	120,013.136	141,878.386	128,210.423 ± 9.6
LIN318	43,826.3060	49,138.9202	46,314.9891 ± 2.1
PCB442	54,538.7924	61,634.0449	58,114.3733 ± 2.5
ATT532	29,258	32,171	30,555.1791 ± 1.1
NRW1379	61,921.4855	65,580.9828	63,745.1455 ± 1.3

Table 3.6. Results for the cheapest insertion and the farthest insertion heuristics starting with the CH of the nodes as initial tour: of course, there is only one solution for each instance

Instance	Cheapest Insertion	Farthest Insertion
BEER127	124,830.995	124,789.471
LIN318	46,903.7692	46,857.7320
PCB442	56,588.8166	59,682.3358
ATT532	30,929	30,835
NRW1379	64,924.9067	63,540.6292

Summarizing, one can conclude that one might find better initialization routines for some insertion heuristic if one has deeper insight into the problem. However, often this additional work might be useless, as the gain depends on the combination of a good initialization and an appropriate insertion heuristic.

3.4 The Savings Heuristic

All construction heuristics introduced above have one thing in common: the item to be inserted is inserted in the best possible way. Thus, each item is somehow egoistic as it looks out only for its own profit, i. e., its own insertion costs, which will be optimal. However, due to this egoistic approach, synergy effects might be lost, such that one looks for another construction heuristic that considers all nodes somehow in parallel.

An example for such a heuristic is the savings algorithm, which was originally developed for the vehicle routing problem (VRP) but is also applied to the TSP in order to obtain rather fast quite good solutions. The TSP is considered to be a VRP and solved in the same way as a corresponding VRP instance. For the TSP case, one starts by randomly selecting one node and calling it "the depot d." (For a VRP, the depot is known in advance, of course.) The algorithm proceeds as shown in Fig. 3.6. $N-1$ traveling salesmen start from depot d to one of the other $N-1$ nodes and return to the depot. This is the initialization routine of the savings heuristic. One now has a tour plan σ with $\sigma(i,j)$ denoting the ith node in the tour of the jth salesman. Thus, intially each column contains one entry, with $\sigma(1,j)$ being the visited customer. One can omit depot d at the beginning and at the end of each tour, but of course the length of the tour of the jth salesman is given by

$$\mathcal{H}_j(\sigma) = D(d,\sigma(1,j)) + \sum_{i=1}^{N_j-1} D(\sigma(i,j),\sigma(i+1,j)) + D(\sigma(N_j,j),d)\,, \quad (3.6)$$

with N_j being the number of customers in tour j. The overall length is given by the sum of these tour lengths. Thus, the TSP is transformed into a multiple traveling salesman problem with $N-1$ salesmen. The task is now to minimize the number of salesmen needed, such that a minimum number of salesmen remains at the end. In the case of the TSP, this number is 1; for a VRP the number might be larger because of the capacity constraint.

This minimization process is performed by uniting the tours of two traveling salesmen into the tour of one traveling salesman by removing the depot from the end of one tour and the beginning of another tour. Thus, one calculates for each pair of tours (a,b) the savings that can be achieved if the tours are united into one tour. Let N_a be the number of nodes in the tour of traveling salesman a, not including the depot at the beginning and at the end, and analogously let N_b be the number of nodes in tour b. In the case of the symmetric TSP, there are four ways to unite the two tours, giving four savings values for uniting the tours:

$$\begin{aligned}
S_1(a,b) &= D(\sigma(N_a,a),d) + D(d,\sigma(1,b)) - D(\sigma(N_a,a),\sigma(1,b))\,,\\
S_2(a,b) &= D(\sigma(1,a),d) + D(d,\sigma(1,b)) - D(\sigma(1,a),\sigma(1,b))\,,\\
S_3(a,b) &= D(\sigma(N_b,b),d) + D(d,\sigma(1,a)) - D(\sigma(N_b,b),\sigma(1,a))\,,\\
S_4(a,b) &= D(\sigma(N_a,a),d) + D(d,\sigma(N_b,b)) - D(\sigma(N_a,a),\sigma(N_b,b))\,.
\end{aligned} \quad (3.7)$$

Then the savings for uniting tours a and b is given by

$$S(a,b) = \max\left\{S_1(a,b),S_2(a,b),S_3(a,b),S_4(a,b)\right\}\,. \quad (3.8)$$

After the savings $S(a,b)$ has been determined for each pair (a,b) of tours, those tours $\tilde{a}$ and $\tilde{b}$ are united for which this savings is maximal:

- If $S_1(\tilde{a},\tilde{b})$ is maximal, then tour $\tilde{b}$ is appended to the end of tour $\tilde{a}$: thus, one gets one sequence $\sigma(1,j),\sigma(2,j),\ldots,\sigma(N_j,j)$, with $\sigma(1,j) = \sigma(1,\tilde{a})$, $\sigma(2,j) = \sigma(2,\tilde{a})$, ..., $\sigma(N_{\tilde{a}},j) = \sigma(N_{\tilde{a}},\tilde{a})$, $\sigma(N_{\tilde{a}}+1,j) = \sigma(1,\tilde{b})$, $\sigma(N_{\tilde{a}}+2,j) = \sigma(2,\tilde{b})$, ..., $\sigma(N_j,j) = \sigma(N_{\tilde{b}},\tilde{b})$, with $N_j = N_{\tilde{a}} + N_{\tilde{b}}$.
- If, however, $S_2(\tilde{a},\tilde{b})$ is maximal, then the direction of tour $\tilde{a}$ is first changed, such that the first customer becomes the last and so on, and after that tour $\tilde{b}$ is appended to the end of tour $\tilde{a}$.
- If $S_3(\tilde{a},\tilde{b})$ is maximal, then tour $\tilde{a}$ is appended to the end of tour $\tilde{b}$.
- If, finally, $S_4(\tilde{a},\tilde{b})$ is maximal, then the direction of tour $\tilde{b}$ must be changed before it is appended to the end of tour $\tilde{a}$.

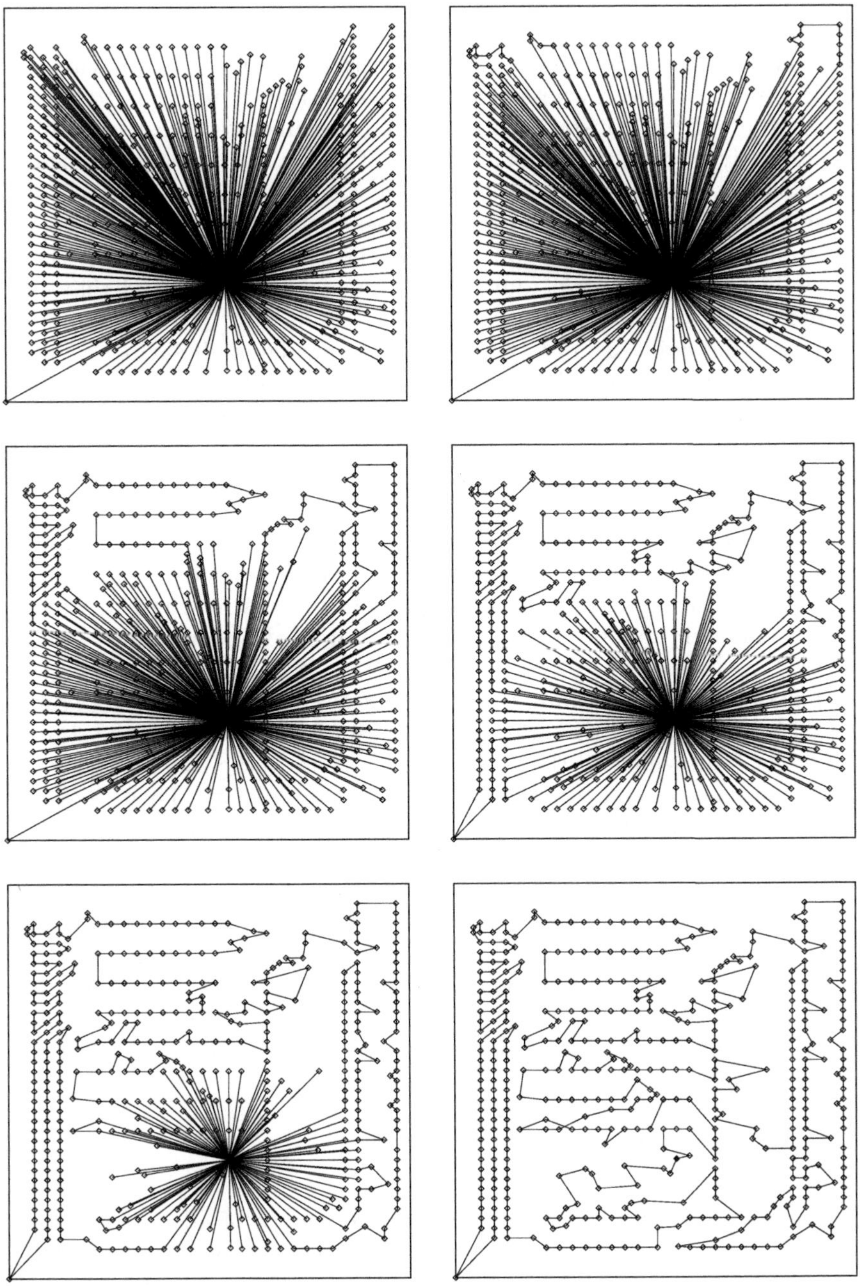

Fig. 3.6. Constructing a solution of the PCB442 instance, using the savings heuristic: the six pictures show various stages of one application of this algorithm, from the initialization via intermediate stages to the final solution

From these descriptions above it follows that the situation is not as easy if the distance matrix is not symmetric. For the ATSP, one must consider the difference in length for changing the direction of a tour. In the case of the VRP, one must check whether a truck is overloaded if it has to serve the united tour. If this constraint is violated, the unification must be forbidden. Analogously, two tours cannot be united if hard time windows of customers are violated after the unification. If such a unification is forbidden due to these additional constraints, then one goes through the list of savings that has to be ordered according to the size of the savings and checks for the next possibility for uniting two tours until a pair of tours is found that can be united.

After this unification of two tours, one calculates all savings values between the united tour and the other tours. One still knows the other savings values from the calculations before. Then one again orders the savings values according to their size and unites those two tours with the largest savings value, if this is allowed.

This procedure is iterated again and again. In the case of the TSP, only one tour remains at the end. In the case of the VRP, it becomes impossible at some step to unite any pair of tours due to the capacity constraints such that the algorithm stops with a number of trucks, which might sometimes be the minimum number of trucks actually needed but usually it is slightly larger.

Looking at Fig. 3.6 again, one sees what actually happens if applying this heuristic: one starts with a starlike structure with the center of the star being the depot from which all the rays originate that are the tours starting at the depot, serving one customer, and ending at the depot. Then neighboring rays with the largest lengths are united such that the radius of the starlike structure around the depot decreases. Automatically, the question arises as to where to put the center of this star, i. e., whether it is better to place the depot in the center or at one of the edges. Of course, one should not consider the geographical center of the area in which the nodes lie as most of the nodes might be clustered in some small region inside the overall area. Thus, the "center of mass" is a preferable measure for the center of the points. The coordinates X_{CM} and Y_{CM} of the center of mass point P_{CM} are simply calculated by

$$X_{\mathrm{CM}} = \frac{1}{N} \sum_{i=1}^{N} x_i \text{ and } Y_{\mathrm{CM}} = \frac{1}{N} \sum_{i=1}^{N} y_i \,. \tag{3.9}$$

Figure 3.7 shows two diagrams in which the length of all possible final solutions is plotted vs. the distance $D(d, P_{\mathrm{CM}})$ between depot d and P_{CM}: for the BEER127 instance, the depot should not be placed in one of the beer gardens far out in the suburbs. Instead, one of the beer gardens rather near the center of mass should be chosen. For the PCB442 problem, the results are just the opposite. Here one gets on average better results if one chooses the

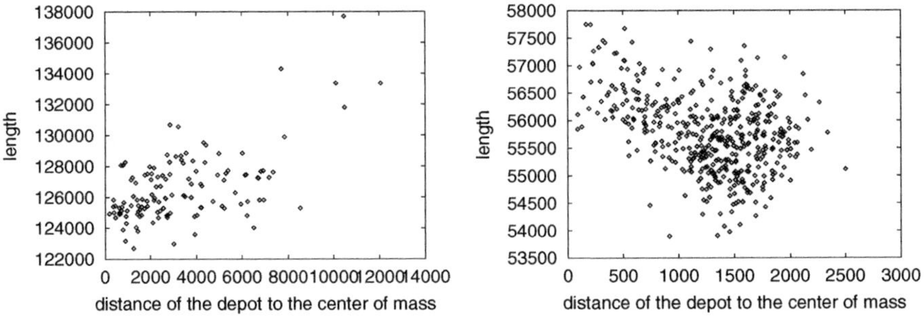

Fig. 3.7. Analysis of the savings heuristic: each *point* stands for one of the N possible applications of the savings algorithm. The length of the final configuration is plotted vs. the distance between the depot node and the center of mass. The data shown on the *left* belong to the BEER127 instance; the data on the *right* to the PCB442 instance

Table 3.7. Results for the savings heuristic: for each instance, each of the nodes was once chosen as the "depot." The mean value denotes the average over N results

Instance	Minimum	Maximum	Mean value ± error
BEER127	122689.400	137693.407	126611.232 ± 198
LIN318	44417.8099	47600.2678	45764.5415 ± 37.6
PCB442	53898.6223	57750.9015	55766.8600 ± 33.2
ATT532	29528	31297	30320.3609 ± 13.7
NRW1379	61167.6236	63451.7975	62397.4092 ± 9.3

location of the depot further out from the center. Thus, there is no general criterion where to optimally place the depot.

Table 3.7 presents all results for the savings algorithm. Comparing them with the results for the insertion type of heuristics, the savings algorithm leads to better results on average. This surely has to do with the integrated view of the savings algorithm. Of course, the savings algorithm also has its weak points, as, e. g., two tours that were already connected cannot be split again; furthermore one tour cannot be inserted but only appended to the end of another tour.

A heuristic that does exactly this and that is related to this savings heuristic is the nearest merger heuristic. In contrast to the savings heuristic, the nearest merger heuristic does not choose a specific node to be a depot. Instead, it initializes the system with N one-node tours. These tours are gradually united. The outline of the algorithm is as follows:

- The heuristic starts with N tours. Each of them contains one of the N nodes such that each node is in one tour.
- While the number of tours is larger than 1, do:

- Select those two tours $\sigma(.,k)$ and $\sigma(.,l)$ with the smallest distance to each other. This distance is given by

$$\min_{\substack{1\le i\le N_k\\1\le j\le N_l}} \{D(\sigma(i,k),\sigma(j,l)\}\ ,$$

with N_k being the number of nodes in tour k and N_l the number of nodes in tour l.
- The tours are united in the cheapest possible way: let i_+ be the successor of tour position i in tour $\sigma(.,k)$ and j_+ the successor of j in $\sigma(.,l)$. Then for all pairs (i,j) one calculates the savings values

$$\begin{aligned}S_1 = {} & D(\sigma(i,k),\sigma(j,l)) + D(\sigma(i_+,k),\sigma(j_+,l))\\ & - D(\sigma(i,k),\sigma(i_+,k)) - D(\sigma(j,l),\sigma(j_+,l))\end{aligned}$$

and

$$\begin{aligned}S_2 = {} & D(\sigma(i,k),\sigma(j_+,l)) + D(\sigma(i_+,k),\sigma(j,l))\\ & - D(\sigma(i,k),\sigma(i_+,k)) - D(\sigma(j,l),\sigma(j_+,l)).\end{aligned}$$

For the pair (i,j) with the minimum savings value, one cuts the edges between $\sigma(i,k)$ and $\sigma(i_+,k)$ and between $\sigma(j,l)$ and $\sigma(j_+,l)$ and forms two new edges: if $S_1 < S_2$, then edges between $\sigma(i,k)$ and $\sigma(j,l)$ and between $\sigma(i_+,k)$ and $\sigma(j_+,l)$ are created. Otherwise, edges between $\sigma(i,k)$ and $\sigma(j_+,l)$ and between $\sigma(i_+,k)$ and $\sigma(j,l)$ are created. Thus, the two possibilities for choosing one of the two directions of the tour to be inserted are considered.
- If the number of tours is larger than 1, then one iterates this procedure again.

Of course, the formulas for the savings values above cover only the general case in which both tours consist of at least three nodes. Otherwise, one must distinguish between one and two nodes for each tour. For each of these cases, one can use even simpler formulas.

Table 3.8 shows the results for applying the nearest merger heuristic to the five considered instances. One finds that the results are very bad, although the heuristic provides more possibilities to connect two tours than the Savings heuristic, which only allows one to append a tour at the end of another tour.

Table 3.8. Results for the nearest merger heuristic: of course, there is only one solution for each instance

Instance	Result
BEER127	141,316.469
LIN318	52,203.8594
PCB442	60,512.9570
ATT532	32,979
NRW1379	66,065.2031

However, due to the depot approach, the savings heuristic makes intelligent use of the angular distribution of the nodes around the depot. This effect is missing here, so that one ends up with far worse results, although the nearest merger heuristic is related to the savings heuristic in its basic philosophy.

As a next step, one might think that a randomization of the algorithm might lead to better results, as, e. g., the bestinsertion heuristic provides better results than the cheapest insertion heuristic. One can analogously define a best merger heuristic in which the tours to be united are chosen at random and the edges to be removed are determined as in the nearest merger heuristic.

Table 3.9. Results for the best merger heuristic: for each instance, 100,000 optimization runs were performed

Instance	Minimum	Maximum	Mean value ± error
BEER127	297,419.569	401,973.722	354,925.723 ± 38.6
LIN318	257,606.980	321,904.772	287,828.522 ± 22.4
PCB442	344,280.234	418,127.578	380,302.261 ± 24.9
ATT532	210,753	251,342	230,491.492 ± 15.4
NRW1379	638,256.993	715,736.532	673,542.708 ± 25.4

Table 3.9 shows the results for this approach, which are all very bad. Here the random approach leads to very long edges that cannot be removed anymore in the subsequent optimization process. Thus, not in every case does a randomization of an algorithm lead to an improvement over the stiff version of the algorithm. One should always first consider the effects of this randomization during the optimization process. In this case, one can even easily predict that the randomized heuristics will not lead to good results if one performs a few steps of this method on a small instance manually.

Generally, construction heuristics can lead very quickly to quite good results that are roughly at least 10% worse than the optimum. They are therefore used in exactly these applications for which a solution has to be found nearly immediately "after pressing the button" and where a loss of roughly 10%, if compared to the unknown optimum, does not hurt. When one wants to reduce this loss to 1% or even 0.1%, more elaborate algorithms must be used. This gain must be paid for with additional calculation time.

4 Local Search Concepts Applied to TSP

4.1 Initialization Routine

In the last chapter, several examples for construction heuristics leading to quite good solutions were given. But there is also another approach that involves starting at some solution and improving this solution step by step until a rather good solution is obtained.

These improvement heuristics usually start with a randomly chosen tour, as shown in Fig. 4.1 for the traveling salesman problem (TSP) instances BEER127 and PCB442. The "construction" of such a random initial configuration is performed as follows:

- Put all nodes of the TSP instance in an unordered bag and set a counter $c = 0$.
- While $c < N$,
  - Increment c by 1,
  - Select one node o from the bag randomly,
  - Set $\sigma(c) = o$,
  - Remove o from the bag.

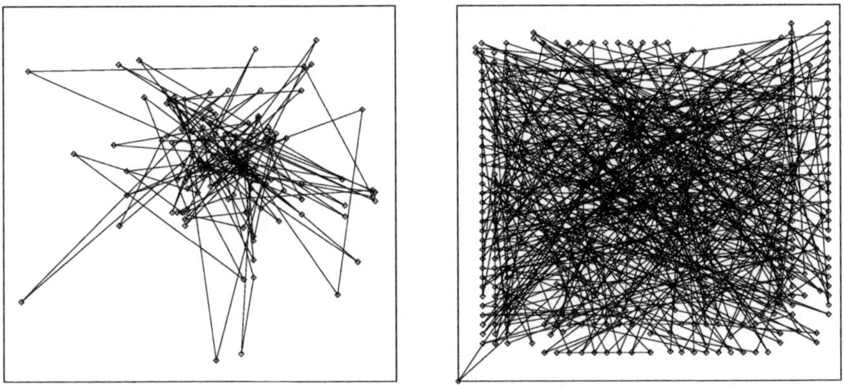

Fig. 4.1. Random configurations of the BEER127 instance (*left*) and the PCB442 instance (*right*)

Each of the $N!$ possible orders of nodes is achieved with equal probability. For the symmetric TSP, the starting node and the direction of the tour does not matter, such that there are $(N-1)!/2$ configurations of the symmetric TSP. For the asymmetric TSP (ATSP), one has $(N-1)!$ configurations.

The average length of all configurations can be computed easily: as the nodes are chosen in a random order, each edge has the same probability to be part of the tour. The mean edge length $\bar{D}$ is given as the average over the lengths of all edges that can occur in a TSP configuration:

$$\bar{D} = \frac{1}{N(N-1)} \sum_{\substack{i,j=1 \\ i \neq j}}^{N} D(i,j)\,. \tag{4.1}$$

(Note that the diagonal elements of the distance matrix never occur as edge lengths.) Thus the average length of a randomly chosen configuration is given by

$$\langle \mathcal{H} \rangle_\infty = N \times \bar{D} = \frac{1}{N-1} \sum_{\substack{i,j=1 \\ i \neq j}}^{N} D(i,j)\,. \tag{4.2}$$

Table 4.1 denotes the results averaged over 100,000 random configurations for five TSP instances. One finds that the mean value of the lengths fluctuates around its expectation value as expected. However, the best solution found is much worse than any solution created by one of the construction heuristics presented in the last chapter. Thus, one cannot hope to find a quite good solution by simply creating configurations at random. There are exponentially more configurations in the energy regime around $\langle \mathcal{H} \rangle_\infty$ than in the regime of the rather good solutions, such that the probability of creating a rather good solution vanishes.

However, a random configuration usually serves as a starting point for improvement heuristics. The task is now to transfer the system from the high-energy unordered regime to a low-energy ordered solution.

Table 4.1. Randomly created solutions: for each instance, the routine for creating an arbitrary configuration was performed 100,000 times. In addition to the results of these runs, the expectation value $\langle \mathcal{H} \rangle_\infty$ for each instance is given

Instance	$\langle \mathcal{H} \rangle_\infty$	Minimum	Maximum	Mean value ± error
BEER127	628,964.464	522,048.692	704,879.121	628,871.220 ± 62.0
LIN318	587,996.659	529,494.856	648,422.848	587,967.946 ± 44.7
PCB442	772,614.928	696,210.110	841,405.018	772,583.835 ± 48.6
ATT532	512,114.712	466,452	556,684	512,161.912 ± 33.5
NRW1379	1,423,598.22	1,337,635.44	1,496,604.12	1,423,553.33 ± 56.0

4.2 Small Moves

For all improvement heuristics , the philosophy is as follows: starting from an arbitrary configuration, one tries to improve this configuration step by step until no further improvement is found. One defines a routine called a move by which a configuration is changed, i. e., a move is performed from a configuration σ to a configuration τ. The question arises as to what such a move would look like. In the concept of local search, a move will only change a configuration slightly, so that one searches somehow in the neighborhood of the current configuration for a better solution.

Thus, the task is now to find small moves that change a TSP configuration only slightly. An obvious move is exchanging two nodes of the tour, as shown in Fig. 4.2. This move is mostly called exchange (EXC), but sometimes also swap, transposition, or or-opt. (Note that in the business of vehicle routing, there is another move sometimes called exchange. This move exchanges two nodes of different tours. Thus, one always has to ask what move is meant.) The procedure is as follows:

- Choose two tour positions i and j with $1 \leq i, j \leq N$ and $i \neq j$ at random. The nodes $\sigma(i)$ and $\sigma(j)$ are to be exchanged.
- Let i_+ be the tour position after tour position i, i. e.,

$$i_+ = \begin{cases} i+1 & \text{if } 1 \leq i \leq N-1 \\ 1 & \text{if } i = N \end{cases} ,$$

  and let i_- be the tour position before i, i. e.,

$$i_- = \begin{cases} i-1 & \text{if } 2 \leq i \leq N \\ N & \text{if } i = 1 \end{cases} .$$

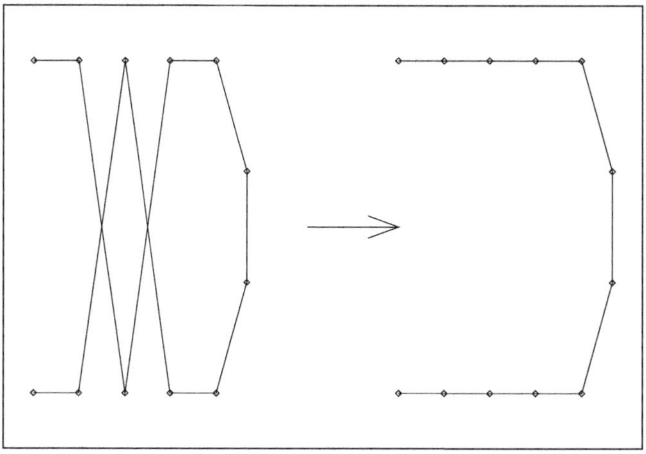

Fig. 4.2. The exchange (EXC)

Analogously, let j_+ be the tour position after j and j_- be the tour position before j.

- Applying the move routine means jumping from the current configuration σ to a new configuration τ. In most acceptance criteria for a move, the energy difference $\Delta\mathcal{H} = \mathcal{H}(\tau) - \mathcal{H}(\sigma)$ is considered. One need not sum up all edge lengths of the configurations σ and τ; instead, one must consider the lengths of the differing edges only:

$$\Delta\mathcal{H} = \begin{cases} \begin{array}{l} \, D(\sigma(i_-),\sigma(j)) + D(\sigma(i),\sigma(j_+)) \\ -\, D(\sigma(i_-),\sigma(i)) - D(\sigma(j),\sigma(j_+)) \end{array} & \text{if } i_+ = j\,, \\ \\ \begin{array}{l} \, D(\sigma(j_-),\sigma(i)) + D(\sigma(j),\sigma(i_+)) \\ -\, D(\sigma(j_-),\sigma(j)) - D(\sigma(i),\sigma(i_+)) \end{array} & \text{if } j_+ = i\,, \\ \\ \begin{array}{l} \, D(\sigma(i_-),\sigma(j)) + D(\sigma(j),\sigma(i_+)) \\ +\, D(\sigma(j_-),\sigma(i)) + D(\sigma(i),\sigma(j_+)) \\ -\, D(\sigma(i_-),\sigma(i)) - D(\sigma(i),\sigma(i_+)) \\ -\, D(\sigma(j_-),\sigma(j)) - D(\sigma(j),\sigma(j_+)) \end{array} & \text{otherwise}\,. \end{cases} \tag{4.3}$$

  Thus, one can easily calculate the energy difference without applying the move. This is also true for the other moves below.
- After the energy difference has been calculated, it is checked whether the move shall be accepted according to the acceptance criterion of the underlying algorithm:
  - If it is accepted, then $\sigma(i)$ and $\sigma(j)$ are exchanged.
  - If it is rejected, then one simply stays with the configuration σ.

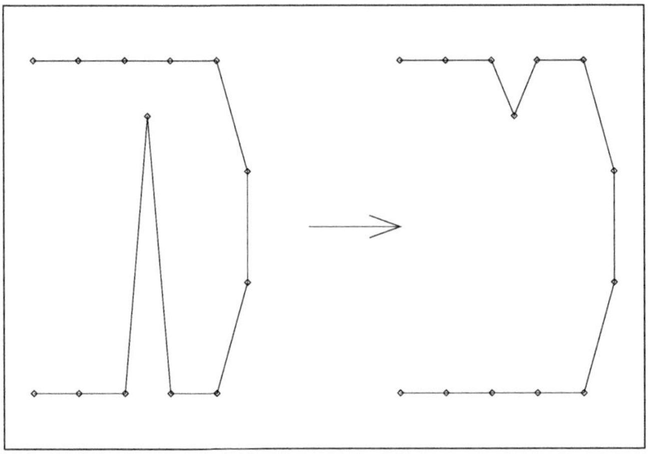

Fig. 4.3. The node insertion move (NIM)

There are also other possibilities for changing the configuration σ slightly. For example, one could shift one node to another position in the tour, as shown in Fig. 4.3. This move is usually called node insertion move (NIM). The procedure for applying the NIM is rather the same as applying the EXC:

- Randomly select a tour position i with $1 \leq i \leq N$ and calculate i_+ as above. The edge between $\sigma(i)$ and $\sigma(i_+)$ shall be removed.
- Randomly select a tour position j with $1 \leq j \leq N$, $j \neq i$, and $j \neq i_+$. $\sigma(j)$ is the node to be inserted between $\sigma(i)$ and $\sigma(i_+)$.
- Determine j_- and j_+ as above.
- The energy difference is given as follows:

$$\Delta\mathcal{H} = \begin{cases} \begin{array}{l} D(\sigma(i),\sigma(j)) + D(\sigma(i_+),\sigma(j_+)) \\ - D(\sigma(i),\sigma(i_+)) - D(\sigma(j),\sigma(j_+)) \end{array} & \text{if } j_- = i_+ \,, \\ \\ \begin{array}{l} D(\sigma(j_-),\sigma(i)) + D(\sigma(j),\sigma(i_+)) \\ - D(\sigma(j_-),\sigma(j)) - D(\sigma(i),\sigma(i_+)) \end{array} & \text{if } j_+ = i \,, \\ \\ \begin{array}{l} D(\sigma(i),\sigma(j)) + D(\sigma(j),\sigma(i_+)) \\ + D(\sigma(j_-),\sigma(j_+)) - D(\sigma(i),\sigma(i_+)) \\ - D(\sigma(j_-),\sigma(j)) - D(\sigma(j),\sigma(j_+)) \end{array} & \text{otherwise}. \end{cases} \tag{4.4}$$

- If the acceptance criterion is fulfilled, then the move has to be applied. In the first two cases, this is easy: $\sigma(j_-)$ has to be exchanged with $\sigma(j)$ in the case $j_- = i_+$; analogously, $\sigma(i)$ is exchanged with $\sigma(j)$ in the case $j_+ = i$. In the other general case, one first stores $\sigma(j)$ in an auxiliary variable h and then distinguishes between $i < j$ and $i > j$: if $i < j$, then one must shift $\sigma(j_-)$, $\sigma(j_- - 1)$, ..., $\sigma(i_+)$ toward the position of the successor, i. e., $\sigma(j_-) \to \sigma(j)$, $\sigma(j_- - 1) \to \sigma(j_-)$, ..., $\sigma(i_+) \to \sigma(i_+ + 1)$, and then insert the node $\sigma(i_+) = h$. In the case $i > j$, one shifts $\sigma(j_+)$, ..., $\sigma(i)$ toward the position of the predecessor, i. e., $\sigma(j_+) \to \sigma(j)$, ..., $\sigma(i) \to \sigma(i-1)$, and then inserts the node $\sigma(i) = h$.
  This description holds for the implementation of a TSP configuration as a 1D array σ, with $\sigma(i)$ being the node at tour position i. However, one can also implement a TSP configuration as a connected list in which there is a pointer from each node to its successor. In this implementation, the update routine for the general case is much easier, as only three pointers have to be updated. One might conclude that one should prefer this implementation as it requires much less computation time for the update of the configuration. However, more computing time is necessary until the energy difference is calculated. Furthermore, except for the random walk (RW), not every move is accepted. Thus, one must check with the acceptance criterion used to see which of these implementations is faster.

A further small move was introduced by Lin [129]. The so-called Lin-2-opt (L2O), which is shown in Fig. 4.4 removes two nonneighboring edges of

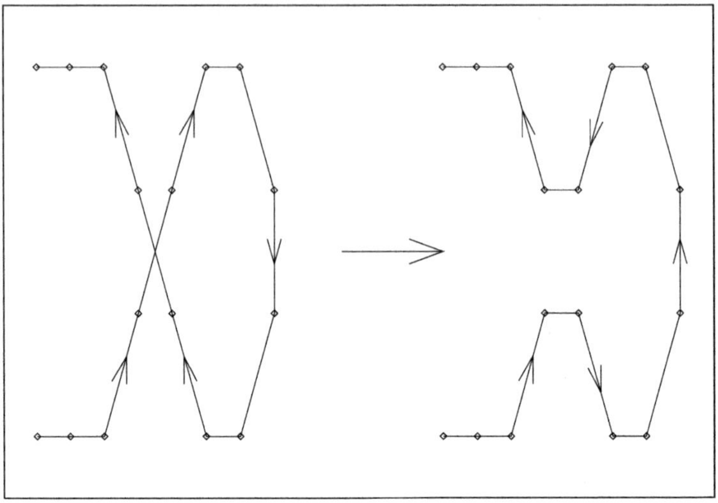

Fig. 4.4. The Lin-2-opt (L2O)

a configuration, changes the direction of one part of the tour, and inserts two new edges connecting the two parts again. The procedure for an array implementation of a configuration of a symmetric TSP is as follows:

- Randomly select a tour position i with $1 \leq i \leq N$.
- Randomly select a tour position j with $1 \leq j \leq N$ and $j \neq i$.
- Sort i and j such that $i < j$.
- Define i_+ and j_+ analogously to the above.
- The cases $i_+ = j$ and $j_+ = i$ are not allowed. In these cases, restart from the very beginning.
- The edges to be removed shall be $(\sigma(i), \sigma(i_+))$ and $(\sigma(j), \sigma(j_+))$. The energy difference is thus given as

$$\Delta\mathcal{H} = D(\sigma(i), \sigma(j)) + D(\sigma(i_+), \sigma(j_+)) - D(\sigma(i), \sigma(i_+)) - D(\sigma(j), \sigma(j_+)) \,. \tag{4.5}$$

- If the acceptance criterion is met, then the part between $\sigma(i)$ and $\sigma(j_+)$ must be turned around. This is done by exchanging pairs of nodes, i. e., $\sigma(i_+) \leftrightarrow \sigma(j)$, $\sigma(i_+ + 1) \leftrightarrow \sigma(j-1)$, …, till $\sigma((i+j)/2) \leftrightarrow \sigma((i+j)/2+1)$ if $i+j$ is even or $\sigma([(i+j)/2]) \leftrightarrow \sigma([(i+j)/2]+2)$ if $i+j$ is odd. ($[x]$ again denotes the Gaussian brackets that take the integer part of x.)

This description must be changed for the ATSP, as one must consider that the length of the tour part that is turned around changes. Furthermore, only in the case of the symmetric TSP is the turning around of one of the two parts arbitrary. Thus, for the ATSP one may not order i and j according to their size, which was done here in order to simplify the update routine.

Thus, there are three small moves for the TSP. A comparison of these moves leads to the following insight:

- These are the smallest possible moves for the TSP, as for each move only two tour positions are selected. It is not possible to perform a tour change if changing something by choosing only one tour position if one is dealing with the symmetric TSP. (For the ATSP, one could turn the whole tour around.)
- The L2O cuts two edges, the NIM three edges, and the EXC four edges.
- The NIM moves one node, the EXC two nodes, and the L2O $|j - i|$ nodes.
- The number of configurations τ that can be reached from one configuration σ is of the order $\mathcal{O}(N^2)$ for all moves: in the case of the EXC, one has exactly $\binom{N}{2}$, in the case of the L2O only $N \times (N-3)/2$ neighbors. Thus, the so-called neighborhood size of all three moves is virtually the same.

The question is now how well these three moves perform.

4.3 Computational Results for Greedy Algorithm

First, one must consider the case that one starts with a random configuration. The task is to end up at a quite good solution. Of course, it makes no sense to apply the moves mentioned above in the RW scenario. As every move is accepted in the RW mode, one would change the configurations at random, thus ending up with random configurations, which are energetically much worse than the optimum solution, as seen above. Thus, another simple acceptance criterion must be found, highlighting the effectiveness of the moves themselves. The simplest criterion provides the greedy algorithm, which rejects any deterioration and accepts all moves with $\Delta\mathcal{H} \leq 0$. Applying this greedy algorithm, the system will freeze after the application of several moves in some locally minimum configuration.

The number of moves needed until the system freezes is proportional to N^2, as the neighborhood size is of order $\mathcal{O}(N^2)$ for every configuration. A few test runs showed that after $1 - 12 \times N^2$ move trials the system was frozen for all three move routines. In order to be on the safe side, we started each simulation with a randomly generated configuration and performed $50 \times N^2$ moves.

The first question of interest is what the final configurations look like for the different types of moves. Figure 4.5 shows final configurations of the BEER127 and the PCB442 instances, which are metastable to either the EXC or the NIM or the L2O. That the configuration is indeed metastable means that there cannot be found a subsequent move of the corresponding type that could improve the solution. This can especially be nicely seen in the case of the L2O, as the configurations no longer contain any crossings: each application of a L2O adds or removes one crossing. Removing a crossing means decreasing the length of the configuration. Thus, if no crossings are

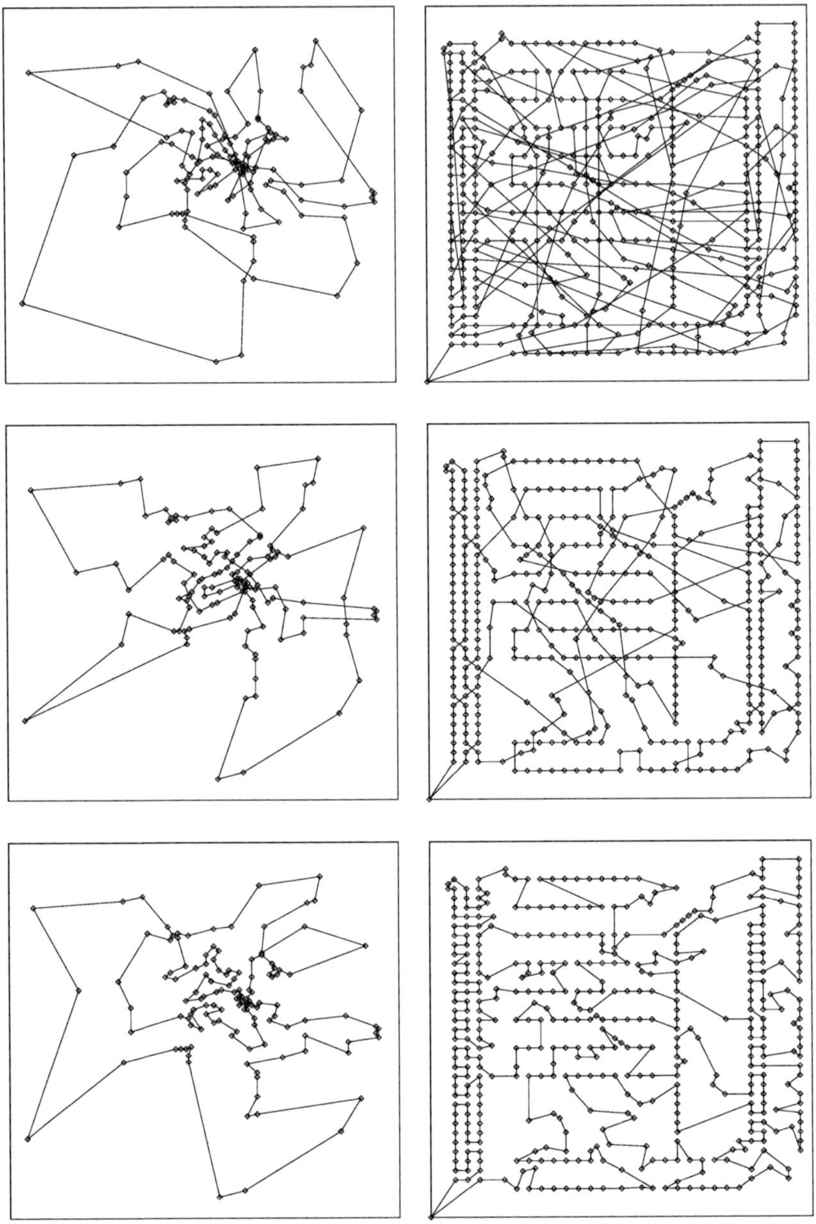

Fig. 4.5. Comparison of solutions produced with the greedy algorithm, using the small moves mentioned in the text. *Left column*: solutions for the BEER127 instance; *right column*: corresponding solutions for the PCB442 instance. The solutions in the *top row* were created using the EXC, those in the *middle* using the NIM, and those at the *bottom* using the L2O

left, the length of the configuration cannot be decreased further with the L2O.

The second question is how well these moves perform and which of these provides the best results. The graphic in Fig. 4.5 already reveals that the L2O leads to the best results. The results for each 100 optimization runs are presented in Table 4.2. One finds that for all instances, the L2O leads to by far the best results, whereas the EXC leads always to the worst results. This enormous difference between the EXC and the L2O was theoretically explained in [201]: as the number of cut edges is two for the L2O and four for the EXC, for large N, the correlation length λ is given by $N/2$ if using the L2O and $N/4$ if using the EXC. As the size of the neighborhood of a configuration is roughly the same for both moves, namely, of order $\mathcal{O}(N^2)$, this difference in λ results in the fact that the L2O must lead to better results than the EXC. Generalizing this approach, one also understands that the quality of the results achieved with the NIM lies in between, as three edges are cut, leading to $\lambda = N/3$, and again the size of the neighborhood of a configuration is of order $\mathcal{O}(N^2)$.

Next, one might wonder whether it would not be best to implement all three moves and start them in a random order, thus getting solutions that are metastable both for the EXC and the NIM and the L2O. Table 4.3 shows the results for this approach, in which each of the moves EXC, NIM, and L2O was called with probability $\frac{1}{3}$. Note that we used $150 \times N^2$ moves here in order

Table 4.2. Results for the greedy algorithm using small moves: for each instance, 100 optimization runs were performed, starting with a random configuration and performing either a EXC $50 \times N^2$ times or a NIM $50 \times N^2$ times or a L2O $50 \times N^2$ times with the greedy acceptance criterion

Instance	Move	Minimum	Maximum	Mean value ± error
BEER127	EXC	159,705.087	208,409.703	182,672.403 ± 977.4
	NIM	132,898.075	161,648.200	146,418.726 ± 596.4
	L2O	121,178.315	138,126.861	129,964.407 ± 325.5
LIN318	EXC	90,116.6104	127,763.120	110,660.750 ± 628.5
	NIM	59,992.7967	78,436.1451	68,407.9569 ± 376.2
	L2O	45,115.9243	49,971.4471	47,276.4710 ± 95.0
PCB442	EXC	112,322.878	142,974.558	129,290.471 ± 708.4
	NIM	64,129.9159	79,939.0300	70,764.4069 ± 296.7
	L2O	54,682.4205	59,026.1119	56,614.6608 ± 82.3
ATT532	EXC	69,294	91,946	80,037.86 ± 449.7
	NIM	37,691	47,862	42,460.43 ± 212.1
	L2O	29,960	32,026	30,867.73 ± 42.2
NRW1379	EXC	164,670.848	196,316.798	181,017.337 ± 607.0
	NIM	89,353.7705	105,344.742	94,486.5234 ± 294.3
	L2O	62,862.6385	64,887.1349	64,046.8071 ± 42.9

Table 4.3. Results for the greedy algorithm using small moves: for each instance, 100 optimization runs were performed, starting with a random configuration and performing $150 \times N^2$ moves with equal probabilities for the move being an EXC, a NIM, or a L2O

Instance	Minimum	Maximum	Mean value ± error
BEER127	119,331.431	134,236.701	126,605.995 ± 305.2
LIN318	43,488.2914	47,875.0510	45,743.4573 ± 78.9
PCB442	52,589.4816	56,840.8241	54,660.7005 ± 76.3
ATT532	29,069	30,530	29,689.64 ± 31.1
NRW1379	60,334.7758	62,208.5391	61,154.0388 ± 39.4

to be on the safe side for the freezing criterion. One indeed finds that this "mixture" leads to a better quality of results than the isolated moves. This result is not surprising when considering the energy landscape: the resulting configurations are now metastable with respect to the EXC, the NIM, and the L2O, whereas the results presented in Table 4.2 were only metastable with respect to one of these moves. If one thus checks all local minima in an energy landscape formed by, e. g., the Exchange only, one would find that most of these configurations could be further improved by a NIM or a L2O. The application of these moves and then also of additional EXCs would lead to better results.

4.4 Local Search as Afterburner for Construction Heuristics

There is also another way of using these moves combined with the greedy acceptance criterion: as seen in the last chapter, the resulting configurations of construction heuristics, such as the bestinsertion heuristic or the savings heuristic, are not locally optimal. Thus, they could be improved by removing crossings, by shifting nodes to other positions, or by exchanging nodes. Thus, one creates a configuration with a construction heuristic and then starts a so-called "afterburner", which consists of an improvement heuristic. This afterburner applies small moves in the greedy mode in order to improve the configuration locally.

Table 4.4 shows the results for starting with configurations produced by the bestinsertion heuristic followed by an afterburner consisting of $50 \times N^2$ EXCs, NIMs, or L2Os in the greedy mode. One finds that the NIM can improve the configurations produced by the bestinsertion heuristic for all instances much more than the EXC or the L2O. Even better results are achieved with a mixture of all three moves.

Table 4.4. Results for the afterburner for the bestinsertion heuristic: for each instance, 100 optimization runs were performed. The "orig" results are those achieved with the bestinsertion heuristic only. Then an afterburner was applied to each of the 100 configurations, consisting of either $50 \times N^2$ EXC or $50 \times N^2$ NIMs or $50 \times N^2$ L2O or $150 \times N^2$ moves with equal probability of an EXC, a NIM, or a L2O ("mixed"). Each of these moves was used in the greedy mode, and thus all deteriorations were rejected

Instance	Move	Minimum	Maximum	Mean value ± error	Mean improvement
BEER127	orig	125,660.270	140,146.531	131,928.823 ± 273.9	
	EXC	122,802.361	137,991.458	130,452.933 ± 277.5	1475.89
	NIM	121,753.542	133,271.145	127,592.701 ± 253.2	4336.12
	L2O	123,103.694	134,933.780	128,469.835 ± 255.8	3458.99
	Mixed	121,147.451	134,091.960	126,540.200 ± 260.8	5388.62
LIN318	Orig	45,069.2572	48,654.3710	46,475.8893 ± 71.76	
	EXC	44,786.9199	48,151.1938	46,030.9755 ± 68.07	444.91
	NIM	43,984.2132	47,134.9966	45,173.8139 ± 62.17	1302.08
	L2O	44,380.9389	46,957.3946	45,427.7567 ± 55.65	1048.13
	Mixed	43,783.6120	46,685.4222	44,771.2695 ± 53.21	1704.62
PCB442	Orig	56,075.4169	60,433.8159	58,186.9793 ± 68.96	
	EXC	55,592.1833	59,534.4677	57,423.6919 ± 68.11	763.29
	NIM	53,365.8901	57,559.2394	55,340.1747 ± 64.72	2846.80
	L2O	54,638.2939	58,231.7207	56,144.9652 ± 71.71	2042.01
	Mixed	53,500.8513	56,206.5877	54,747.9606 ± 59.60	3439.02
ATT532	Orig	29,725	31,489	30,612.67 ± 33.71	
	EXC	29,508	31,074	30,251.67 ± 29.63	361.00
	NIM	28,720	30,151	29,570.17 ± 31.17	1042.50
	L2O	29,257	30,493	29,916.42 ± 26.91	696.25
	Mixed	28,650	29,995	29,354.41 ± 28.44	1258.26
NRW1379	Orig	62,661.6777	64,754.9425	63,594.9357 ± 42.93	
	EXC	61,794.5966	63,818.7890	62,833.7421 ± 41.28	761.19
	NIM	60,126.8920	61,632.9720	60,968.9513 ± 27.21	2625.98
	L2O	61,381.6769	63,133.1927	62,193.8470 ± 33.59	1401.09
	Mixed	59,965.4997	61,185.0856	60,631.2822 ± 28.59	2963.65

A look at Table 4.5 reveals that also in the case where the initial solutions are produced with the savings heuristic, the NIM leads to the largest improvements among the three moves. Again, the EXC performs worst. A mixture of all three moves leads to even better results, which are on average only 3 to 5% worse than the optimum.

Table 4.5. Results for the afterburner for the savings heuristic: the procedure is the same as for the results in Table 4.4, with the exceptions that the initial configurations were produced not with the bestinsertion but with the savings heuristic and that N optimization runs were performed

Instance	Move	Minimum	Maximum	Mean value ± error	Mean improvement
BEER127	Orig	122,689.400	137,693.407	126,611.232 ± 197.04	
	EXC	120,405.089	134,417.855	124,612.765 ± 178.36	1998.47
	NIM	119,411.536	132,192.833	122,650.917 ± 161.18	3960.32
	L2O	119,936.472	130,000.871	123,781.982 ± 166.50	2829.25
	Mixed	118,444.675	128,040.274	122,372.420 ± 143.14	4238.81
LIN318	Orig	44,417.8099	47,600.2678	45,764.5415 ± 37.55	
	EXC	43,199.9493	46,689.5988	44,938.3216 ± 38.18	826.22
	NIM	43,008.0510	45,572.6886	44,206.8113 ± 30.67	1557.73
	L2O	43,098.0751	45,773.1190	44,431.7113 ± 30.83	1332.83
	Mixed	42,857.2567	45,379.7298	43,943.4939 ± 28.04	1821.05
PCB442	Orig	53,898.6223	57,750.9015	55,766.8600 ± 33.18	
	EXC	53,184.7074	56,554.6051	54,872.4913 ± 30.78	894.37
	NIM	51,701.5554	54,684.4690	53,021.0843 ± 25.53	2745.78
	L2O	52,422.0055	55,591.6721	54,110.9355 ± 27.31	1655.92
	Mixed	51,622.9015	54,165.6020	52,843.2454 ± 23.79	2923.61
ATT532	Orig	29,528	31,297	30,320.36 ± 13.61	
	EXC	28,951	30,852	29,729.47 ± 14.00	590.89
	NIM	28,465	30,057	29,088.00 ± 12.50	1232.36
	L2O	28,637	30,416	29,447.88 ± 11.65	872.48
	Mixed	28,213	29,918	28,921.26 ± 11.54	1399.10
NRW1379	Orig	61,167.6236	63,451.7975	62,397.4092 ± 9.31	
	EXC	60,518.0125	62,547.8736	61,504.1214 ± 8.63	893.29
	NIM	59,339.9513	61,199.3219	60,196.2318 ± 7.47	2201.18
	L2O	60,118.8580	62,266.0126	61,139.2124 ± 8.36	1258.20
	Mixed	59,192.6633	61,031.2744	60,011.4728 ± 7.41	2385.94

5 Next Larger Moves Applied to TSP

In the last chapter, the smallest possible moves for applying the local search approach to the traveling salesman problem (TSP) were introduced. The results were already quite promising but not very good. Thus, one wonders whether a weakening of the local search approach, i. e., using larger moves, might lead to better results. These larger moves can be composed by smaller ones. As the best results for the smallest moves were achieved with the Lin-2-opt (L2O), we will only consider constructions of higher-order moves based on this L2O.

5.1 Lin-3-Opts

For the L2O, two tour positions i and j were considered after which the tour was cut. Analogously, one can fix three tour positions, i, j, and k, after which the tour is cut. Thus, one gets three sequences that have to be reconnected in such a way that a new closed tour is built. Let i, j, and k be ordered according to their size such that $i < j < k$, and let i_+, j_+, and k_+ be the tour positions after the positions i, j, and k, respectively. Thus, one can write the cut tour as follows:

$$\ldots\ \sigma(i)\big|\sigma(i_+)\ \ldots\ \sigma(j)\big|\sigma(j_+)\ \ldots\ \sigma(k)\big|\sigma(k_+)\ \ldots$$

Note that the tour is, of course, closed such that the partial tour starting at $\sigma(k_+)$ ends with $\sigma(i)$. If $j \neq i_+$, $k \neq j_+$, and $i \neq k_+$, there are eight possibilities to reconnect the three parts:

1. One can trivially reconnect $\sigma(i)$ with $\sigma(i_+)$, $\sigma(j)$ with $\sigma(j_+)$, and $\sigma(k)$ with $\sigma(k_+)$, thus ending up with the old configuration.
2. One can connect $\sigma(i)$ with $\sigma(i_+)$, $\sigma(j)$ with $\sigma(k)$, and $\sigma(j_+)$ with $\sigma(k_+)$, thus ending up with

   $$\ldots\ \sigma(i)\big|\sigma(i_+)\ \ldots\ \sigma(j)\big|\sigma(k)\ \ldots\ \sigma(j_+)\big|\sigma(k_+)\ \ldots$$

   However, this move could be performed with a L2O cutting the tour after $\sigma(j)$ and $\sigma(k)$. Thus, this possibility does not lead to a larger move.

3. One can connect $\sigma(i)$ with $\sigma(j)$, $\sigma(i_+)$ with $\sigma(j_+)$, and $\sigma(k)$ with $\sigma(k_+)$. Then one gets the tour

$$\dots \sigma(i)\big|\sigma(j) \dots \sigma(i_+)\big|\sigma(j_+) \dots \sigma(k)\big|\sigma(k_+) \dots$$

However, this reconnection could also be done with a L2O cutting the tour after $\sigma(i)$ and $\sigma(j)$. Again this is not an L3O.
4. Then one can reconnect the three parts as follows:

$$\dots \sigma(i)\big|\sigma(j) \dots \sigma(i_+)\big|\sigma(k) \dots \sigma(j_+)\big|\sigma(k_+) \dots$$

This is the first true possibility for a L3O as there are three new edges in the tour. For this move, two L2Os would have to be performed to simulate it, namely, one turning around $\sigma(i_+), \dots, \sigma(j)$ and one turning around $\sigma(j_+), \dots, \sigma(k)$.
5. Then one can reconnect the three parts as follows:

$$\dots \sigma(i)\big|\sigma(j_+) \dots \sigma(k)\big|\sigma(i_+) \dots \sigma(j)\big|\sigma(k_+) \dots$$

This is a further true possibility for a L3O as there are again three new edges in the tour. However, here three L2Os would be needed to simulate this L3O: first a L2O cutting the tour after positions i and k would have to be performed, then one after i and $i + (k - j)$, and finally one after $i + (k - j)$ and k.
6. The reconnection

$$\dots \sigma(i)\big|\sigma(j_+) \dots \sigma(k)\big|\sigma(j) \dots \sigma(i_+)\big|\sigma(k_+) \dots$$

is also a true possibility for a L3O. Here, only two L2Os would be necessary to simulate the L3O, the first one after tour positions i and k, the second one after i and $i + (k - j)$.
7. Analogously, the reconnection

$$\dots \sigma(i)\big|\sigma(k) \dots \sigma(j_+)\big|\sigma(i_+) \dots \sigma(j)\big|\sigma(k_+) \dots$$

leads to three new edges and is thus a true L3O. Here also only two L2Os are needed to simulate it, namely, one cutting after tour positions i and k and one after $k - (j - i)$ and k.
8. In contrast, the reconnection

$$\dots \sigma(i)\big|\sigma(k) \dots \sigma(j_+)\big|\sigma(j) \dots \sigma(i_+)\big|\sigma(k_+) \dots$$

can also be performed with only one L2O, which cuts the tour after $\sigma(i)$ and $\sigma(k)$.

Thus, four possibilites for a Lin-3-opt (L3O) remain: one can either exchange two succeeding parts of the tour without changing their direction (L3O1), or one can change the directions of two succeeding parts of the tour (L3O2), or

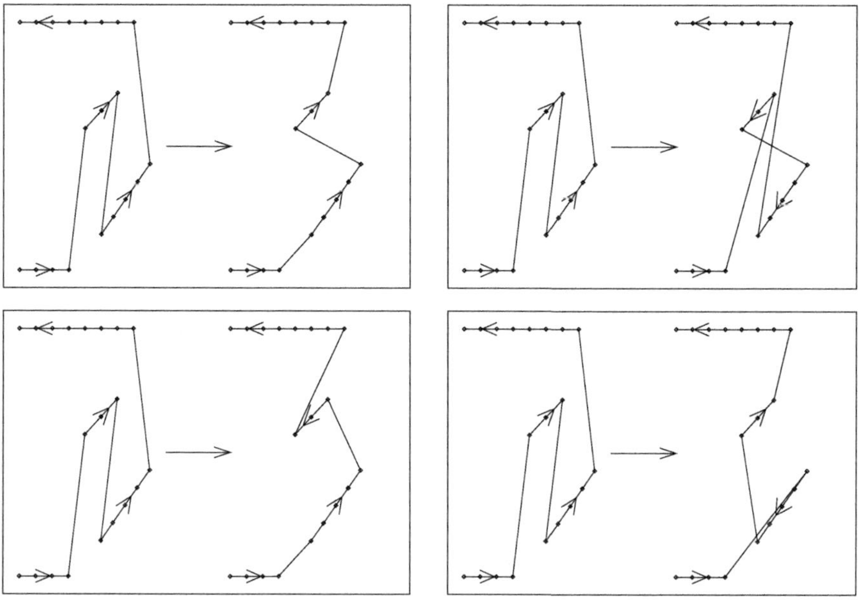

Fig. 5.1. The Lin-3-opts: L3O1 (*top left*), L3O2 (*top right*), L3O3 (*bottom left*), and L3O4 (*bottom right*)

one can exchange two succeeding parts of the tour and change the direction of one of these (L3O3 and L3O4). These moves are presented in Fig. 5.1.

However, if one of the two parts consists of only one node, i. e., either $i_+ = j$ or $j_+ = k$, then the number of possible moves leading to three new edges reduces to one. The only possible L3O in this case is the node insertion move (NIM), which was introduced in the last chapter. As the NIM is a special case of the L3O, but has a neighborhood size of order $\mathcal{O}(N^2)$ like the L2O, it is sometimes called Lin-2.5-opt. Even worse, if both $i_+ = j$ and $j_+ = k$, then the only possible move is the nearest neighbor exchange, which can be considered as a special case of either the exchange or the L2O as the nodes $\sigma(j)$ and $\sigma(k)$ are exchanged and the tour is cut after $\sigma(i)$ and $\sigma(k)$. In this case, there is no longer any true L3O.

5.2 Higher-Order Lin-n-Opts

One can go on to moves of even higher order, for which four, five, six, or more edges have to be cut. Analogously to the considerations for the L3O, one can determine the number of possible L4Os to be 25, the number of possible L5Os to be 208, and so on. The number of possible Lin-n-opts can be calculated as follows [203]:

Assume that there are n edges cut in the current tour. Thus, it is split into n sequences containing $A(i)$ nodes with

$$\sum_{i=1}^{n} A(i) = N\,. \tag{5.1}$$

Depending on the neighborhood relation of these n cuts to each other, the cutting of the tour into partial sequences leads to the following structures that have to be distinguished in the analysis below:

- There are pieces containing two or more nodes, i. e., $A(i) \geq 2$. Let α_0 be the number of these parts.
- Then there are isolated pieces that contain only one node, and that are surrounded by two pieces containing at least two nodes. Thus, this case is given for a one-node piece $A(i)$ by
$$A(i-1) \geq 2 \ \wedge \ A(i) = 1 \ \wedge \ A(i+1) \geq 2\,. \tag{5.2}$$
Let α_1 be the number of such pieces.
- Then there are tuples of directly succeeding pieces with each one containing only one node. Thus, here three directly succeeding edges were cut and the neighboring edges around them remained uncut. Thus, one has a tuple $(i, i+1)$ with
$$A(i-1) \geq 2 \ \wedge \ A(i) = 1 \ \wedge \ A(i+1) = 1 \ \wedge \ A(i+2) \geq 2\,. \tag{5.3}$$
Let α_2 be the number of these tuples.
- Furthermore, there are triples, quadruples, quintuples, and so on of neighboring pieces, each one consisting of only node node. Let the number of these parts be α_3, α_4, ..., α_N.

Then the number of cuts n can be expressed as follows:

$$n = \alpha_0 + \sum_{i=1}^{N} i\alpha_i\,. \tag{5.4}$$

In what follows, the assumption shall hold that not every edge of the tour is cut, so that $n < N$ and thus $\alpha_0 > 0$. Thus, one can always choose a part consisting of two or more nodes and fix its direction. We will only consider the symmetric TSP here. For the asymmetric TSP (ATSP), some of the numbers given have to be multiplied by 2.

Next the number of reconnections leading to feasible configurations are determined. Feasible means that there is a closed roundtrip touching each node exactly once. For this purpose, first the special case that each part consists of at least two nodes will be considered such that $n = \alpha_0$. Let $\mathcal{M}$ be the number of possibilities for creating a feasible configuration. In general, this number depends on the vector $\boldsymbol{\alpha} = (\alpha_0, \alpha_1, \ldots, \alpha_N)$. $\mathcal{M}$ is given by

$$\mathcal{M}(\alpha_0, 0, \ldots, 0) = 2^{n-1}(n-1)! = \mathcal{M}_n \tag{5.5}$$

and therefore depends only on n.

In the general case, there are also other $\alpha_i > 0$. The calculation of $\mathcal{M}$ does not depend on whether parts consisting of only one node are isolated or come in tuples or triples. Thus, one can rewrite the dependency in the form

$$\mathcal{M}(\boldsymbol{\alpha}) = \mathcal{M}(\alpha_0, n - \alpha_0, 0, \ldots, 0)\,. \tag{5.6}$$

Starting from the part consisting of two nodes and with a fixed direction, one has $n - \alpha_0$ possibilities to connect a piece consisting of only one node, thus reducing their number $n - \alpha_0$ by 1, and $2(\alpha_0 - 1)$ possibilities to connect a piece consisting of at least two nodes, thus also reducing their number by 1. One can solve this recursion to obtain

$$\mathcal{M}(\boldsymbol{\alpha}) = 2^{-(n-\alpha_0)}\mathcal{M}(n, 0, \ldots, 0) = 2^{\alpha_0 - 1}(n-1)!\,. \tag{5.7}$$

However, $\mathcal{M}$ is only the number of possibilities for creating a feasible configuration. The number $\mathcal{N}$ of true Lin-n-opts is much smaller. As was already shown for the L3O, many possibilities turn out to be Lin-k-opts, with $k < n$, if cut edges are used again. These former edges must be excluded in the calculation of the true Lin-n-opts as otherwise the numbers of Lin-$(n-i)$-opts might be added several times.

Again the starting point is the special case in which each part consists of at least two nodes. In this case, the number $\mathcal{M}(n, 0, \ldots, 0)$ must be reduced by the number of possibilities for a smaller Lin-k-opt. There are $\binom{n}{k}$ possibilities to choose old edges if there are $n-k$ cuts. Thus, the number of true Lin-n-opts is given by

$$\mathcal{N}(n, 0, \ldots, 0) = 2^{n-1}(n-1)! - \sum_{k=0}^{n-1} \binom{n}{k} \mathcal{N}(k, 0, \ldots, 0)\,, \tag{5.8}$$

with

$$\mathcal{N}(\mathbf{0}) = 1\,. \tag{5.9}$$

Thus, there is 1 Lin-0-opt, namely, the identity, 0 L1Os, as by cutting only one edge no new tour can be formed, 1 L2O, 4 L3Os, 25 L4Os, 208 L5Os, 2121 L6Os, and so on.

The next more general case is that there are not only pieces consisting of at least two nodes but also isolated pieces consisting of one node and surrounded by pieces of at least two nodes. For each of these isolated pieces, one can proceed as follows: one extends this isolated piece to two nodes by doubling the node. Thus, one gets $\mathcal{N}(\alpha_0 + 1, \alpha_1 - 1, \ldots)$ instead of $\mathcal{N}(\alpha_0, \alpha_1, \ldots)$ possibilities. By changing its direction, one can connect it (in contrast to before when it consisted of only one edge) to those edges of the neighboring parts to which it was connected before. There are two possibilities to connect it in this way to one of the two neighboring parts and one way to connect it in this way to both neighboring parts. These cases must be forbidden now. The resulting number must be divided by 2, as a partial sequence containing only one node does not have two different directions, such that one gets the recursive formula:

$$\begin{aligned}\mathcal{N}(\boldsymbol{\alpha}) = \tfrac{1}{2}\Big(& \mathcal{N}(\alpha_0+1,\alpha_1-1,\alpha_2,\ldots,\alpha_{N-2}) \\ & -2\cdot\mathcal{N}(\alpha_0,\alpha_1-1,\alpha_2,\ldots,\alpha_{N-2}) \\ & -\mathcal{N}(\alpha_0-1,\alpha_1-1,\alpha_2,\ldots,\alpha_{N-2})\Big).\end{aligned} \tag{5.10}$$

Analogously, one can derive a formula if there are tuples of neighboring parts with only one node each. Here one expands one of the two parts to two nodes, such that there is one tuple less, but one isolated part more and one longer part more. Analogously to the above, the false possibilities must be subtracted and the result divided by 2, such that one gets the following formula:

$$\begin{aligned}\mathcal{N}(\boldsymbol{\alpha}) = \tfrac{1}{2}\Big(& \mathcal{N}(\alpha_0+1,\alpha_1+1,\alpha_2-1,\alpha_3,\ldots,\alpha_{N-2}) \\ & -\mathcal{N}(\alpha_0,\alpha_1+1,\alpha_2-1,\alpha_3,\ldots,\alpha_{N-2}) \\ & -\mathcal{N}(\alpha_0+1,\alpha_1,\alpha_2-1,\alpha_3,\ldots,\alpha_{N-2}) \\ & -\mathcal{N}(\alpha_0,\alpha_1,\alpha_2-1,\alpha_3,\ldots,\alpha_{N-2})\Big).\end{aligned} \tag{5.11}$$

The approach is the same for triples, quadruples, and so on. Here it is appropriate to blow up a single-node part at the frontier such that the following recursive formula is obtained:

$$\begin{aligned}\mathcal{N}(\boldsymbol{\alpha}) = \tfrac{1}{2}\Big(& \mathcal{N}(\alpha_0+1,\alpha_1,\ldots,\alpha_{i-1}+1,\alpha_i-1,\alpha_{i+1},\ldots,\alpha_{N-2}) \\ & -\mathcal{N}(\alpha_0,\alpha_1,\ldots,\alpha_{i-1}+1,\alpha_i-1,\alpha_{i+1},\ldots,\alpha_{N-2}) \\ & -\mathcal{N}(\alpha_0+1,\alpha_1,\ldots,\alpha_{i-2}+1,\alpha_{i-1},\alpha_i-1,\alpha_{i+1},\ldots,\alpha_{N-2}) \\ & -\mathcal{N}(\alpha_0,\alpha_1,\ldots,\alpha_{i-2}+1,\alpha_{i-1},\alpha_i-1,\alpha_{i+1},\ldots,\alpha_{N-2})\Big).\end{aligned} \tag{5.12}$$

Generally, one should proceed with the recursion in the following way: first, those nonzero α_i for which i is maximal should vanish. This approach should be iterated with decreasing i until one ends up with a formula for tours with pieces consisting of at least two nodes.

So far, however, we have only calculated the number of possibilities for reconnecting several partial sequences, which were created after the former

Table 5.1. Number of possibilities for cutting the tour of a traveling salesman if each partial sequence shall contain at least two nodes: n denotes the number of cuts of the Lin-n-opt, $C(n)$ the number of possibilities

n	$C(n)$
2	$N \times (N-3)/2$
3	$N \times (N-4)(N-5)/3!$
4	$N \times (N-5)(N-6)(N-7)/4!$
5	$N \times (N-6)(N-7)(N-8)(N-9)/5!$

tour had been cut, to a new tour. We must still determine the number of possibilities for cutting the tour in order to create these partial sequences.

We again start out with the special case in which every partial sequence to be created contains at least two nodes. By empirically going through all possibilities, we found the formulas given in Table 5.1. From this result we deduce a general formula for the possibilities $C(n)$ for the Lin-n-opt:

$$C(n) = N \times \prod_{i-1}^{n-1} (N-n-i) \times \frac{1}{n!} = \frac{N}{N-n} \times \binom{N-n}{n} . \tag{5.13}$$

Thus we find here that the neighborhood created by a Lin-n-opt is of order $\mathcal{O}(N^n)$.

In the general case, an arbitrary Lin-n-opt can also lead to partial sequences containing only one node. Here we have to distinguish between various types of cuts: The cuts introduced by a Lin-n-opt can be isolated, i. e., they are between two sequences with more than one node each. Then they can lead to isolated nodes that are between two partial sequences with more than one node each, and so on. Let us view this here from the point of view of the cuts of the tour. All in all, a Lin-n-opt generally leads to n cuts in the tour. Let us denote an i-type multicut (with $1 \leq i \leq n$) at position j (with $1 \leq j \leq N$) as the scenario where the tour is cut at i successive positions after the node with the tour position number j. Thus, the tour is cut by an i-type multicut successively between pairs of nodes with the tour position numbers $(j, j+1), (j+1, j+2), \ldots, (j+i-1, j+i)$.

Furthermore, let β_i be the number of i-type multicuts: β_1 is the number of isolated cuts, and β_2 is the number of 2-type multicuts by which the tour is cut after two successive tour positions such that a partial sequence containing only one node is created, surrounded by two sequences containing more than one node. Thus, β_2 is also the number of isolated nodes surrounded by longer sequences and is thus identical to α_1. Analogously, 3-type multicuts lead to tuples of nodes that are surrounded by partial sequences with more than one node; thus β_3 is the number α_2 of these tuples. Analogously, 4-type

Table 5.2. Number of possibilities for cutting the tour of a traveling salesman: n denotes the overall number of cuts performed by the Lin-n-opt, β_1, β_2, β_3, and β_4 denote the number of 1-, 2-, 3-, and 4-type multicuts as defined in the text. Only β_i values for $i \leq 4$ are considered here in our examples. For the L6O, only some special cases are considered. The number C of possibilities depends on all these numbers. All these formulas for $C(\boldsymbol{\beta})$ were found manually

n	β_1	β_2	β_3	β_4	C
2	2	0	0	0	$N \times (N-3)/2$
	0	1	0	0	N
3	3	0	0	0	$N \times (N-4)(N-5)/3!$
	1	1	0	0	$N \times (N-4)$
	0	0	1	0	N
4	4	0	0	0	$N \times (N-5)(N-6)(N-7)/4!$
	2	1	0	0	$N \times (N-5)(N-6)/2$
	0	2	0	0	$N \times (N-5)/2$
	1	0	1	0	$N \times (N-5)$
	0	0	0	1	N
5	5	0	0	0	$N \times (N-6)(N-7)(N-8)(N-9)/5!$
	3	1	0	0	$N \times (N-6)(N-7)(N-8)/3!$
	1	2	0	0	$N \times (N-6)(N-7)/2$
	2	0	1	0	$N \times (N-6)(N-7)/2$
	0	1	1	0	$N \times (N-6)$
	1	0	0	1	$N \times (N-6)$
6	2	2	0	0	$N \times (N-7)(N-8)(N-9)/2/2$
	0	3	0	0	$N \times (N-7)(N-8)/3!$
	3	0	1	0	$N \times (N-7)(N-8)(N-9)/3!$
	1	1	1	0	$N \times (N-7)(N-8)$
	0	0	2	0	$N \times (N-7)/2$
	2	0	0	1	$N \times (N-7)(N-8)/2$
	0	1	0	1	$N \times (N-7)$

multicuts lead to triples of sequences containing only one node each, and so on. Generally, we have for all $i \geq 1$ that

$$\alpha_i = \beta_{i+1} \,, \tag{5.14}$$

but for $i = 0$ the situation is different: each i-type multicut produces a further sequence consisting of at least two nodes, such that we have

$$\alpha_0 = \sum_{i=1}^{n} \beta_i \,. \tag{5.15}$$

Please note that α_0 is both the number of these longer partial sequences and the number of all i-type multicuts. The overall number n of cuts can be expressed as

$$n = \sum_{i=1}^{n} i\beta_i \,. \tag{5.16}$$

In this general case, the number C of ways of cutting the tour also depends not only on n but on the entries of the vector $\boldsymbol{\beta} = (\beta_1, \beta_2, \ldots, \beta_n)$ as well. As Table 5.2 shows, the order of the neighborhood size is now given as $\mathcal{O}(N^{\alpha_0})$. From these examples in the table we empirically derive the formula

$$C(\boldsymbol{\beta}) = N \times \prod_{i=1}^{\sum_{j=1}^{n} \beta_j - 1} (N - n - i) \times \frac{1}{\prod_{i=1}^{n} \beta_i!} \tag{5.17}$$

for the number of possibilities for cutting a tour with a Lin-n-opt, leading to β_i many i-type multicuts. Note that if the upper index of a product is smaller than the lower index, then this so-called empty product is 1. This formula can be rewritten as

$$C(\boldsymbol{\beta}) = \frac{N}{N-n} \times \binom{N-n}{\alpha_0} \times \frac{\alpha_0!}{\prod_{i=1}^{n} \beta_i!} \tag{5.18}$$

making use of α_0 as the sum of all β_i. Note that Eq. (5.13) for the special case with each sequence containing at least two nodes is a special case of Eq. (5.18).

5.3 Computational Results for the Greedy Algorithm

Now the quality of these larger moves shall be investigated, starting with the four types of the L3O. Table 5.3 shows the results for the greedy algorithm using one of these four move types. The names of these types are the same as in Fig. 5.1. In a few test runs, the system was frozen after $1 - 2 \times N^3$ moves, so that we spent $10 \times N^3$ moves to be on the safe side. One finds that the results for any of these L3Os are much better than those for the L2O in Table 4.2, except for the NRW1379 instance. These results are in nice agreement with the remark in [201] on the results in [116] that the L3Os lead to better results than the L2O. This is surely an effect of the neighborhood size of a L3O, which is of order $\mathcal{O}(N^3)$, whereas the neighborhood size of the L2O is only $N(N-3)/2$ and thus of order $\mathcal{O}(N^2)$.

Furthermore, the quality of the results differs between the various types of L3Os: L3O3 and L3O4, both of which exchange two successive parts of the tour and change the direction of one of these parts, provide better results than the other two types of the L3O. Mostly, the variant L3O2, which changes the directions of two successive parts of the tour simultaneously, provides the worst results among the L3Os. Please note that different results can be achieved if an other optimization algorithm, like simulated annealing, is used.

Table 5.3. Results for the greedy algorithm using L3Os: for each instance, 100 optimization runs were performed, starting with a random configuration and performing $10 \times N^3$ times one of the four L3Os with the greedy acceptance criterion. In the "L3Oall" scenario, each of the four L3Os was implemented and called with the same probability; here $20 \times N^3$ move trials were performed

Instance	Move	Minimum	Maximum	Mean value ± error
BEER127	L3O1	119,218.371	131,235.436	125,118.494 ± 254.3
	L3O2	120,935.053	133,226.981	125,888.805 ± 260.4
	L3O3	118,589.344	127,980.167	121,981.992 ± 187.4
	L3O4	118,899.537	127,588.470	121,928.533 ± 187.4
	L3Oall	118,629.043	127,881.507	121,396.175 ± 194.1
LIN318	L3O1	43,883.1117	46,992.0997	45,210.0271 ± 63.9
	L3O2	44,209.8503	47,368.0175	45,466.4980 ± 66.1
	L3O3	42,863.7951	44,822.3800	43,632.8142 ± 39.5
	L3O4	42,626.6703	45,173.6630	43,534.8133 ± 44.4
	L3Oall	42,401.9352	44,514.0820	43,518.6735 ± 40.7
PCB442	L3O1	53,547.9002	56,348.0251	54,850.7759 ± 60.7
	L3O2	52,975.6709	57,838.1568	54,847.6013 ± 85.7
	L3O3	51,658.9173	54,163.2202	52,689.2427 ± 47.9
	L3O4	51,627.2806	53,519.6159	52,545.4050 ± 41.0
	L3Oall	51,480.2480	53,892.2666	52,493.4278 ± 44.5
ATT532	L3O1	28,716	30,639	29,732.22 ± 33.7
	L3O2	29,229	30,817	30,031.61 ± 30.7
	L3O3	28,275	29,369	28,692.05 ± 22.8
	L3O4	28,076	29,159	28,673.47 ± 21.4
	L3Oall	28,281	29,102	28,718.27 ± 19.6
NRW1379	L3O1	64,624.4812	66,712.2867	65,611.2830 ± 35.6
	L3O2	63,267.3838	64,788.3074	64,008.2578 ± 37.1
	L3O3	58,730.9852	59,768.7048	59,199.8516 ± 25.4
	L3O4	58,727.9973	59,656.5893	59,173.3947 ± 23.5
	L3Oall	58,650.2000	60,077.9761	59,307.8550 ± 26.1

One can proceed to moves of even greater size, namely, the L4Os. As mentioned before, there are 25 different types of a L4O. One of these is the so-called two-bridge move: it cuts the tour after four positions i, j, k, and l, i. e.,

$$\dots \sigma(i)\big|\sigma(i_+) \dots \sigma(j)\big|\sigma(j_+) \dots \sigma(k)\big|\sigma(k_+) \dots \sigma(l)\big|\sigma(l_+) \dots,$$

and connects the parts as follows:

$$\dots \sigma(i)\big|\sigma(k_+) \dots \sigma(l)\big|\sigma(j_+) \dots \sigma(k)\big|\sigma(i_+) \dots \sigma(j)\big|\sigma(l_+) \dots$$

Thus, it exchanges two nonsuccessive parts of the tour. This move is called L4O1 in Table 5.4. Furthermore, we implemented another variant of the L4O,

Table 5.4. Results for the greedy algorithm using two types of L4Os: for each instance, 100 optimization runs were performed, starting with a random configuration and performing N^4 times one L4O with the greedy acceptance criterion. As the computing time is rather large, only small instances are considered

Instance	Move	Minimum	Maximum	Mean value ± error
BEER127	L4O1	123,767.073	134,986.024	129,588.275 ± 244.2
	L4O2	121,407.255	131,300.030	125,622.863 ± 214.7
LIN318	L4O1	45,184.3246	48,569.5839	46,740.9896 ± 72.2
	L4O2	44,539.2015	46,932.1063	45,451.7793 ± 48.3
PCB442	L4O1	55,673.1534	58,855.2662	57,381.3213 ± 71.3
	L4O2	53,793.3942	56,843.4938	55,209.2331 ± 64.0

which additionally changes the direction of the middle part of the tour:

$$\dots\sigma(i)\big|\sigma(k_+)\dots\sigma(l)\big|\sigma(k)\dots\sigma(j_+)\big|\sigma(i_+)\dots\sigma(j)\big|\sigma(l_+)\dots$$

The L2O is a special case of this L4O2, if both $i_+ = j$ and $k_+ = l$.

As the neighborhood size of the L4Os is of order $\mathcal{O}(N^4)$, it is not surprising that the number of L4O moves needed until the system is frozen in the greedy mode is also of order $\mathcal{O}(N^4)$. Table 5.4 provides the results for some smaller TSP instances. Comparing these results to those in Table 5.3, one finds that the L4O1 provides worse results and the L4O2 provides results of roughly the same quality. As the calculation time needed is increased so strongly and as there is no payback with a significant improvement of the results, we do not consider these L4Os or even higher moves any further.

5.4 Combination of Moves of Various Sizes

Table 5.5 shows the results for a mixture of the three smallest moves and the four variants of the L3O. We find that this mixture leads to better results than a mixture of the smallest moves only. As already discussed above, the calculation time t needed until the system freezes in some locally minimum solution is given by

$$t = \alpha(N, n) \times N^n\,, \tag{5.19}$$

with n being the number of positions to choose in the move routine for selecting an edge to cut or a node to move and $\alpha(N, n)$ being some constant depending strongly on n. However, the considerations till now have been rather theoretical: first, one is interested in achieving a solution that is as good as possible but also in a calculation time as short as possible. In practical applications, one does not care whether the system is actually frozen or not; thus one does not want to spend the calculation time to ensure that the system is frozen. Secondly, when following the approach above, the best

Table 5.5. Results for the greedy algorithm using the small moves EXC, NIM, and L2O and the four variants of the L3O with equal probability: for each instance, 100 optimization runs were performed, starting with a random configuration and performing $50 \times N^3$ times a move with the greedy acceptance criterion

Instance	Minimum	Maximum	Mean value ± error
BEER127	119,173.772	128,031.879	122,507.497 ± 210.9
LIN318	42,750.1022	45,225.3364	43,842.1611 ± 48.5
PCB442	51,584.0933	55,147.4393	52,957.8833 ± 53.2
ATT532	28,341	29,305	28,863.08 ± 19.3
NRW1379	59,322.0434	61,053.6215	60,068.1042 ± 31.8

way would be to perform a bunch of Lin-N-opts, which are guaranteed to lead to the optimum configuration as all configurations are neighbors of each other and, thus, only one minimum remains in the energy landscape. But this contradicts the local search approach, which requires that neighboring configurations be rather similar to each other. Furthermore, it requires much calculation time to find that special Lin-N-opt that actually leads to the optimum.

Thus, one usually considers the smallest possible moves with order n_0 and the moves with the next higher order $n_0 + 1$ only. Only seldom are moves of the order $n_0 + 2$ used. Furthermore, one measures the calculation time in sweeps linearly in the system size N: if N moves are performed, then a sweep is performed. Of course, this definition does not consider that two tour positions are involved in the L2O, three in the L3Os, and so on. Therefore, it is neither a true measure in terms of interactions nor a true measure in times of using a specific tour position in a move. Despite that, we will use this term from now on and will study the quality of the results that can be achieved depending on the number of sweeps invested.

6 Ruin & Recreate Applied to TSP

6.1 Application of Ruin & Recreate

As already pointed out in Chap. 10, Part I, it is sometimes not sufficient to work with the local search paradigm, i. e., to use only small moves in order to change a configuration. Instead, large moves must be invented that destroy the current configuration to some degree and rebuild the remaining subsystem to a solution for the whole problem according to a given rule set. Such a ruin & recreate (R & R) move thus consists of two parts, a ruin part that removes some parts of the system and a recreate part that reinserts these parts into the system.

For the traveling salesman problem (TSP), there are two obvious possibilities for performing the ruin part of the move: one could remove either edges or nodes from the roundtrip of the traveling salesman. Let us consider here the case of the nodes only. Some nodes are removed from the tour and thus no longer served by the traveling salesman. There are generally at least the following two possibilities for choosing the nodes to be removed:

- One could always randomly select some nodes and remove them from the tour.
- But one could also use some neighborhood relation between the nodes. In the case of the TSP, this neighborhood could be given by the distances between the nodes or by the sequence of the tour.

We will thus consider these three Ruin types:

- Radial ruin:
  A node c is randomly selected. Furthermore, a number n of nodes to be removed from the system is chosen. Then the n nearest neighbors of c (including c) are removed from the roundtrip of the traveling salesman; the predecessors and successors of the removed nodes are connected such that a shorter tour consisting of the remaining $N - n$ nodes is built.
  Although the neighborhood relation might have nothing in common with the radial appearance, the radial ruin can easily be imagined for a geographic neighborhood relation with an underlying Euclidean metric, and thus this terminus technicus should be used for this type as one can always visualize a neighborhood relation by introducing such a metric that one gets a sphere.

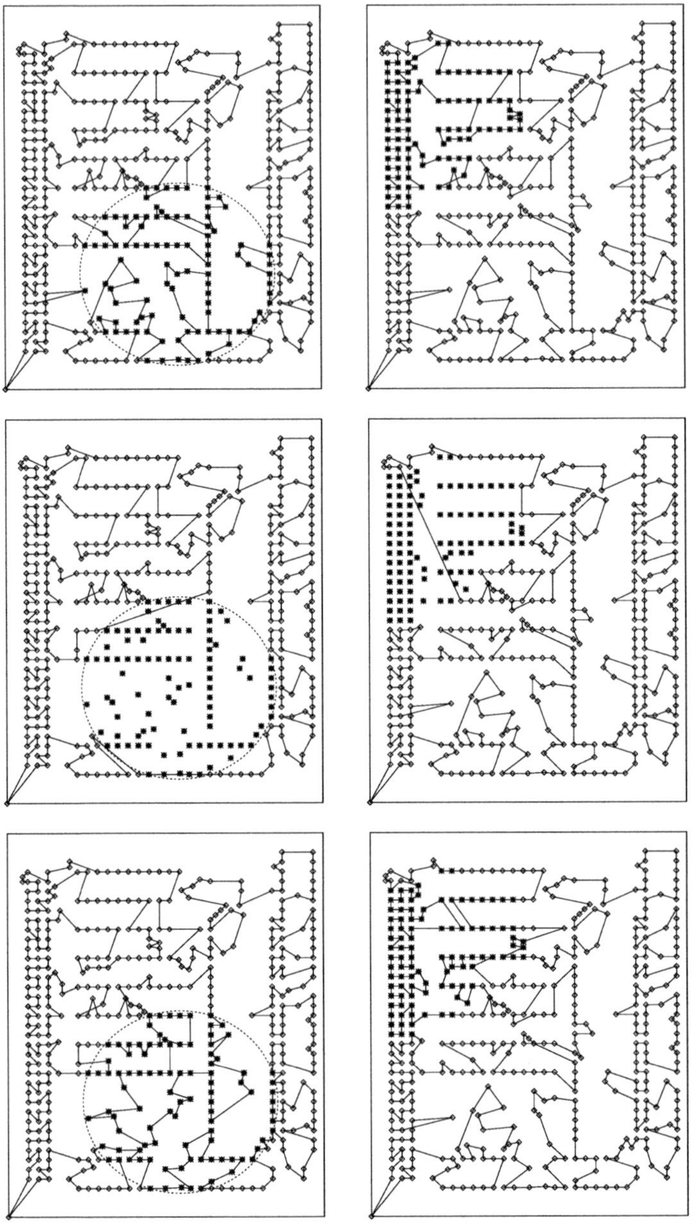

Fig. 6.1. R & R move applied to the PCB442 instance using the radial ruin (*left*) and the sequential ruin (*right*). *Top row*: current configurations in which nodes to be removed are already selected. (For the radial ruin, these nodes lie within a circle around the central node *c*.) *Middle row*: these nodes are removed and a roundtrip through the other nodes remains. *Bottom row*: the removed nodes are already reinserted by the bestinsertion heuristic

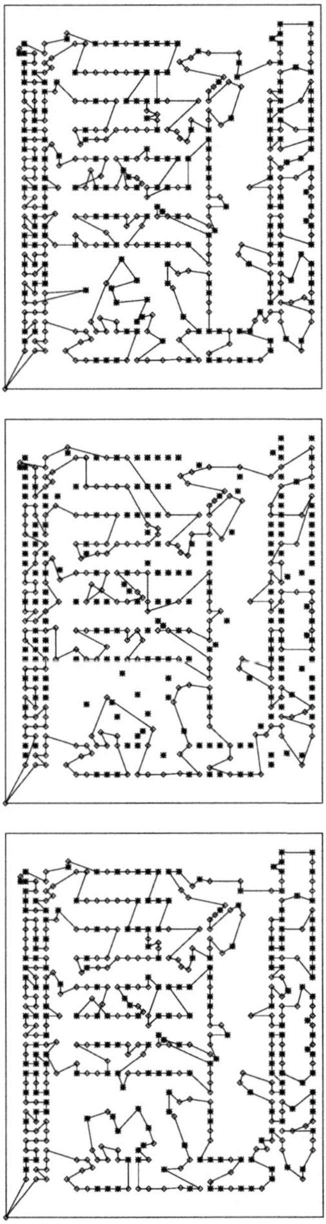

Fig. 6.2. R & R move using the random ruin: as in Fig. 6.1, the *top graphic* shows the initial configuration, the *middle graphic* the tour through the remaining subsystem, and the *bottom graphic* the recreated configuration

- Sequential ruin:
  Here a sequence of n successive nodes in the tour is removed and the predecessor and the successor of the sequence are connected. Thus, like the radial ruin, the sequential ruin uses some neighborhood relation; however, here the neighborhood is given by the relative positions of the individual nodes in the current solution; thus, this is not a static neighborhood.
- Random ruin:
  Here n nodes are randomly selected. Again the roundtrip through the remaining nodes in the system is closed by inserting edges between the predecessors and successors of the removed nodes.

Let us assume that the removed nodes are put in a bag one by one. Then they have to be reinserted into the system, so that a complete configuration for the original problem consisting of all N nodes is built. This recreate part of the move may not be done randomly, as then the probability that a good configuration will be created will be very small. Instead, something "intelligent" must be done. Furthermore, the recreation of a configuration should not take too much time. Thus, a construction heuristic is used that starts with the already existing partial solution, takes the nodes from the bag, and reinserts them into the system. For example, one could use the bestinsertion heuristic.

Thus we have defined three R & R moves: R & $\mathrm{R_{rad}}$ combines the radial ruin and the bestinsertion heuristic to a move, R & $\mathrm{R_{seq}}$ combines the sequential ruin and the bestinsertion heuristic, and R & $\mathrm{R_{ran}}$ consists of the random ruin and the bestinsertion heuristic. How these moves work is illustrated in Figs. 6.1 and 6.2. In the case of the radial ruin, the nodes to be removed lie in a circle because of the Euclidean metric. For a Manhattan or Tschebyscheff metric, they would lie in a square. In the case of the sequential ruin, the tour can be folded in a way such that the nodes also lie in some local region. Contrarily, the tour sequence can also be topologically extended through the whole region of the problem. Generally, these R & R moves are never local but have an impact on the whole system, as removed nodes may be placed in an other part of the tour by the bestinsertion heuristic.

As already mentioned in Chap. 10, one must consider a maximum fraction $\mathcal{P}$ up to which the system is destroyed. One finds that one should not remove too large fractions from the radial and sequential ruins, as then the area that must be recreated would become too large. However, one may of course remove even 50% of the system when using the random ruin, as an "overall skeleton" always remains from which a tour for the complete system can be reconstructed very well.

6.2 Analysis of R & R Moves in RW and GRE Modes

First, the question arises as to how to start the optimization run with these moves. It is a rather natural choice not to start with a randomly created configuration but with one that was fully created by the construction heuristic that is used by the recreate part of the R & R move. Thus, a "ruin & recreate all" move, i. e., the bestinsertion heuristic, is performed at the beginning. Furthermore, a suitable maximum fraction $\mathcal{P}$ must be selected for each R & R move. Then a sequence of R & R moves can be performed, e. g., in the random walk (RW) or in the greedy (GRE) mode.

However, one must be aware of the fact that the RW is not the type of RW that was performed with the small moves. For the small moves, new edges were selected at random, such that every configuration was created with the same probability and a real RW through the configuration space was performed by accepting each of these small moves. Here, however, also every move is accepted according to the RW acceptance criterion but the resulting configurations are not random because of the construction heuristic that tries to place every node to be reinserted on the best position in the tour.

If one plotted a histogram of the lengths of randomly created configurations and then a histogram of the lengths of configurations that were then altered by a series of small moves, one would get the same histogram as some edges are replaced by other randomly selected edges. However, here the question arises as to whether the histogram of configurations created by the bestinsertion heuristic is changed by R & R moves with $\mathcal{P} < 1$.

Figure 6.3 shows probability distributions of the PCB442 instance, to which we will restrict ourselves in this chapter, for constructing a solution with the bestinsertion heuristic (R & R_{all}) and then altering it by either 100 R & R_{rad} moves or by 100 R & R_{seq} moves or by 100 R & R_{ran} moves, depending on the maximum fraction $\mathcal{P} = 0.01, 0.02, 0.05, 0.1, 0.2,$ or 0.5 of the system to be destroyed. One finds that the probability distribution is shifted. For small fractions $\mathcal{P}$, the solutions are always improved by the R & R moves. This is quite obvious, as in this case the R & R moves that try to reinsert each removed node in the best possible way work in a quasigreedy mode. For large $\mathcal{P}$, the results for the three move types differ: R & R_{ran} clearly provides the best results and R & R_{rad} changes the distribution only slightly, whereas R & R_{seq} obviously worsens the configurations. These results can be easily explained: R & R_{ran} can best make use of the roundtrip through the remaining nodes as the nodes that were removed are uniformly distributed over the whole system, such that an overall frame remains on which the recreate part of the move can work very well. R & R_{rad}, however, constructs a part of the system in a sphere in a completely new way, with only a small number of hints from the surrounding area. The ruin part of R & R_{seq}, however, produces one long edge in the system; the removed nodes are then often inserted on other edges such that after the recreate this long edge might remain. Thus, this move is an example of a bad combination of a ruin and a recreate. Please note that this sequential ruin cannot be said to be generally bad. Only the combination with the bestinsertion heuristic as the recreate part of the move

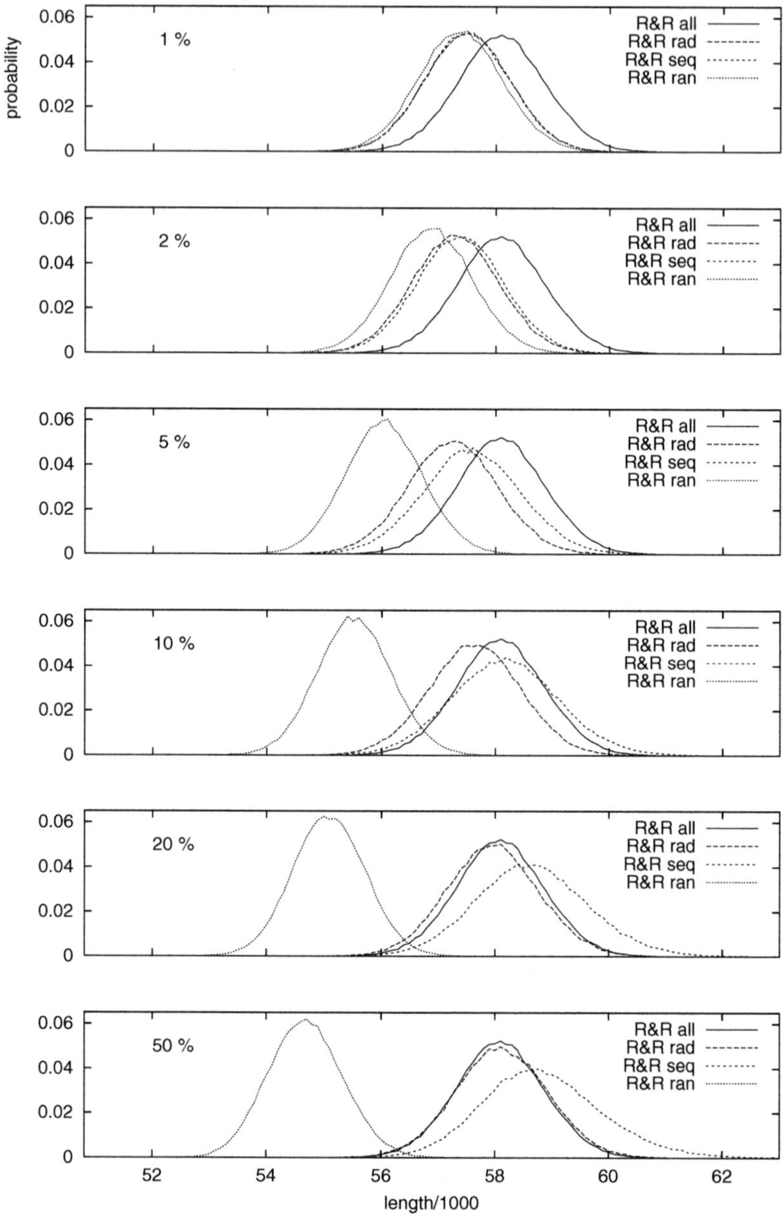

Fig. 6.3. Distribution of the lengths of 100,000 solutions created with the bestinsertion heuristic and then altered with 100 R & R moves in a random walk: in each graphic, the results are shown for a specific maximum fraction $\mathcal{P}$ up to which the system is destroyed. For not too large $\mathcal{P}$ the configuratons are improved. For small $\mathcal{P}$, the peaks of the distributions are always shifted toward smaller lengths. For the random ruin, the results get even better if $\mathcal{P}$ is large, whereas they get worse for the sequential ruin

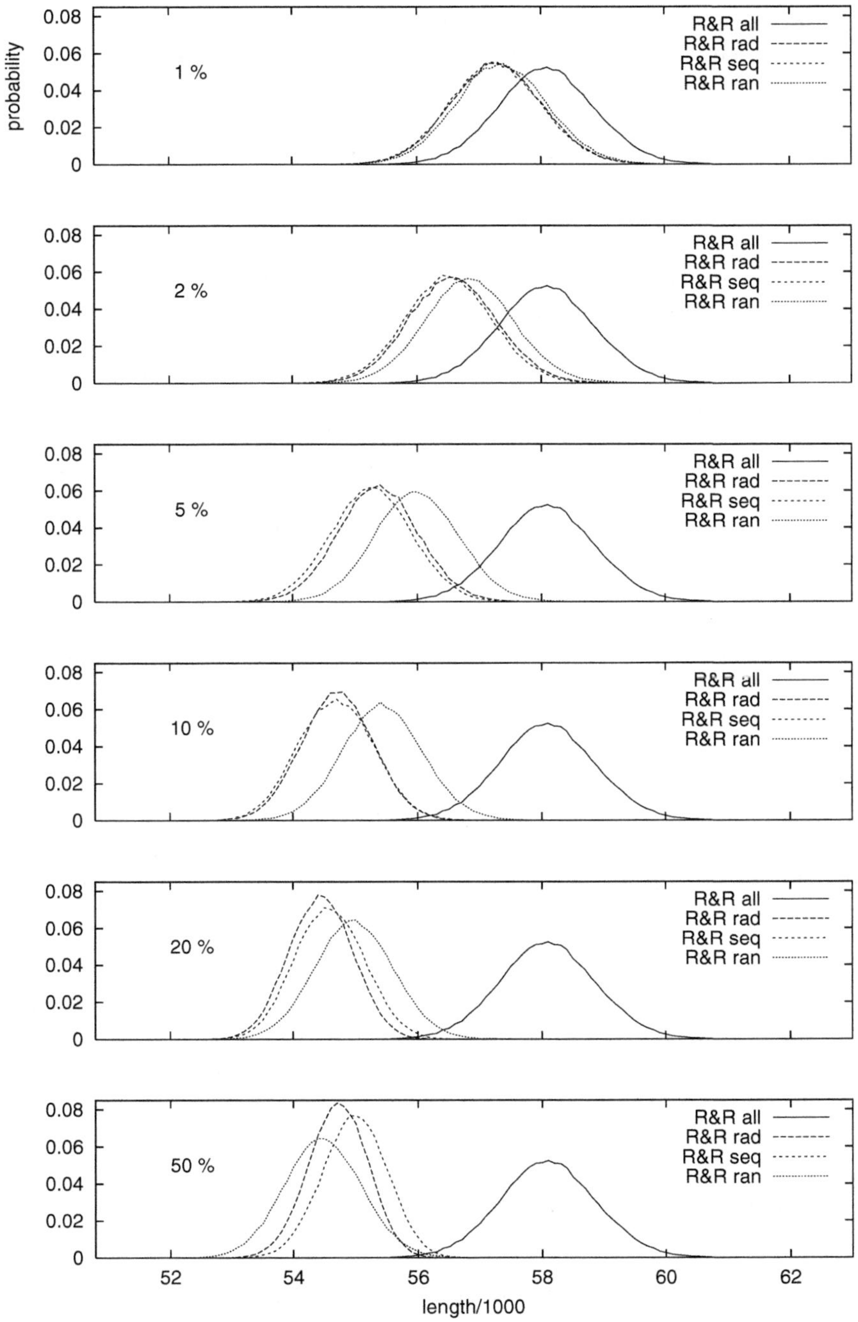

Fig. 6.4. Distribution of the lengths of 100,000 solutions, first created with the bestinsertion heuristic and then altered with 100 move trials in the greedy mode, for various $\mathcal{P}$

leads to bad results. Other recreates that, e. g., would reinsert all removed nodes between the end nodes of the long edge might lead to better results.

For comparison, analogous results for these moves shall be shown for the greedy algorithm if again applied to the PCB442 instance. Figure 6.4 shows that the results differ from those for the RW. First, the results are mostly better if using the greedy algorithm instead of the RW, which one would, of course, expect. The greatest differences are achieved for the radial and the sequential ruin, whereas the gain is relatively small for the random ruin. The results differ for small $\mathcal{P}$ only slightly as for these small fractions it is nearly equivalent to using RW or greedy: if, e. g., only one node is removed and bestinsertion looks for the cheapest possibility for reinserting the node, then one gets either the previous solution or a better solution. Here RW and greedy coincide. If a small number of nodes is removed, only small deviations from the greedy algorithm are possible. For larger $\mathcal{P}$, the distributions achieved with the greedy algorithm exhibit a sharper peak, and the peak is generally transferred to smaller lengths. For $\mathcal{P} = 0.5$ one gets worse results than for $\mathcal{P} = 0.2$ when the sequential or radial ruin is used. The optimum fraction up to which the system should be destroyed thus depends on the type of ruin here, too.

Thus far, only some trend has been observed for the various moves. Of course, these changes do not stop after 100 move trials, as can be observed in Fig. 6.5. The mean value of 10 optimization runs converges for large $\mathcal{P}$ toward a constant after roughly 1000 move trials if using the RW and working with the radial or sequential ruin. For small $\mathcal{P}$, the various runs do not differ much between the RW mode and the greedy mode. If using the random ruin, one gets a different behavior: the mean value continues with its decrease for all $\mathcal{P}$. Furthermore, it is remarkable that all curves decrease in a sigmoidal way. The best results are achieved if a large $\mathcal{P}$ is used because these moves lead to larger reorderings in the system on a larger scale and thus to more improvements. Again the radial and sequential ruins are more closely related to each other than to the random ruin. Additionally, one must consider that after the large amount of calculation time that was invested in these graphics, most curves indicate that even after such a huge number of move trials the system is not frozen in the greedy mode for larger and medium $\mathcal{P}$. Further improvements could be found especially for the random ruin.

6.3 Ruin & Recreate as Self-Contained Algorithm

All the Markovian improvement heuristics that start with some solution and try to improve it gradually and are more elaborate than the RW or the greedy have a property in common, namely, a control parameter. This control parameter is gradually reduced during the optimization run such that a transition between the RW mode and the greedy mode is performed. Of course, one can

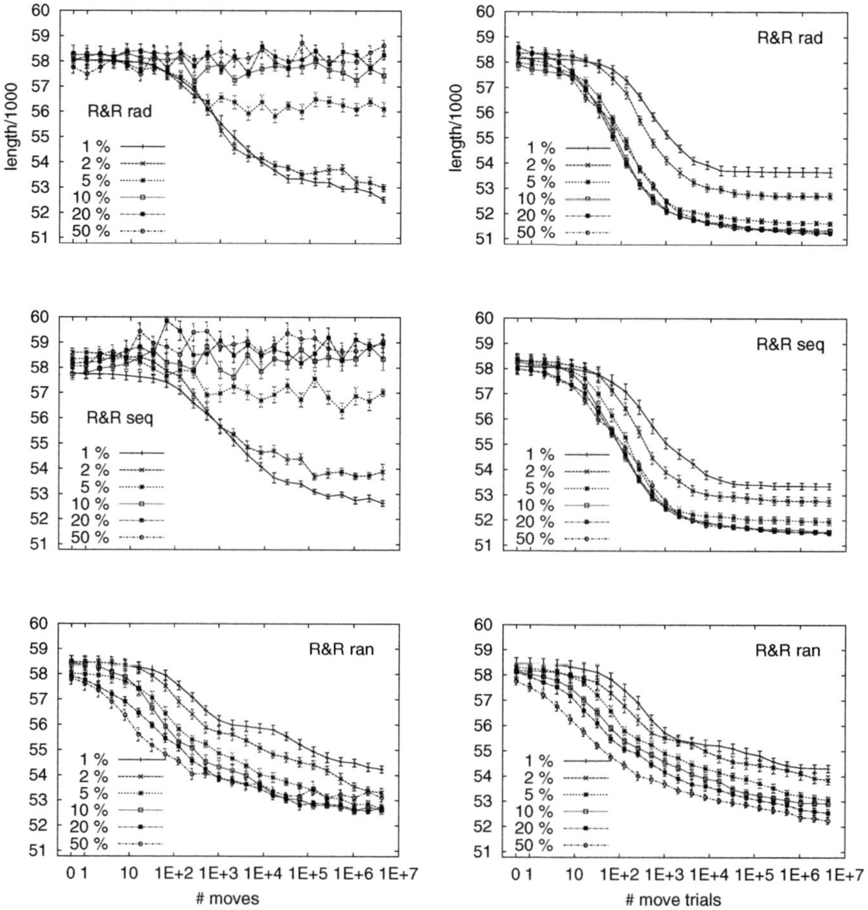

Fig. 6.5. Mean length averaged over ten runs as a function of the number of R & R moves for R & R_{rad}, R & R_{seq}, and R & R_{ran} with various $\mathcal{P}$. Zero moves corresponds to the initialization with the bestinsertion heuristic. *Left column*: results for RW; *right column*: results for greedy

use R & R moves also in optimization runs with, e. g., simulated annealing (SA), thus achieving even better results than with the greedy algorithm [194].

However, here we want to consider that R & R already contains the control parameter $\mathcal{P}$ up to which the system may be destroyed at a maximum. As shown above, the results achieved strongly depend on this parameter. One could also vary this parameter during the optimization run.

Figure 6.6 shows the results for an optimization run for the PCB442 instance with each of these three move types. The acceptance criterion used is that of the RW. The maximum number n of nodes to be removed is decreased linearly from $N-1$ to 1 in steps of 1. The decrease of $\mathcal{P}$ and of n leads

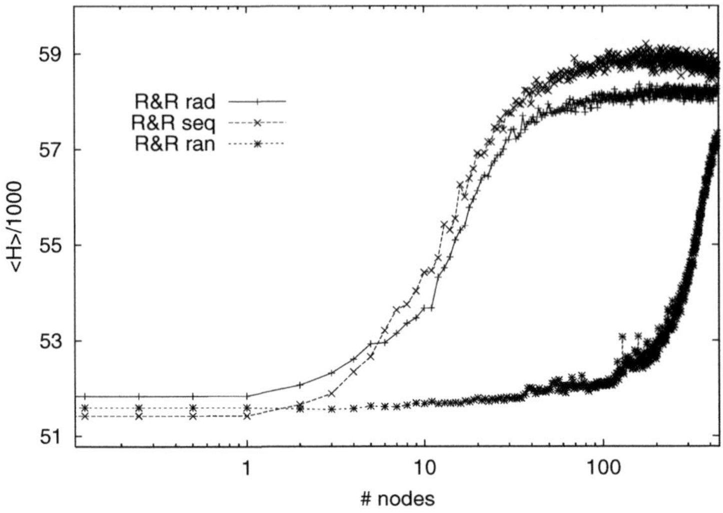

Fig. 6.6. Decrease of the length of the PCB442 instance with a decreasing maximum number of nodes to be removed in the ruin part of a R & R move: for each move type, one optimization run was performed. Whereas R & R_{rad} shows a nice sigmoidal decrease, R & R_{seq} clearly exhibits a knee at about $n = 200$ and R & R_{rad} drops immediately. The points which seem to be at $n < 1$ are still using $n = 1$

to a transition from the RW to the greedy regime. For the radial and the sequential ruins, one gets a sigmoidal decrease of the mean energy. Similar transitions will be achieved in the following chapters when working with SA and related algorithms. The curve for the sequential ruin exhibits a knee between $100 \leq n \leq 300$, which is in agreement with the results above.

6.4 Discussion of Application Possibilities of Ruin & Recreate

There are also other possibilities for defining R & R moves. For example, one could replace the bestinsertion heuristic by any other construction heuristic mentioned above. But one could also think of an improvement heuristic doing the work of the recreate part of the move: thus, such a large move would make use of either small moves or of other large moves of a smaller range. These moves are then accepted or rejected according to the acceptance criterion of the underlying optimization algorithm. If using, e. g., SA, one could think of a two-temperature concept: for the overall system, there is a global temperature T. But while applying a large move to a part of the system, a local temperature T_l overwrites T in the local part of the system. This local temperature is gradually reduced from a high value to zero, thus leading to a good solution of the local part of the problem. The rebuilt configuration of the overall system is then either accepted or rejected with the global temperature T [165].

One can show for both the above-mentioned R & R moves that make use of construction heuristics and for moves that use a local optimization run as the recreate part of the move that this R & R paradigm leads to better results in the same amount of calculation time than the standard local search approach [194, 165]. Furthermore, when not using the greedy algorithm, one should not mix small moves and these R & R moves, as this leads to worse results. It was shown above that the R & R moves, in contrast to small moves, lead not to random configurations if the RW criterion is applied but to much better configurations. If now additionally the Lin-2-opt, e. g., were applied in the RW, it would introduce two new randomly selected edges, leading on average to a worse configuration and in consequence worsening the R & R result. Analogously, the local and the large moves would feel different degrees of freedom if a more elaborate algorithm with a control parameter were applied at each value of this control parameter.

We will now return to the seven small moves and study the quality of algorithms based on these moves, as otherwise it would be unclear as to what improvement is due to the algorithm and what is due to the intelligence in the R & R moves. Furthermore, for large moves, one would always have to test which ruin and recreate combination is the best for the underlying optimization algorithm and to what extent the system should be destroyed in order to get the best results.

7 Application of Simulated Annealing to TSP

7.1 Simulated Annealing for the TSP

The application of simulated annealing (SA) to the traveling salesman problem (TSP) is rather straightforward. One usually starts with a random configuration and performs a series of moves changing the configuration gradually. One uses the same moves as for the greedy algorithm, but now the acceptance probability function is given by the Metropolis criterion [Eq. (11.6) in Part I]. The temperature is decreased from a high value to a value at which it is so small that the system is frozen. For the TSP, the energy difference is given by the difference between the lengths of the current configuration and the tentative new configuration, and the decrease of the temperature must be performed in an exponential way, i. e., by $T_{\text{new}} = f \times T_{\text{old}}$. We will generally work with $f = 0.99$ in what follows so as not to quench the system.

Thus, the outline of the algorithm is as follows:

- First, the system must be initialized by determining the distance matrix D between the N nodes and by creating a random configuration as initial configuration for the simulation.
- Second, an appropriate start temperature must be determined that is large enough, i. e., above the freezing temperature T_{f}. This can be done by, e. g., performing a random walk (RW) in which each move is accepted and by measuring the energy differences $\Delta\mathcal{H}_i$ occurring during this RW. The initial temperature can then be set to, e. g., $10 \times \max\{\Delta\mathcal{H}_i\}$, as described in Chap. 15 in Part I. Then the main part of the optimization process can be started.
- As long as the temperature is larger than some end value, do:
  - Perform a series of sweeps at each temperature step. Each move is accepted or rejected according to the Metropolis criterion.
  - Reduce the temperature T by a constant factor, i. e., $T = f \times T$.
- At last, the final configuration is printed.

Figures 7.1 and 7.2 show snapshots of an optimization run with SA applied to the PCB442 instance. The optimization run starts out with a randomly created configuration as shown in Fig. 4.1. Then it applies the Metropolis criterion at the high temperature value $T = 10{,}000$. At this temperature, one does not see any difference between these configurations when walking over them.

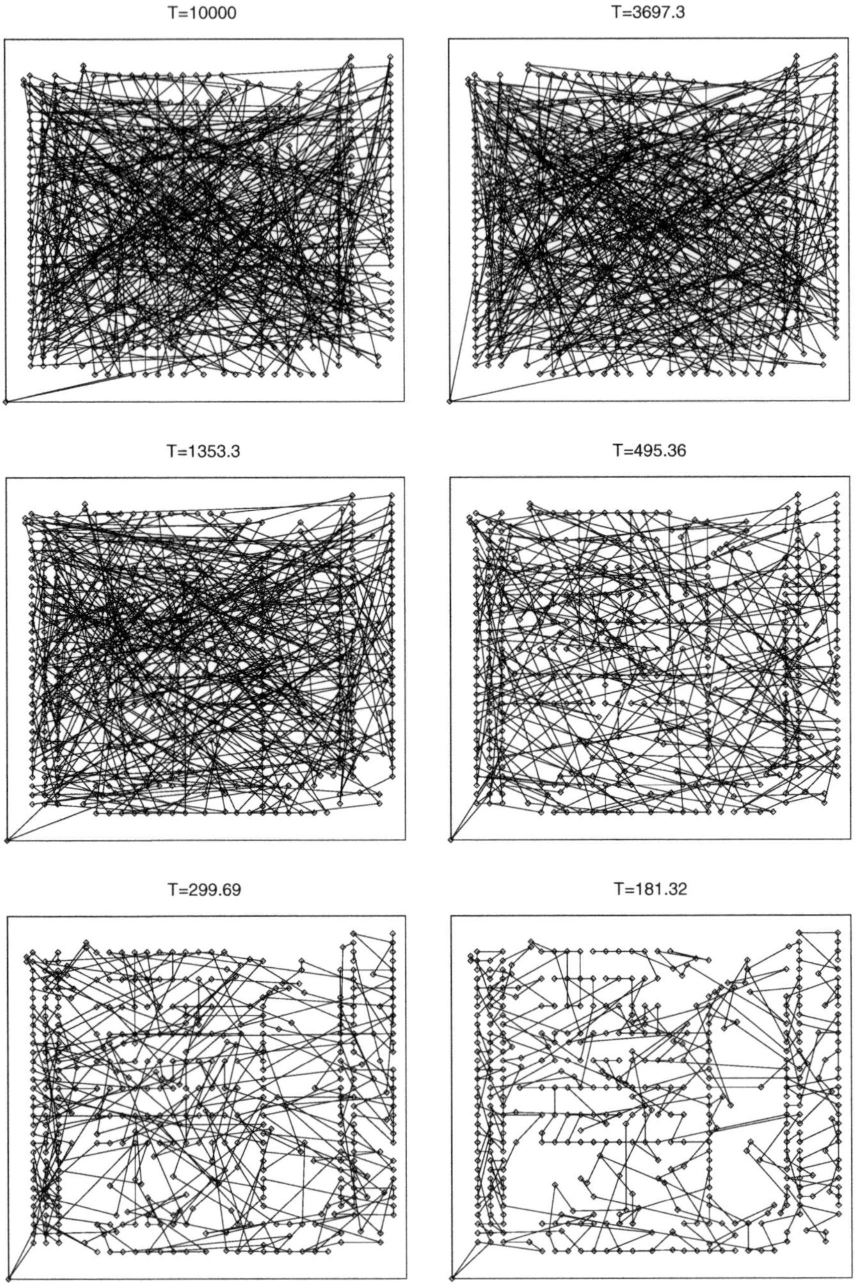

Fig. 7.1. Application of SA to the PCB442 instance at large temperatures $T > T_{\mathrm{f}}$: with decreasing temperature, the system leaves the range of the random configurations and gets to configurations with smaller energies

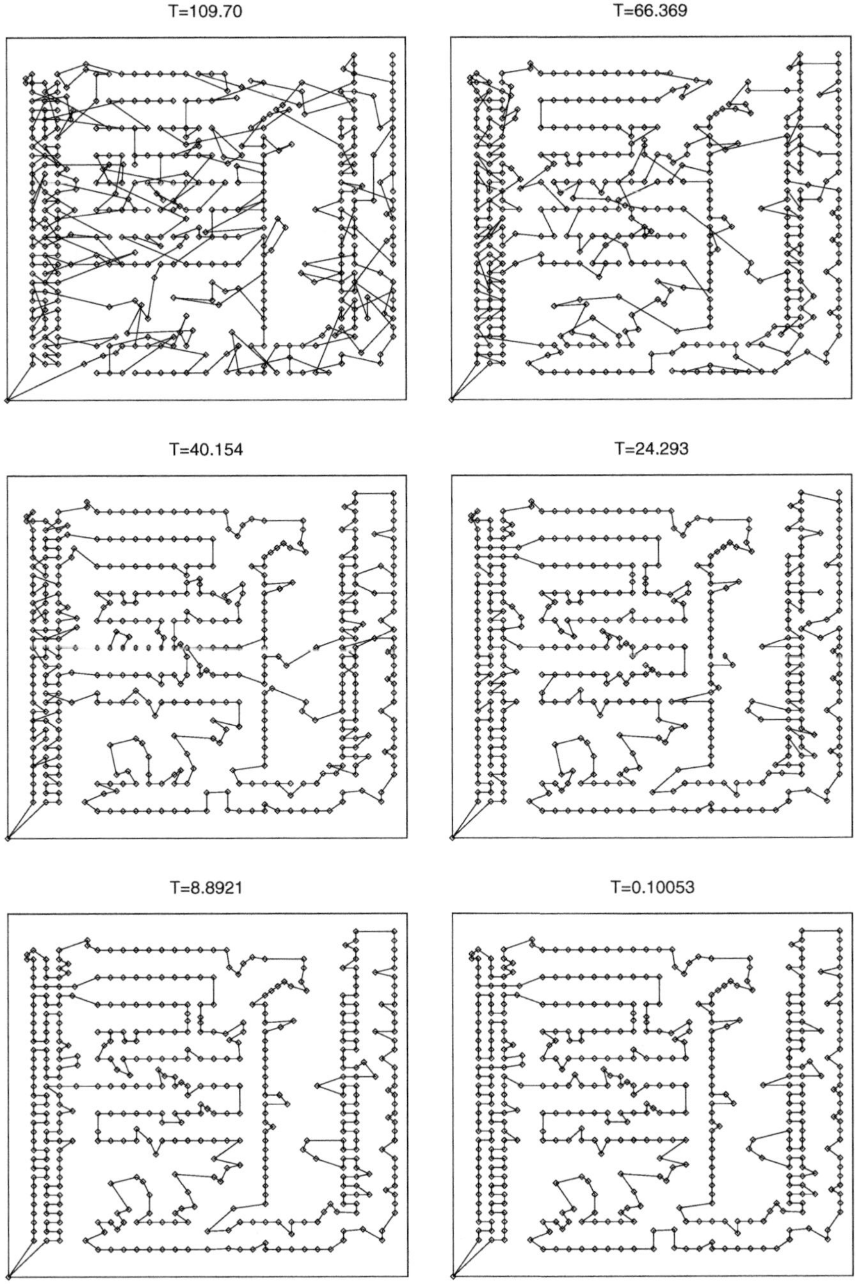

Fig. 7.2. Application of SA to the PCB442 instance at small temperatures $T < T_{\mathrm{f}}$: with decreasing temperature, the system freezes, orders itself, and, finally, gets stuck in a local or like here in the global optimum

Decreasing the temperature further, more and more short edges become part of the roundtrip, and the original spider net is thus thinned out. At a temperature $T \approx 180$, the system undergoes its transition from the unordered to the ordered phase. Then, at smaller temperatures, the system undergoes some local rearrangements and accepts some remaining improvements. But the system is no longer able to leave the local valley at these small temperatures. Summarizing, when the temperature is decreased, SA drives the system to be optimized gradually from the range of random configurations to the regime of ordered solutions by accepting changes in the configuration with the Metropolis criterion.

7.2 Computational Results for Observables of Interest

From a physicist's point of view, the first questions arising are what the observables of interest are and how they change during the cooling process. Two observables, which are always available for closer investigation, are the expectation value of the mean energy $\langle \mathcal{H} \rangle$ and the specific heat C.

Figure 7.3 shows the results for these observables of optimization runs for three TSP instances. In each run, the temperature was decreased from its initial value to the final value by a factor of 0.99, so that the number of temperature steps is in each case given by $[-5/\log_{10}(0.99)] = 1145$. In each temperature step, 1100 measurements were performed. The first 100 measurements were thrown away as the system needs some time to equilibrate at the new temperature value. Between two measurements, 30 sweeps were performed, such that the configurations whose measurements were taken are quite independent of each other. Each sweep contains N move trials. The seven small moves [exchange (EXC), node insertion move (NIM), Lin-2-opt (L2O), and the four variants of the Lin-3-opt (L3O)] are called with equal probability. Note that we present in this figure and in the following figures raw data, i. e., the real measurements are shown without any binning or otherwise smoothing of the data. Nicer looking figures can of course be obtained if the number of measurements is strongly increased or if the measured data are binned, for example by averaging over 5 to 25 successive data points.

For all three instances, the mean energy $\langle \mathcal{H} \rangle$ decreases smoothly in a sigmoidal way with decreasing temperature T on a logarithmic scale. Finally, it becomes constant at some small value of T. From that temperature on, the system is frozen. The right half of Fig. 7.3 shows the specific heat C, which is calculated according to the relation $C = \mathrm{Var}(\mathcal{H})/T^2 = (\langle \mathcal{H}^2 \rangle - \langle \mathcal{H} \rangle^2)/T^2$. One finds a nice rounded peak around the so-called freezing temperature T_{f} of the system. One notices that the fluctuations on the left leg of the peak are much larger than those on the right leg. This is simply due to the fact that the temperature is smaller such that fluctuations in the variance of the energy are not so suppressed as in the high-temperature regime. Here in all cases the specific heat vanishes at small temperatures. However, due to rounding

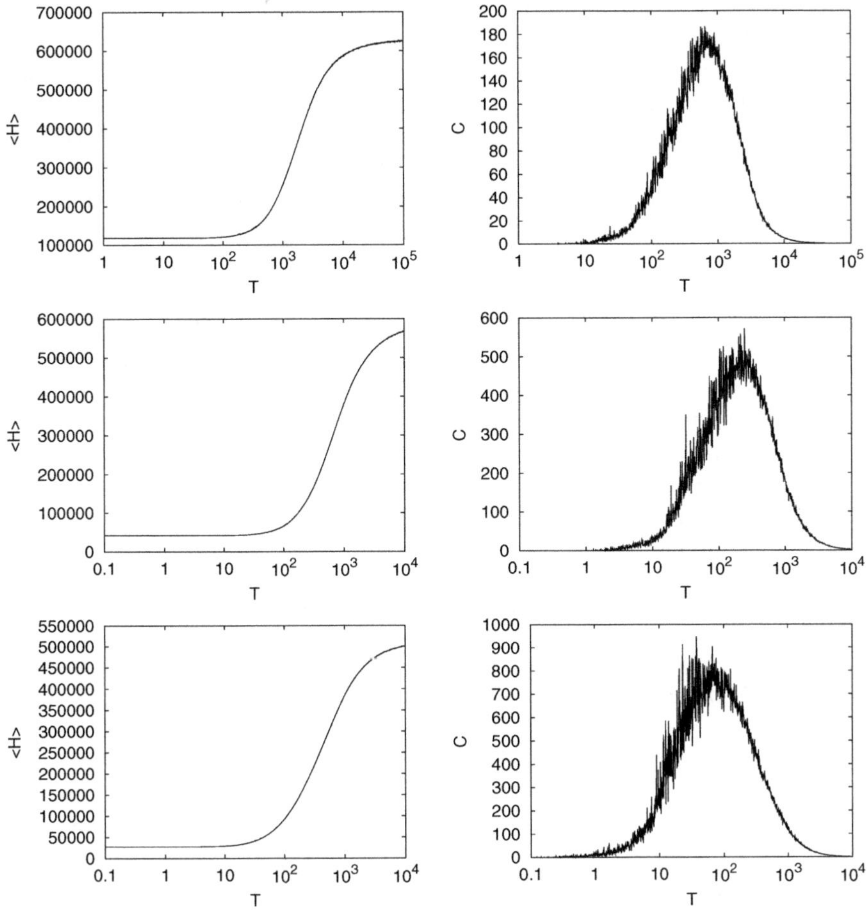

Fig. 7.3. Computational results for the mean energy $\langle \mathcal{H} \rangle$ (*left*) and the specific heat C (*right*) of the three TSP instances using SA: BEER127 (*top*), LIN318 (*middle*), ATT532 (*bottom*)

errors, it could sometimes happen that a small positive or negative value for the variance remains, so that one gets an increase in the specific heat at very small temperatures that is proportional to $1/T^2$.

The specific heat is very important for the optimization process when using SA. One must start the optimization run with an initial temperature larger than or at least at the freezing temperature T_f in order to get to good results. Furthermore, if the peak looks rather symmetric on a logarithmic scale, as it is here, then one should decrease the temperature in an exponential way as the system obviously freezes over several orders of magnitude of the temperature. Otherwise, if the peak appears rather symmetric if plotted with a linear T-axis, one should prefer a linear cooling schedule. Finally, one should always have a look at the figure of the specific heat. If it does not show a nice peak but large fluctuations in the form of several sharp spikes, then the calculation time must be increased as otherwise the system will be simply quenched down. There is also another possibility for measuring the specific heat, namely, by using the definition of the specific heat, $C = \partial\langle\mathcal{H}\rangle/\partial T$. One can thus calculate numerically this derivative using the results for the mean energy values. If the specific heat looks exactly the same as if calculated via the variance of the energy, one can say that the simulation is in equilibrium. However, this is only a weak condition for having a true equilibrium. Mostly, however, one only refers to the method of calculating the specific heat via the energy variance, as it is numerically more stable, so that one sees a real peak with a much smaller amount of calculation time as would be necessary if one were to use the derivative definition which, if the calculation time was not long enough, only shows quasirandom fluctuations.

Mostly, only the mean energy and the specific heat are available for a closer investigation of the problem. However, sometimes one has deeper insight into the problem and is able to define an appropriate order parameter ξ. Then one can virtually extend the Hamiltonian by

$$\tilde{\mathcal{H}}(\sigma) = \mathcal{H}(\sigma) - \lambda\xi(\sigma) \tag{7.1}$$

with some Lagrange multiplier λ with $\lambda \to 0$. Thus, one can consider the mean value $\langle\xi\rangle$ of this control parameter and also the corresponding susceptibility

$$\chi = \frac{\partial\langle\xi\rangle}{\partial\lambda} = \frac{\mathrm{Var}(\xi)}{T} \,. \tag{7.2}$$

This susceptibility can be considered the answer by the system to the force applied by the Lagrange multiplier λ via the Zeeman term $-\lambda\xi$ in the Hamiltonian, such that it will order itself according to the considered order parameter.

For the TSP, an appropriate order parameter was already introduced, namely, the projection of the current configuration on the ground state by comparing the edges used in the configuration, as sketched in Sect. 1.9. The results for this order parameter and its corresponding susceptibility are shown in Fig. 7.4. The order parameter increases with decreasing temperature in a sigmoidal way. But only if the optimization run ends up in the global optimum of the problem does the order parameter converge to 1, as is the case for the BEER127 instance. For the other two instances, the optimization runs ended up at some quasioptimum configurations that have many edges in common with the corresponding global optimum configurations, so that the final value is about 0.7 in both cases.

The corresponding susceptibilities, which are shown in the right half of Fig. 7.4, exhibit (despite fluctuations) a nice peak at the system's critical temperature T_{c}. This peak is much sharper than the peak of the specific heat and, in addition, is at a lower temperature, the temperature at which

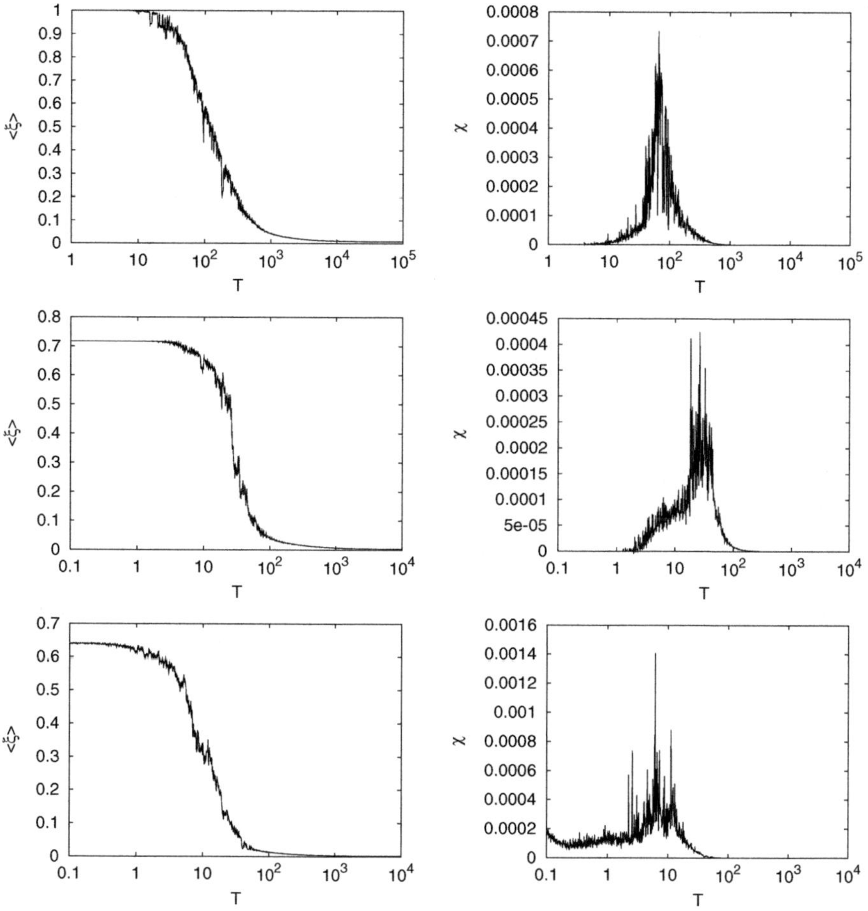

Fig. 7.4. Computational results for the expectation value $\langle \xi \rangle$ of the order parameter χ (*left*) and the corresponding susceptibility χ (*right*) of three TSP instances using SA: BEER127 (*top*), LIN318 (*middle*), ATT532 (*bottom*)

the ordering transition really takes place. Thus, if one has a good order parameter for some problem, it is advantageous to use the susceptibility at least in addition to the specific heat. Looking closer at the susceptibility of the LIN318 instance, one suspects that the peak might be a double peak, which might come from an ordering and clustering effect as in [110], so that one achieves first an ordering in the large length scale, whereas the small parts are ordered inside themselves at smaller temperatures. However, one must perform more measurements in order to verify this assumption. Investigating the graphics for the susceptibility of the ATT532 instance more closely, one finds that it increases for very small temperatures proportional to $1/T$. This comes from a rounding error in the calculation of the variance of the order parameter. This variance should vanish; however, numerically a small positive value remains such that the susceptibility increases hyperbolically.

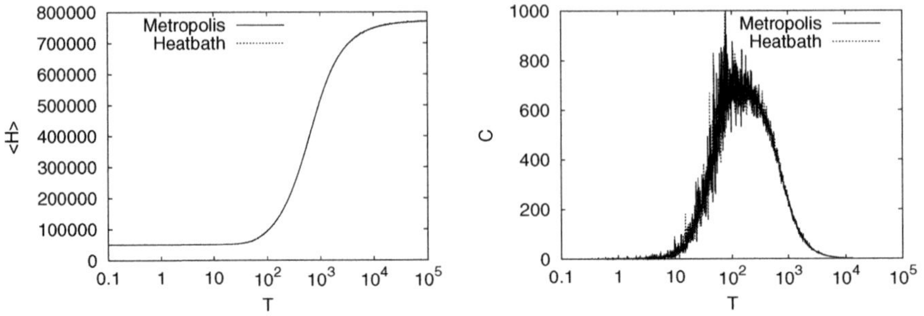

Fig. 7.5. Computational results for the mean energy $\langle \mathcal{H} \rangle$ (*left*) and the specific heat C (*right*) of the PCB442 instance using SA with either the Metropolis criterion or the heat bath criterion

Finally, the question arises as to whether we get the same results for these observables when using other acceptance criteria for SA. As mentioned in the derivation of SA in Chap. 11 in Part I, some arbitrariness in the explicit choice of the transition probability remains. Above, we used the most widely used Metropolis criterion. Figure 7.5 shows a comparison of the results for the mean energy and the specific heat of the PCB442 instance when using either the Metropolis criterion or the heat bath criterion. We find that the use of either of these two criteria leads to the same results. This is not surprising as both criteria are intended to lead the system to the Boltzmann equilibrium, such that the expectation values for the observables must be identical.

7.3 Computational Results for Acceptance Rates

Next the freezing behavior of the optimization process if using SA shall be studied. For this purpose, first of all the total acceptance rate A of the moves must be considered. Figure 7.6 shows the results for the acceptance rate for the three TSP instances. In all three cases, one finds a nice sigmoidal decrease of the acceptance rate with decreasing temperature. Note that the peak of the specific heat is located at a temperature at which less than 10% of the moves are accepted. As it is sufficient to start at the peak of the specific heat, one should always consider that most moves are rejected during the optimization run. Thus, one should program the move routines accordingly. For example, one should always calculate the energy difference without explicitly applying the move if possible and appropriate. Furthermore note that the total acceptance rate nearly vanishes at the order transition.

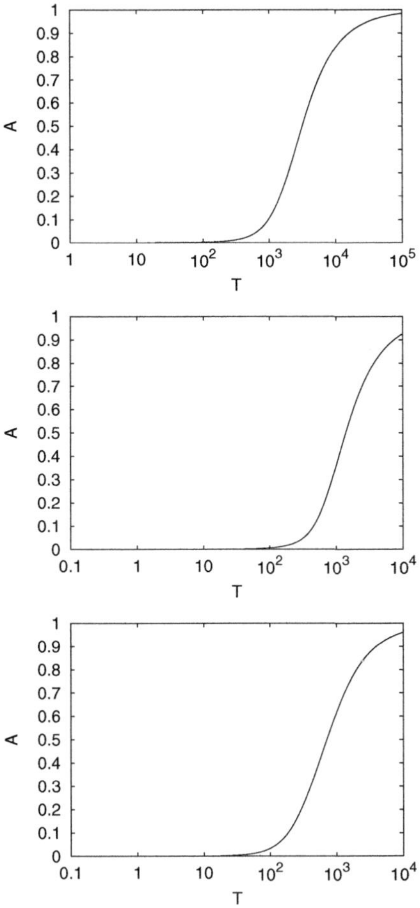

Fig. 7.6. Computational results for the total move acceptance rate A of the three TSP instances using SA: BEER127 (*top*), LIN318 (*middle*), ATT532 (*bottom*)

However, not only the total acceptance rate A for all moves but also the partial acceptance rates A_i of the individual moves are of interest. Figure 7.7 shows the results for the seven applied moves. One finds at first glance that the acceptance rates for the four variants of the L3O coincide with each other. However, there are small differences between the sigmoidal decreases in the acceptance rates for the moves EXC, NIM, and L2O. The acceptance rate for the EXC decreases at slightly higher temperatures than the acceptance rate for the NIM and this again at slightly higher temperatures than the acceptance rate of the L2O. This can be easily explained: as the EXC changes four, the NIM three, and the L2O only two edges and as the edge lengths of the new and of the old edges are summed up, when calculating the energy difference $\Delta\mathcal{H}$, the occurring absolute energy differences for the EXC are

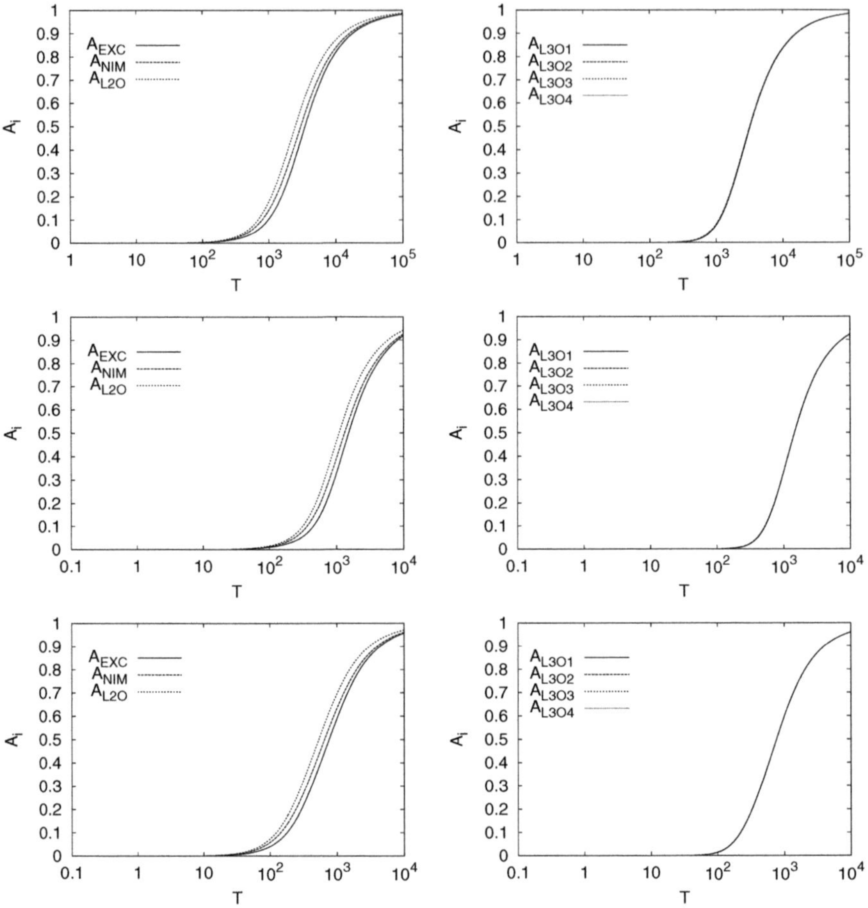

Fig. 7.7. Computational results for the partial acceptance rates A_i of the moves EXC, NIM, and L2O (*left*) and of the four variants of the L3O (*right*) for three TSP instances using SA: BEER127 (*top*), LIN318 (*middle), ATT532 (*bottom*)*

on average larger than those for the NIM and even more than those of the L2O. This difference in the average deteriorations leads to different transition temperatures. The acceptance rates of the L3Os nearly coincide with the acceptance rate for the EXC, but their decreases are a bit steeper.

However, one could also have implemented various moves for which acceptance rates decrease at very different temperatures. If one has thus a move a with an acceptance rate decreasing at rather high temperatures and a move b with an acceptance rate decreasing much later in the optimization process, it usually does not make any sense to call the move a often at very small temperatures. Generally, one could observe the acceptance rates of the individual moves and then decide for the next temperature step how often each move will be called. This decision can also be influenced by the averaged (absolute) size of the energy changes a move causes, i. e., a move is also worthwile even if seldom accepted, if it leads to large improvements.

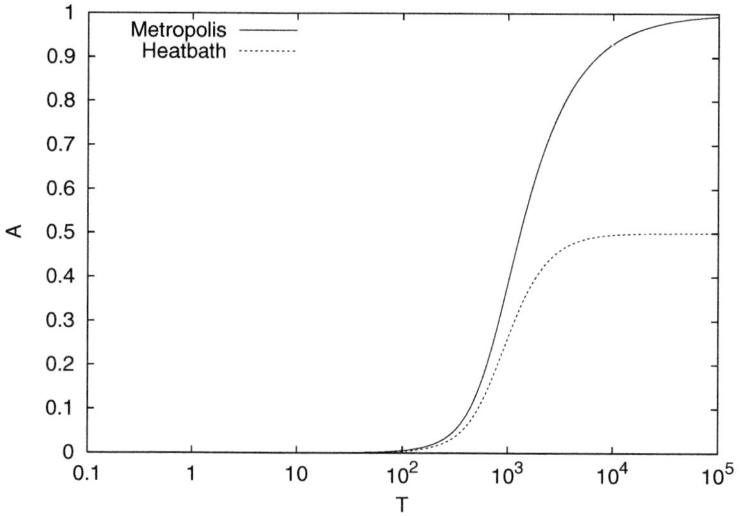

Fig. 7.8. Computational results for the total move acceptance rate A of the PCB442 instances using SA with either the Metropolis criterion or the heat bath criterion

Again we ultimately want to compare the results achieved with the Metropolis criterion and the heat bath criterion. Figure 7.8 shows the curves for the total acceptance rate A when using one of these two acceptance criteria. We find that the heat bath acceptance probability starts out at 0.5 at very high temperatures and decreases sigmoidally to zero with decreasing temperatures. It is always smaller than the Metropolis acceptance probability.

The system performs in both cases the same transition from the high-energy random configurations to the low-energy ordered solutions. The starting value of the acceptance rate of 0.5 simply denotes that 50% of the moves are accepted. But the system is indeed in a quasi-RW mode at these high temperatures. Trivial moves with $\Delta\mathcal{H} = 0$ are accepted with a probability of 0.5 by the heat bath criterion. Also, the other moves are accepted with probabilities of ≈ 0.5. Thus, the heat bath criterion leads at high temperatures basically to a quasi-RW in which on average every second move is accepted, nearly independently of its energy difference. Using the Metropolis criterion at high temperatures, the improving moves are always accepted and the worsening moves are nearly always accepted, so that the total acceptance rate is slightly smaller than 1. Using the heat bath criterion, the total acceptance rate cancels out to 0.5 at high temperatures, as the acceptance probability for any move is roughly 0.5. Thus, the dynamics is slowed down if replacing the Metropolis criterion by the heat bath criterion, but the results for the expectation values of the observables remain, of course, the same.

7.4 Quality of the Results Achieved with Various Computing Times

Finally, we want to consider the quality of the results that can be achieved by spending a certain amount of computing time for a SA run in comparison to the quality of the results achieved with the greedy algorithm when spending exactly the same amount of time. Figure 7.9 shows the quality of the results vs. the calculation time spent. For each TSP instance, 100 optimization runs were performed; the mean value is thus averaged over 100 final configurations, so that the error bars are the same size as the symbols. Analogously, the minima and maxima are also taken from these 100 runs each. (Henceforth, the terms minimum, mean, and maximum length will always refer to the results of 100 optimization runs unless it is clearly stated otherwise.) The calculation time is denoted as the number of Monte Carlo sweeps spent per temperature step of SA. For the greedy algorithm, corresponding calculation times were used; the Metropolis criterion, which is used for SA, was simply replaced by the greedy criterion.

In the SA runs, the BEER127 instance was cooled down from the initial temperature $T_\mathrm{i} = 10^5$ to the final temperature $T_\mathrm{f} = 1$ in each optimization run. Additionally, a final greedy step was performed. Analogously, the LIN318, PCB442, and ATT532 instances were cooled down from $T_\mathrm{i} = 10^4$ to $T_\mathrm{f} = 0.1$, then a final greedy step was performed. For the NRW1379 instance, the parameters were $T_\mathrm{i} = 10^3$ and $T_\mathrm{f} = 10^{-2}$; also, a greedy step at $T = 0$ was performed at the end. These fixed values were set in order to be able to compare the individual results better as a function of the calculation time spent and in order to allow each system the transition from the unordered to the ordered regime.

The cooling schedule was exponential with a cooling factor of $f = 0.99$. Thus, all in all, 1147 temperature steps were performed in each optimization run for each instance. The number of sweeps per temperature step differed: they were 1, 3, 10, 30, 100, ..., or 30000. Summarizing, the overall number of sweeps spent is given by the number of temperature steps multiplied by this number of sweeps per temperature step.

In order to have slightly more space in the graphics, the lengths of the final configurations were divided by a factor of 1000, such that the labeling at the y-axis takes up less space.

Figure 7.9 clearly shows that the lengths achieved with SA decrease with increasing calculation time, at first rather fast, but then the improvement diminishes if more calculation time is invested. For the larger instances, one finds that the minimum and maximum qualities found for a certain amount

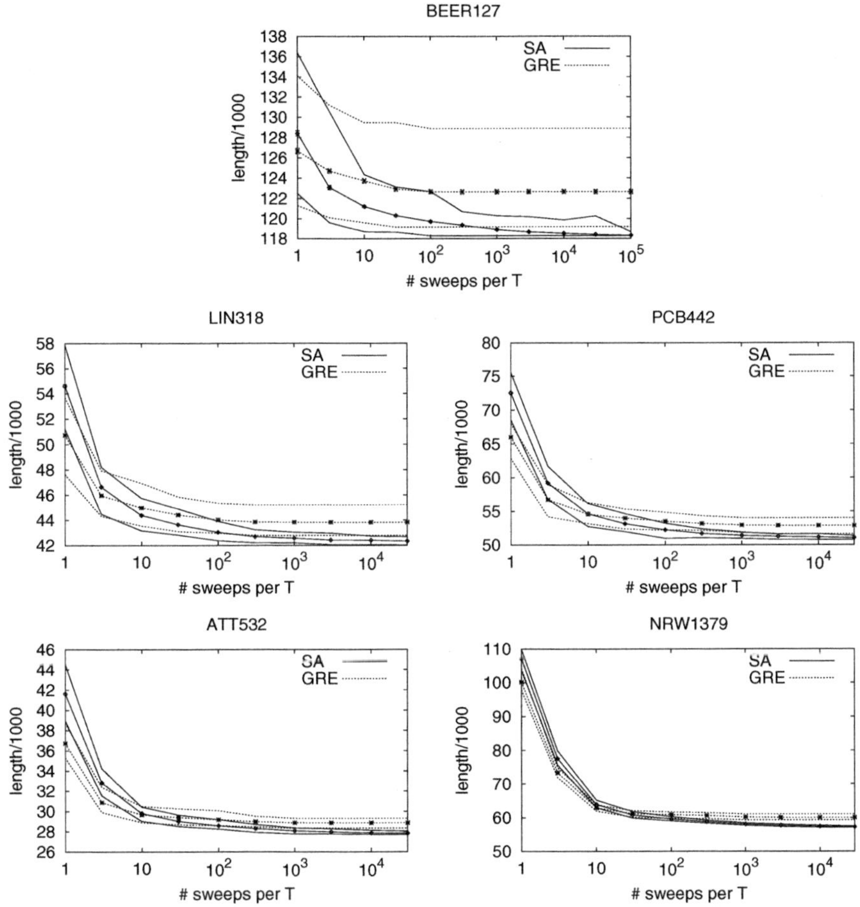

Fig. 7.9. Quality of the results achieved with SA (*straight lines*) and the greedy algorithm (*dotted lines*) for five TSP instances (BEER127, LIN318, PCB442, ATT532, and NRW1379) vs. number of sweeps per temperature step: the three *lines* show the minimum lengths, mean lengths (*error bars* of the size of the *symbols*), and maximum lengths

of calculation time are not far from each other if the number of sweeps is increased beyond 1000 sweeps per temperature step. Furthermore, if the calculation time is increased by a factor of 10, the worst result out of 100 runs is then nearly as good as the best result previously. Thus, when working with SA, it is generally best to spend the whole calculation time available on one long optimization run. Performing many short optimization runs clearly leads to worse results. Comparing these results for SA with those for the greedy algorithm, one clearly finds that the greedy algorithm is superior for very short calculation times as it spends the whole calculation time at $T = 0$, where it only accepts improvements, whereas SA gradually quenches the system down to this regime but initially also allows deteriorations. However, as more calculation time is spent, SA surpasses the greedy algorithm. One finds for all instances that the greedy algorithm, which leads to a freezing of the system in some local minimum at $\approx 3 \times 10^3$ sweeps per temperature step, i. e., after $\approx 3 \times 10^6$ overall sweeps, cannot lead to such good configurations as SA is able to produce for longer calculation times. Furthermore, one finds for the greedy that the variation between results of different runs is greater than that was seen for SA.

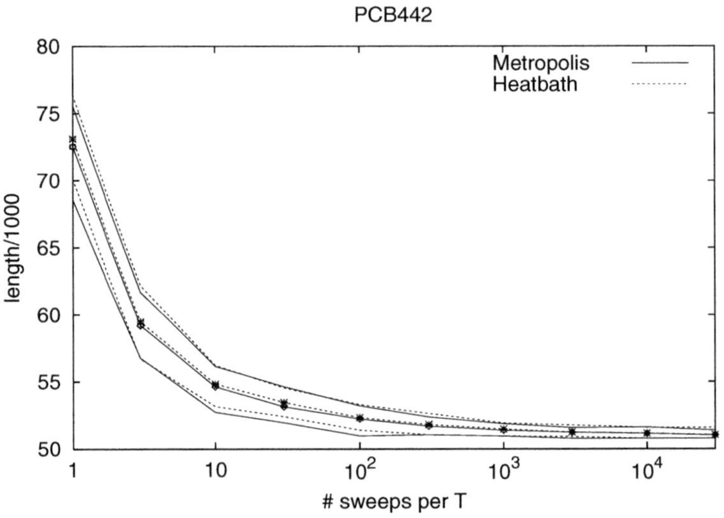

Fig. 7.10. Quality of the results achieved with SA for the PCB442 instance using either the Metropolis criterion (*straight lines*) or the heat bath criterion (*dotted lines*) vs. number of sweeps per temperature step. The three *lines* show the minimum lengths, mean lengths (*error bars* of the size of the *symbols*), and maximum lengths. The results for the Metropolis criterion are replotted from Fig. 7.9

Next we again investigate whether it makes a difference to use the heat bath criterion instead of the Metropolis criterion. Figure 7.10 compares the results of these two acceptance criteria for the PCB442 instance. In both cases, the same parameters for the starting value of the temperature and so on were used. Looking closely, we find that the Metropolis criterion always leads on average to better results. For short computing times, this difference is significant, as can be seen at the error bars; for long computing times, it is only marginal. This result can easily be explained by the fact that the dynamics is faster if using the Metropolis criterion. However, it is not always advantageous to use the Metropolis criterion: if it takes a large amount of calculation time to actually update a configuration if a move is accepted and only a comparatively small amount to calculate the energy difference of the proposed move, then it might sometimes be better to use the heat bath criterion and perform more sweeps per temperature step. We, however, will stay with the Metropolis criterion here. This scenario of working with SA using the Metropolis criterion and using the parameters mentioned above will become our standard case with which we will usually compare the results for other algorithms in the following chapters.

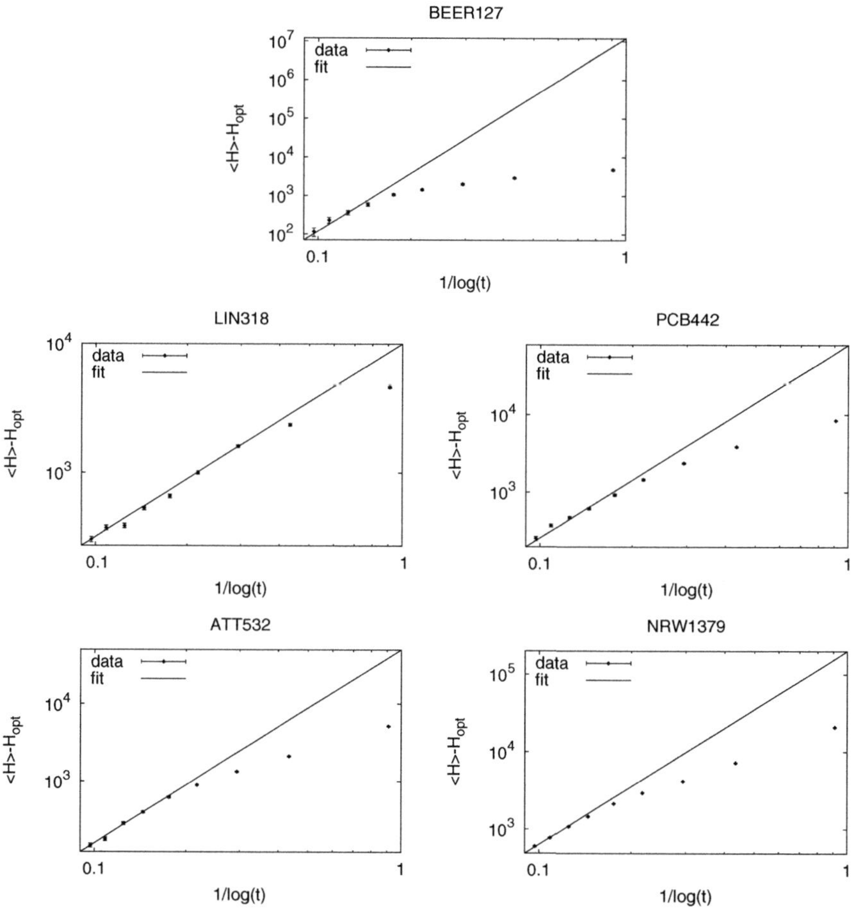

Fig. 7.11. Investigating the Grest hypothesis on data for SA shown in Fig. 7.9: the deviation of the mean length to the optimum is plotted vs. the inverse of the logarithm of the calculation time, which is simply measured in number of sweeps per temperature step. A hand-made fit leads to a dependency as $1/\log(t)^\zeta$ with $\zeta \approx 5$ for the BEER127 instance, $\zeta \approx 1.5$ for the LIN318 instance, and $\zeta \approx 2.5$ for the other three instances

Based on the results for SA in Fig. 7.9, we investigate the Grest hypothesis [Eq. (16.1) in Part I] for these five TSP instances: according to Grest, the mean deviation of the results achieved with SA after some time t from the globally optimum solution $\mathcal{H}_{\text{opt}}$ is proportional to $1/\log(t)^{\zeta}$ with some power ζ. Figure 7.11 shows the data of Fig. 7.9 replotted in such a way as to be able to verify the Grest hypothesis. In a double logarithmic plot, the exponent ζ is clearly not roughly 1. Furthermore, it is impossible to get all points on one straight line; there is even a slight curvature for those points that correspond to long computing times. We used only those for fitting as they are the more important points. We get $\zeta \approx 5$ for the BEER127 instance and $\zeta \approx 1.5$ for the LIN318 instance. Only for the three largest instances do we clearly find an exponent of $\zeta \approx 2.5$. Thus, for large TSP instances, the Grest hypothesis might hold for long computing times.

8 Dependencies of SA Results on Moves and Cooling Process

8.1 Results for Various Small Moves

In this chapter, we want to study which things have an impact on the quality of the results. These "things" are the moves and the cooling schedule.

First we want to start with an investigation into what level of quality can be reached with a particular move type. Thus, instead of calling a "one move trial routine", which then again calls with an equal probability of 1/7 one of the seven moves exchange (EXC), node insertion move (NIM), Lin-2-opt (L2O), and the four variants of the L3O, one could, e. g., simply call only one of these moves all the time.

Here we present only results for the PCB442 instance, but the conclusions can be made generally. The optimization runs were performed in the same way as in Sect. 7.4, i. e., the PCB442 instance was cooled down from the initial temperature 10^4 to 0.1 with a cooling factor of 0.99. Finally, one greedy step was performed, such that 1147 temperature steps were performed. Thus, the move routine was called 1147 (the number of temperature steps) times the number of sweeps per temperature times 442 (the number of TSP nodes).

Figure 8.1 shows the results for various calculation times depending on the move used. In the "all" scenario, every move was called with equal probability, such that these results are identical to those in Fig. 7.9 for the PCB442 instance. At first glance, we find that the results differ strongly for the various moves: the worst results are clearly achieved with the EXC for all calculation times. For short calculation times, the L2O leads to the best results, followed by the "all moves" scenario and the NIM. Thus, with only a rather small amount of calculation time, it is best to work with a good small move only and not to use larger moves. However, usually one does not know in advance which of the small moves will lead to good results. Furthermore, in this time regime, which lasts only fractions of seconds, it is better first to use a construction heuristics and then, if there is still time available, to proceed with an afterburner in the greedy mode.

For longer calculation times, however, the curves for the various moves except the EXC seem to coincide. In order to better be able to study the differences between the moves, we enlarge them by looking at the relative deviation

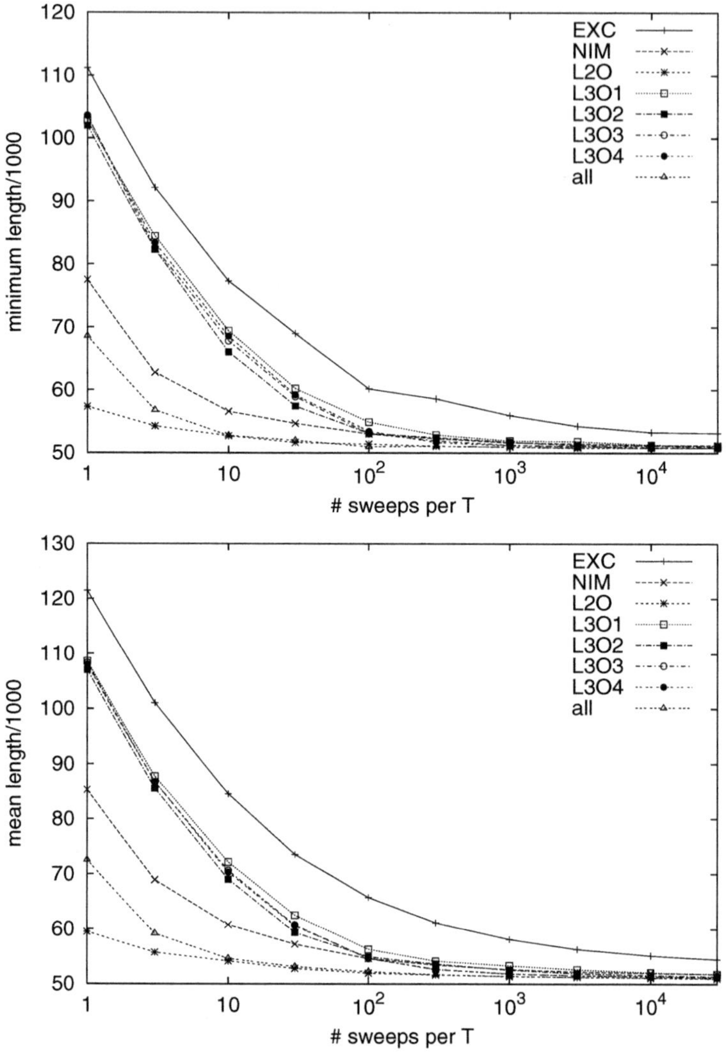

Fig. 8.1. Minimum and mean lengths vs. calculation time when using only one of the seven moves or all moves when applying SA to the PCB442 instance

$$\delta_x = \frac{x - \mathcal{H}_{\text{opt}}}{\mathcal{H}_{\text{opt}}} \tag{8.1}$$

of the results to the optimum value of the TSP instance. x denotes either the mean value $\langle \mathcal{H} \rangle$ or the minimum value found. A deviation of 0.1 thus means that the value is 10% worse than the optimum value.

Figure 8.2 shows the results for the relative deviations derived from the data shown in Fig. 8.1. The differences are enlarged with a logarithmic y-axis.

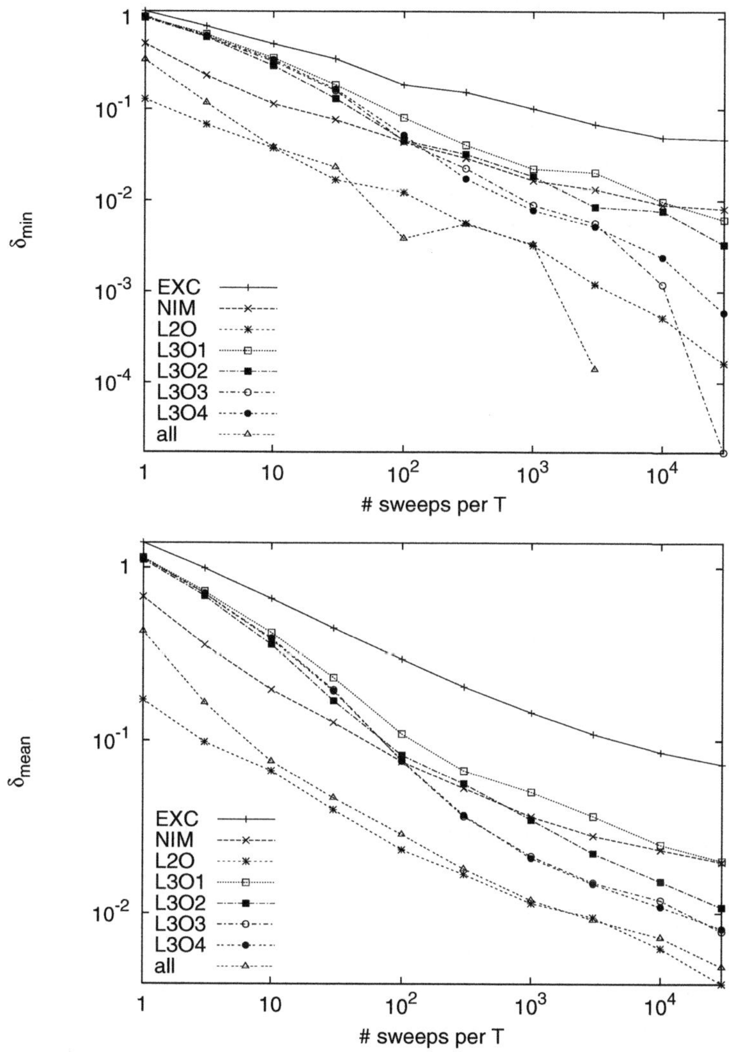

Fig. 8.2. Minimum and mean deviation δ [as defined in Eq. (8.1)] vs. calculation time (based on data of Fig. 8.1)

Points not shown are thus exactly 0. One finds that the mean deviation is approximately the same for the L2O and the "all moves" scenario, in the long time range. Furthermore, these two scenarios lead to the best results, followed by the L3Os and the NIM. Looking at δ_{min}, we find that the "all moves" scenario finally finds the optimum if the number of sweeps per temperature step is at least 10^4, whereas the L2O alone does not lead to it.

Summarizing, it is best to implement as many smallest-order and next-higher-order moves as possible. The often heard sentence "You only need to implement one small move as the algorithm will do it all" is simply false or at least not generally true. Surely, one might find a move like the L2O for a specific problem, which leads to rather the same quality of results as the ensemble of moves, but then it must be the right move and not, e. g., the analog to the EXC. This additional implementation work pays off in the quality of the results, especially if more calculation time is invested.

Thus, we will return to our "all moves" scenario, in which each of these seven small moves is called with equal probability. Of course, one could think of even more complex move routines, in which some moves are called more often than others. Furthermore, these calling rates could be changed adaptively due to the acceptance rates of each type of move and the energy changes or gains to which they lead at each temperature. One could even move on to some system with artificial intelligence or a neural network that then adapts the calling rates. However, as far as we know, there is no general rule set for all possible problems on how to choose the calling rates of the individual moves: one move type could be more important than another type if it is accepted more often, if it leads to larger improvements, or if it leads to a larger acceptance rate for another move after an accepted move trial. There are many possibilities for implementing an adaptive calling algorithm based on rules of thumb containing these importance measures. Of course, such an algorithm could lead to even better results, especially for short calculation times [180]. We, however, want to stay with our simple calling procedure.

8.2 Results for Monotonous Cooling Schedules

The next question is how strong the influence of the cooling schedule is on the quality of the results. First, we will consider the standard exponential cooling schedule as usual, not with a cooling factor of $f = 0.99$ but with smaller cooling factors of $0.1 \leq f \leq 0.98$. Again the system is cooled down from $T_{\mathrm{i}} = 10^4$ to $T_{\mathrm{f}} = 0.1$.

Figure 8.3 shows the results for these cooling factors. Note that here the results are not plotted vs. the number of sweeps per temperature step but vs. the overall number of sweeps. As a smaller cooling factor is used, the number of temperature steps decreases. The number of temperature steps for each cooling factor is given in Table 8.1. We find that the curves for the different cooling factors seem to coincide totally. Thus, at first sight, the results seem to be independent of the cooling factor f, as the curves for the various f scale nicely.

We again blow up the differences by looking at the relative deviation as defined in Eq. (8.1). Figure 8.4 clearly shows that for short calculation times, the results are independent of the cooling factor used. In the long time period, however, the results for the smallest cooling factors $f = 0.1$ and $f = 0.2$ are

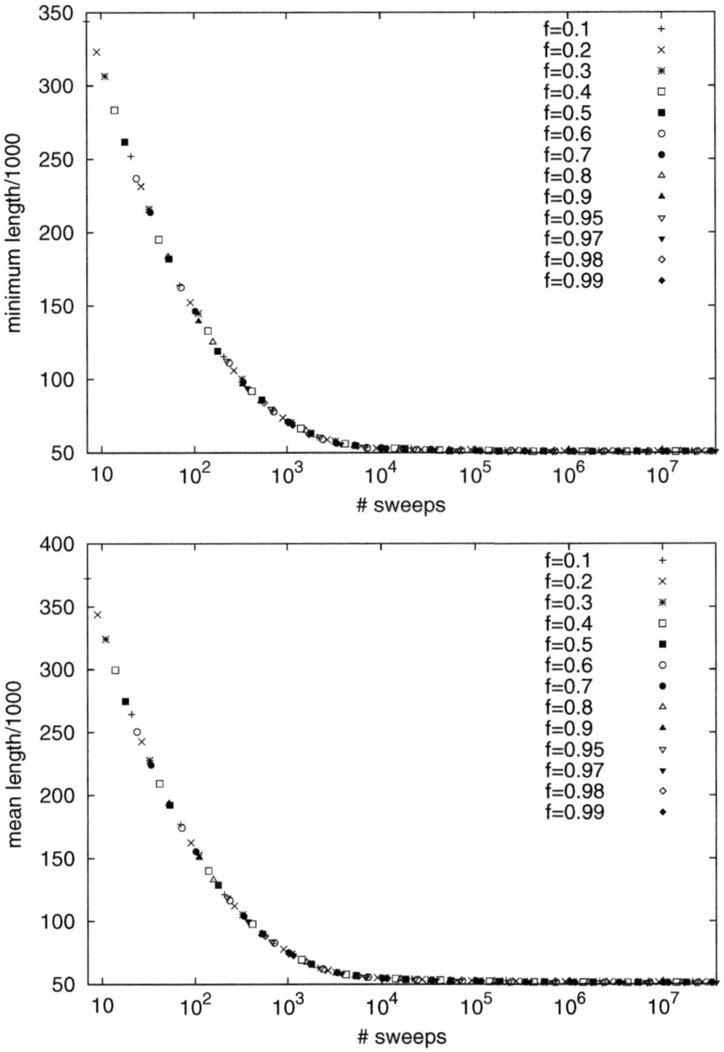

Fig. 8.3. Minimum and mean lengths vs. number of sweeps for various cooling factors when applying SA to the PCB442 instance

Table 8.1. Number of temperature steps for various cooling factors for the results shown in Fig. 8.3

Cooling factor	0.1	0.2	0.3	0.4	0.5	0.6	0.7	0.8	0.9	0.95	0.97	0.98	0.99
Temperature steps	7	9	11	14	18	24	34	53	111	226	379	571	1147

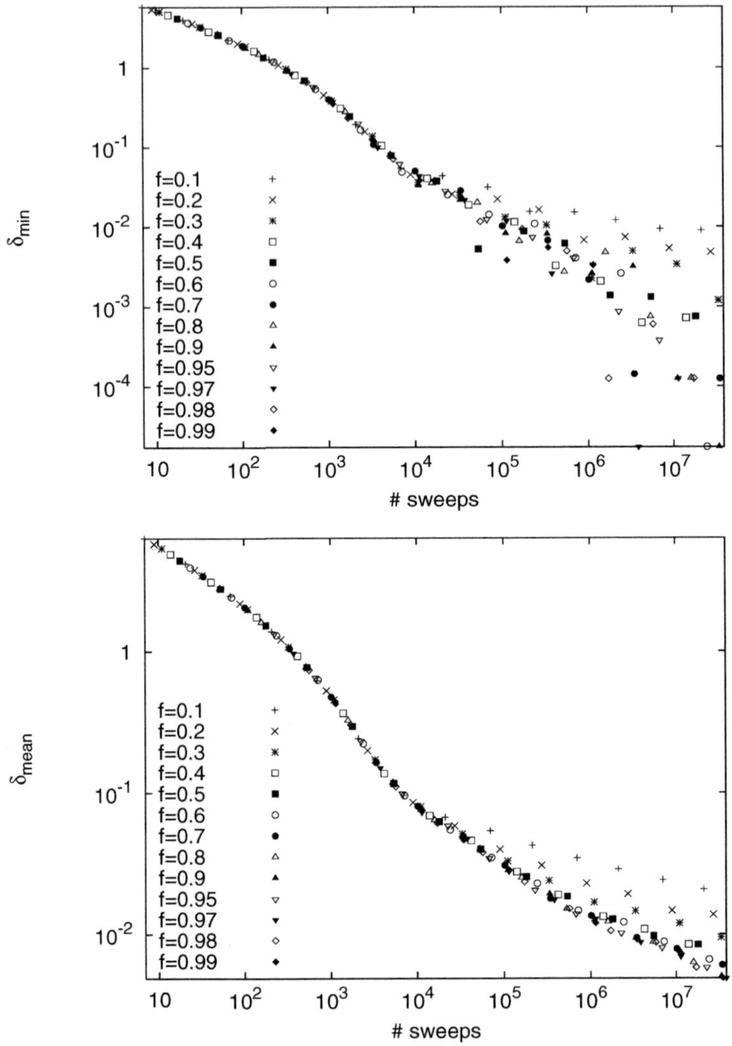

Fig. 8.4. Minimum and mean deviation δ [as defined in Eq. (8.1)] vs. calculation time (based on the data of Fig. 8.3)

significantly worse than those for the larger cooling factors. First, the curve for $f = 0.1$ deviates from the other curves, then that of $f = 0.2$, and, finally, even the curve for $f = 0.3$ starts to move away from the other curves that lead to better results.

Summarizing, we wish to state that it is better to use a larger cooling factor. According to our experience also with other problems, one should generally work with $f \geq 0.8$.

The second question when restricting oneself to monotonic cooling schedules, as one usually does, is at what initial temperature T_i should the optimization run be started and at what final temperature T_f should it end. As already mentioned in Chap. 11 in Part I, the initial temperature T_i must be large enough such that the system is not quenched down in a greedylike way. Additionally, the final temperature T_f must be so small that the system is virtually frozen. Here we want to study the quenching effect only, as it does not make much sense to study the quality of "high-temperature solutions". Thus, we use the "good" cooling factor $f = 0.99$, which leads to good results, as shown above, and the final temperature $T_\mathrm{f} = 0.1$, at which the PCB442 system is frozen. Again we cool the system from an initial temperature T_i, which is to be varied here, exponentially down to T_f and finally add a greedy step, such that we get the overall number of temperature steps shown in Table 8.2.

Table 8.2. Number of temperature steps for various initial temperatures T_i

Initial temperature	0.3	1	3	10	30	100	300	1000	3000	10000	30000	100000
Temperature steps	111	231	340	460	569	689	798	918	1027	1147	1256	1376

Figure 8.5 shows the results for various initial temperatures T_i. As the figure makes clear, the results differ between the small initial temperatures $T_\mathrm{i} \leq 10$ and the large $T_\mathrm{i} \geq 100$. The points for $T_\mathrm{i} = 30$ lie in between these two scenarios. In the short time period, the results for small T_i are better. This result is in accordance with results shown in Fig. 7.9 and can be explained like them: for very short computing times, the greedy algorithm leads to better results than SA. Starting at a rather small initial temperature means quenching the system. One can interpret the greedy as some kind of superquench of the system. Thus, these results for small T_i must be better for short calculation times as no time is lost at high temperatures where also large deteriorations are accepted. For long calculation times, however, we clearly find that the results for small T_i are worse than those for $T_\mathrm{i} \geq 100$, which seem to be rather identical.

To see these differences better, we blow them up by looking at the deviation from the optimum as defined in Eq. (8.1). Figure 8.6 shows the relative deviation of the minimum result found and of the mean results. We clearly find for long computing times that, for $3 \leq T_\mathrm{i} \leq 100$, the smaller the initial temperature, the worse the results. The results for $T_\mathrm{i} \leq 3$ lie close together, so obviously it does not make any difference whether $T_\mathrm{i} = 3$ or $T_\mathrm{i} = 0.3$ as the system is in the greedy mode anyway. On the other hand, for $T_\mathrm{i} \geq 100$, we cannot see significant differences between the runs with different T_i but the same number of sweeps per temperature step. (These results form the groups of seven points in the graphic for δ_mean.) Thus, increasing the initial tem-

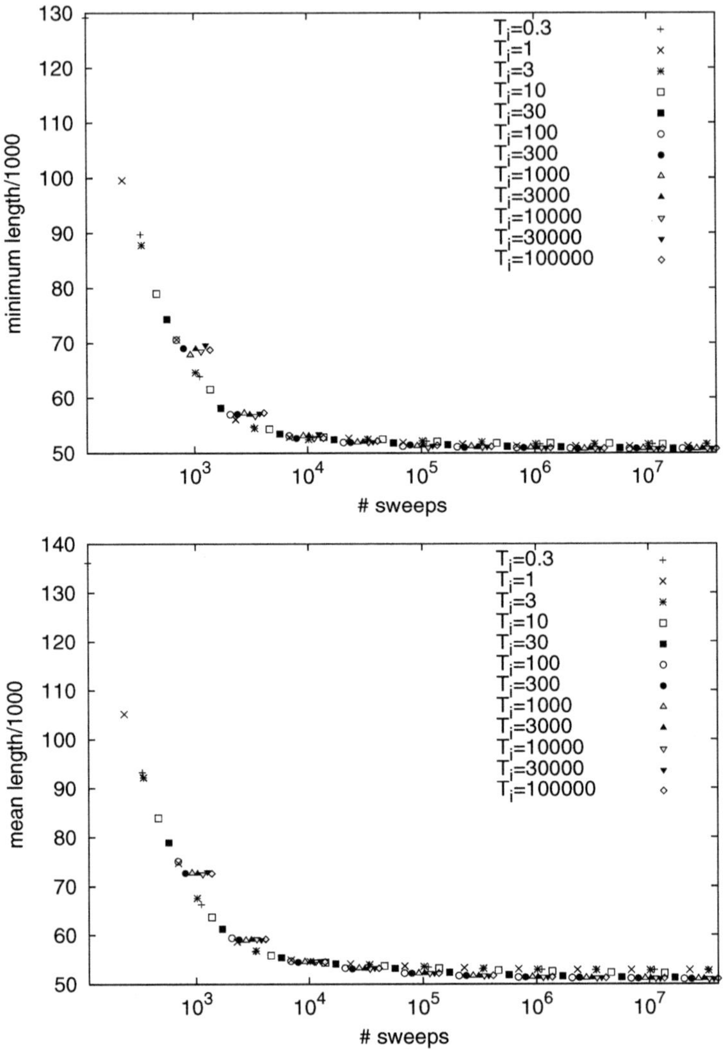

Fig. 8.5. Minimum and mean lengths vs. number of sweeps for various initial temperatures T_{i} when applying SA to the PCB442 instance

perature T_{i} too much means wasting calculation time at high temperatures as we find that we get rather equally good results for $10^2 \leq T_{\mathrm{i}} \leq 10^5$ here: cooling down the system exponentially from 10^5 to 0.1 means trying twice as many moves as cooling it down from 10^2 to 0.1 if the number of move trials per temperature step stays the same. Furthermore, more moves are accepted at larger temperatures, such that also more computing time for updating the configuration and thus really applying the move must be invested. Note that

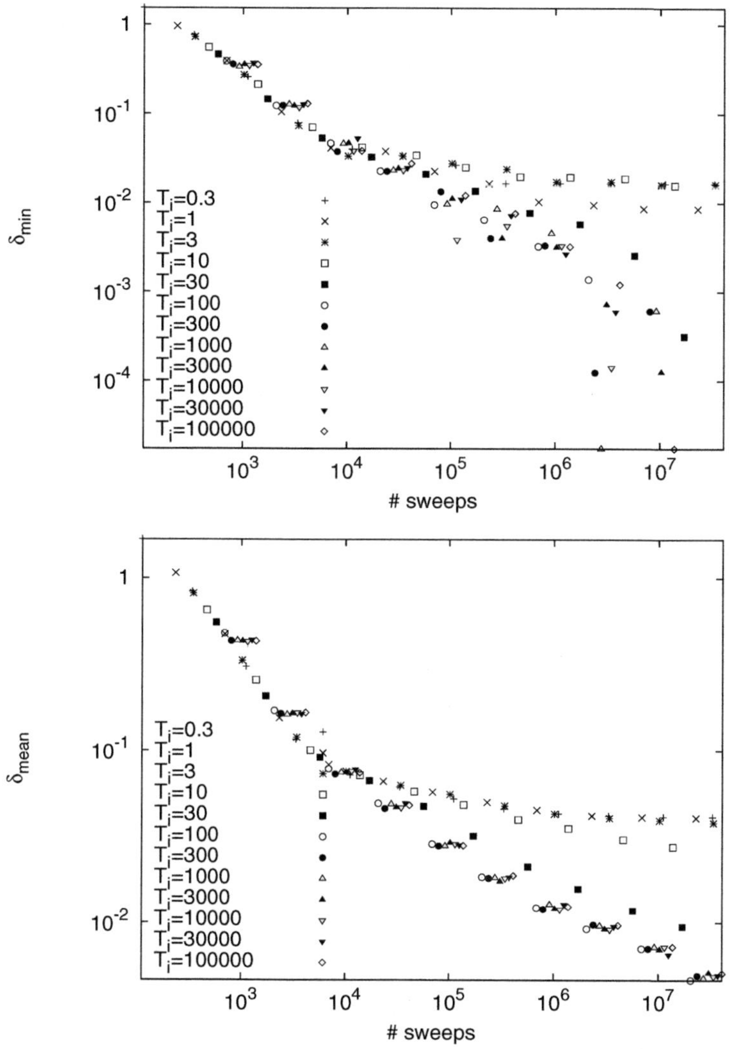

Fig. 8.6. Minimum and mean lengths vs. number of sweeps for various initial temperatures $T_{\rm i}$ when applying SA to the PCB442 instance

we do not want to make a plaedoyer for using small initial temperatures. If working on more complex problems than the TSP, it might be that the system is restricted to some subspace in the search space if $T_{\rm i}$ is chosen too small. Thus, we recommend using the criteria introduced in Chap. 15, Part I, even if some calculation time might be wasted. This is still much better than starting at too small temperatures and thus getting worse results as shown here.

Of course, one could also work with more elaborate monotonous cooling schedules in which the cooling factor is changed adaptively, which also leads to better results, if the calculation time is concentrated on the important temperature range. But again, although optimization libraries like the TopC library developed at the IBM Scientific Center Heidelberg containing such elaborate techniques have been developed, adapting each of the many parameters to a given problem instance in a perfect way still requires much testing [180].

8.3 Results for Bouncing

The next question is whether the results can be improved if a nonmonotonous cooling schedule is used. We want to investigate this question by a rather straightforward approach to nonmonotonous cooling called bouncing: in a first iteration, the system is cooled down from a high initial temperature T_i to a freezing temperature T_f as usual. For this first iteration, which will be called iteration No. 1 in this section, we will use parameters as before: we cool down the PCB442 instance from $T_\mathrm{i} = 10{,}000$ to $T_\mathrm{f} = 0.1$ with an exponential cooling schedule using a cooling factor of $f = 0.99$—which makes 1146 temperature steps—and finish the first iteration with a greedy step at $T = 0$. Then we perform a series of bouncing iterations, in which

- We take the final configuration of the previous iteration as the starting configuration,
- We set the starting temperature to a specific value T_B,
- We cool the system down to $T_\mathrm{f} = 0.1$ in an exponential schedule with 1146 cooling steps, and
- Finally, we perform one greedy step.

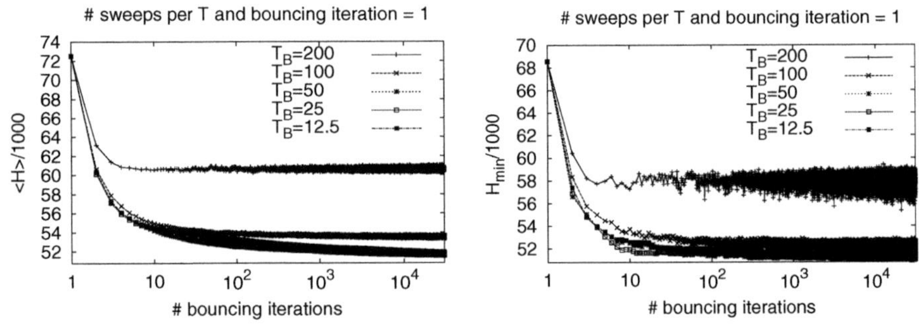

Fig. 8.7. Bouncing the PCB442 instance: mean and minimum lengths achieved with 100 optimization runs for various bouncing temperatures T_B ($T_\mathrm{B} = 200, 100, 50, 25, 12.5$) and for various numbers of sweeps per temperature step in each bouncing iteration vs. the number of bouncing iterations. Note that the lower bound in each graphic is exactly at the optimum value 50,783.5 ... for the PCB442 instance

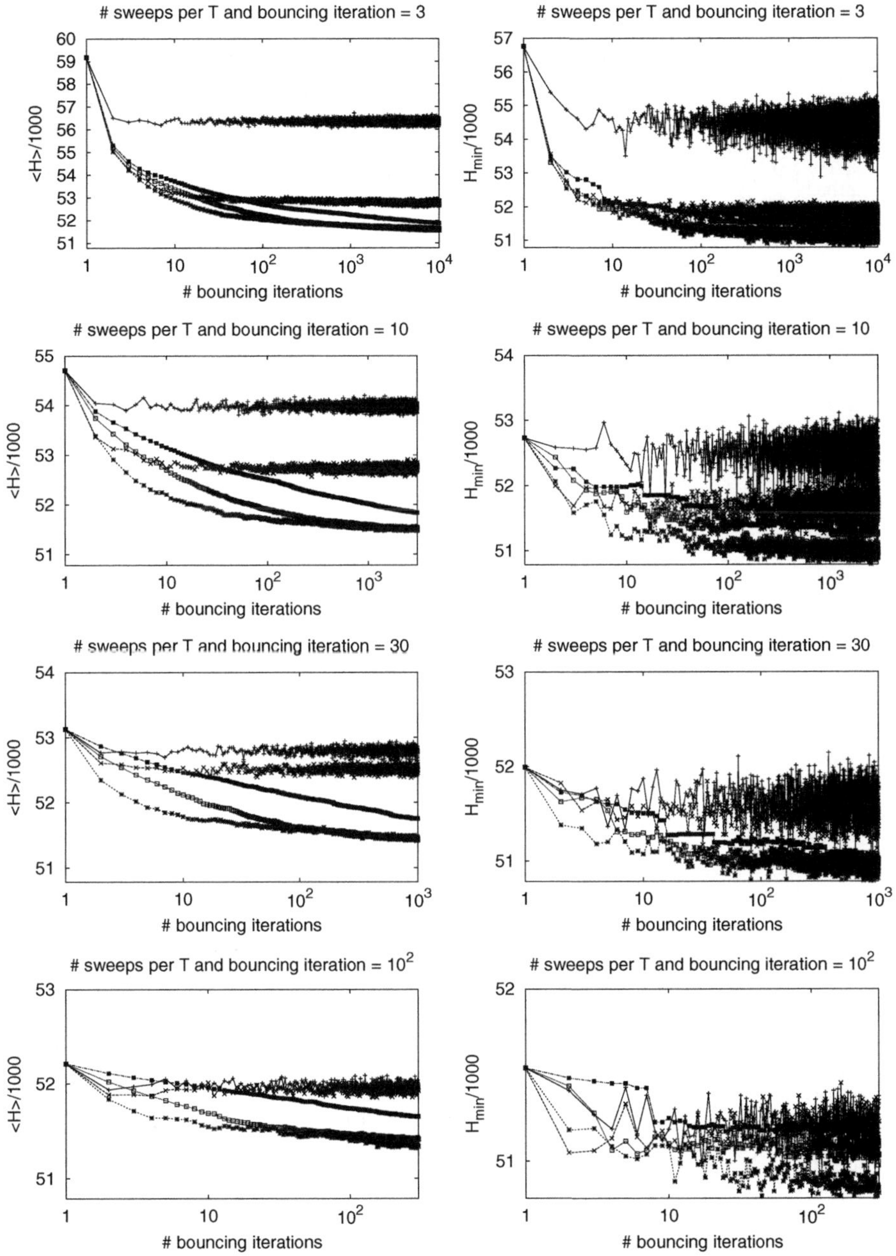

Fig. 8.7. (continued)

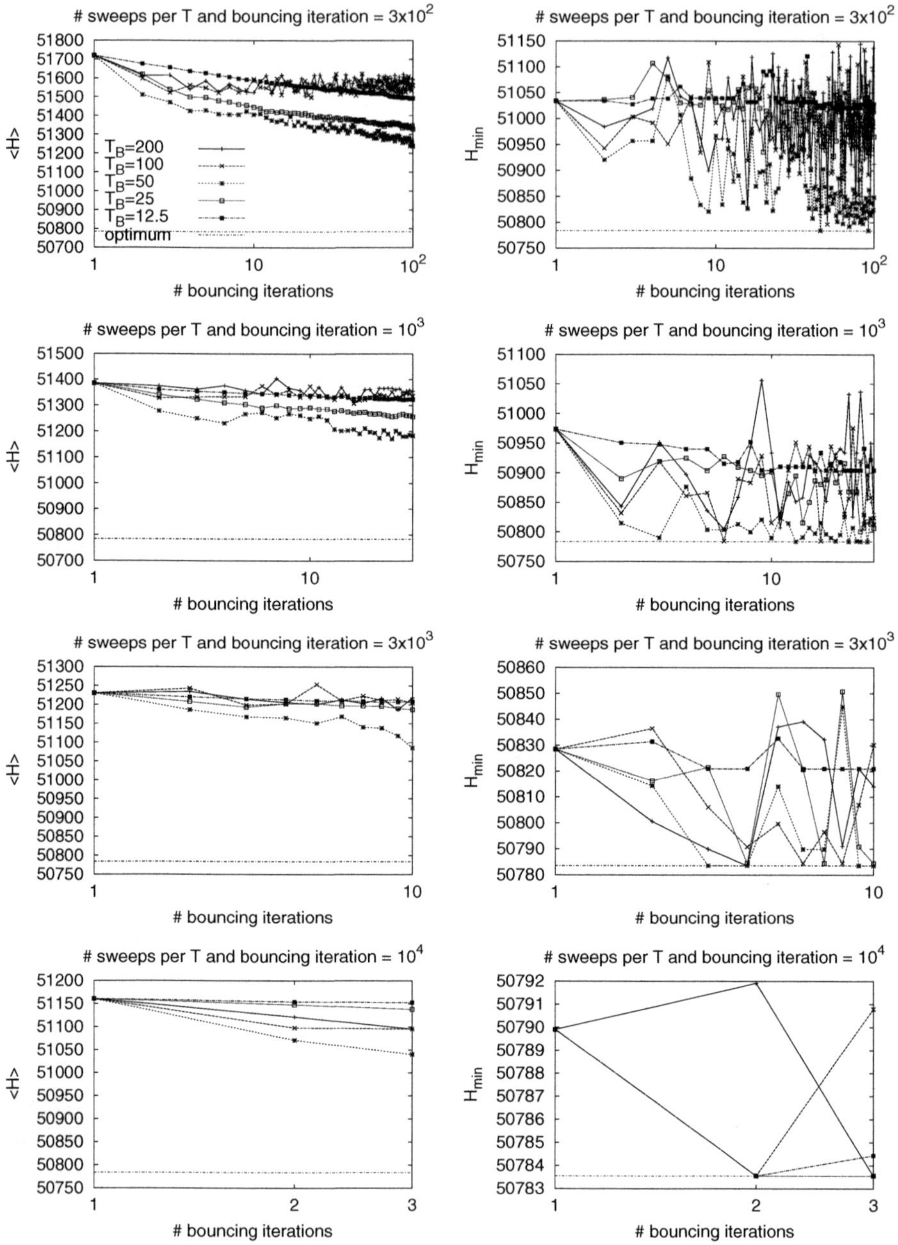

Fig. 8.8. Bouncing the PCB442 instance: results analogous to those in Fig. 8.7, but for larger numbers of sweeps per temperature step and bouncing iteration, the optimum value is now shown separately

In each temperature step, the same amount of sweeps is performed as usual. As we want to compare the bouncing results with the results for SA with a monotonous cooling schedule, the number of sweeps per temperature step is as usual $1, 3, 10, 30, \ldots$. Note that in each bouncing iteration, the same number of move trials is performed as in the usual optimization run, which now serves as iteration No. 1. Furthermore, note that the individual temperature values are closer to each other than in the first iteration, i. e., the cooling factor in the bouncing iterations is larger than 0.99. We bounce those systems with less calculation time in each bouncing iteration more often, so that the overall calculation time measured in number of sweeps is the same for all parameter sets and thus we can really compare the results at the end.

Figures 8.7 and 8.8 show the mean and minimum results for various values of the bouncing temperature T_B and for various numbers of sweeps per temperature step. For cases with less calculation time, all curves decrease strongly in the first few iterations. Then they either fluctuate around some constant value in the case of larger bouncing temperatures T_B or continue to decrease for smaller T_B.

We clearly find that the bouncing process is generally able to gradually improve the quality of the solutions: the mean lengths decrease with an increasing number of bouncing iterations. For a small number of sweeps per temperature step, first the runs with $T_B = 50$ lead to the best results, but later on the results of the runs with $T_B = 25$ are better than these. For larger numbers of sweeps, the runs with $T_B = 50$ lead on average to the best results in all bouncing iterations. Furthermore, we notice that by far the worst results are always achieved if we bounce the system up to $T_B = 200$, followed by $T_B = 100$ and $T_B = 12.5$.

Looking at the specific heat in Fig. 8.9, we find that the bouncing temperatures used cover the three bouncing regimes mentioned in Sect. 15.2 in Part I:

- $T_B = 200$ lies above the freezing temperature T_f of the system, $T_B = 100$ only slightly below. Using these bouncing temperatures, one "melts" the system, so to speak. In the case of very short calculation times, one can still improve the initial configuration, as the system is quickly quenched down to low temperatures and the final configuration of the previous iteration is surely not fully optimized, so that one can find easily some improvements. For long calculation times in each bouncing iteration, the quality of the solutions stays roughly the same as in the first iteration.
- At $T_B = 12.5$, the specific heat starts to vanish. The system is only slightly warmed up if reheated only up to this temperature. Again, for short calculation times, improvements can be found as the calculation time for each iteration is simply too short to really get in a local minimum in the first iteration. Thus, the curves for $T_B = 12.5$ decrease slowly in this case. For long calculation times per iteration, however, the bouncing iterations do not lead to larger improvements as the Monte Carlo walker is not able to

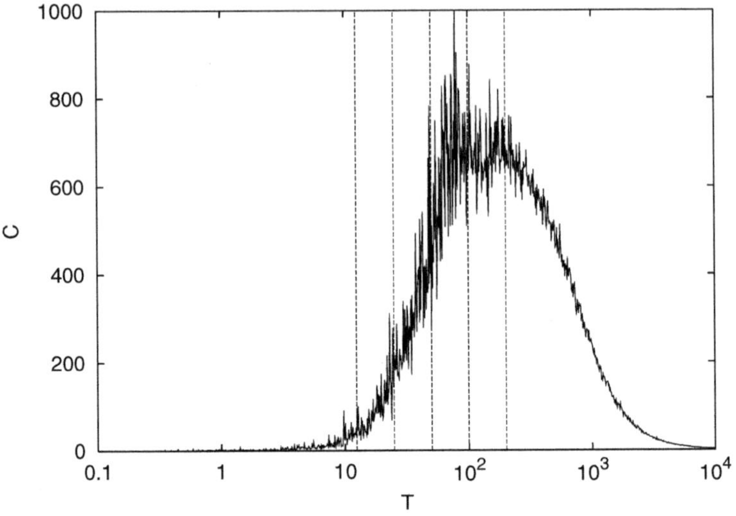

Fig. 8.9. Specific heat of the PCB442 instance with the five bouncing temperatures used in this investigation

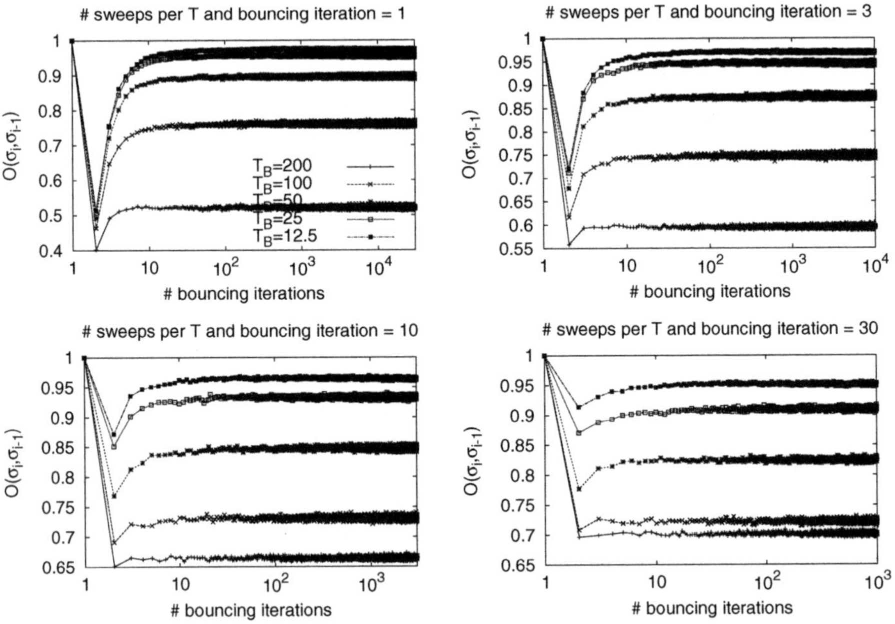

Fig. 8.10. Bouncing the PCB442 instance: here the symmetric overlap of the final configuration σ_i of each bouncing iteration i with the final configuration of the previous bouncing iteration, which serves as the initial configuration for the next bouncing iteration, is shown

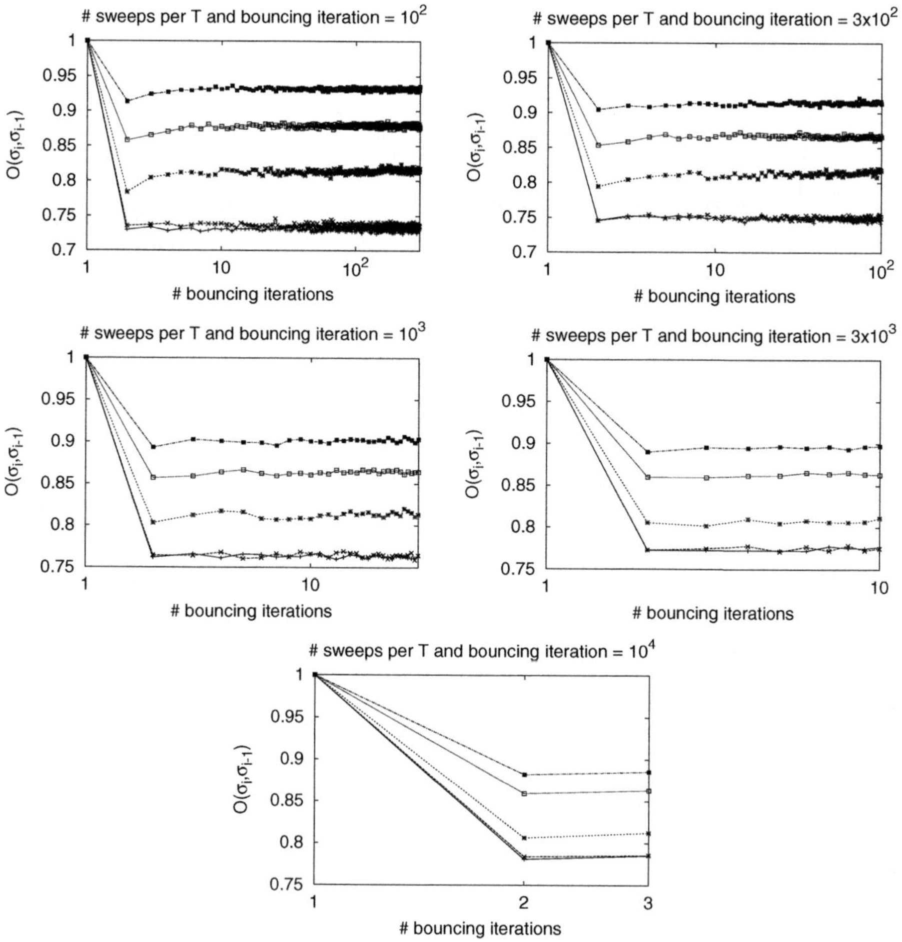

Fig. 8.10. (continued)

leave the local valley of the initial configuration due to the small value of T_{B}.

- The bouncing temperatures $T_{\mathrm{B}} = 25$ and $T_{\mathrm{B}} = 50$ are on the left leg of the specific heat. Thus, the system is considerably warmed up in these cases, but it is not melted. Thus, these bouncing temperatures "lie in the right temperature range", as can also be seen from the fact that the results for these T_{B} are generally the best.

Note that these numbers of a "good" bouncing temperature strongly depend on the instance. One should first perform a conventional SA run in which the specific heat of the system is measured. Then one knows the temperature

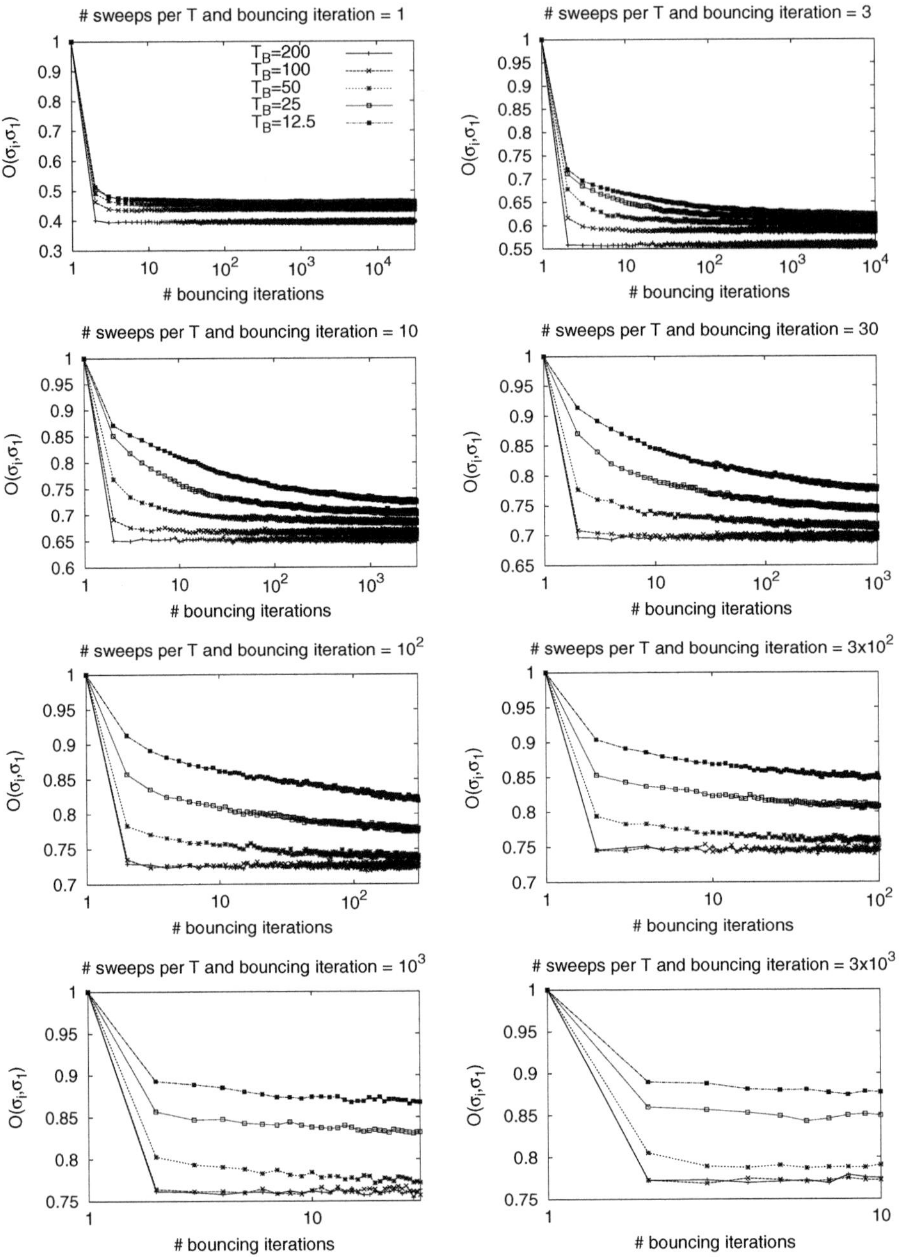

Fig. 8.11. Bouncing the PCB442 instance: here the symmetric overlap between the final configuration σ_i of the bouncing iteration i with the final configuration σ_1 of the first iteration, i. e., of the conventional monotonically cooled optimization run, is shown

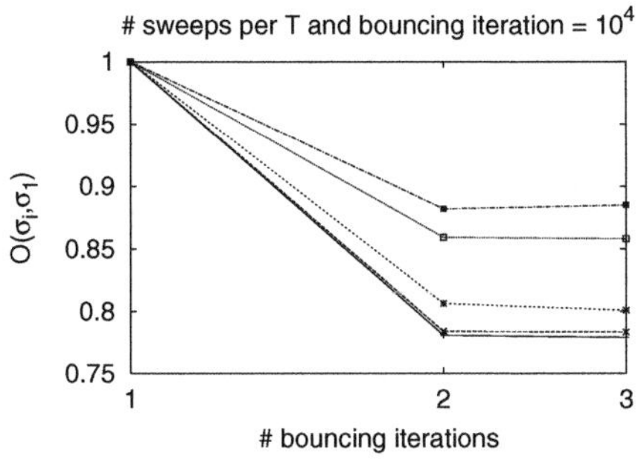

Fig. 8.11. (continued)

range in which one must reheat the system in order to strongly improve the results with a bouncing process.

These insights are in good agreement with the results shown in Figs. 8.10 and 8.11. Here we consider the overlap between the final configurations σ_i of the successive bouncing iterations i. The overlap $O(\sigma, \tau)$ between configurations σ and τ is simply given by the number of edges occurring both in σ and in τ in relation to the overall number of edges: let

$$\eta_\sigma(i,j) = \begin{cases} 1 & \text{if } \sigma^{-1}(i) = \sigma^{-1}(j) \pm 1 \bmod N\,, \\ 0 & \text{otherwise}\,, \end{cases} \tag{8.2}$$

i. e., $\eta_\sigma(i,j) = 1$ if j is either the predecessor or the successor of i in configuration σ. Then the overlap is given as

$$O(\sigma, \tau) = \frac{1}{2N} \sum_{i,j=1}^{N} \eta_\sigma(i,j) \times \eta_\tau(i,j)\,. \tag{8.3}$$

We first consider the overlap between the final configuration σ_i of the bouncing iteration i and the final configuration σ_{i-1} of the previous bouncing iteration $i-1$, which serves as the initial configuration for the next bouncing iteration. For the first bouncing iteration, we simply set $\sigma_0 = \sigma_1$ such that the overlap has the value 1 there. We find that for small calculation times the overlap increases strongly in the first few iterations. Then it becomes nearly a constant in all cases. Furthermore, we find that these final constants strongly depend on the corresponding bouncing temperature T_B: the smaller T_B is, the larger is the overlap with the initial configuration of the bouncing iteration. This result is not surprising at all, as the possibilities for moving for the Monte Carlo walker are enlarged if the system is reheated to a higher temperature. However, the agreement with the results for the lengths is very good:

- The results for the lengths coincide for a number of sweeps ≥ 100 for $T_\mathrm{B} = 200$ and $T_\mathrm{B} = 100$ in Figs. 8.7 and 8.8. Analogously, the overlaps coincide

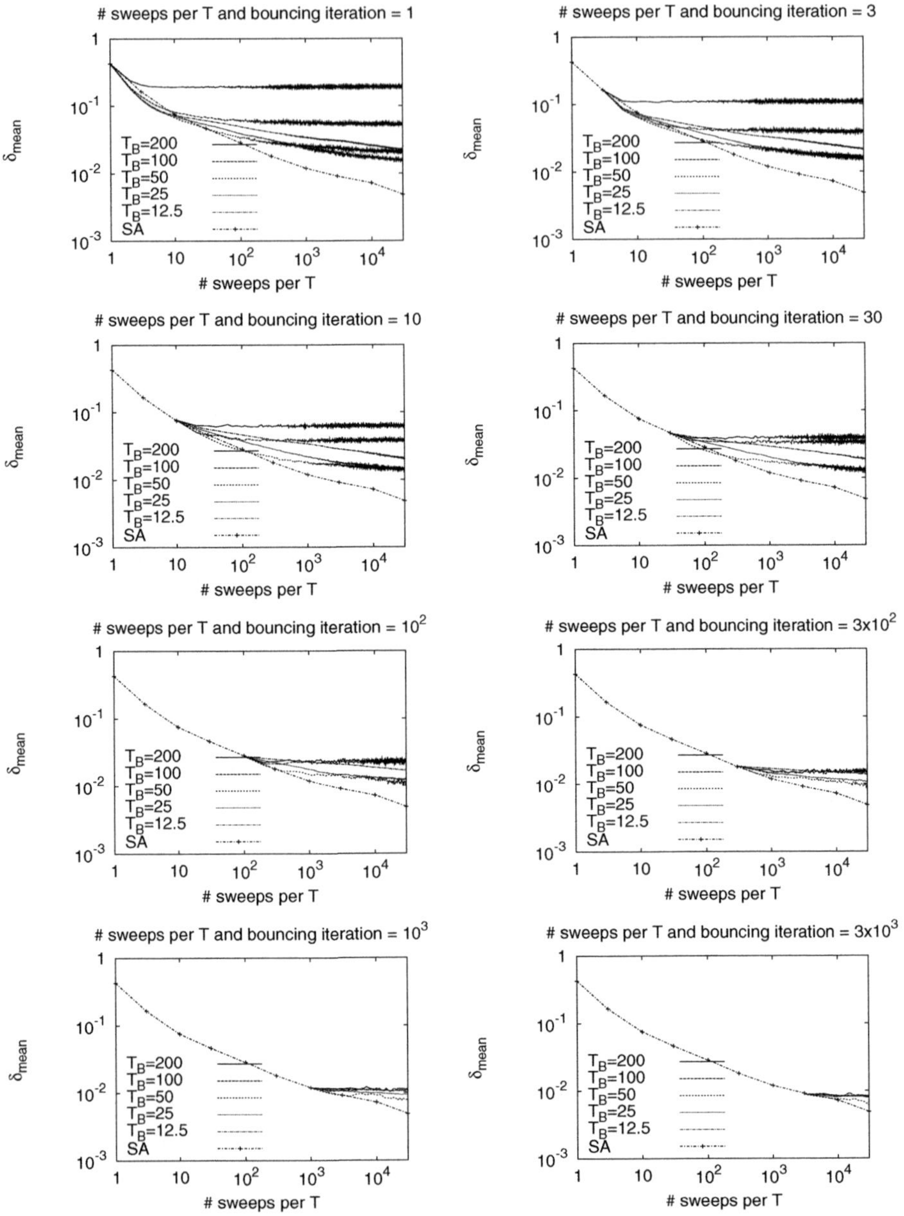

Fig. 8.12. Bouncing the PCB442 instance: the results of Figs. 8.7 and 8.8 for the mean length are replotted as the mean deviation from the optimum, as defined in Eq. (8.1). For comparison, the results of monotonically cooled optimization runs with SA of Fig. 7.9 for the PCB442 instance are shown with the label "SA"

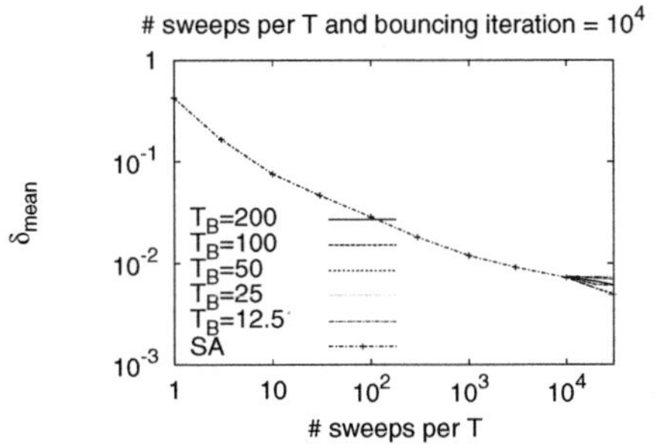

Fig. 8.12. (continued)

for these sweep numbers. Furthermore, they stay roughly constant at ≈ 0.76 if the number of sweeps is increased. This number is also the overlap between rather good configurations that are generated independently of each other for this specific instance.

- The best results can, however, be achieved if this overlap to the initial configuration is larger and is $\gtrapprox 0.8$ for longer computing times.
- If the system is only slightly warmed up, then the overlap has values > 0.9 and thus is simply too large. In this case, the Monte Carlo walker does not leave the local valley and is thus not able to get to better solutions.

Instead of projecting the final configuration σ_i of the bouncing iteration i on the final configuration σ_{i-1} of the previous bouncing iteration $i-1$, one can also consider the projection of σ_i on σ_1, i. e., on the result of the initial optimization run, which is now successively bounced. Figure 8.11 shows the results for this projection. We find that this projection decreases more strongly for very short computing times. Furthermore, the decrease again depends on the bouncing temperature T_{B}: for $T_{\mathrm{B}} = 200$, this overlap drops to its final value and stays rather constant there. This is also the case for $T_{\mathrm{B}} = 100$ and longer computing times. For smaller T_{B}, the overlap with the original solution gradually decreases. The lower T_{B} is, the slower is this decrease.

Note that these values for this specific PCB442 instance might differ strongly from values for other instances and other problems. In particular, the PCB442 instance exhibits a highly degenerate ground state and thus also higher-energy states are degenerate. There are trivial moves by which one can jump in the greedy mode between various configurations as the energy of some neighboring configurations is exactly the same.

Till now, we have only considered the relative improvement due to the bouncing process. However, looking at Figs. 8.7 and 8.8 again, we find that there are large differences in the quality of the final configurations σ_1 of the first iteration due to the different calculation times spent. The longer this first monotonously cooled iteration lasts, the better these results are, which are to be improved by the bouncing process. However, the question arises as to whether it is better to perform a short initial cooling run and then a long bouncing process with many iterations or to spend all the available calculation time in one monotonically cooled optimization run.

Figure 8.12 shows the results for the mean relative deviations from the optimum as defined in Eq. (8.1), both for the results of the bouncing process shown in Figs. 8.7 and 8.8 and for the results for monotonically cooled optimization runs that are taken from Fig. 7.9. The results are plotted in such a way that one can clearly see what quality of results one gets as a function of the computing time.

In this figure, the results for SA occur as a line from which the results for the bouncing processes take off after the first monotonically cooled iteration. For a small number of sweeps per temperature step in each bouncing iteration one finds that the results achieved after 1–10 bouncing iterations can indeed be better for short computing times. Investing more computing time, bouncing cannot lead to as good results as a long monotonically cooled optimization run.

However, the bouncing idea has still its place, not only in the case that the available computing time is very short, and it can be very useful: after having invested some computing time on several computers, one can check the final configurations that one likes most and then bounce the best one several times in order to improve its quality but not lose too many of the nice properties one wants to have. There are also other applications in which one needs to get a first preliminary solution in a rather short time in order to have a solution at all, but then one can often use additional calculation time when available in order to improve the result while not altering the solution too much.

8.4 Results for Parallel Tempering

The next cooling approach we want to investigate is parallel tempering (PT). Here one considers not just one Monte Carlo walker walking through the energy landscape but a set of Monte Carlo walkers. Each of these Monte Carlo walkers i has assigned a fixed temperature T_{i}. They accept each move according to the Metropolis criterion. After some number of sweeps, they are allowed to switch their positions in the energy landscape if a Metropolis-like criterion [Eq. (15.30) in Part I] is fulfilled. The exact implementation of the PT algorithm is as follows:

1. Set up a set of processes, each one starting with a randomly generated configuration and assign a temperature T_{i} to each process i, which is held constant during the simulation. These temperature values correspond to the temperatures of the conventional SA algorithm. Actually, we use 1147 processes, assigning to each of them one of the temperature values of the SA simulations, i. e., $T_{\mathrm{i}} = 10{,}000 \times 0.99^{i-1}$ for $1 \leq i \leq 1146$ and $T_{1147} = 0$.

2. Then each process performs the same number of sweeps. This leads to a final configuration σ_i for each process i.
3. After that pairs of processes with neighboring temperature values check according to Eq. (15.30) in Part I whether or not they should exchange their configurations. As the number of processes is odd, here the process pairs $(1,2), (3,4), \ldots, (1145, 1146)$, and the process pairs $(2,3), (4,5), \ldots,$ $(1146, 1147)$ are used in an alternating way. Note that there is always one marginal process that does not take part in this exchange process. Furthermore, note that the exchange with process 1147, which is run in the greedy mode, is performed according to the greedy acceptance criterion, such that the better of the two configurations is moved to the greedy Markov chain.
4. If some stopping criterion is not met (here a given calculation time), the algorithm returns to step 2.

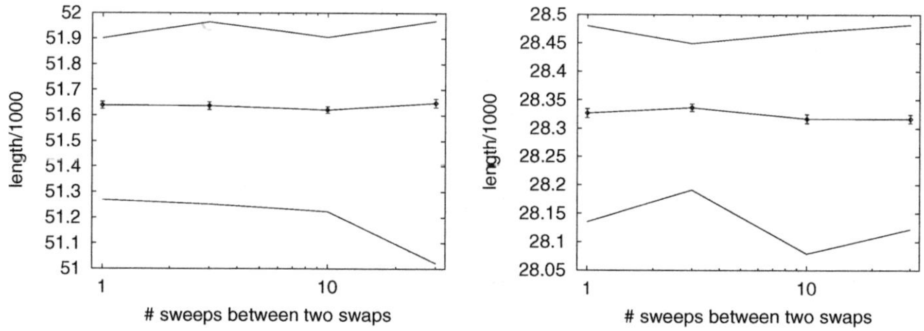

Fig. 8.13. Minimum, mean, and maximum lengths vs. number of sweeps between two swaps achieved with PT applied to the PCB442 instance (*left*) and to the ATT532 instance (*right*): for each of the two instances, 100 optimization runs were performed

We decided to use the same overall time as for the longest SA simulations, namely, an overall number of 30,000 sweeps per temperature step, in order to compare the results with those of standard SA. This overall number of 30,000 is split as a product of the number of sweeps between two swaps and the number of swaps. Now the question arises as to how often the swap of the configurations shall be performed. Figure 8.13 shows the results averaged over 100 simulations using this PT with various numbers of sweeps between two swaps. Note that as a result of the simulation, the length of the final configuration of process 1147, which was run at the lowest temperature $T = 0$, was returned.

The amount of 30 sweeps is already so large that the configuration at the highest temperature cannot tunnel to the last process, as here only 1000 swaps are performed, whereas there are 1147 processes. On the other hand, it does not make sense to use less than at least one sweep between two swaps, as the transferred configuration has to get used to the new temperature value. We find that the mean results are rather independent of the number of sweeps between two swaps, but the minimum results that can be achieved for the PCB442 instance become better if there is more time between two swaps.

The question is now whether this method is better or worse than SA. Neglecting the time for the swaps, the same overall number of move trials was performed as with SA using 30000 sweeps per temperature step and 1147 temperature steps. SA leads on average to results of $51{,}032 \pm 16$ for this scenario if applied to the PCB442 instance and to results of $27{,}834 \pm 9$ if applied to the ATT532 instance. SA reaches sometimes even the optimum configuration. On the other hand, PT leads on average to values larger than 51,600 for the PCB442 instance and larger than 28,300 for the ATT532 instance. The best result achieved for the PCB442 instance is 51,021.375, for the ATT532 instance 28,079. However, this is arguing from the point of view of a single-processor machine, on which all these processes of PT must be run. One can also argue from the point of view of people working on large workstation clusters or on supercomputers with a huge number of processors: here the different processes of PT can be run in parallel, such that—if we neglect again the time for swapping the configurations—each processor only performs 30,000 sweeps and has thus finished its work within 1/1147 of the time the single-processor machine needs. If we now compare this scenario with SA, the nearest value to that is where we perform 30 sweeps for each of the 1147 temperature steps. Here we get $53{,}146 \pm 51$ for the PCB442 instance and $29{,}016 \pm 23$ for the ATT532 instance, so that in this sense SA leads to much worse results than PT. On the other hand, one can also argue that if one uses a parallel computer search for the optimum of a given problem with SA, one can start the same SA program on each processor and finally take the best result. For 1000 simulations performed, with 30 sweeps per temperature step, this best result was 51730.8471 for the PCB442 instance and 28419 for the ATT532 instance, which is in both cases still much worse than the mean result achieved with PT.

A further question is whether this PT method can really be interpreted as a type of cooling process for SA. For this reason, we show in Fig. 8.14 the lengths of the final configurations of the processes in a PT optimization run drawn vs. the temperature values of the individual processes. We find that these lengths fluctuate nicely around the curve for the expectation value of the energy as a function of the temperature. Thus, after some time, each process has moved the system starting at a random configuration into the correct range of energies at its system temperature. The swapping of the configurations between the processes can speed up adjusting the systems to the temperatures.

The question now is whether the results achieved with PT can be further improved. The philosophy of SA is not to quench the system down by

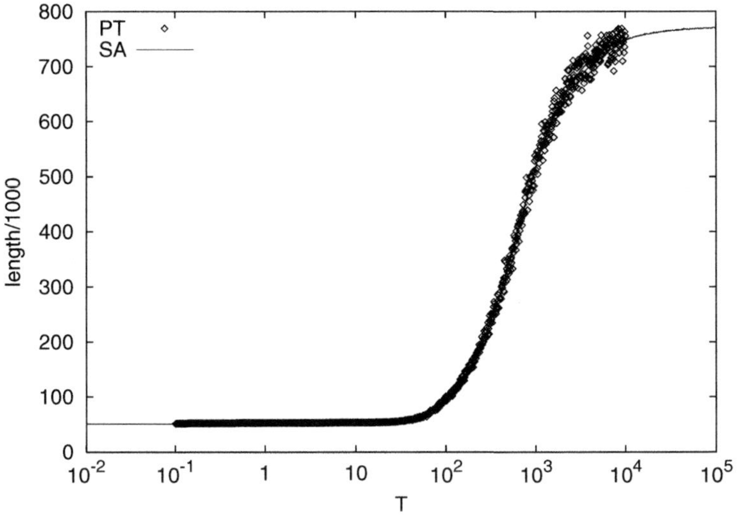

Fig. 8.14. Lengths of the final PCB442 configurations of the processes in a PT optimization run vs. temperature values of each process. For comparison, the expectation value $\langle \mathcal{H} \rangle_T$ of the energy as a function of the temperature, which was measured in a SA run, is drawn as a straight line

cooling it somewhat fast but to cool it slowly. If we consider the processes of a PT simulation, then we find that those processes with small temperatures quench their configurations down, which are initialized with randomly generated configurations. Thus the idea is already to start with much better configurations, at least at the low temperatures. Much better configurations like these can be easily and quickly generated by some construction heuristic, like the bestinsertion heuristic. Figure 8.15 shows results that are analogous to those in Fig. 8.13, but now all initial configurations are generated with the bestinsertion heuristic. We find that the results for the PCB442 instance do not improve—on the contrary, they are slightly worse—whereas the results for the ATT532 instance become better. As there is no general advantage of starting with preoptimized solutions, we will return to the random initialization.

The next question concerning the quality of the results as well is how many processes will be used. On the one hand, the temperatures of the processes must be "dense" enough so that the acceptance ratio for the swapping of the configurations is large enough. For each process, the lengths of the occurring configurations fluctuate after some time around the expectation value of the energy. The widths of these distributions differ for the various temperature values. But these distributions should overlap; otherwise one can only use that part of the PT criterion in which the acceptance probability decreases exponentially with the difference between the inverse temperature values. As

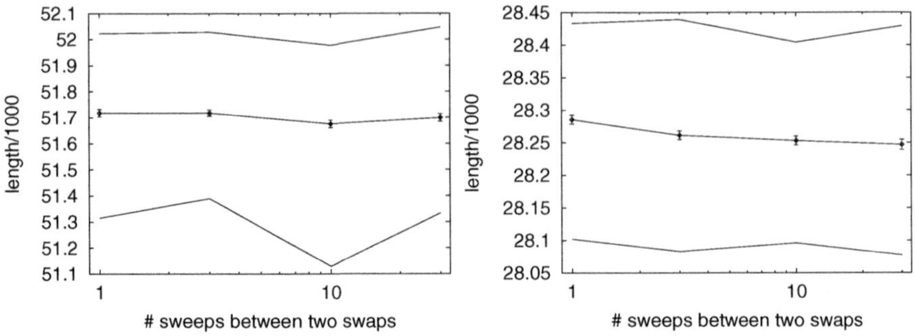

Fig. 8.15. Minimum, mean, and maximum lengths vs. number of sweeps between two swaps if applying PT to the PCB442 instance (*left*) and to the ATT533 instance (*right*) and starting with the bestinsertion heuristic: like the results shown in Fig. 8.13, the results are averaged over 100 optimization runs

we see for our approach in Fig. 8.14, we are well on the safe side, according to this criterion, as the final lengths shown there are not monotonically decreasing. On the other hand, let us consider an argument sometimes heard for this PT method: let us consider the swapping of the configurations as a RW process, i. e., the configurations perform a RW in the temperature space [208]. (Of course, this is not the case, as the swap is not always accepted but only with a Metropolis-like criterion, but we want to use this picture here.) Then, according to the Einstein relation, the range of temperature steps a configuration can trespass is proportional to the square root of the time that passed since the RW was started. Thus, the time a configuration needs to visit all the occurring processes increases at least quadratically with the number of processes. Summarizing, we need a large number of dense temperature values but a small number of processes.

In order to solve this contradiction, an approach can be used in which there are fewer ensemble members and the individual ensemble members of the PT process gradually decrease their temperature values $T_{\rm i}$. In our implementation, we use $1146/n+1$ ensemble members with $n = 1, 2, 3, 191, 382, 573$, and 1146, which are the factors of the number 1146. The last member of the ensemble is always set to the temperature $T = 0$ from the very beginning. The other members start with temperature values of $T_{\rm i} = 10^4 \times 0.99^{i-1}$. The temperatures are then gradually decreased exponentially with the cooling factor 0.99, such that the final temperatures for these ensemble members are given by $T_{\rm i} = 10^4 \times 0.99^{1146\times(n-1)/n+(i-1)}$. Thus the whole temperature range of the previous simulations is covered. This decrease of the temperature is performed uniformly during the entire optimization process. As there is only a fraction of the previous number of ensemble members left, we increase the overall number of sweeps performed by each ensemble member by this factor n in order to use the same overall time for the simulation runs.

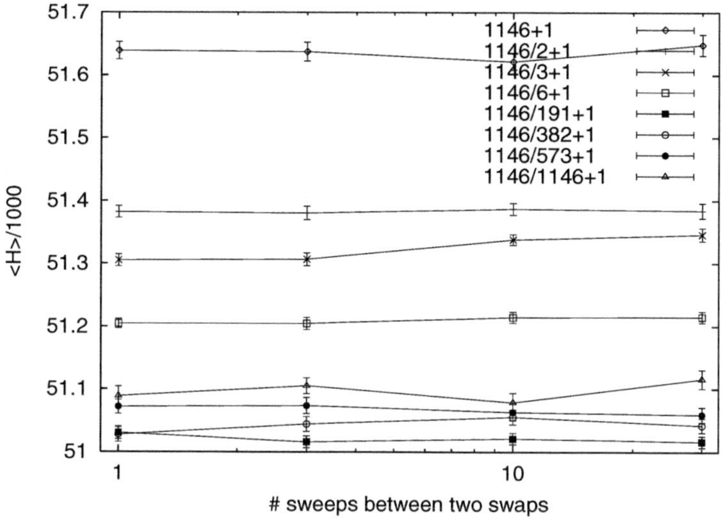

Fig. 8.16. Average lengths vs. number of sweeps between two swaps when applying PT with a cooling process to the PCB442 instance: each data point shows the mean value and the error bar of 100 optimization runs

Figure 8.16 shows the results for various n, averaged over 100 PT simulations each. One clearly finds that the results strongly improve if n is increased from 1 to 191 and thus the number of ensemble members decreases from 1147 to 7. (Even the optimum of the PCB442 problem can be reached with this approach.) However, if n is increased even further, the results get worse again, but stay generally much better than the original PT approach without cooling and better than the results for very small n. Thus, there is an optimum intermediate number of processes when using this PT approach. The results are then even able to compete with those of SA if PT must be carried out on a single processor.

Summarizing, PT is an interesting alternative to SA. However, it is only useful if there is a parallel computer or a workstation cluster one can work on. A small or medium-sized cluster is sufficient as the number of processes should not be too large.

9 Application to TSP of Algorithms Related to Simulated Annealing

9.1 Computational Results for Threshold Accepting

The way to apply threshold accepting (TA) to the traveling salesman problem (TSP) is exactly the same as for simulated annealing (SA) except that the acceptance criterion must be changed. Instead of the Metropolis criterion, the threshold criterion is used, i. e., the tentative new configuration is accepted if it is either better or up to a threshold T worse than the current configuration. Thus, the acceptance function of TA can only take one of the two discrete values 0 and 1 such that there is never the need to calculate a random number in order to decide whether to accept or reject a move. Furthermore, no exponential must be calculated. Thus, the computing time for the accept function is much shorter than for SA. For some problems, the energy difference $\Delta\mathcal{H}$ can be calculated very quickly. For instance, when treating the TSP with the seven small moves, we usually work with, one simply adds up up to eight numbers that are looked up from the distance matrix. Thus, in this case, it was advantageous to use TA instead of SA some years ago in order to save the calculation time for calculating the exponential. (Because of this calculation time problem, many models developed in physics exhibit only a very small number of possible energy differences, such that the exponentials of these differences $\Delta\mathcal{H}_i$, i. e., the values $\exp(-\Delta\mathcal{H}_i/(k_\mathrm{B}T))$, can be easily calculated at the beginning of each temperature step, then stored in a table, and finally looked up from this table if needed.)

As TA can be considered a deterministic variant of SA, also the range of the control parameter in which the transition from the high-energy configurations to the low-energy solutions occurs can be expected to be rather the same. Thus, we use the same initial and final values for the threshold here as we used in Chap. 7 for the temperature. Indeed, we simply copied the SA program and replaced the Metropolis criterion by the threshold criterion. There is no more to do if switching from SA to related algorithms like TA.

Investigating TA, we first want to compare the results for some observables with the curves obtained for SA. First, Fig. 9.1 shows the decrease in the mean energy $\langle\mathcal{H}\rangle$ with decreasing threshold and temperature T and the curves for the specific heat C of three TSP instances both for TA and for SA. One clearly sees that the decrease in $\langle\mathcal{H}\rangle$ is steeper in the case of TA.

Furthermore, at large thresholds, the mean energy is virtually constant. The "specific heat" is calculated via $C = \mathrm{Var}(\mathcal{H})/T^2$ both for SA and TA. Of course, this formula is only correct for SA. But we find that this observable also gives some insight into the behavior of TA, as it measures the fluctuations of the energy related to the control parameter. However, one must be aware of the fact that this is then not specific heat in the strict physical sense in the case of TA. One finds for all three instances that the peak of this specific heat lies at slightly larger values of T in the case of TA. Furthermore, the height of the peak is much smaller. Thirdly, one finds that the specific

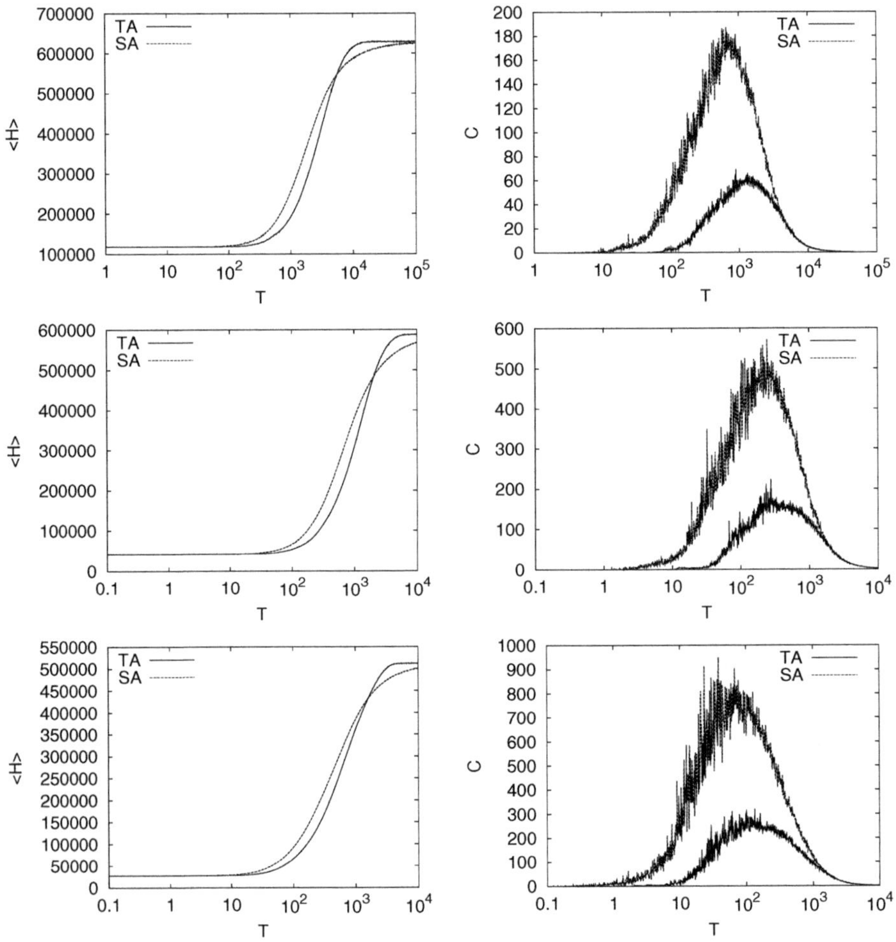

Fig. 9.1. Computational results for the mean energy $\langle \mathcal{H} \rangle$ (*left*) and the specific heat C (*right*) of three TSP instances [BEER127 (*top*), LIN318 (*middle*), ATT532 (*bottom*)] using TA. For comparison, the results for SA of Fig. 7.3 are shown as *dashed lines*

heats for SA and TA are rather identical at large values of T and that the curves differ from each other when approaching the peak of the specific heat. In conclusion, we get the result that the range of the control parameter T in which the transition between the high-energy and the low-energy regime occurs is indeed roughly the same for SA and TA. Thus, one can really consider TA to be an approximation of SA.

Next we compare TA and SA with the curves for the mean value of the order parameter ξ, which was introduced in Sect. 1.9, and the corresponding susceptibility χ, which are both shown in Fig. 9.2. The susceptibility is calculated as $\chi = \mathrm{Var}(\mathcal{H})/T$ both for SA and for TA, although the formula

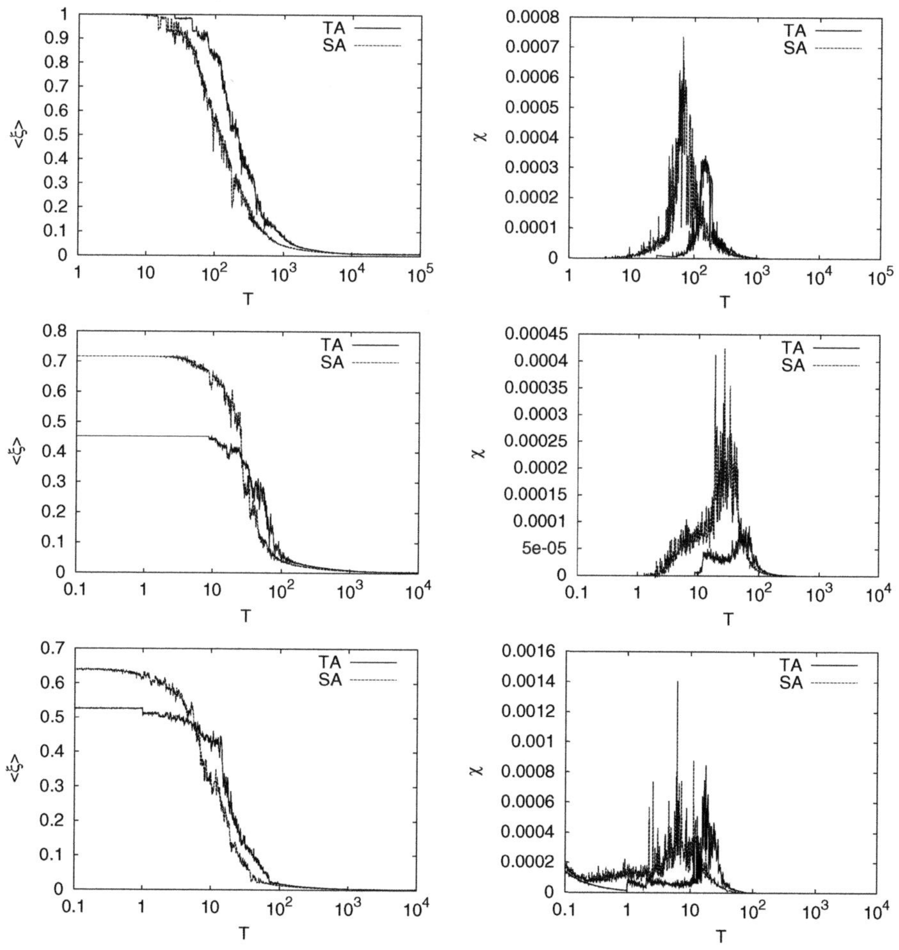

Fig. 9.2. Computational results for the expectation value $\langle\xi\rangle$ of the order parameter (*left*) and the corresponding susceptibility χ (*right*) of three TSP instances [BEER127 (*top*), LIN318 (*middle*), ATT532 (*bottom*)] using TA. For comparison, the results for SA from Fig. 7.4 are shown as *dashed lines*

is only strictly correct for SA. Again the figure contains the curves for SA for comparison, too. The curves for the order parameter also increase from slightly above zero to their final values, which depend on the locally minimum configurations in which the simulation runs finally get stuck, at roughly the same values of the control parameter, with the curve for TA increasing a little bit earlier. Thus, we see the same transition range of the control parameter here as well. The peak of the susceptibility χ lies, like the peak of the specific heat, at slightly larger values of the control parameter T and its height is smaller again than in the case of SA. Looking more closely at the peaks of the susceptibilities of TA at small thresholds, one finds that χ increases $\propto 1/T$

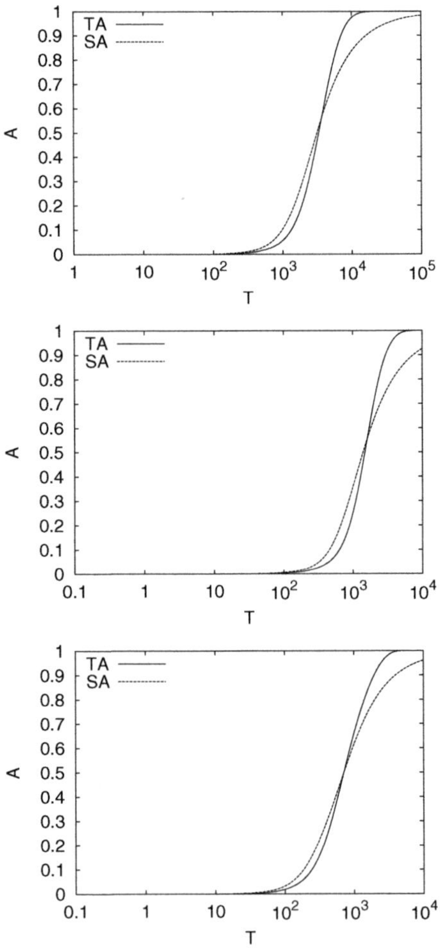

Fig. 9.3. Computational results for the total move acceptance rate A of three TSP instances [BEER127 (*top*), LIN318 (*middle*), ATT532 (*bottom*)] using TA. For comparison, the results for SA from Fig. 7.6 are shown as *dashed lines*

for small thresholds T until the minimum positive energy difference of the system is crossed. After that the susceptibilities might increase again, due to a constant rounding error, such that the variance of the control parameter is not exactly zero, as it should be, but some small constant value, such that χ increases again $\propto 1/T$.

For other systems, we have already found more pronounced steps on the left legs of the specific heat and the susceptibility [182, 185, 189]: first these observables increase $\propto 1/T^2$ and $\propto 1/T$, respectively, then there is a sharp step downwards, then they increase again. This can be iterated several times, so that the sequence of the smallest energy differences of a system can be determined.

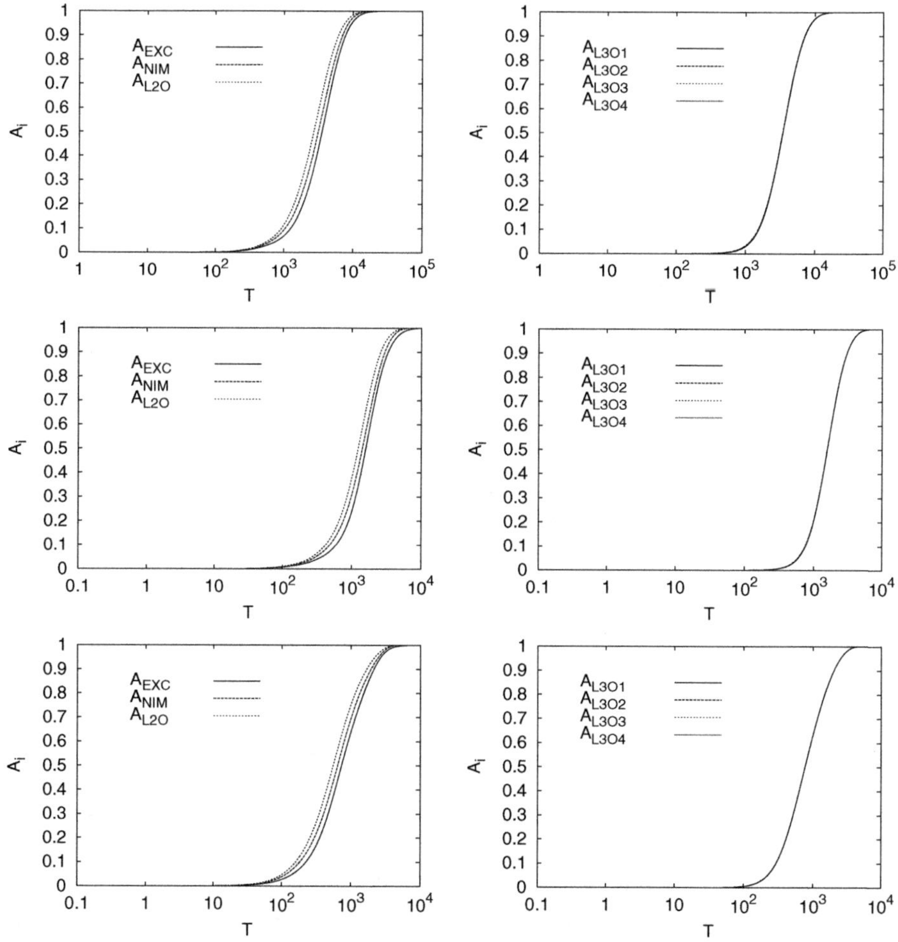

Fig. 9.4. Computational results for the partial acceptance rates A_i of the moves EXC, NIM, and L2O (*left*), and of the four variants of the L3O (*right*) for three TSP instances using TA: BEER127 (*top*), LIN318 (*middle*), ATT532 (*bottom*)

Next we compare the curves for the total acceptance rate of all moves between TA and SA. Figure 9.3 shows that the acceptance rates of TA start with a value of exactly 1. Thus, at the beginning, the system is in the random walk (RW) mode, as every move is accepted. Then the acceptance rate decreases more steeply than for SA. Again the transition range is roughly the same for SA and TA.

Now we consider the partial acceptance rates of the individual moves, shown in Fig. 9.4. We find that they all decrease nicely in a sigmoidal way from 1 to some small value. Just as for SA, we find for TA that the partial acceptance rates of the four variants of the Lin-3-opt (L3O) are identical.

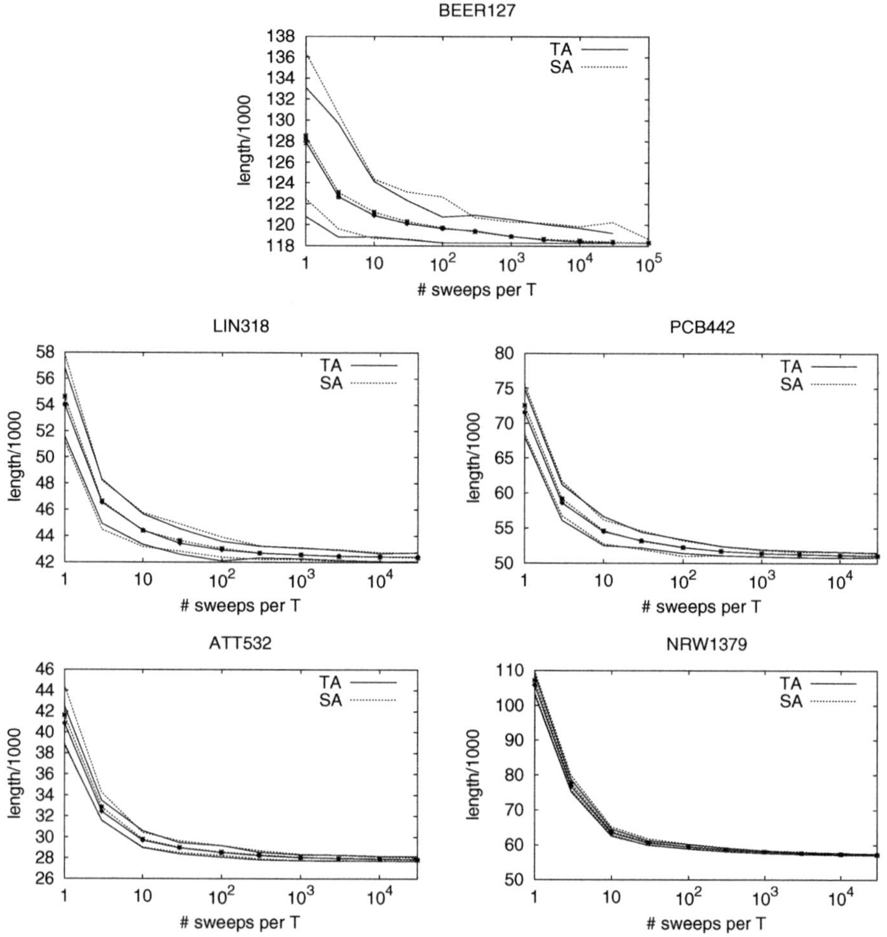

Fig. 9.5. Quality of the results achieved with TA (*straight lines*) compared to those achieved with SA (*dotted lines*, replotted from Fig. 7.9), for five TSP instances (BEER127, LIN318, PCB442, ATT532, and NRW1379) vs. number of sweeps per threshold step: the three *lines* show the minimum, mean (*error bars* are the size of the *symbols*), and maximum lengths

Some years ago the claim was made that TA is a superior method to SA due to the better quality of the results [53]. Thus, we compare the results we get for various calculation times measured in sweeps for five TSP instances from SA and TA. In both cases, the same exponential cooling schedule with a factor of $f = 0.99$ was used, the system was cooled from the initial value $T_\mathrm{i} = 10^5$ to the final value $T_\mathrm{f} = 1$ in the case of the BEER127 instance, from $T_\mathrm{i} = 10^3$ to $T_\mathrm{f} = 0.01$ in the case of the NRW1379 instance, and from $T_\mathrm{i} = 10^4$ to $T_\mathrm{f} = 0.1$ for all other instances. Finally, one greedy step was added. In each temperature and threshold, step, a number of sweeps $(1, 3, 10, 30, \ldots, 30{,}000)$ was performed. Figure 9.5 shows the quality of the results achieved vs. the number of sweeps used per temperature step. For each instance, 100 optimization runs were performed for both SA and TA.

We find that the results for SA and for TA are roughly of the same quality; sometimes SA is better, sometimes TA. Thus, the two algorithms can be considered equally good. Only for short calculation times, generally the results achieved with TA are better than those achieved with SA. On the other hand, when investing a lot of calculation time, one is more likely to get a better result with SA. For very large amounts of calculation times, SA is then clearly superior to TA.

9.2 Computational Results for Penna Criterion

The next acceptance criterion that is related to and competing with the Metropolis criterion and that we investigate here is the Penna criterion, as an example of the criteria based on the Tsallis statistic. The Penna criterion is given as

$$p(\sigma \to \tau) = \begin{cases} 1 & \text{if } \Delta\mathcal{H} \leq 0\,, \\ 0 & \text{if } q < 1 \text{ and } \Delta\mathcal{H} \geq \dfrac{kT}{1-q}\,, \\ \left(1 - (1-q)\dfrac{\Delta\mathcal{H}}{kT}\right)^{\frac{1}{1-q}} & \text{otherwise}\,. \end{cases} \tag{9.1}$$

Note that the Penna criterion is a "Tsallis-inspired" criterion but does not converge to the Tsallis probability distribution. A special case of this Penna criterion is the Metropolis criterion for $q = 1$. In this case, we use the original Metropolis criterion instead.

First, the question arises as to what the curves for the observables look like when using this criterion. We had to split the results for $q \leq 1$ and $q \geq 1$, as they behave quite differently. Figure 9.6 shows the results for the mean value of the energy and the specific heat C, which was again measured as $C = \mathrm{Var}(\mathcal{H})/T^2$, for the PCB442 instance, to which we want to restrict the discussion in this section. We find that the temperature range in which the transition from the high-energy to the low-energy regime occurs is shifted toward smaller temperatures with increasing q. For $q < 1$, we get curves looking rather similar to that of SA; the curves for the energy seem to be simply shifted sideward. The amount of shift roughly depends in a logarithmic way on the absolute value of q if q is strongly negative. The curves for $q < 1$ are only slightly steeper than that for SA. However, we find, that the decrease in the energy is strongly slowed down when using values of q with $q > 1$: the transition range is extended by some orders of magnitude of the temperature if $q = 2$ is used.

Looking at the specific heat, we find that the height of the peak strongly increases and is shifted toward smaller values of the temperature with increasing q.

Next we want to consider the total acceptance rate A as a measure of the extent to which the system is frozen at a particular temperature. Figure 9.7 shows nice sigmoidal decreases in the acceptance rate. We find behavior similar to that of the decrease in the mean energy shown in Fig. 9.6: for negative q, the curve is simply shifted sideward. For $q > 1$, the decrease in the acceptance rates lasts more orders of magnitude of the temperature than for $q = 1$. The question is now whether this slower transition behavior pays off with better results.

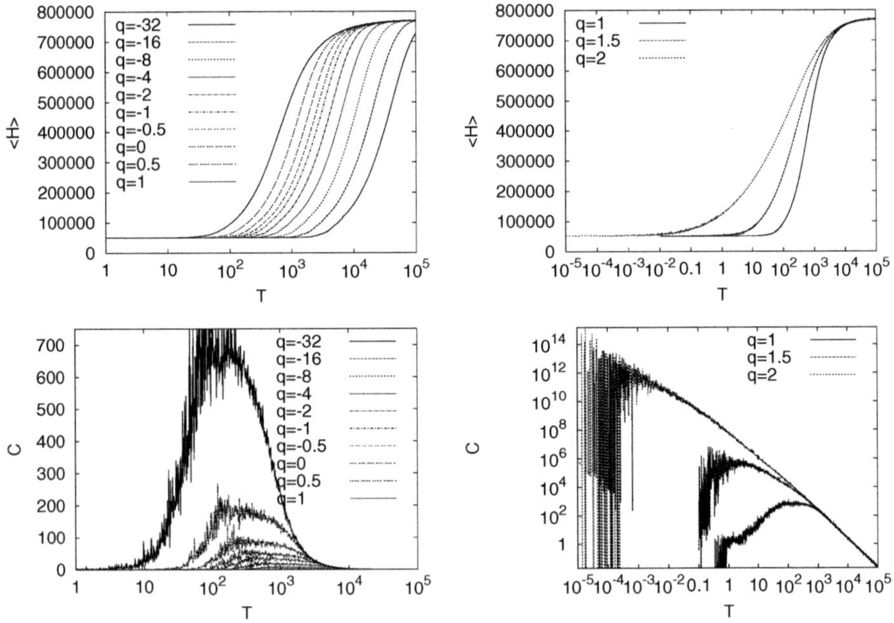

Fig. 9.6. Applying the Penna criterion to the PCB442 instance: *top*: decrease of the mean energy $\langle \mathcal{H} \rangle$ with decreasing temperature; *bottom*: specific heat C for various q-values ($q \leq 1$ *left*, $q \geq 1$ *right*). The results for $q = 1$ are actually results for SA and replotted from Fig. 7.3

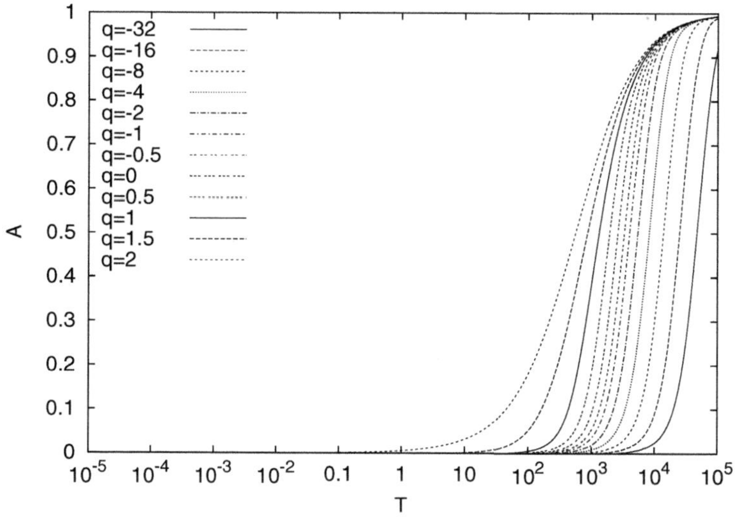

Fig. 9.7. Applying the Penna criterion to the PCB442 instance: decrease in the acceptance rate A with decreasing temperature T for various q-values

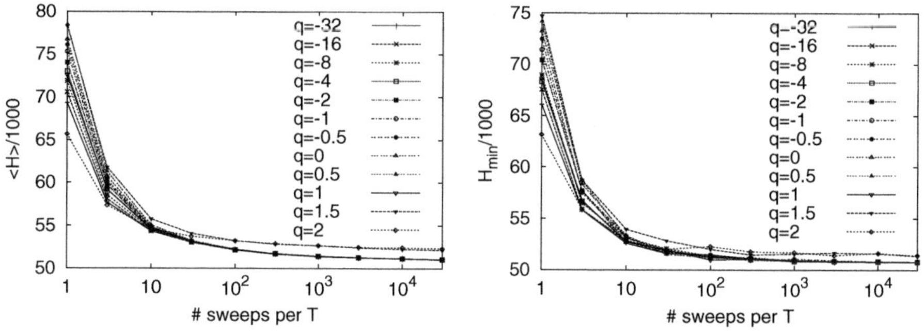

Fig. 9.8. Applying the Penna criterion to the PCB442 instance: mean (*left*) and minimum (*right*) length vs. number of sweeps per temperature step

Figure 9.8 shows the quality of the results achieved with the Penna criterion achieved with various values of q. For $-32 \leq q < 1$, the PCB442 instance was cooled down from $T_\mathrm{i} = 10^5$ to $T_\mathrm{f} = 1$, for $q = 1$ and $q = 1.5$, from $T_\mathrm{i} = 10^4$ to $T_\mathrm{f} = 0.1$, and for $q = 2$, from $T_\mathrm{i} = 10^4$ to 10^{-5}. Finally, a greedy step with $T = 0$ was performed. Note that the simulation runs with $q = 2$ thus got much more calculation time than the other optimization runs. For each q and for each number of sweeps per temperature step, 100 optimization runs were performed. We find that the results for $-2 \leq q \leq 1$ are the best. The worst results are achieved with $q > 1$.

In Fig. 9.9, the differences in the quality are magnified using the relative deviation δ_mean from the optimum. We again find that there are no large

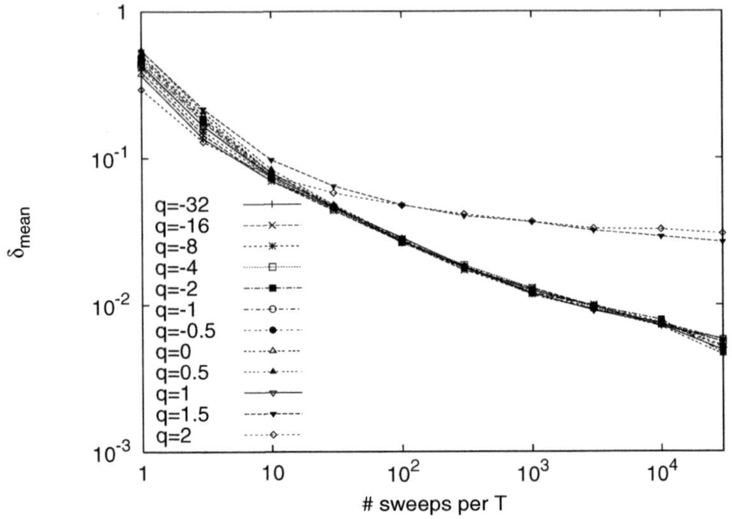

Fig. 9.9. Applying the Penna criterion to the PCB442 instance: deviation δ_{mean} from the optimum for various q-values (data of Fig. 9.8 replotted)

differences for $q \leq 1$. Again one finds that the achieved results are much worse for $q > 1$. Summarizing, one should use either small negative values or values up to $q = 1$ when working with the Penna criterion. The Penna criterion does not lead to results better than SA for the TSP.

9.3 Computational Results for Great Deluge Algorithm

The next algorithm we investigate is the great deluge algorithm (GDA). Unlike the previous algorithms, the acceptance probability of this algorithm depends not on the energy difference between the current configuration and the tentative new configuration but only on the energy value of the tentative new configuration. If this energy is smaller than the control parameter $\mathcal{T}$, then the new configuration is accepted. The Monte Carlo walker can thus be considered to be a bottom-dwelling fish in a lake and $\mathcal{T}$ to be the water level. The fish can perform a RW on the bottom of the lake but cannot leave the lake. With decreasing water level, the possible moves of the fish drive it to deeper and deeper regions of the lake. Finally, he/she is stranded at some local minimum, which is hopefully rather deep.

The question is now at what value of the control parameter to start and how to decrease it. We start with a random initial configuration σ_0 and set the initial water level $\mathcal{T}_i = 1.5 \times \mathcal{H}(\sigma_0)$ such that the system is safely in the RW mode at the beginning. Then $\mathcal{T}$ is decreased exponentially with a factor $f = 0.99$. However, a problem occurs if the fish is then above the decreased water level. Thus, we reduce the water level $\mathcal{T}$ iteratively with this cooling factor, but if the fish then moves above the water level, we set the water level to the energy value of the current configuration in order to avoid this problem. If this occurs often, then the cooling of the system is slowed down considerably. Generally, the system could even choose its own cooling schedule.

Figure 9.10 shows the decrease in the water level and the mean energy with increasing time. We find that there are five regimes:

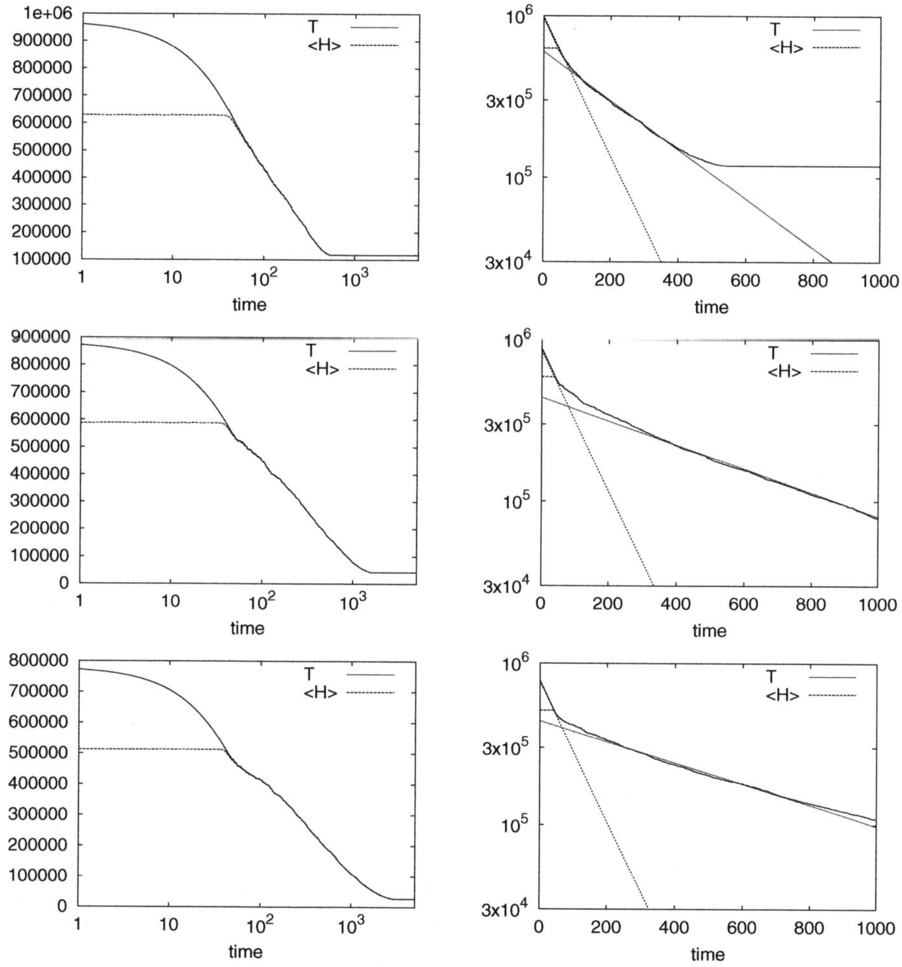

Fig. 9.10. Water level $\mathcal{T}$ and mean energy $\langle\mathcal{H}\rangle$ decreasing with increasing time measured in cooling steps: computational results for three TSP instances [BEER127 (*top*), LIN318 (*middle*), ATT532 (*bottom*)] when applying the GDA; graphic on *right* is a blowup of graphic on *left* and shows additional fit *curves*

- Initially, the water level decreases exponentially with a factor of $f = 0.99$ until it reaches that range of areas in which the system oscillates in its RW.
- Then the water level starts to press the system down to smaller energies.
- After this short transition period, the cooling schedule that the system adaptively creates seems to become roughly exponential again: for the BEER127 instance, one gets on average $f = 0.9966$ between the 150. and 350. time steps, for the LIN318 instance, $f = 0.9983$ between the 400. and 1000. steps, and for the ATT532 instance, $f = 0.9985$ between the 250. and 750. time steps.
- Then a second transition occurs in which the freezing behavior is even more slowed down.
- Finally, the system freezes in some local minimum, so that further time steps do not lead to further improvements.

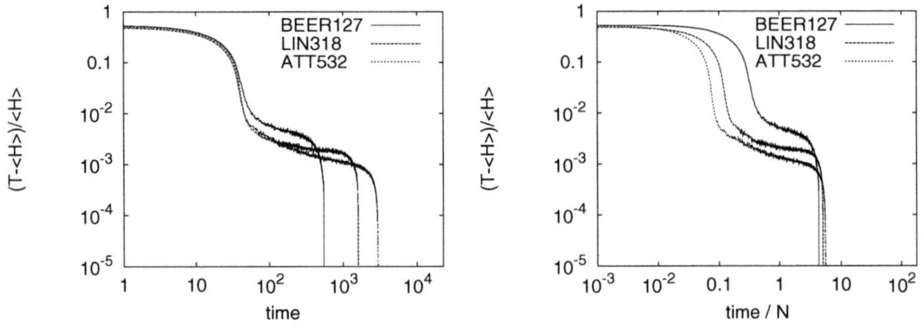

Fig. 9.11. Normalized difference between the water level $\mathcal{T}$ and the mean energy $\langle\mathcal{H}\rangle$: computational results for three TSP instances applying the GDA (based on data shown in Fig. 9.10)

One finds nice scaling behaviors common to all investigated TSP instances, as Fig. 9.11 shows: when plotting the normalized difference between the water level and the mean energy vs. the time measured in water level steps, the first transition regime occurs at the same time for all instances. This is not surprising, as we set the initial water level to 1.5 times the length of the random initial configuration. However, if we divide the number of water level steps by the system size, then the curves coincide roughly with each other at the second breakdown, at which the system starts to get stuck in some local valley and finally freezes in a local minimum.

Plotting the mean energy vs. the water level, which plays the role of the temperature in the GDA, provides further insight. Figure 9.12 shows that one does not obtain the same picture as if using SA or TA: first the mean energy is virtually constant at high water levels, i. e., the system is in a RW mode initially. After a short transition range, the mean energy decreases linearly as $y = x$ with the decreasing water level. The small differences between the water level and the mean energy are too small to be visible here.

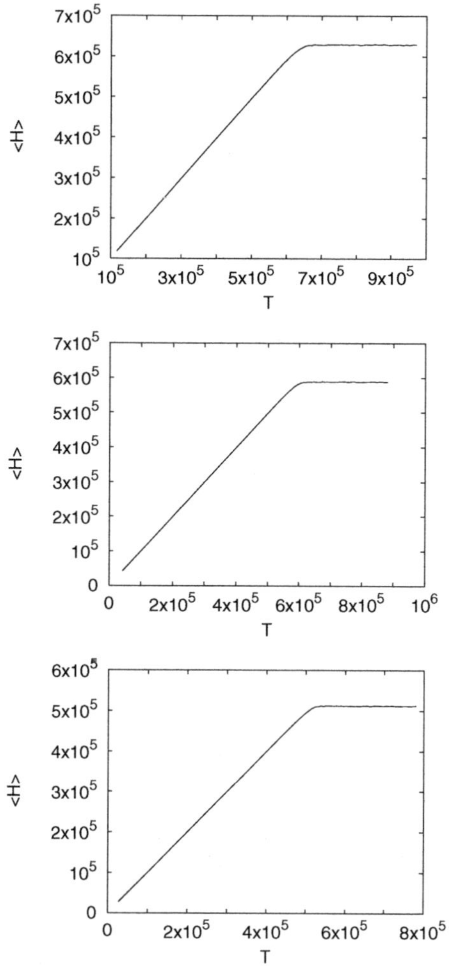

Fig. 9.12. Mean energy $\langle\mathcal{H}\rangle$ vs. water level $\mathcal{T}$ for the TSP instances BEER127 (*top*), LIN318 (*middle*), and ATT532 (*bottom*): results are replotted from Fig. 9.10

Next we want to investigate the fluctuations of the energy at some water level. A useful means for that is $C = \mathrm{Var}(\mathcal{H})/\mathcal{T}^2$, although this formula for the specific heat is only valid in the case of SA. Figure 9.13 shows C for three TSP instances. We find that at large water levels, the curves increase $\propto 1/\mathcal{T}^2$ with decreasing water level, as expected, as the system performs a RW at these large water levels, such that the variance of the energy remains constant. After the peak C breaks down rather fast, as the water level restricts the system to the configurations in some half space of the energy landscape.

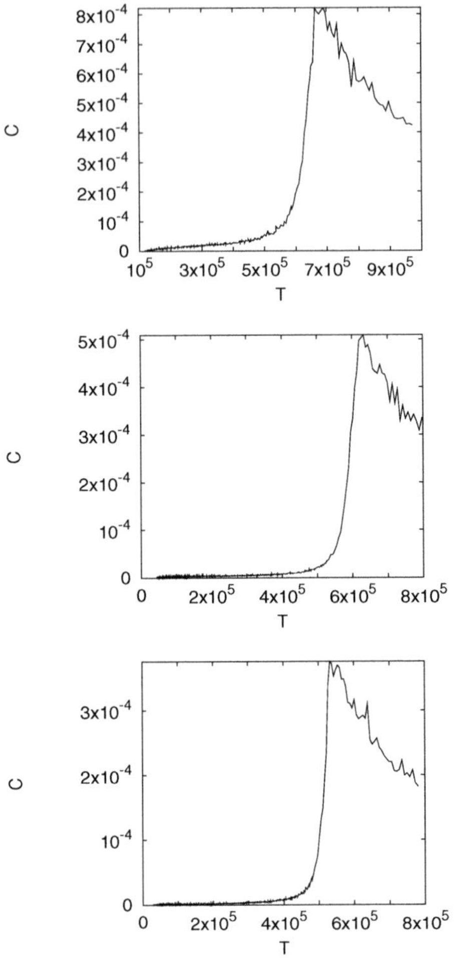

Fig. 9.13. C vs. water level T of the GDA for three TSP instances [BEER127 (*top*), LIN318 (*middle*), ATT532 (*bottom*)]

Although the system could perform a restricted RW due to the acceptance criterion, it obviously likes to stay slightly below the water level, as the results of the last figures show. As the system does not prefer a specific configuration, we can only conclude that there are many more configurations slightly below the water level than far below the water level. Due to the self-chosen cooling schedule of the algorithm, we can even say that the number of configurations increases exponentially with increasing water level, i. e., with increasing energy. One could also interpret these results in such a way that each good configuration has exponentially more neighboring configurations with worse energy values than with better energy values. However, these two statements are equivalent to each other. Summarizing, the GDA is a good means to measure the relative density of states at a given energy value.

Next we again investigate more closely the order parameter ξ and the corresponding susceptibility χ [measured as $\chi = \mathrm{Var}(\mathcal{H})/\mathcal{T}$], which are shown for three TSP instances in Fig. 9.14. The order parameter increases strongly at the freezing transition. The behavior is rather different from that of SA and TA, where a sigmoidal increase could be observed in which the order parameter reached its final value rather smoothly. Here, however, the order parameter increases to its final value. This increase stops when the system is frozen in some local minimum.

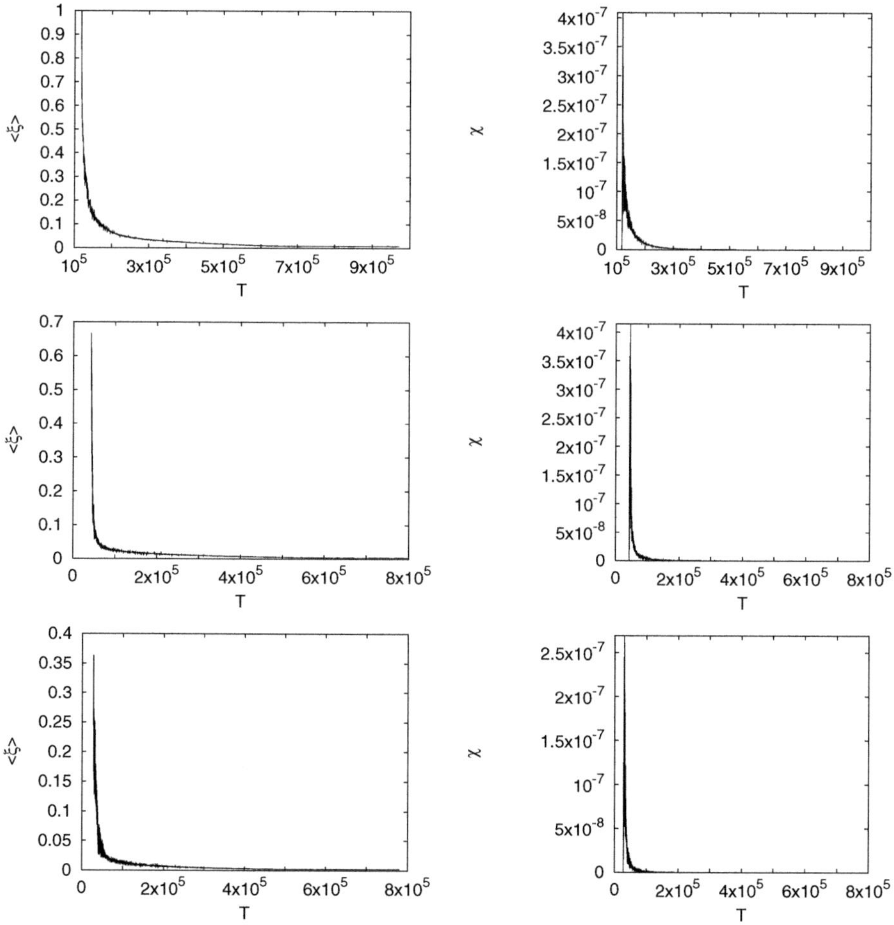

Fig. 9.14. Expectation value of the order parameter ξ (*left*) and the corresponding susceptibility χ (*right*) vs. water level $\mathcal{T}$ of the GDA for three TSP instances [BEER127 (*top*), LIN318 (*middle*), ATT532 (*bottom*)]

At high water levels, at which the system performs an unrestricted RW, the susceptibility increases $\propto 1/T$, as expected. But after a short transition range in which the water level nearly meets the mean energy value, the susceptiblity starts on average to increase $\propto 1/T^3$, as one can see when plotting the susceptibility in a log-log plot. However, one can even better fit two $1/T^5$ functions to it between which a transfer occurs. Then the susceptibility increases even faster than that, before it breaks down when the system freezes in a local minimum.

The total acceptance rates shown in Fig. 9.15 do not look familiar, either. For large T, the acceptance rates are equal to 1, as the system performs

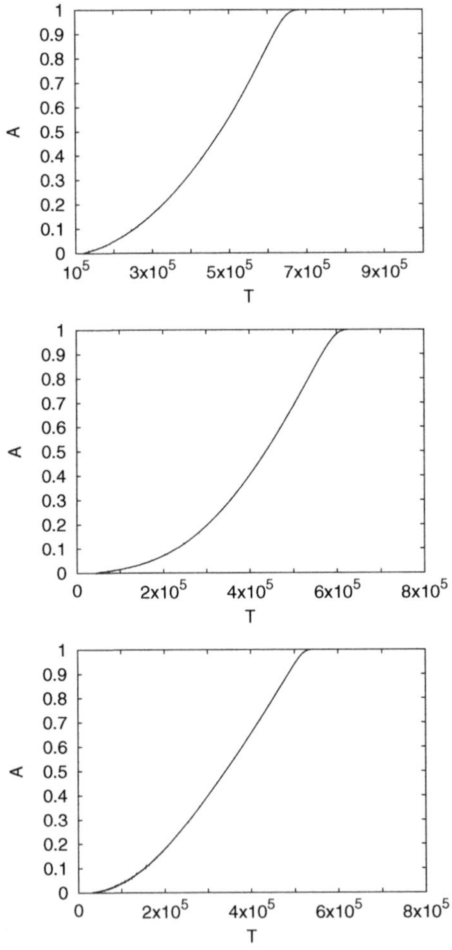

Fig. 9.15. Acceptance rate vs. water level T of the GDA for three TSP instances [BEER127 (*top*), LIN318 (*middle*), ATT532 (*bottom*)]

a RW there. Lowering $\mathcal{T}$, the acceptance rates decrease $\propto \mathcal{T}^{2-3}$. Finally, the acceptance rate breaks down as the system gets stuck in some local valley and finally freezes at its bottom.

Figure 9.16 shows the partial acceptance rates for the individual moves. Here, too, we find that the acceptance rates for the four types of the L3O are again rather identical. All acceptance rates start to decrease from 1 at roughly the same value of $\mathcal{T}$. The decrease in the acceptance rate of the Lin-2-opt (L2O) is the slowest one. The acceptance rate for the EXC first decreases fastest but then crosses the acceptance rates for the four L3Os.

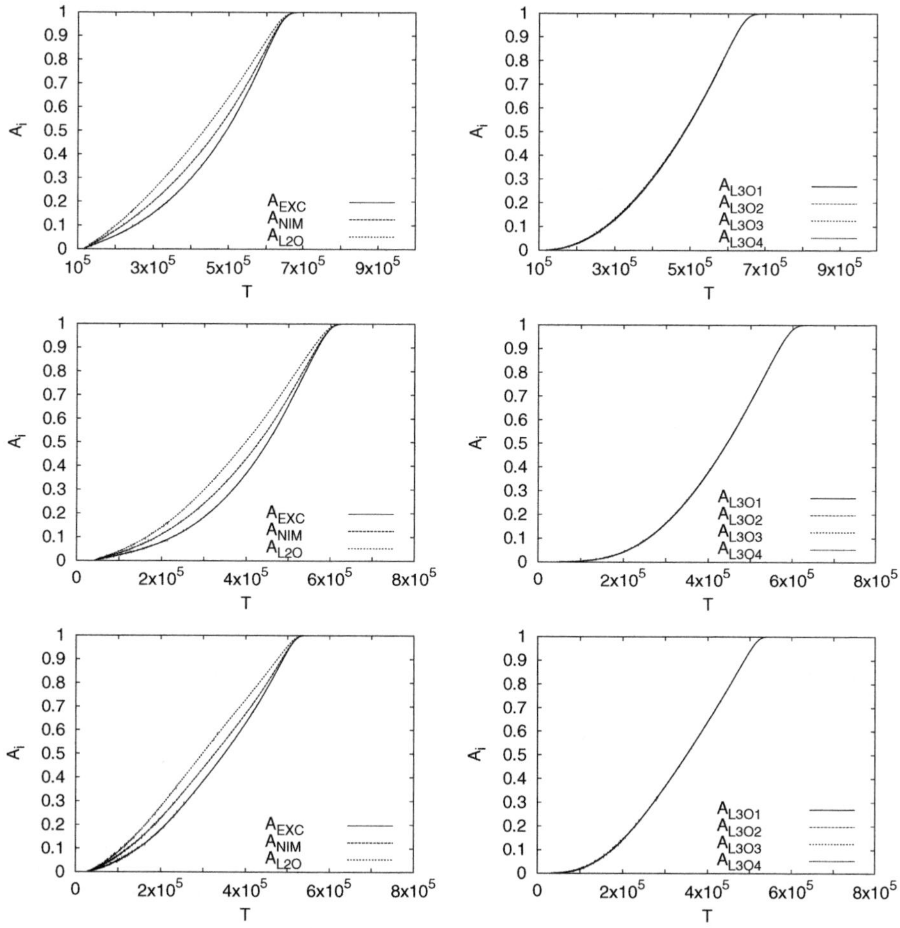

Fig. 9.16. Partial acceptance rates of the individual moves (EXC, NIM, L2O *left*, and the four variants of the L3O *right*) vs. water level T of GDA for three TSP instances [BEER127 (*top*), LIN318 (*middle*), ATT532 (*bottom*)]

Till now, the GDA has been allowed to select the new water level on its own, as $\mathcal{T}$ could be set to the energy of the current configuration if that was larger than the desired next value of the water level. However, in practical applications, one usually avoids this slowing down of the cooling schedule. But then the simulation might run into the problem that the system is stuck in a configuration above the water level and cannot get below the water level if every neighboring configuration is also above the water level. Thus,

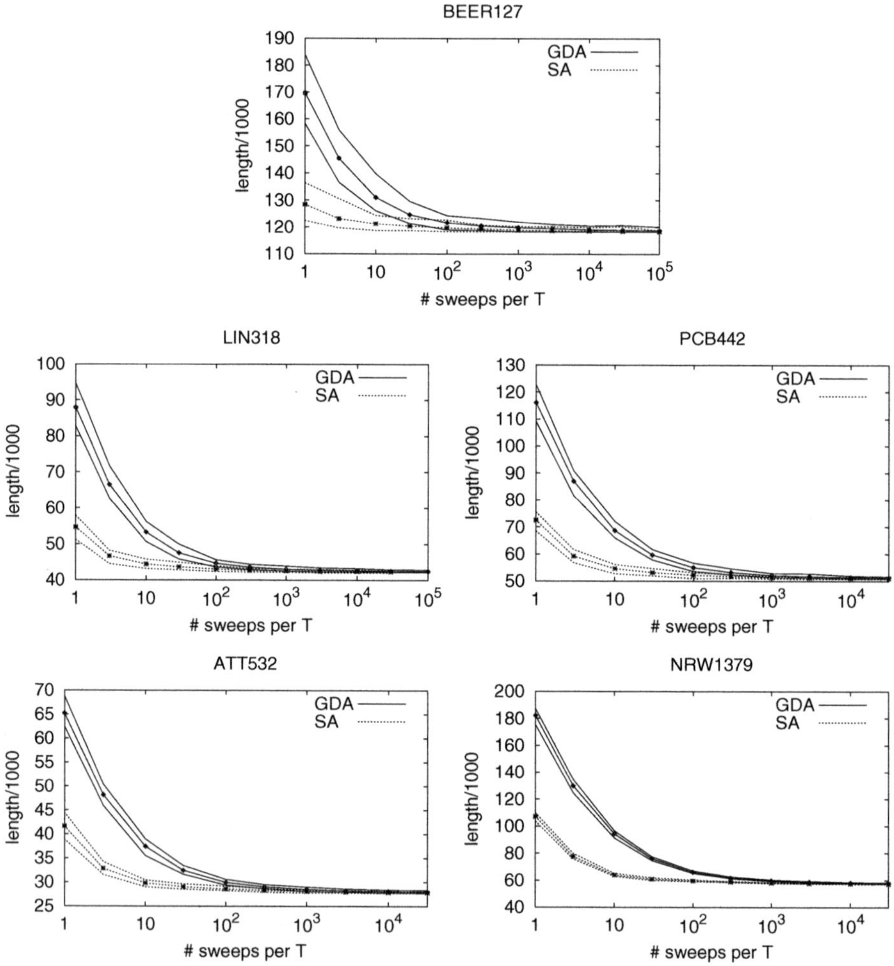

Fig. 9.17. Quality of the results achieved with GDA (*straight lines*) compared to those achieved with SA (*dotted lines*, replotted from Fig. 7.9), for five TSP instances (BEER127, LIN318, PCB442, ATT532, and NRW1379) vs. number of sweeps per water level step: three *lines* show the minimum, mean (*error bars* are the size of the *symbols*), and maximum lengths

even paths leading iteratively downhill and into the water cannot be used. Therefore, the acceptance criterion of the GDA is altered such that each move improving the current configuration is also accepted. This extension is only important for the case mentioned. Here the system is quenched into some local valley as if by a greedy descent, but basically the algorithm is still a GDA as it allows for a restricted RW once the Monte Carlo walker is below the surface of the water.

Figure 9.17 shows the results achieved with the GDA for five TSP instances. For comparison, the results achieved with SA are shown, too. SA was selected as there are proofs both for SA and the GDA that they may lead to the optimum configuration in an infinite amount of time.

Looking at Fig. 9.17 we find that the GDA leads to much worse results than SA for short calculation times. Increasing the calculation time, the difference between the GDA and SA becomes smaller. However, the mean value of the results provided by SA always stays significantly better than that of the GDA. Summarizing, SA is superior to the GDA and should thus be preferred. A further point to make is that only the energy difference must be calculated if using SA. Of course, one can also calculate the energy difference and add it to the energy of the current configuration to get the energy of the tentative new configuration that is to be accepted or rejected by the GDA. But after the acceptance of a series of moves, the energy of the current configuration must be recalculated in order to reduce the amount of rounding errors. This takes some additional calculation time.

9.4 Computational Results for Record-to-Record Travel

The record-to-record travel (R2R) algorithm is, given its acceptance criterion, closely related to TA. In contrast to TA, R2R compares the tentative new configuration not to the current one but to the best configuration found so far. Thus, a move is accepted either if it leads to an improvement or if the tentative new configuration is only up to some threshold T worse than the best solution found so far. As this criterion is technically rather similar to that of TA, we will compare the results for R2R with the results for TA here.

Again we start out with the decrease in the mean energy with decreasing threshold (Fig. 9.18). The decrease in the energy looks rather similar to that of TA. But the transition regime in which the energy decreases from the regime of the random configuration to the regime of ordered solutions is shifted to larger values of the threshold by roughly one magnitude. The pseudo specific heat, which is calculated as $C = \mathrm{Var}(\mathcal{H})/T^2$ for both algorithms and which is shown on the right half of Fig. 9.18, shows, however, a completely different behavior: for the R2R algorithm, three parts of C can be distinguished: a small peak is observed at large T. In the intermediate threshold range, C fluctuates rather menacingly. Finally, at small thresholds,

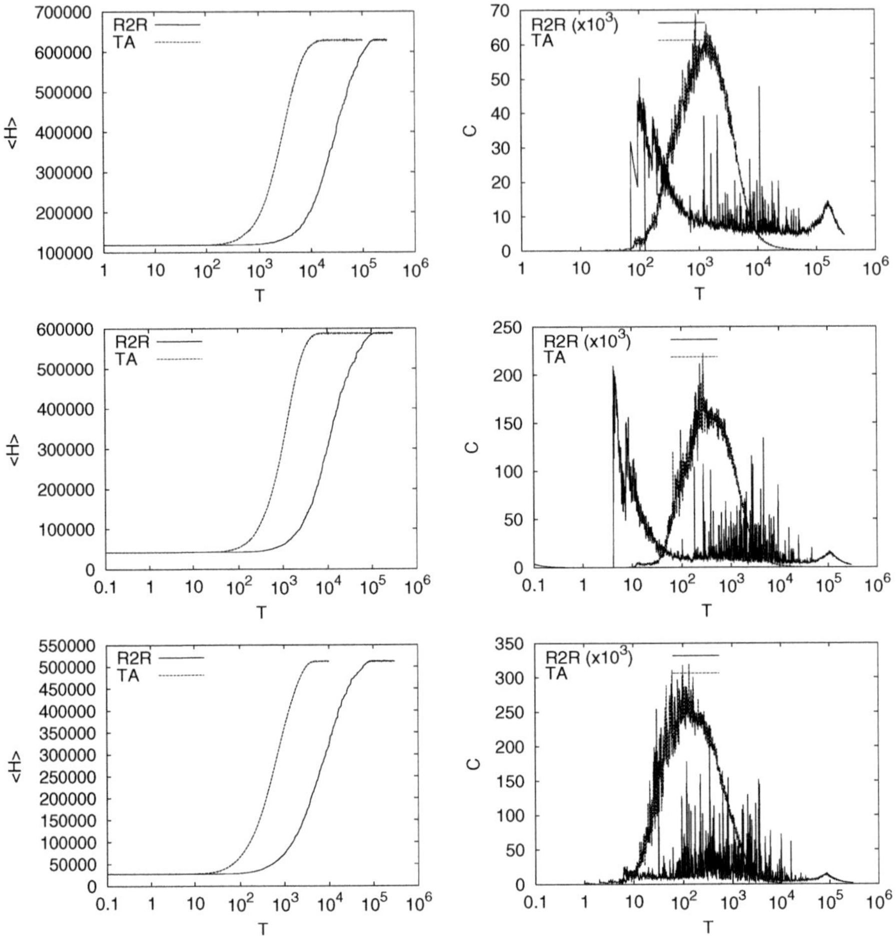

Fig. 9.18. Computational results for the mean energy $\langle \mathcal{H} \rangle$ (*left*) and the pseudo specific heat C (*right*) of three TSP instances [BEER127 (*top*), LIN318 (*middle*), ATT532 (*bottom*)] using R2R. For comparison, the results for TA of Fig. 9.1 are shown

there is a second peak that is much larger than the one at high thresholds. Such fluctuations as seen in the intermediate part here could also occur when working with SA or TA and spending only a very small amount of calculation time. In that case, such fluctuations indicate that the system is not equilibrated as the measurements are taken, so that one gets fluctuations instead of a round or sharp peak. Here, however, the calculation time for the R2R algorithm was the same as for TA. It could be that this is a sign that the system has not yet reached equilibrium, whatever that is in the case of R2R. Even more interesting are the left large peaks. They differ strongly between simulations as each Monte Carlo walker meets other best-so-far solutions on his way at different times.

Let us now take another look at the order parameter and its corresponding susceptibility (Fig. 9.19). We get a nice sigmoidal increase of the order parameter with decreasing threshold with the R2R algorithm as well. This increase is shifted by roughly one order of magnitude of the threshold toward larger threshold values compared to TA. This result is in accordance with the result for the susceptibility, which also increases earlier. The peaks of the susceptibilities of the various instances for R2R look otherwise rather the same as for TA: if there is a wide peak for a specific instance with TA, there

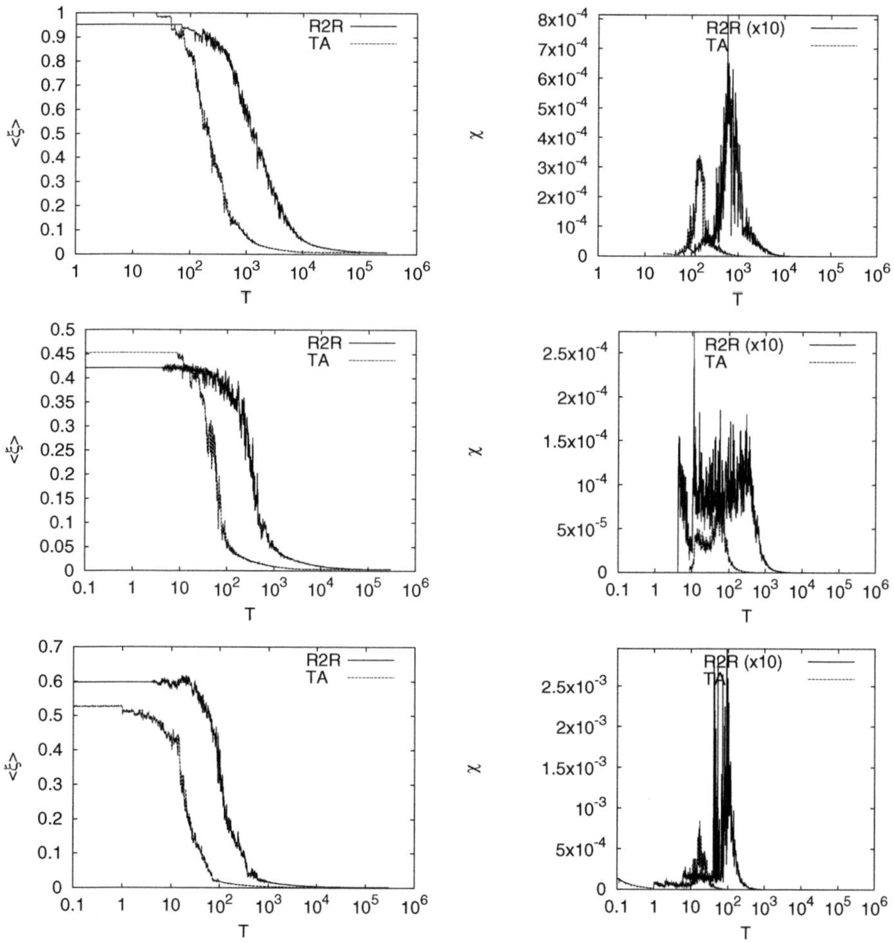

Fig. 9.19. Computational results for the order parameter $\langle \xi \rangle$ and the corresponding susceptibility χ of three TSP instances [BEER127 (*top*), LIN318 (*middle*), ATT532 (*bottom*)] using R2R. For comparison, the results for TA from Fig. 9.2 are shown

is also a wide peak with R2R. Analogously, sharp peaks are shifted on the threshold axis.

The same shift can be observed in the total acceptance rates A (Fig. 9.20). In addition, the decrease in the acceptance rate is slightly steeper for TA than for the R2R, which is also the case for the decrease in the energy. Thus, the transition range of this R2R algorithm is wider than the transition range of TA.

Next we consider the partial acceptance rates of the individual moves for the R2R algorithm. Figure 9.21 shows curves similar to those of TA (Fig. 9.4).

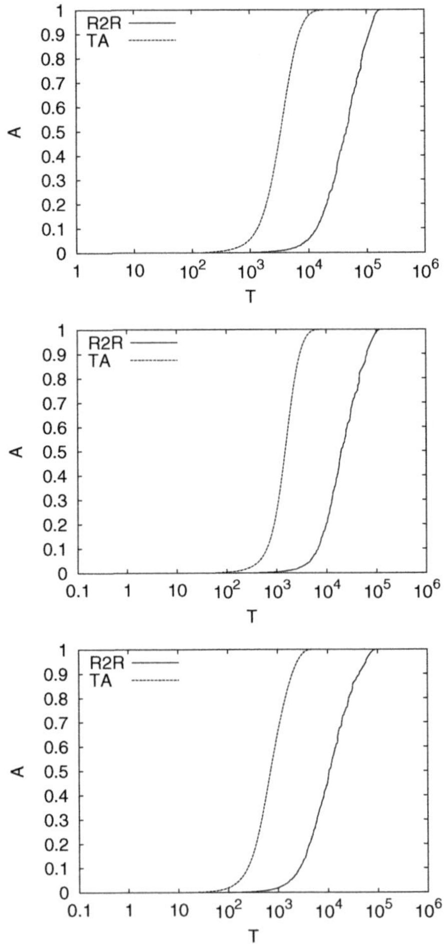

Fig. 9.20. Computational results for the total move acceptance rate A of three TSP instances [BEER127 (*top*), LIN318 (*middle*), ATT532 (*bottom*)] using R2R. For comparison, the results for TA from Fig. 9.3 are shown

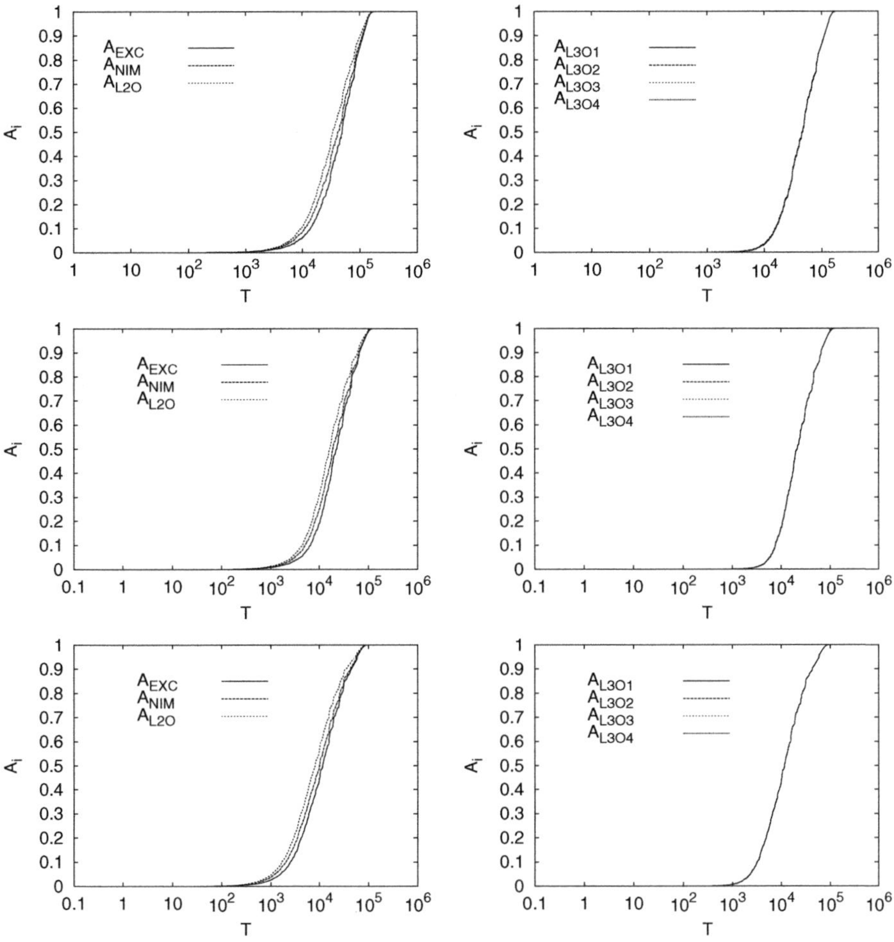

Fig. 9.21. Computational results for the partial acceptance rates A_i of the moves EXC, NIM, and L2O and of the four variants of the L3O for three TSP instances using R2R: BEER127 (*top*), LIN318 (*middle*), ATT532 (*bottom*)

Again we notice that the curves for the four variants of the L3O coincide with each other. The curves for the smaller moves again differ only slightly.

Finally, we consider the quality of the results that can be achieved using the R2R algorithm and compare it to the results for TA again. Figure 9.22 shows the results as usual for the five TSP instances. We find that for short calculation times, TA is clearly superior to the R2R algorithm. However, the R2R algorithm produces slightly better results on average when a large amount of calculation time is spent. Reconsidering the calculation time to be spent one again must consider that the total energy of the tentative new configuration must be determined as in the GDA, such that one must—at

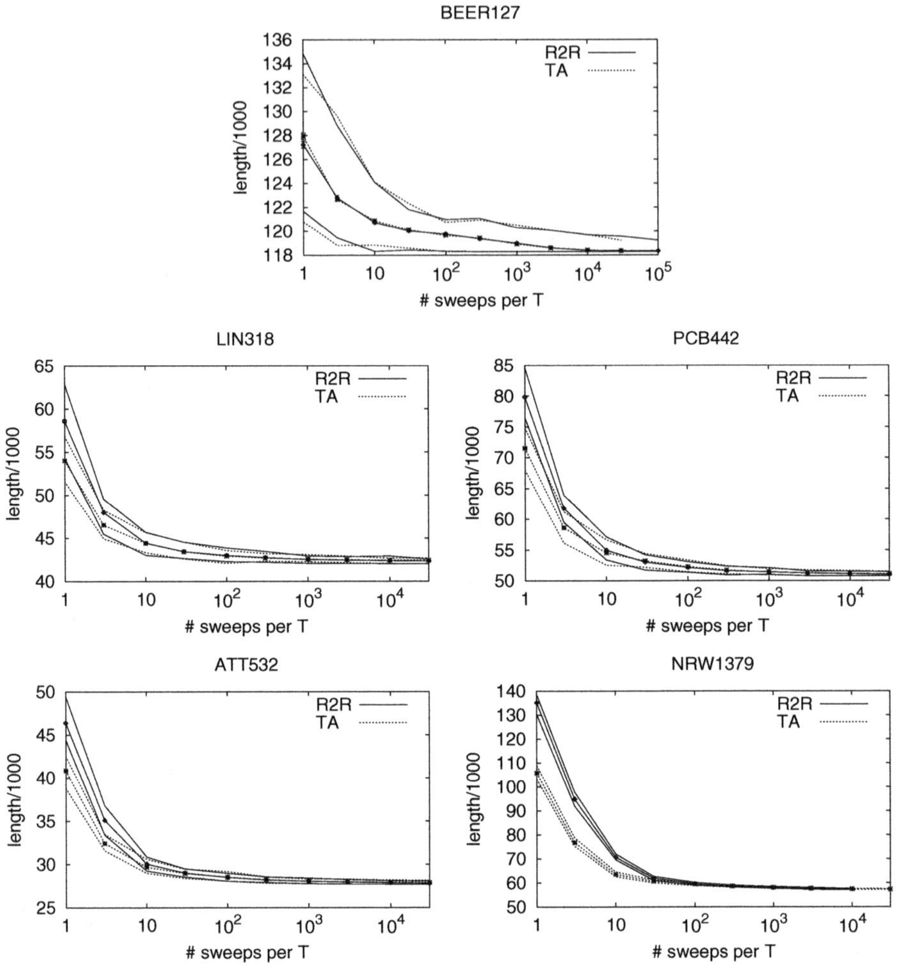

Fig. 9.22. Quality of the results achieved with R2R (*straight lines*) vs. those achieved with TA (*dotted lines*, replotted from Fig. 9.5), for five TSP instances (BEER127, LIN318, PCB442, ATT532, and NRW1379) vs. number of sweeps per threshold step: the three *lines* show the minimum, mean (*error bars* are the size of the *symbols*), and maximum lengths

least sometimes—calculate not only the energy difference but the total new energy.

Summarizing, we have shown some results for the application of algorithms closely related to SA. One might argue that the results were mostly the same or very similar time after time. However, there are also a few small differences among these algorithms. Thus, we showed these curves over and over in order to see what stays the same and where the differences lie.

Generally speaking, however, as the concept of SA to introduce some control parameter in order to enable the system to climb over barriers in the energy landscape and to reduce this control parameter in order to drive the system from the regime of unordered high-energy configurations to ordered low-energy solutions is transferred to similar algorithms, one cannot expect anything completely new here. One cannot even say which of these algorithms is the best. Some are better at shorter calculation times, others at longer times. It may even depend on the computer being used, e. g., how much time it takes to calculate an exponential of a number.

The algorithms related to SA shown here simply demonstrate that this concept of cooling a system down with a temperaturelike control parameter and thus getting (quasi) optimum solutions is very general and does not depend very much on a specific acceptance criterion.

10 Application of Search Space Smoothing to TSP

10.1 A Small Toy Problem

Till now, more elaborate iterative improvement heuristics have been studied that make use of a control parameter by which the Monte Carlo walker is partially enabled to climb over barriers in the energy landscape. But there are also algorithms that try to change the energy landscape by removing the barriers, such that a greedy Monte Carlo walker who never climbs upward might be able to reach the global minimum.

Let us start with the small toy instance given in Table 10.1. It consists of four nodes on the unit square but the distance $D(2,1)$ is $\sqrt{2}$ instead of 1, such that the TSP instance becomes asymmetric. The distance matrix D is

Table 10.1. Small toy instance of an asymmetric TSP: as shown in the picture, it consists of four nodes located at the edges of a unit square such that the distances are already given. The distance $D(2,1)$ will be, however, $\sqrt{2}$, so that the TSP instance becomes asymmetric. Furthermore, only one move will be implemented by which two neighboring nodes in the tour are exchanged. *Left table*: asymmetric distance matrix of this instance; *right table*: sequences, lengths, and neighbors of individual states

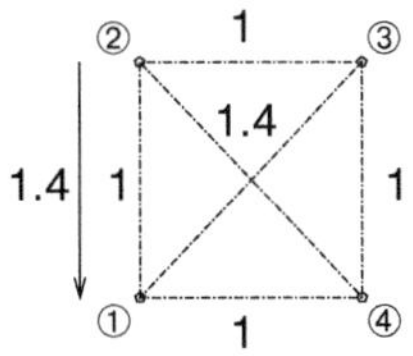

Distance matrix

	1	2	3	4
1	0	1	$\sqrt{2}$	1
2	$\sqrt{2}$	0	1	$\sqrt{2}$
3	$\sqrt{2}$	1	0	1
4	1	$\sqrt{2}$	1	0

$\longrightarrow$

State	Sequence	Length	Neighboring states
# 1	(1 2 3 4)	4	# 2, # 3, # 4, # 5
# 2	(1 2 4 3)	$2+2\sqrt{2}$	# 1, # 3, # 5, # 6
# 3	(1 3 2 4)	$2+2\sqrt{2}$	# 1, # 2, # 4, # 6
# 4	(1 3 4 2)	$1+3\sqrt{2}$	# 1, # 3, # 5, # 6
# 5	(1 4 2 3)	$2+2\sqrt{2}$	# 1, # 2, # 4, # 6
# 6	(1 4 3 2)	$3+\sqrt{2}$	# 2, # 3, # 4, # 5

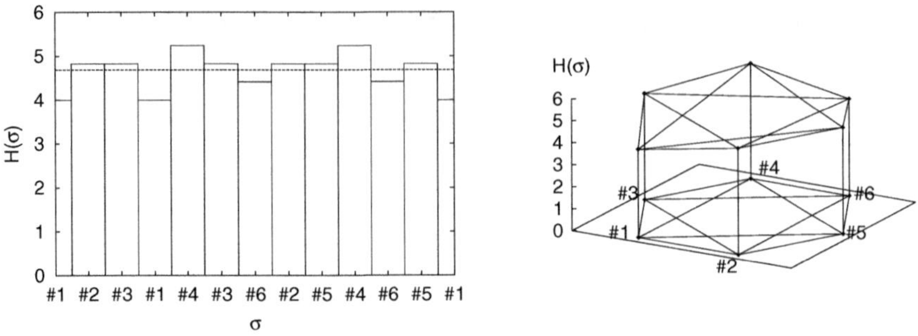

Fig. 10.1. Two representations of the energy landscape of the toy instance introduced in Tab. 10.1. *Left*: closed Markov chain of minimum length containing all edges in the search space, i. e., all possible moves between various configurations, exactly once. *Dotted line*: mean length of all states, which is roughly 4.69. *Right*: the search space is projected in the 2D space in such a way that the six states and the links between them form both a hexagon and a hexagram. On *top* of this plane, which stands for the search space, a further dimension is introduced that refers to the energies of the states. Thus, we see in this projection on the 3D space what the energy landscape looks like for this problem

shown in the left table. The only move allowed will be the nearest neighbor exchange, a special case of the Lin-2-opt (L2O) and the exchange (EXC), by which two neighboring nodes in the sequence are exchanged. Thus, the energy landscape contains six states, among them the global optimum (1234) and a local minimum (4321) with lengths 4 and $3 + \sqrt{2} \approx 4.41$, respectively. The other four states are neighbors of these minima, as shown on the right in Table 10.1. A closed Markov chain containing all edges in the energy landscape is shown in Fig. 10.1. The mean length of the states is $\langle \mathcal{H} \rangle = (7 + 5\sqrt{2})/(4 - 1) \approx 4.69$. Furthermore, this figure shows a projection of this energy landscape into the 3D space, which gives an even better insight into this problem than looking only at the closed Markov chain, which represents a cut through the landscape along the path a Monte Carlo walker has gone.

We now introduce an easy smoothing rule:

- First, all states are assumed to have the same length $\langle \mathcal{H} \rangle$. Thus, the energy landscape is flat at the very beginning.
- In the second step, the state whose length deviates most from $\langle \mathcal{H} \rangle$ resumes its original length. Thus, the energy landscape retains its previous form, except that there is now a hole or a peak in the surface.
- In the next steps, the other states return to their original lengths. The order in which they return is determined by the deviation of their lengths to the mean length. Thus, the one with the second largest deviation returns, then the one with the third largest deviation, and so on. Finally, all states return to their original lengths.

In the toy model, the following sequence of "desmoothing" steps is applied:

1. First, the energy landscape forms a plain, as all states are of equal length.
2. Then state 1 returns to its original length. The other states remain on the plateau. Here state 1 becomes the global minimum of the system.
3. State 4 returns to its original position in the energy landscape and becomes the global maximum.
4. State 6 becomes a local minimum.
5. States 2, 3, and 5 return to their original lengths at the same time.

When performing a greedy walk over the smoothed energy landscape at each step, the Monte Carlo walker will fall into the global minimum in step 2.

Of course, this is only a simple example, but it demonstrates the potential of this idea. However, more elaborate techniques for search space smoothing (SSS) must be used for practical applications as a complete enumeration of all configurations and thus an exact knowledge of the energy landscape is usually impossible for larger system sizes. Thus, rules must be found for smoothing the entries in the distance matrix in order to smooth the energy landscape in an indirect way.

10.2 Gu and Huang Approach

Gu and Huang introduced the following approach to smoothing the distances with a smoothness-control parameter α [76]:

- First, normalize all distances by dividing them by the maximum distance: let $D_{\max}$ be the maximum of all distances and

$$d(i,j) = D(i,j)/D_{\max}\,. \tag{10.1}$$

  This linear transformation rescales all lengths to the interval [0; 1]; however, the shape of the energy landscape is not changed.
- Secondly, the mean normalized distance

$$\bar{d} = \frac{1}{N(N-1)} \sum_{\substack{i,j=1 \\ i\neq j}}^{N} d(i,j) \tag{10.2}$$

  is calculated. Due to the normalization of the distances, this mean distance lies between 0 and 1. Note that the edges from a node to itself cannot be part of a feasible TSP configuration. The diagonal elements of the distance matrix are left out both in the sum and in the normalization factor.
  Figure 10.2 shows mean distances for various TSP scenarios. First, the TSP nodes could be placed randomly, e. g., in the unit square. We considered instances with 3, 10, 30, 100, 300, 1000, and 3000 nodes. For each TSP size, 100 random instances were generated. The lines show the minimum, maximum, and mean value (with error bars) of $\bar{d}$. We find that the mean

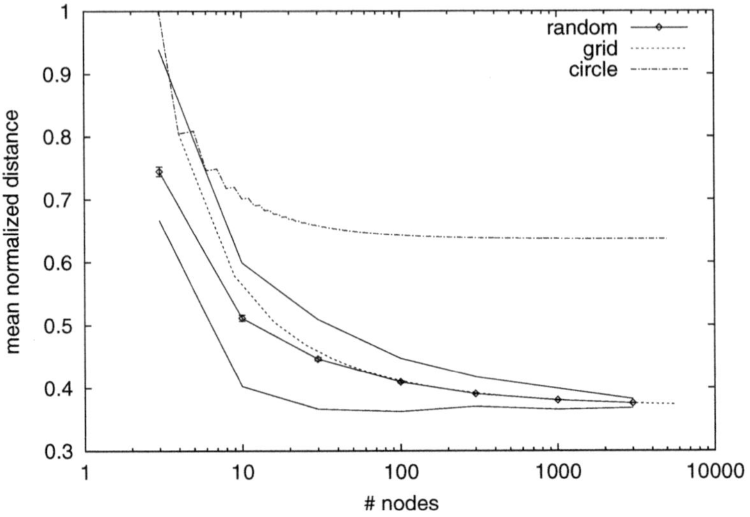

Fig. 10.2. Mean distances $\bar{d}$ as defined in Eq. (10.2) of TSP instances with the nodes placed randomly in a unit square ("random"), placed on a regular quadratic grid ("grid"), and placed on a circle ("circle") with equal distance to the nearest neighbors

Table 10.2. Mean distances $\bar{d}$ as defined in Eq. (10.2) of five TSP benchmark instances

Instance	$\bar{d}$
BEER127	0.254739...
LIN318	0.380025...
PCB442	0.361045...
ATT532	0.345025...
NRW1379	0.348021...

distance decreases monotonously with increasing system size. Secondly, we consider instances in which the nodes are placed on a quadratic grid. Thus, the number of nodes N is always given by $N = L^2$ with L being the linear dimension of the grid, i. e., the number of columns and rows. Here, too, the curve decreases monotonously. Furthermore, we find that the curve for the random instances and for the grid instances coincide at large N, which is not surprising as these two scenarios can be considered in the limit of large N as a Monte Carlo and a deterministic measurement of the integral

$$\frac{1}{D_{\max}} \int \mathrm{d}\boldsymbol{r} \int \mathrm{d}\boldsymbol{r}' |\boldsymbol{r} - \boldsymbol{r}'| \times \varrho(\boldsymbol{r}) \times \varrho(\boldsymbol{r}')$$
$$= \frac{1}{\sqrt{2}} \int_0^1 \mathrm{d}x \int_0^1 \mathrm{d}y \int_0^1 \mathrm{d}a \int_0^1 \mathrm{d}b \sqrt{(x-a)^2 + (y-b)^2} \tag{10.3}$$

with the point density $\varrho(\boldsymbol{r}) \equiv 1$. For the largest TSP instance we used for Fig. 10.2, we get a value of ≈ 0.37 for this integral. On the other hand, we consider the case of the nonfrustrated TSP in which all nodes lie on a circle, such that the nodes are the corners of a regular "N-gon". Here the mean distance $\bar{d}$ also decreases monotonously with increasing N, in the sense that it decreases monotonously both for even N and for odd N. If N is even, its mean normalized distance $\bar{d}$ is slightly smaller than the value for $N+1$. These values converge to the limiting value

$$\begin{aligned} &\frac{1}{D_{\max}} \int \mathrm{d}\boldsymbol{r} \int \mathrm{d}\boldsymbol{r}' |\boldsymbol{r}-\boldsymbol{r}'| \times \varrho(\boldsymbol{r}) \times \varrho(\boldsymbol{r}') \\ &= \frac{1}{2} \int\limits_0^{2\pi} \mathrm{d}\varphi \int\limits_0^{2\pi} \mathrm{d}\psi \left| \begin{pmatrix} \cos(\varphi)-\cos(\psi) \\ \sin(\varphi)-\sin(\psi) \end{pmatrix} \right| \times \frac{1}{2\pi} \times \frac{1}{2\pi} \\ &= \frac{1}{8\pi^2} \int\limits_0^{2\pi} \mathrm{d}\varphi \int\limits_0^{2\pi} \mathrm{d}\psi \sqrt{2(1-\cos(\varphi-\psi))} \\ &= 16\pi/(8\pi^2) = 2/\pi \approx 0.6366\ldots \,. \end{aligned} \tag{10.4}$$

This limiting value is of course much larger than in the previous two cases as there is a large hole in the midst of the spatial distribution of the nodes. In practical applications, nodes are mostly placed neither according to a special geometry nor purely at random. However, the results for random instances are usually much closer to those for real-life instances than those of a special geometry.
Finally, we consider the values for $\bar{d}$ in the five benchmark instances we work with here. The values are given in Table 10.2. The very small value for the BEER127 instance is due to the special distribution of the nodes where many nodes are close together in the center and only a few nodes are outside. The values of the other instances agree nicely with those for random and grid instances. Note that the values for $\bar{d}$ are significantly smaller than 0.5 in all five benchmark instances, which will be important later on for the explanation of the results of the SSS technique.

- Furthermore, the deviations $\Delta(i,j)$ of the distances to the mean distance,

$$\Delta(i,j) = d(i,j) - \bar{d}\,, \tag{10.5}$$

hold also the condition $0 \leq |\Delta(i,j)| < 1$ for all distances.

- Gu and Huang introduced a power law for the smoothed distances:

$$d_\alpha(i,j) = \begin{cases} \bar{d} + \Delta(i,j)^\alpha & \text{if } \Delta(i,j) \geq 0\,, \\ \bar{d} - (-\Delta(i,j))^\alpha & \text{otherwise}\,. \end{cases} \tag{10.6}$$

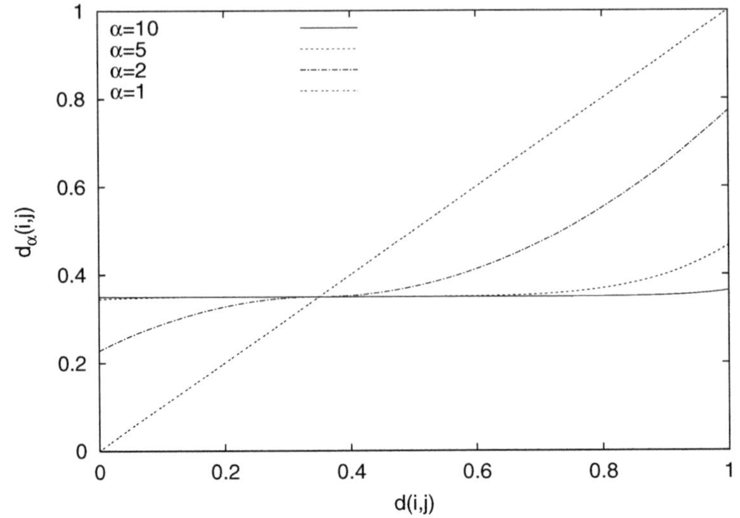

Fig. 10.3. Distances smoothed with the power law smoothing formula (10.6). *Curves*: values for smoothed distances $d_\alpha(i,j)$ as functions of the original normalized distances $d(i,j)$ for the assumed mean normalized distance $\bar{d} = 0.35$ and for various values of the smoothness control parameter α

Using this formula, α must be decreased gradually from a rather high value at which all distances are roughly the same (Fig. 10.3) such that the energy landscape is rather flat, to a final value of 1 at which the distances return to their original values and the energy landscape returns to its original shape. For $\alpha \gg 1$, one gets $d_\alpha(i,j) \approx \bar{d}$, as the deviation is smaller than 1. This is the reason for the normalization of the distances. At each value of α, several greedy sweeps are applied to the system in order to lead the system into a local or even the global minimum of the smoothed energy landscape. At large α, the system will thus get into the global optimum of the smoothed energy landscape if there is only one minimum left. With decreasing α, the landscape is deformed and the minima are shifted to configurations near those configurations that are the minima in the previous landscape. With new greedy runs, the system follows these changes in the landscape. If α is decreased slowly enough, one can usually trust in a guidance effect [76], i. e., the system does not lose the trail of its old local or global minimum.

As in the investigation of simulated annealing (SA) and its related algorithms, we first want to investigate the behavior of SSS by looking at some observables of the TSP instances. Figure 10.4 shows in its left half the decrease in the mean energy for three TSP instances. However, one must consider that there are two energy functions when working with this SSS technique: there is of course the original Hamiltonian $\mathcal{H}_{\text{orig}}$ that has to be considered. The greedy optimization, however, works at each value of the smoothness-control

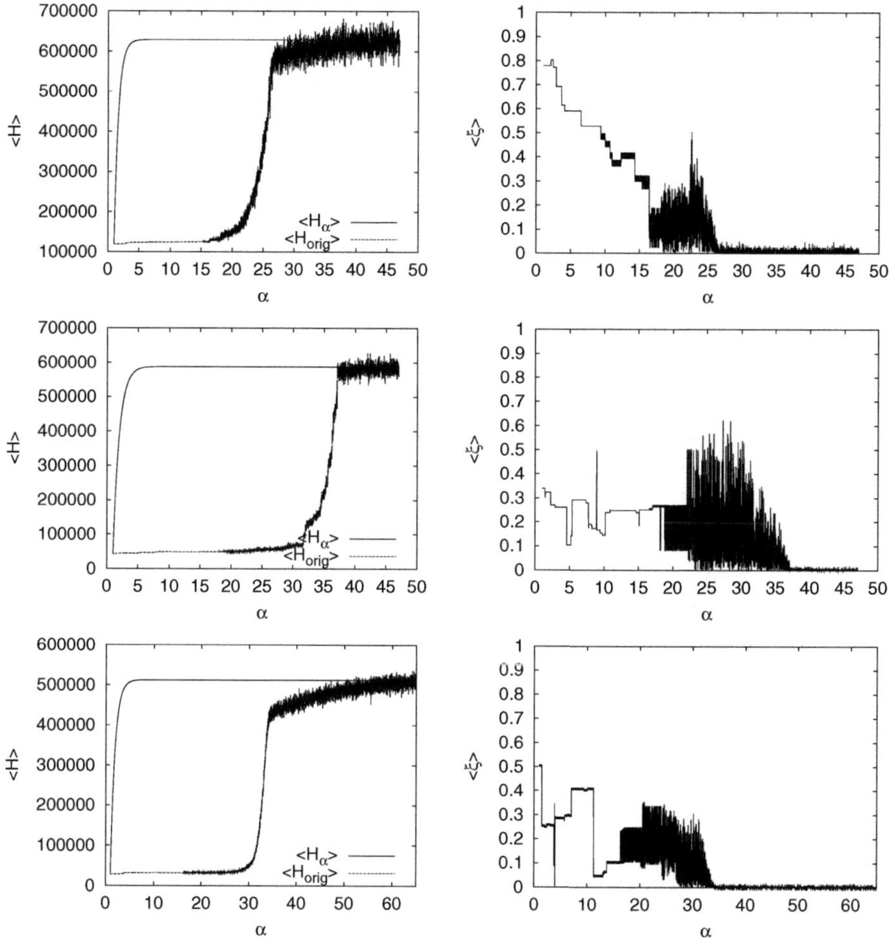

Fig. 10.4. Computational results for the mean energy $\langle \mathcal{H} \rangle$ (*left*) and the order parameter $\langle \xi \rangle$ (*right*) of three TSP instances using the power law formula for SSS: BEER127 (*top*), LIN318 (*middle*), ATT532 (*bottom*). The mean energy $\langle \mathcal{H}_\alpha \rangle$ according to the smoothed distance values $D_{\max} \times d_\alpha(i,j)$ and the mean energy $\langle \mathcal{H}_{\text{orig}} \rangle$ calculated with the original distances $D(i,j)$ are shown

parameter α with a smoothed Hamiltonian $\mathcal{H}_\alpha$, which only coincides with $\mathcal{H}_{\text{orig}}$ at the final value $\alpha = 1$. Thus, first, we must consider $\langle \mathcal{H}_\alpha \rangle$: it stays exactly constant for very large α and then decreases monotonously. Over a wide α-range, $\langle \mathcal{H}_\alpha \rangle$ stays virtually constant. Only for small α does the decrease in $\langle \mathcal{H}_\alpha \rangle$ become dramatic. Thus, looking only at $\langle \mathcal{H}_\alpha \rangle$, one would conclude that the optimization is performed at small α-values.

However, the mean energy $\langle \mathcal{H}_{\text{orig}} \rangle$, calculated with the original distances $D(i,j)$, which is simply measured at each α-step but has no influence on the optimization process, first fluctuates around the curve for $\langle \mathcal{H}_\alpha \rangle$, then drops to a narrow range of the control parameter α, and finally decreases slightly in a gradual way. The fluctuation at the beginning hints that the greedy technique here nearly coincides with the random walk (RW): obviously, most or all configurations are of equal length, so that most, or even all, moves are accepted. $\langle \mathcal{H}_{\text{orig}} \rangle$ drops to a range where $\langle \mathcal{H}_\alpha \rangle$ is still exactly or virtually constant. This means that the energy landscape is no longer flat at these values of α. As the mean value $\bar{d}$ is smaller than 0.5, the first edges not of length $\bar{d}$ must be those with $D(i, j) \approx D_{\max}$. Thus, the system first starts to avoid taking longer edges into the configuration. This leads to the large drop of $\langle \mathcal{H}_{\text{orig}} \rangle$, whereas $\langle \mathcal{H}_\alpha \rangle$ is still rather constant as the other edges are still of the same length $\bar{d}$. This type of landscape can be considered a monument valley landscape, which is shown in the color picture: in such a landscape, there is a plain but also some isolated hills sticking out of the plain. The hills are made of the configurations containing the longest edges.

Finally, at smaller α, $\langle \mathcal{H}_{\text{orig}} \rangle$ looks like the development of the mean value of the energy in the greedy case: the system is quasifrozen, but sometimes improvements are found. But here these improvements are due to the guidance effect of SSS: by reducing the smoothness-control parameter α, the smoothed energy landscape is deformed, so that a configuration that had previously been the local minimum might then be on a crooked wall. Applying the greedy algorithm, the system is able to move to the new local minimum nearby, in the slightly desmoothed energy landscape. Hopefully, this guidance effect will lead to good results in the end, where the landscape is finally turned into the energy landscape of the original problem.

The right half of Fig. 10.4 shows the expectation value of the order parameter ξ. At large α, $\langle \xi \rangle$ is slightly larger than zero and thus indicates that the system is in a RW mode in which the configurations sometimes contain one or more edges of the optimum configuration. Then in the α-range in which $\langle \mathcal{H}_{\text{orig}} \rangle$ drops, $\langle \xi \rangle$ fluctuates rather strongly. Depending on the exact value of α, the system likes more or less the valley in which the global optimum lies. At small α, where $\langle \mathcal{H}_{\text{orig}} \rangle$ shows already a mostly frozen behavior, $\langle \xi \rangle$ is also practically frozen. But large jumps in $\langle \xi \rangle$ occur between some successive α-steps. Obviously, sometimes the system falls in a quite different valley in the energy landscape when viewed from the point of view of the original landscape. Figure 10.4 shows that this can either gradually lead to a better approximation of the optimum, as it does for the BEER127 instance, or it can be more up and down, as it is for the other two instances.

A further measure for the freezing behavior of this SSS approach is the total acceptance rate of the moves, which is shown in Fig. 10.5. We find that the curves for the total acceptance rate are rather similar to those for $\langle \mathcal{H}_{\text{orig}} \rangle$ in Fig. 10.4 and strengthen the points made in the discussion of $\langle \mathcal{H}_{\text{orig}} \rangle$ above. At large α, the system is in a quasi-RW mode in which most moves are accepted. With decreasing α, more and more moves are rejected. Then in a very narrow α-range, a transition occurs in which both the total acceptance rate and $\langle \mathcal{H}_{\text{orig}} \rangle$ break down. For small α, the acceptance rate vanishes.

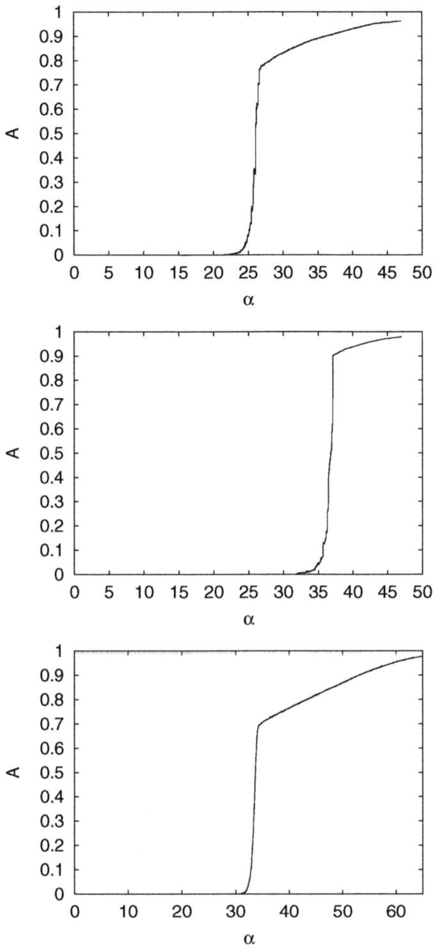

Fig. 10.5. Computational results for the total acceptance rate A of three TSP instances using the power law formula for SSS: BEER127 (*top*), LIN318 (*middle*), ATT532 (*bottom*)

We get the same picture for the partial acceptance rates of the moves in Fig. 10.6. As in the graphics for SA and its related algorithms, we find again that the acceptance rates for the four variants of the Lin-3-Opt (L3O) coincide with each other for all values of the control parameter. On the other hand, there are some differences between the curves for the smaller moves: for large α, the acceptance rate of the L2O is larger than that for the node insertion move (NIM), which is in turn larger than that for the EXC. In some narrow "critical range" of α, these curves break down. The acceptance rates for the four variants of the L3O and for EXC vanish first, followed by the acceptance rate of the NIM, and again followed by the L2O.

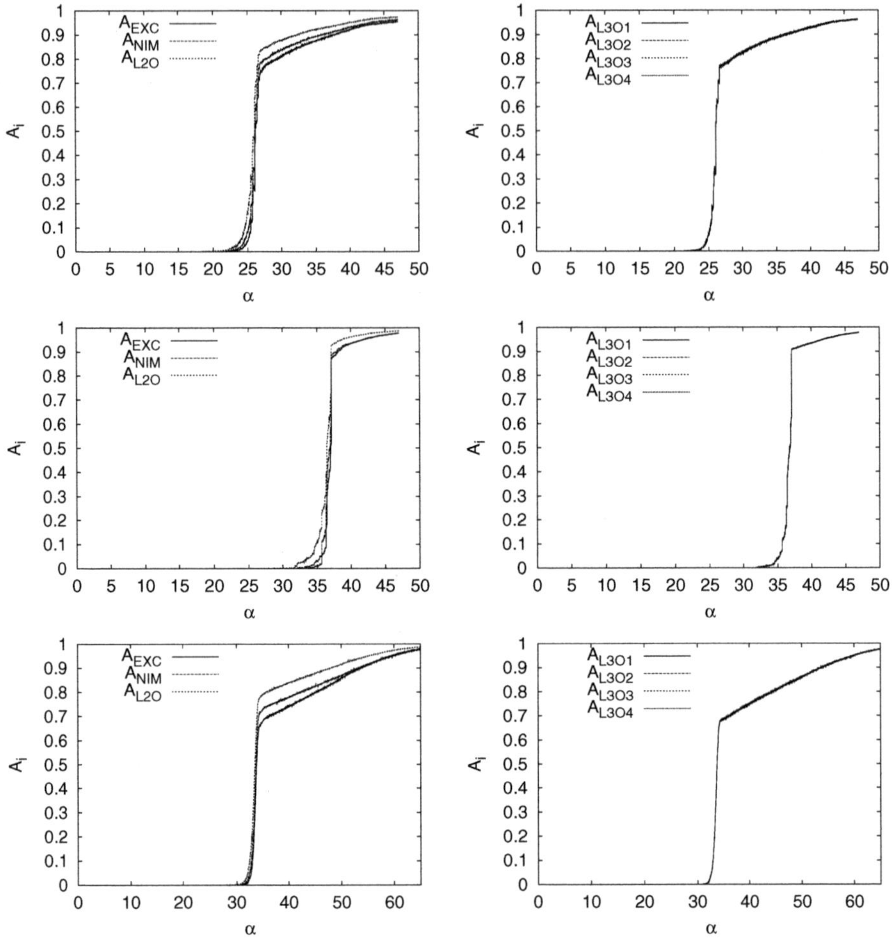

Fig. 10.6. Computational results for the partial acceptance rates A_i of the small moves EXC, NIM, and L2O (*left*) and of the four variants of the L3O (*right*) of three TSP instances using the power law formula for SSS: BEER127 (*top*), LIN318 (*middle*), ATT532 (*bottom*)

The most astounding difference between SSS with the power law formula by Gu and Huang and SA-type optimization algorithms is that obviously here the system orders itself on a linear and not a logarithmic α-scale. This is clearly not only due to the construction of the α-decrease schedule, which must end at a final value of 1 instead of the final temperature value of 0, because the successive α-values could also be chosen as $1+(\alpha_i-1)\times f^n$ with an initial α-value α_i and a "cooling factor" f. Instead, we find nearly sigmoidal decreases of $\langle\mathcal{H}_{\text{orig}}\rangle$ in Fig. 10.4 and of the total and partial acceptance rates on a linear α-scale. Thus, we decrease α linearly when using the power law formula by Gu and Huang, from a large initial value to 1 in steps of 0.01. Finally, we add one greedy step to the original energy landscape in order to end up at exactly $\alpha = 1$.

Finally, but most important for optimization purposes, the question arises as to whether this SSS approach leads to better results than the SA-type optimization algorithms. Thus, we compare the results achieved with SSS using the power law formula with those achieved with SA in Fig. 10.7. The ini-

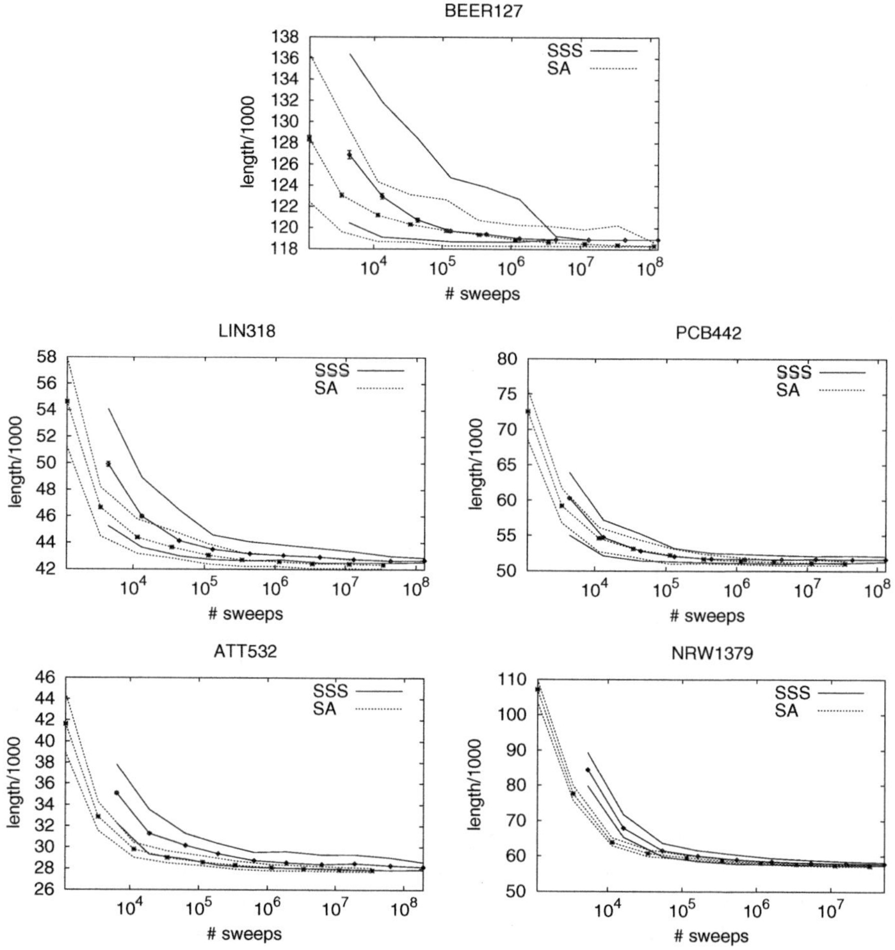

Fig. 10.7. Quality of the results achieved with SSS using the power law formula (*straight lines*) and SA (*dotted lines*) for five TSP instances (BEER127, LIN318, PCB442, ATT532, and NRW1379) vs. overall number of sweeps: the three *lines* show minimum lengths, mean lengths (*error bars* are mostly the size of the *symbols*), and maximum lengths

tial value of the smoothness-control parameter was chosen to be 45.001 for the BEER127, LIN318, and PCB442 instances, 65.001 for the ATT532 instance, and 55.001 for the NRW1379 instance. Thus, we performed 4402 α-steps for the BEER127, LIN318, and PCB442 instances, 6402 for the ATT532 instance, and 5402 for the NRW1379 instance. At each α-step, some number of sweeps (1, 3, 10, 30, ...) was performed. These results are now compared to those for SA shown in Fig. 7.9. The number of temperature steps when using SA was for all instances 1147. Generally, we find that—when using our parameters, which allow the transition between the unordered high-energy and the ordered low-energy regime without wasting too much calculation time at large values of the control parameter—SA generally leads to better results for these five TSP instances than SSS with this power law formula. However, from this comparison, one may not conclude that SA is generally better than SSS for all problems and for all imaginable smoothing formulas and cooling schedules. As the number of smoothing formulas is infinite, one might also find for some problem a smoothing formula that leads to better results than SA.

Of course, further Gu–Huang-like smoothing formulas can be developed with more complicated smoothing formulas. These should be based on the normalized distances and their deviation from the mean distance $\bar{d}$. Furthermore, they must lead to a flat landscape for large α and to the original landscape for some small value of α. Here we give only a few examples [181, 186, 182]:

- Exponential smoothing:
  Here a Gu–Huang-like formula containing an exp(...) term is used:

$$d_\alpha(i,j) = \begin{cases} \bar{d} + \dfrac{\alpha}{\exp\left(\dfrac{\alpha}{\Delta(i,j)}\right) - 1} & \text{if } \Delta(i,j) \geq 0\,, \\ \bar{d} - \dfrac{\alpha}{\exp\left(\dfrac{\alpha}{-\Delta(i,j)}\right) - 1} & \text{otherwise}\,. \end{cases} \tag{10.7}$$

  For large α, the deviation from the mean distance $\bar{d}$ vanishes much more strongly than for the original power law formula. For $\alpha \to 0$, considering the relation $\exp(x) \to 1 + x$ for $x \to 0$, one gets the original landscape. Using this exponential smoothing formula, α is decreased linearly in steps of 0.01 from the initial value of 18.001 to 0.001. Finally, a greedy step on the original landscape corresponding to $\alpha = 0$ is added. Thus, 1802 α-steps are performed.
  Figure 10.8 shows the results for the PCB442 instance for the mean energy $\langle \mathcal{H} \rangle$ calculated according to the smoothed metric and according to the original metric and for the total acceptance rate A. We find that the results are rather similar to those for the power law smoothing formula: again, the system is in a quasi-RW mode at large values of α. $\langle \mathcal{H}_{\text{orig}} \rangle$ and A drop in

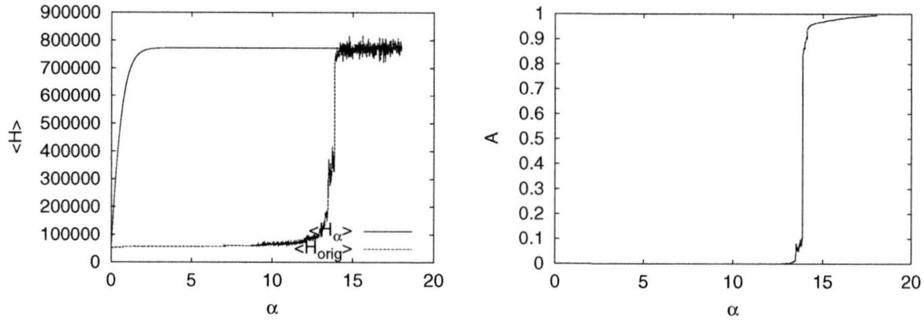

Fig. 10.8. Computational results for the mean energy $\langle\mathcal{H}\rangle$ (*left*) and the total acceptance rate A (*right*) of the PCB442 instance using the exponential formula (10.7) for SSS: *left*: both the mean energy $\langle\mathcal{H}_\alpha\rangle$ according to the smoothed distance values $D_{\max} \times d_\alpha(i,j)$ and the mean energy $\langle\mathcal{H}_{\text{orig}}\rangle$ calculated with the original distances $D(i,j)$ are shown

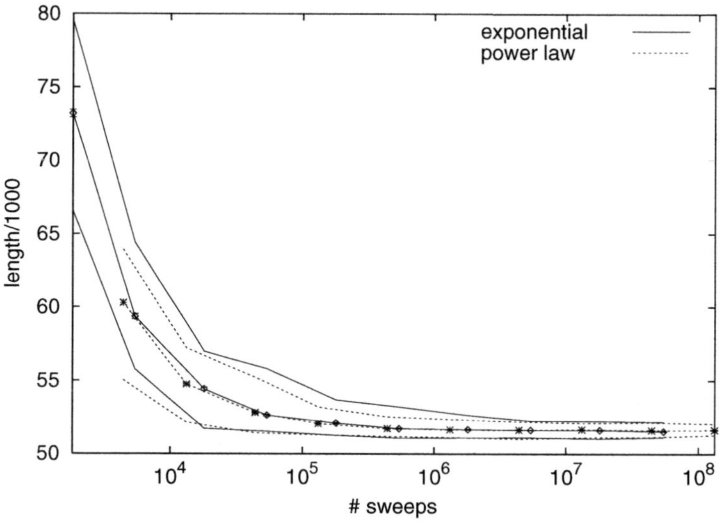

Fig. 10.9. Quality of the results achieved with SSS using the exponential formula (*straight lines*) and using the power law formula (*dotted lines*) for the PCB442 instance vs. overall number of sweeps: the three *lines* show minimum lengths, mean lengths (*error bars* are mostly the size of the *symbols*), and maximum lengths

the same narrow range of α while $\langle\mathcal{H}_\alpha\rangle$ is still rather constant. At small α, $\langle\mathcal{H}_\alpha\rangle$ drops significantly and finally coincides with the curve for $\langle\mathcal{H}_{\text{orig}}\rangle$ at $\alpha = 0$, while the system is already frozen. Again we find that the system orders itself on a linear scale of α.

For optimization purposes, it is even more important to study the quality of the results achieved with this exponential smoothing formula. Figure 10.9 shows that the results for both formulas are roughly of the same quality.

- Sigmoidal smoothing:
  As an example of a sigmoidal smoothing formula, the hyperbolic tangent shall be used as a smoothing function:

$$d_\alpha(i,j) = \bar{d} + \frac{\tanh\left(\alpha\Delta(i,j)\right)}{\alpha} . \tag{10.8}$$

  These sigmoidal smoothing functions like the tangens hyperbolicus have in common that they are rather flat for large argument values and that they are linear around the origin. Thus, for very large α, the sigmoidal function can be replaced by its limiting values, which are small compared to α, such that all distances are equal to $\bar{d}$. At $\alpha \to 0$, however, the relation $\tanh(x) \approx x$ can be used, such that the original landscape is received at the end.

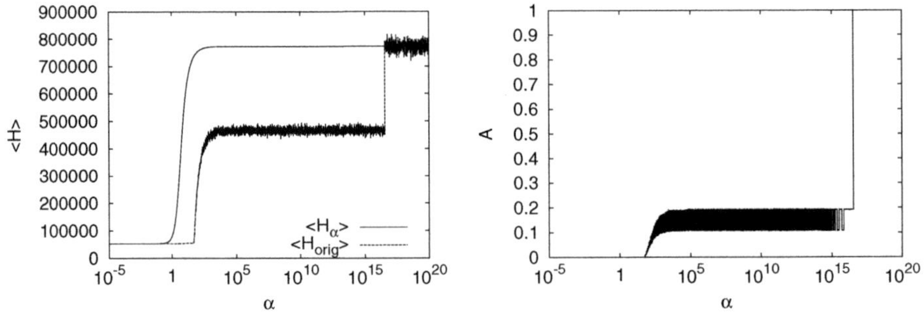

Fig. 10.10. Computational results for the mean energy $\langle\mathcal{H}\rangle$ (*left*) and the total acceptance rate A (*right*) of PCB442 instance using the sigmoidal formula (10.8) for SSS: *left*: both the mean energy $\langle\mathcal{H}_\alpha\rangle$ according to the smoothed distance values $D_{\max} \times d_\alpha(i,j)$ and the mean energy $\langle\mathcal{H}_{\text{orig}}\rangle$ calculated with the original distances $D(i,j)$ are shown

As Fig. 10.10 shows, this sigmoidal smoothing formula (10.8) leads to a new behavior of both $\langle\mathcal{H}_{\text{orig}}\rangle$ and the total acceptance rate A: as usual, $\langle\mathcal{H}_{\text{orig}}\rangle$ fluctuates around $\langle\mathcal{H}_\alpha\rangle$ at large α, while the acceptance rate is 1. Then both $\langle\mathcal{H}_{\text{orig}}\rangle$ and A drop to an intermediate value. Obviously, the system performs a restricted RW on some plateau in the energy landscape, which is completely surrounded by some hills. One could imagine this type of landscape as the Great Basin in the USA, which is also surrounded by hills. We additionally find that the acceptance rate fluctuates here in contrast to results in [186]: this has to do with the different move set. The small moves L2O, EXC, and NIM only show one constant value of the acceptance rate in this intermediate range of α. But the variants of the L3O jump between a value of ≈ 0.15 and 0, such that we get an overall fluctuation of the total acceptance rate A. After this intermediate range, both A and $\langle\mathcal{H}_{\text{orig}}\rangle$

decrease again, and the system freezes. In this range of their decrease, $\langle \mathcal{H}_\alpha \rangle$ also starts to decrease. The curves of $\langle \mathcal{H}_{\text{orig}} \rangle$ and $\langle \mathcal{H}_\alpha \rangle$ also coincide later at smaller α. But, most interesting of all, we find that the system here orders on a logarithmic α-scale, such that we must plot the data vs. a logarithmic α-axis and decrease α exponentially by a factor of 0.99 from 2×10^{17} to 10^{-6}. After that, a greedy step in the original landscape corresponding to $\alpha = 0$ is added, so that we have 5340 overall α-steps.

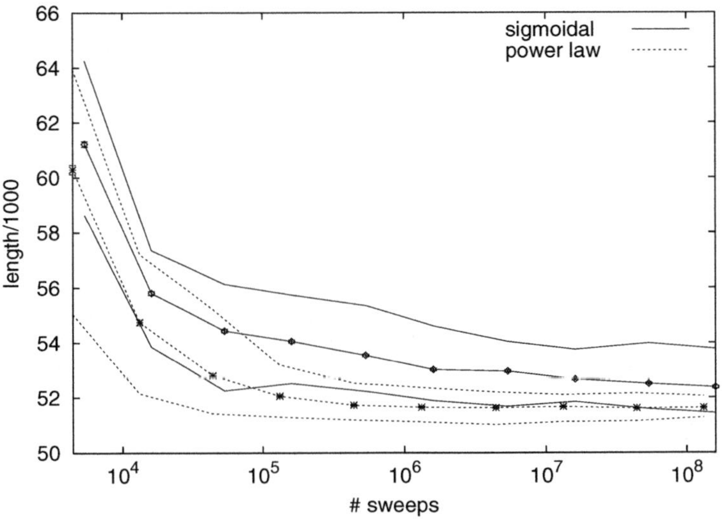

Fig. 10.11. Quality of the results achieved with SSS using the sigmoidal formula (*straight lines*) and the power law formula (*dotted lines*) for the PCB442 instance vs. overall number of sweeps: the three *lines* show minimum lengths, mean lengths (*error bars* are mostly the size of the *symbols*), and maximum lengths

We again compare the quality of the results achieved with this sigmoidal smoothing formula with those for the original power law smoothing formula. Figure 10.11 shows that the sigmoidal smoothing formula (10.8) leads to significantly worse results than the power law formula (10.6).

- Logarithmic smoothing:
  The logarithm can also be used as a smoothing function:

$$d_\alpha(i,j) = \begin{cases} \bar{d} + \dfrac{\log(1+\alpha\Delta(i,j))}{\alpha} & \text{if } \Delta(i,j) \geq 0\,, \\ \bar{d} - \dfrac{\log(1-\alpha\Delta(i,j))}{\alpha} & \text{otherwise}\,. \end{cases} \tag{10.9}$$

  Here one makes use of the fact that the logarithm increases very slowly for large increasing arguments. Furthermore, when considering $\log(1+x) \approx x$ for small $|x|$, one finds that one gets the original landscape for small α.

Just as for sigmoidal smoothing, we find for logarithmic smoothing that α must be decreased in an exponential way: Fig. 10.12 shows pictures similar to those of Fig. 10.10; however, the extended medium range of $\langle \mathcal{H}_{\text{orig}} \rangle$ and of the acceptance rate A are missing. Instead we find that after the first sharp decrease in $\langle \mathcal{H}_{\text{orig}} \rangle$ and A, both quantities show fluctuations.

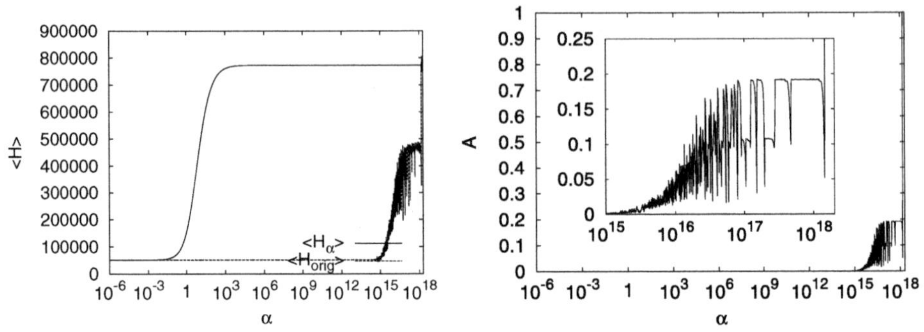

Fig. 10.12. Computational results for the mean energy $\langle \mathcal{H} \rangle$ (*left*) and the total acceptance rate A (*right*) of the PCB442 instance using the logarithmic formula (10.9) for SSS: *left*: both the mean energy $\langle \mathcal{H}_\alpha \rangle$ according to the smoothed distance values $D_{\max} \times d_\alpha(i,j)$ and the mean energy $\langle \mathcal{H}_{\text{orig}} \rangle$ calculated with the original distances $D(i,j)$ are shown. The *inset* in the right graphic shows a blowup of the acceptance rate in order to show the fluctuations better

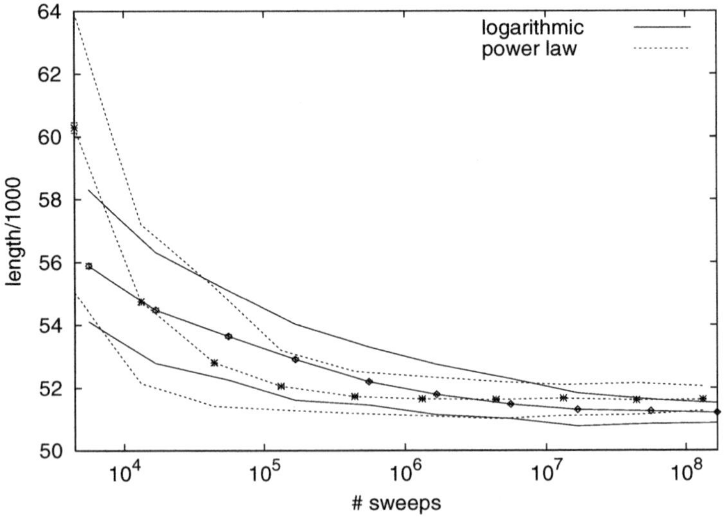

Fig. 10.13. Quality of the results achieved with SSS using the logarithmic formula (*straight lines*) and the power law formula (*dotted lines*) or for the PCB442 instance vs. overall number of sweeps: the three *lines* show minimum lengths, mean lengths (*error bars* are mostly the size of the *symbols*), and maximum lengths

These fluctuations, which are shown enlarged in the inset, can even be self-similar depending on the moves used [186]. Then both $\langle \mathcal{H}_{\text{orig}} \rangle$ and A vanish while $\langle \mathcal{H}_\alpha \rangle$ is still constant. At much smaller values of α, $\langle \mathcal{H}_\alpha \rangle$ decreases and finally becomes identical with $\langle \mathcal{H}_{\text{orig}} \rangle$ at $\alpha = 0$. Working with this logarithmic smoothing formula (10.9), we decrease α from 2×10^{18} to 10^{-6} exponentially with a factor of 0.99. Again we add a greedy step in the original landscape at the end, so that we have overall 5569 α-steps.

Finally, we compare the results achieved with the logarithmic smoothing formula (10.9) again with those obtained with the power law formula (10.6). Figure 10.13 shows that power law smoothing leads to significantly better results for short computing times, whereas the logarithmic smoothing formula provides better results when investing more computing time. Of all the smoothing formulas studied here, the logarithmic smoothing formula is the only one able to lead to the global optimum of the PCB442 instance and to other quasioptimum results.

10.3 Effect of Numerical Precision on Smoothing

When increasing the smoothness-control parameter α to even larger values than used above, the acceptance rate becomes exactly 100% for all smoothing formulas, i. e., every move is accepted, the greedy and the RW coincide totally. Thus, the smoothed energy landscape must be completely flat at such large values of α. This is not an effect of the formulas themselves, as these would always lead to very small deviations of the smoothed distances to the mean normalized distance $\bar{d}$, as long as α took finite positive values. Thus, this effect is only due to the finite numerical precision on a computer: each real number is represented by some number of bits. Nowadays, usually one of the two following representations is used: the real number is either stored in four bytes, i. e., 32 bits (in which case it is called a `REAL` or `REAL*4` in Fortran and usually a `float` in C, C++, and Java) or it is stored in eight bytes, i. e., 64 bits (in which case it is called a `DOUBLE PRECISION` or a `REAL*8` in Fortran and usually a `double` in C, C++, and Java. However, these precisions are only fixed for Fortran and Java; one must check the precision with the `sizeof` command when working with C and C++.) Within these bits, both the mantisse and the exponent are stored. Thus, not every real number can be stored on a computer but only a small set of them, and these are only represented with some finite precision. Thus, if a very small number b is added to a comparatively large number a, the result for $a + b$ is a. This explains why on the computer all lengths have the value $\bar{d}$ for very large α, so that the lengths of the configurations are identical and the energy landscape is completely plain.

However, the next question is whether this finite numerical precision on a computer also affects the smoothing steps at smaller α. Figure 10.14 shows with the example of the power law smoothing formula that the curves for our observables of interest stay rather the same if another precision for the real numbers is used. However, the transition from the random configurations at large values of α to the quasifrozen system occurs at much smaller values of α if the precision of the real numbers is decreased. Here we present only results for real numbers with 4-byte or 8-byte precision, as these are the two most often used numerical precisions.

In both scenarios, α was decreased linearly in steps of 0.01. The final value of α must be 1, of course. However, the initial value of α can be chosen much smaller if fewer bits are used for the representation of real numbers as the transition range is shifted to smaller values of α. Thus, an initial value of 20.001 was used for α when working with four bytes, whereas an initial value of 45.001 was used when working with eight bytes. In both cases, a greedy step in the original landscape corresponding to $\alpha = 1$ was added in the end. Thus, we have 4402 α-steps when working with eight bytes, whereas we have

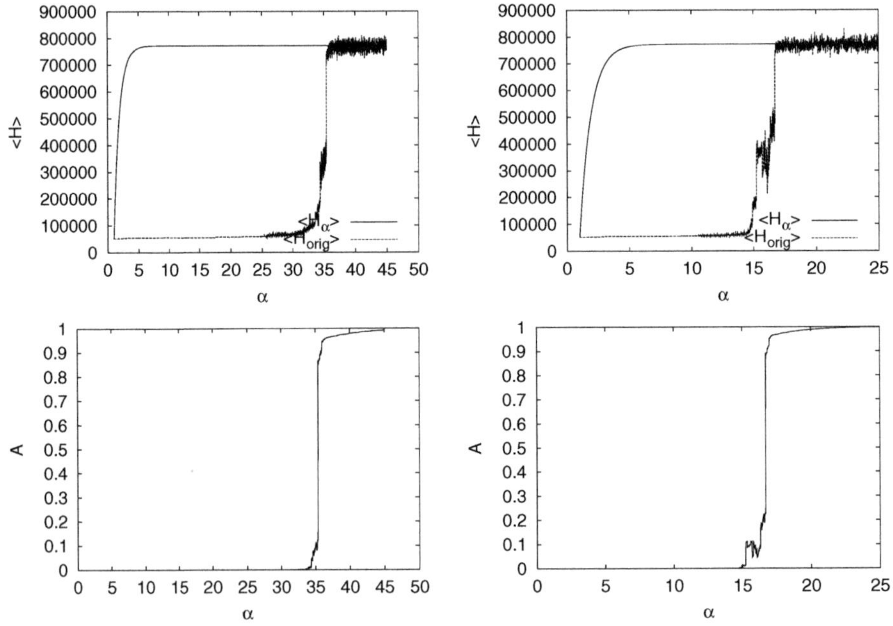

Fig. 10.14. Computational results for the mean energy $\langle \mathcal{H} \rangle$ (*top*) and the total acceptance rate A (*bottom*) of the PCB442 instance using the power law formula (10.6) for SSS: *left*: results from using 8-byte-long REALs; *right*: results from using 4-byte-long REALs. As usual, both the mean energy $\langle \mathcal{H}_\alpha \rangle$ according to the smoothed distance values $D_{\max} \times d_\alpha(i,j)$ and the mean energy $\langle \mathcal{H}_{\text{orig}} \rangle$ calculated with the original distances $D(i,j)$ are shown

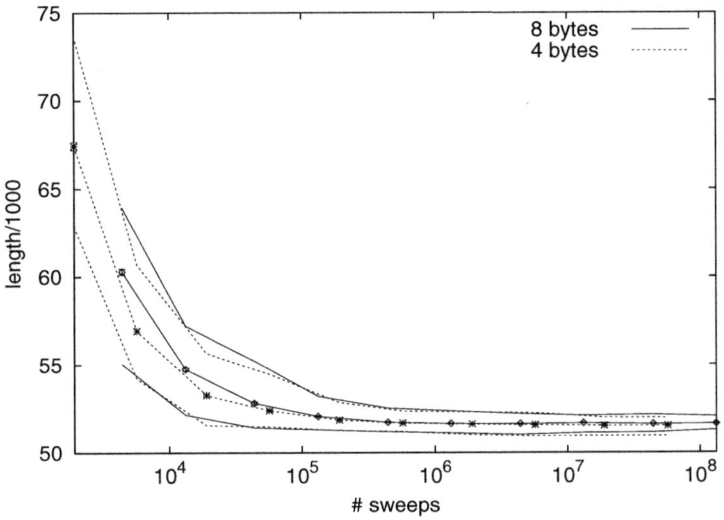

Fig. 10.15. Quality of the results achieved with SSS using the power law formula (10.6) with 8-byte precision (*straight lines*) and 4-byte precision (*dotted lines*) for the PCB442 instance vs. overall number of sweeps: the three *lines* show minimum lengths, mean lengths (*error bars* are mostly the size of the *symbols*), and maximum lengths

only 1902 when working with four. The question arises as to which of the two scenarios leads to the better results. Figure 10.15 shows that, although less calculation time was invested in the REAL*4 runs, they led to slightly better results. Thus, more precision does not necessarily lead to better results. Here the transition range is only shifted but not widened so that the different precision of the numbers mainly leads to the difference in the initial value of α, which can be chosen to be much smaller.

Note that such numerical rounding effects play no such role for the transition regimes of SA and its related algorithms. There the energy differences are not changed by changing the control parameter. They are always related to the control parameter as "$\Delta\mathcal{H}/T$". The acceptance functions can be written, e. g., as $\Theta(\exp(-\Delta\mathcal{H}/T) - r)$ in the case of SA, with r being the random number and $\Theta(x)$ the Heaviside function returning a 1 if $x \geq 0$ and 0 otherwise and written as $\Theta(1 - \Delta\mathcal{H}/T)$ in the case of threshold accepting (TA). Of course, the rounding errors are of different sizes for various precisions. But the calculated number $\Delta\mathcal{H}/T$ stays rather the same. Thus, the results for, e. g., the location of the critical transition range stay the same.

10.4 Smoothing with Finite Numerical Precision Only

As finite numerical precision plays such a dominant role in SSS, the question arises as to whether it is also possible to smooth the energy landscape by rounding errors only on the basis of a linear smoothing function. For this purpose, we define the following hyperbolic smoothing function, with which we will now work:

$$d_\alpha(i,j) = \bar{d} + \frac{1}{\alpha}\Delta(i,j)\,. \tag{10.10}$$

Looking at this formula from an exact mathematical point of view, it would simply lead to a linear rescaling of the energy landscape without changing its shape. Thus, changing the smoothness-control parameter α and performing a greedy step at each value of α would be identical to running the greedy algorithm on the original landscape. However, as mentioned above, we can make use of smoothing effects due to rounding errors if we choose an initial value of α large enough such that all distances become equal to $\bar{d}$ at the beginning. The optimization run ends at $\alpha = 1$ in the original energy landscape.

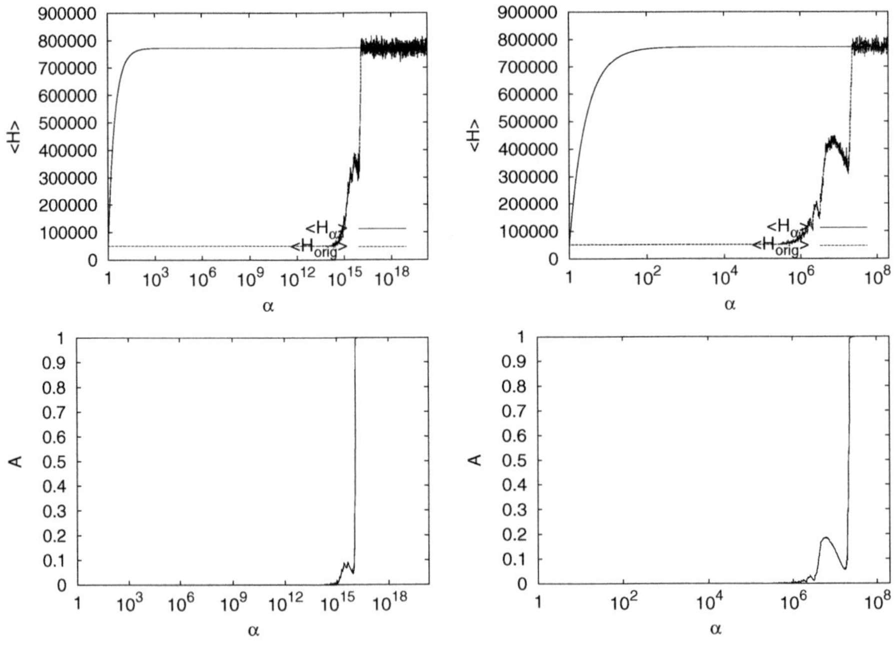

Fig. 10.16. Computational results for the mean energy $\langle\mathcal{H}\rangle$ (*top*) and the total acceptance rate A (*bottom*) of the PCB442 instance using the hyperbolic formula (10.10) for SSS: *left*: results for using 8-byte-long REALs; *right*: results for using 4-byte-long REALs. As usual, both the mean energy $\langle\mathcal{H}_\alpha\rangle$ according to the smoothed distance values $D_{\max} \times d_\alpha(i,j)$ and the mean energy $\langle\mathcal{H}_{\text{orig}}\rangle$ calculated with the original distances $D(i,j)$ are shown

Figure 10.16 shows results for using this hyperbolic smoothing formula working with `REAL*8` and `REAL*4`. We find that we get smoothing effects due to rounding errors in both cases at large values of α: at very large values, all distances are equal to $\bar{d}$ as expected, such that the system performs a RW in the flat energy landscape. Then the curves for $\langle \mathcal{H}_{\text{orig}} \rangle$ and the total acceptance rate A drop significantly in a narrow range of α. We then find an intermediate increase of these two curves, such that the system is obviously able to move more and get to worse configurations again. But after that both curves decrease again, and the system freezes. Over this whole range of α where this happens, $\langle \mathcal{H}_\alpha \rangle$ stays (rather) constant. Finally, $\langle \mathcal{H}_\alpha \rangle$ decreases at comparatively small α and unites with the curve of $\langle \mathcal{H}_{\text{orig}} \rangle$ at $\alpha = 1$. Just as for sigmoidal and logarithmic smoothing, we find that the system orders itself on a logarithmic scale such that we decrease α exponentially with a factor of 0.99 from 2×10^{16} to 1 in the case of 8-byte precision and from 5×10^7 to 1 in the case of 4-byte precision in the production runs. Adding a greedy step at the end in both cases, we have 3736 α-steps when working with REAL*8 and 1765 α-steps when working with REAL*4.

Figure 10.17 shows the quality of the results achieved with the hyperbolic smoothing formula (10.10) both for 8-byte-long and 4-byte-long real numbers. We find again that, although much less calculation time was invested in the runs with 4-byte precision, they lead to slightly better results, such that it seems to be preferrable to work with REAL*4 when using the SSS approach.

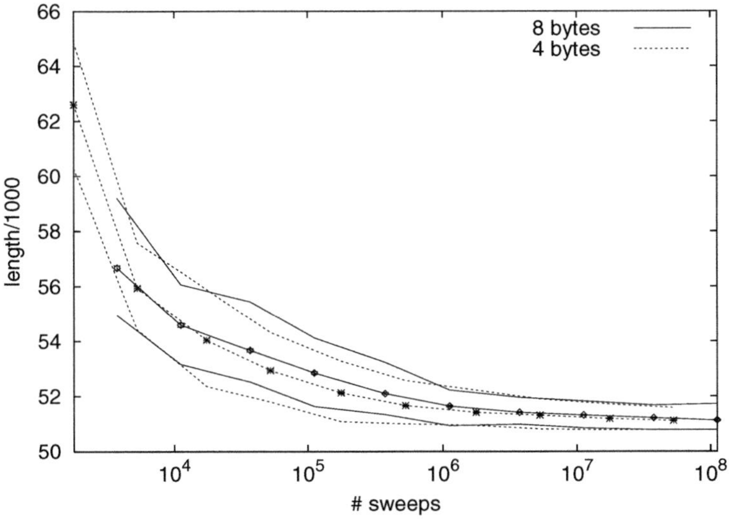

Fig. 10.17. Quality of the results achieved with SSS using the hyperbolic formula (10.10) with 8-byte precision (*straight lines*) and 4-byte precision (*dotted lines*) for the PCB442 instance vs. overall number of sweeps: the three *lines* show the minimum lengths, mean lengths (*error bars* are mostly the size of the symbols), and maximum lengths

When comparing the results for hyperbolic smoothing to those for the other smoothing formulas, we find that we get the best results with this hyperbolic approach. Perhaps this has to do with the fact that we work here indeed all the time much more with the original landscape than if using one of the other formulas: smoothing with this linear formula here does not lead to a shifting of the local minima. The main effect when desmoothing the landscape is that neighboring states that were degenerate at larger values of α suddenly have differing energy values. Thus, more and more degeneracies are desolved while decreasing α until the original landscape is received, which is then only linearly transformed at the end.

Summarizing, this SSS approach is an interesting ansatz when working with iterative improvement heuristics. Basically, it is an approach strongly related to the SA-type optimization algorithms: again a control parameter is introduced that transfers the system gradually from a RW phase to a greedy phase. The results we got here for SSS are generally slightly worse than those we got for SA and TA. However, it might be that when using other parameters or working on other problems they could lead to better results.

11 Further Techniques Changing the Energy Landscape of a TSP

11.1 The Convex–Concave Approach to Search Space Smoothing

Based on the search space smoothing (SSS) approach by Gu and Huang [76], discussed in the previous chapter 10, Coy et al. introduced a related approach [42] based on considerations of the shape of the smoothing functions introduced in [76] and [186] and discussed in the last chapter 10. All these smoothing functions $f(d(i,j))$ except the hyperbolic one can be split at the mean normalized distance $\bar{d}$ into a convex and a concave part: a real function $f(x)$ is said to be convex in some interval $[a;b]$ if

$$f(t \times x_1 + (1-t) \times x_2) \leq t \times f(x_1) + (1-t) \times f(x_2)\,, \tag{11.1}$$

with $0 \leq t \leq 1$ and $x_1, x_2 \in [a;b]$. $f(x)$ is said to be concave in some interval if $-f(x)$ is convex there.

Figure 10.3 shows that the power law smoothing formula (10.6) is convex for $d(i,j) > \bar{d}$ and concave for $d(i,j) < \bar{d}$. On the other hand the sigmoidal smoothing formula, e. g., is concave for $d(i,j) > \bar{d}$ and convex for $d(i,j) < \bar{d}$. Thus, the question can be asked whether the convex or the concave part of the smoothing function is the important one for the smoothing process and, based on that, whether a completely convex or a completely concave smoothing function would lead to even better results [42].

A straightforward approach to investigating this idea is to use the approach of Gu and Huang: introduce a smoothness control parameter α, work with normalized distances $d(i,j) = D(i,j)/D_{\max}$, i. e., $0 \leq d(i,j) \leq 1$, and define the convex smoothing formula

$$d_\alpha(i,j) = d(i,j)^\alpha \tag{11.2}$$

and the concave smoothing formula

$$d_\alpha(i,j) = d(i,j)^{1/\alpha} \tag{11.3}$$

with $\alpha \geq 1$ [42]. Note that these formulas are nearly identical with the power law smoothing formula (10.6) by Gu and Huang, but the concept of using the mean distance $\bar{d}$ is omitted here in order to get an overall convex or concave

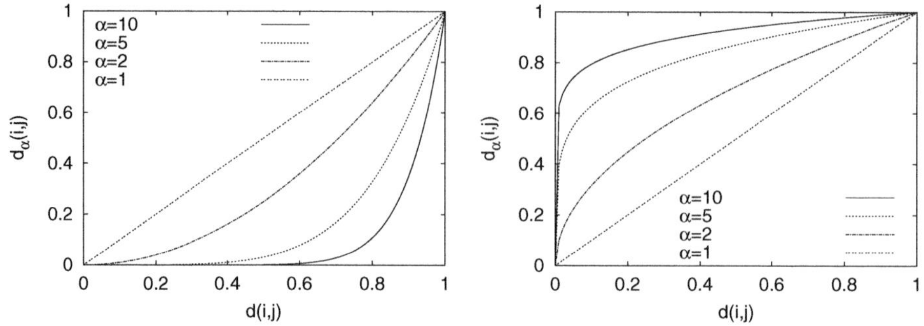

Fig. 11.1. Distances smoothed with the convex smoothing formula (11.2) (*left*) and with the concave smoothing formula (11.3) (*right*): the *curves* show the values for the smoothed distances $d_\alpha(i,j)$ as functions of the original normalized distances $d(i,j)$ for various values of the smoothness-control parameter α. One gets $d_\alpha(i,j) = d(i,j)$ for $\alpha = 1$

function (Fig. 11.1). Thus, all nonzero distances converge to 1 for $\alpha \to \infty$ when using the concave formula. Analogously, all smoothed distances except those for $d(i,j) = 1$ converge against 0 for $\alpha \to \infty$ when using the convex formula. Thus, we get flat landscapes at large α and the original landscape at $\alpha = 1$ for both formulas.

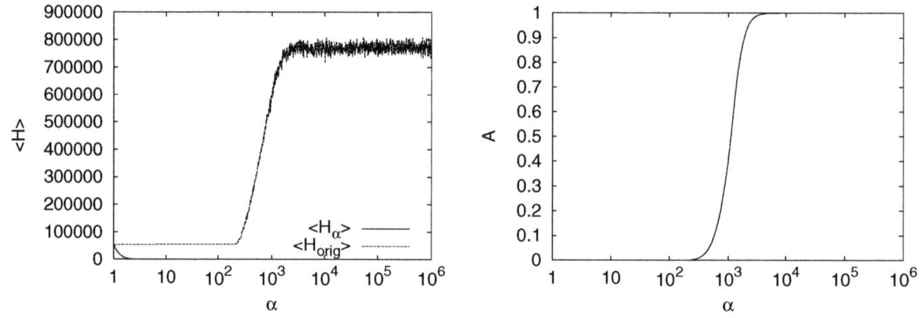

Fig. 11.2. Computational results for the mean energy $\langle \mathcal{H} \rangle$ (*left*) and the total acceptance rate A (*right*) of the PCB442 instance using the convex formula (11.2) for SSS. *Left*: both the mean energy $\langle \mathcal{H}_\alpha \rangle$ according to the smoothed distance values $D_{\max} \times d_\alpha(i,j)$ and the mean energy $\langle \mathcal{H}_{\text{orig}} \rangle$ calculated with the original distances $D(i,j)$ are shown

Again we first study the developments of the mean energy and of the acceptance rate with decreasing smoothness control parameter α (Fig. 11.2) for the convex smoothing formula (11.2). We find that $\langle \mathcal{H}_\alpha \rangle$ is hardly visible in the left graphic as it remains zero until small values of α are reached. On the other hand, $\langle \mathcal{H}_{\text{orig}} \rangle$ shows the behavior we are used to. It decreases from the energy range of the random configurations in some range of the control parameter α to the regime of the ordered configurations. In the same range, the total acceptance rate A drops from slightly below 1 to values slightly larger than 0. Note that when using this convex smoothing formula, the greedy approach can never completely coincide with the random walk (RW), as the longest edge is always of length 1, whereas the other lengths are decreased to 0. Thus, the longest edge will never be accepted by the greedy algorithm. We find that at α-values smaller than 5×10^4, the system leaves this quasi-RW mode. In the production runs, we thus decrease α exponentially from 5×10^4 to 1 with a factor of 0.99. Furthermore, we add one greedy step at $\alpha = 1$ in the original landscape. At small α, the curve for $\langle \mathcal{H}_\alpha \rangle$ increases toward the curve for $\langle \mathcal{H}_{\text{orig}} \rangle$, while the curve for $\langle \mathcal{H}_{\text{orig}} \rangle$ shows that sometimes improvements can be found even at small α.

Although the convex smoothing formula (11.2) is related to the power law smoothing formula (10.6), we find that here the system orders itself on a logarithmic α-scale (α must be decreased over several orders of magnitude), whereas it ordered itself on a linear scale for the power law smoothing formula. Thus, we want to compare the results achieved with the convex smoothing formula with those for another smoothing formula that also orders the system over several orders of magnitude of the control parameter. As the hyperbolic smoothing formula (10.10) also orders the system in a logarithmic way and as it is analytically a special case of a convex function, we will compare the results for the convex smoothing formula with those for the hyperbolic formula here. Figure 11.3 shows the results for both smoothing formulas.

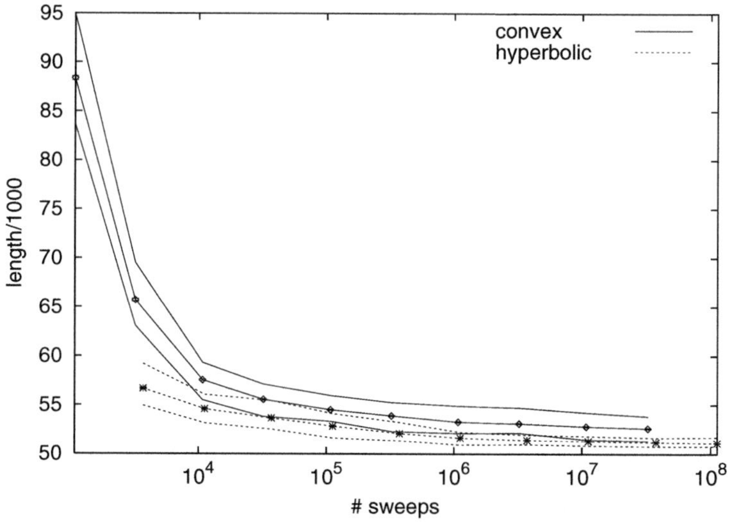

Fig. 11.3. Quality of the results achieved with SSS using the convex formula (11.2) (*straight lines*) and the hyperbolic formula (10.10) (*dotted lines*) for the PCB442 instance vs. overall number of sweeps: the three *lines* show minimum lengths, mean lengths (*error bars* are mostly the size of the *symbols*), and maximum lengths

Note that $1, 3, 10, 30, \ldots$, or 30,000 sweeps were performed per α-step in both cases, but that 1078 α-steps were performed using the convex function and 3736 α-steps when working with the hyperbolic function. We find that the hyperbolic smoothing function leads to significantly better results than the convex smoothing function.

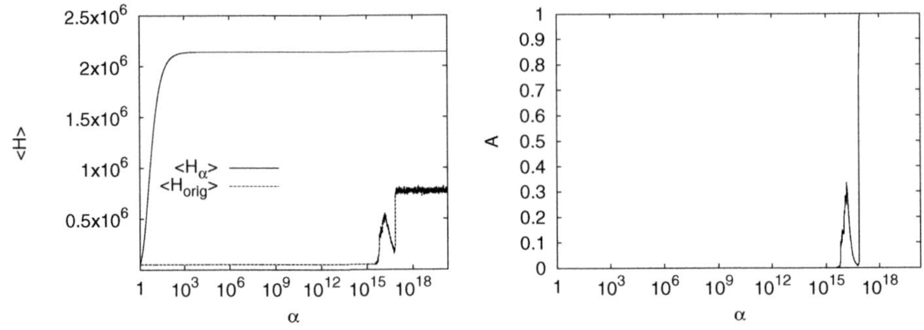

Fig. 11.4. Computational results for the mean energy $\langle \mathcal{H} \rangle$ (*left*) and the total acceptance rate A (*right*) of the PCB442 instance using the concave formula (11.3) for SSS. *Left*: both the mean energy $\langle \mathcal{H}_\alpha \rangle$ according to the smoothed distance values $D_{\max} \times d_\alpha(i,j)$ and the mean energy $\langle \mathcal{H}_{\mathrm{orig}} \rangle$ calculated with the original distances $D(i,j)$ are shown

Next we consider the results for the concave smoothing function (11.3). Figure 11.4 shows the development of the mean energy and of the total acceptance rate with decreasing α. Again $\langle \mathcal{H}_\alpha \rangle$ is rather constant over a wide α-range at a value of $D_{\max} \times N \approx 4841.487 \times 442 \approx 2.14 \times 10^6$, as all smoothed normalized distances are equal to 1 for very large α and deviate from 1 only slightly with decreasing α. These starting deviations lead to the breakdown of both $\langle \mathcal{H}_{\mathrm{orig}} \rangle$ and A at $\alpha \approx 7 \times 10^{16}$. Then the curves for these observables gradually increase again before a second decrease occurs. At small α, the curve for $\langle \mathcal{H}_\alpha \rangle$ decreases strongly and tends toward the curve for $\langle \mathcal{H}_{\mathrm{orig}} \rangle$. The last improvements for the system can be found at $\alpha \approx 2$. Thus, this smoothing formula also orders the system over several orders of magnitude of the smoothness control parameter, such that we use again an exponential cooling schedule.

In the production runs, we decreased α from 9×10^{16} exponentially to 1 with a factor of 0.99 and added a greedy step in the original landscape at the end, so that we had overall 3886 α-steps here. Figure 11.5 shows the results for the concave smoothing formula (11.3) compared to those for hyperbolic smoothing. We find that the results for concave smoothing are on average significantly better than those for hyperbolic smoothing, so that the concave smoothing formula may be considered to be the best smoothing formula found so far. As the results do not spread as widely as for hyperbolic smoothing,

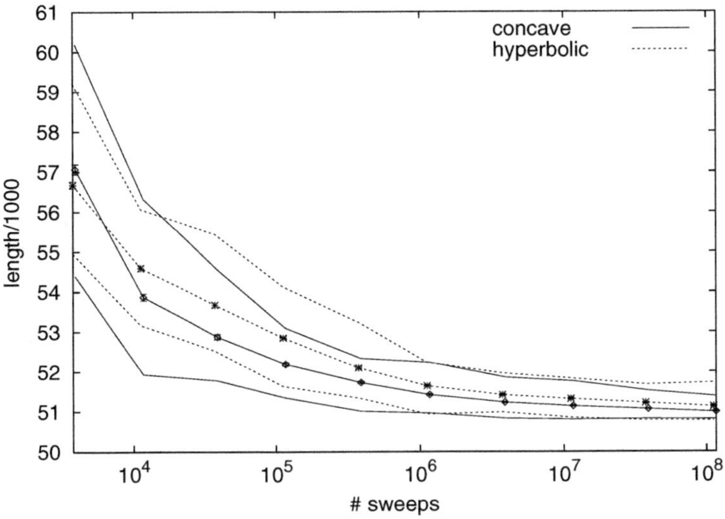

Fig. 11.5. Quality of the results achieved with SSS using the concave formula (11.3) (*straight lines*) and the hyperbolic formula (10.10) (*dotted lines*) for the PCB442 instance vs. overall number of sweeps: the three *lines* show minimum lengths, mean lengths (*error bars* are the size of the *symbols*), and maximum lengths

we did not obtain a result here with optimum length using it in contrast to the hyperbolic and the logarithmic smoothing formulas.

Finally, one might ask whether one could combine the convex smoothing formula (11.2) with the concave smoothing formula (11.3). This can be be done, e. g., by first performing some sweeps using the concave smoothing formula and then performing the same number of sweeps using the convex smoothing formula. Figure 11.6 shows the results for this approach. For each α-step, we perform the measurements separately for the concave and for the convex part. Thus, the data points jump between the curves for convex and concave smoothing shown in Figs. 11.2 and 11.4. For large α, the system is in the RW mode for the convex smoothing formula but already in the greedy mode for the concave smoothing formula. Thus, the system performs a RW if the convex formula is applied; then it is quenched down in some local valley after switching to the concave formula. With the next switch to the convex formula, the system leaves the local minimum again and walks freely through the energy landscape.

As Fig. 11.7 shows, switching between the convex and the concave formula does not lead to a further improvement in the results. Here we decrease α from 5×10^4 to 1 exponentially with a factor of 0.99. At each α-step with $\alpha > 1$, we perform two steps, one in the convex and one in the concave mode. At $\alpha = 1$, we perform a greedy step in the original landscape. Thus, we multiply here the number of sweeps per α-step with $2 \times 1078 - 1 = 2155$,

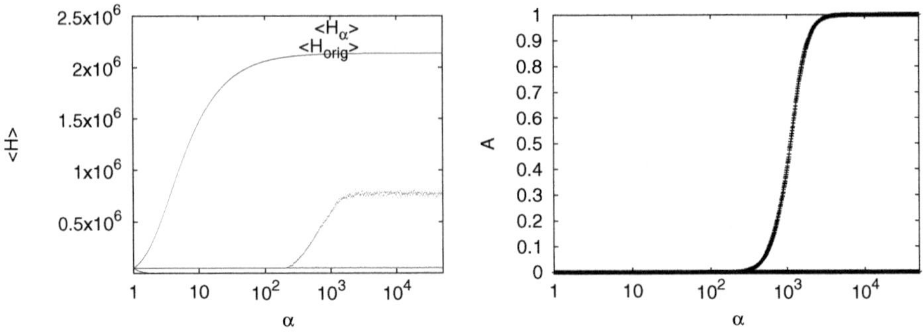

Fig. 11.6. Computational results for the mean energy $\langle \mathcal{H} \rangle$ (*left*) and the total acceptance rate A (*right*) of the PCB442 instance using both the convex formula (11.2) and the concave formula (11.3) for SSS. *Left*: both the mean energy $\langle \mathcal{H}_\alpha \rangle$ according to the smoothed distance values $D_{\max} \times d_\alpha(i,j)$ and the mean energy $\langle \mathcal{H}_{\text{orig}} \rangle$ calculated with the original distances $D(i,j)$ are shown. At each value of α, we first perform the sweeps in the concave mode, then print the values for $\langle \mathcal{H}_\alpha \rangle$, $\langle \mathcal{H}_{\text{orig}} \rangle$, and A, and then perform the same number of sweeps in the convex mode and print the results from that. Thus, the data *points* cannot be connected: since they jump between two *curves*, the *lines* would simply fill the *area* between both *curves*

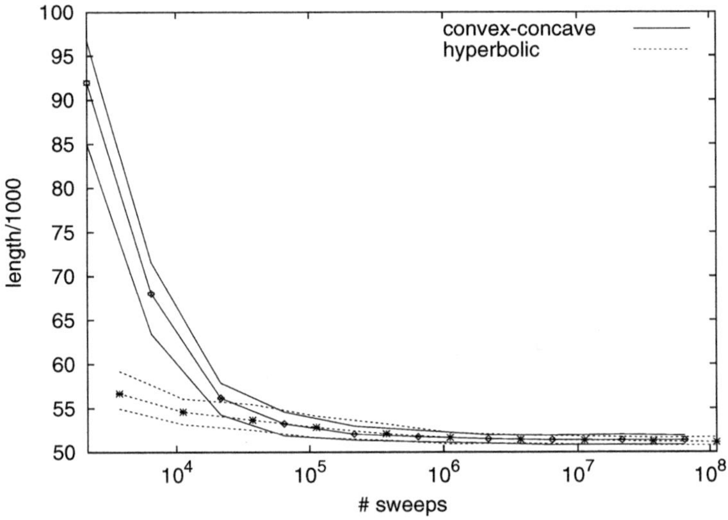

Fig. 11.7. Quality of the results achieved with SSS switching between the concave formula (11.3) and the convex formula (11.2) (*straight lines*) and using the hyperbolic formula (10.10) (*dotted lines*) for the PCB442 instance vs. overall number of sweeps: the three *lines* show minimum lengths, mean lengths (*error bars* are mostly the size of the *symbols*), and maximum lengths

the number of steps, in order to get the overall number of sweeps. We find that the results for this combined technique are again rather good but not as good as when using the concave formula alone.

Coy et al. changed the original power law smoothing formula by Gu and Huang to make it completely convex or concave. This can, of course, also be done with the other smoothing formulas introduced in the last chapter. For example, the sigmoidal smoothing formula (10.8) can be altered to

$$d_\alpha(i,j) = \frac{\tanh(\alpha d(i,j))}{\alpha}, \tag{11.4}$$

thus resembling an overall concave formula in the interval $[0; 1]$. Figure 11.8 shows results for the mean energy and for the acceptance rate. The picture for the mean energy looks similar to that for the convex formula shown in Fig. 11.4: at large α, the mean value calculated with the smoothed distance matrix vanishes, while the mean value calculated with the original distance matrix shows that the system is in a RW mode. $\langle \mathcal{H}_\alpha \rangle$ continuously increases over the whole α-range. The mean value calculated with the original distance matrix exhibits a breakdown, after which both curves start to coincide. Much more interesting is the sight of the acceptance rate: for large α, the total acceptance rate jumps irregularly between total acceptance of all moves and a value of $0.428\ldots$. Both values start to decrease at the same α-values.

Figure 11.9 shows the results achieved with the sigmoidal–concave smoothing formula in comparison with results achieved using the original sigmoidal smoothing formula. We decreased α from 2000 to 0.01 by a factor of 0.99 when working with the sigmoidal–concave formula and added a greedy step in the original landscape at the end, thus having only 1216 α-steps. Just as with the power law smoothing formula, we find here also that this convex–

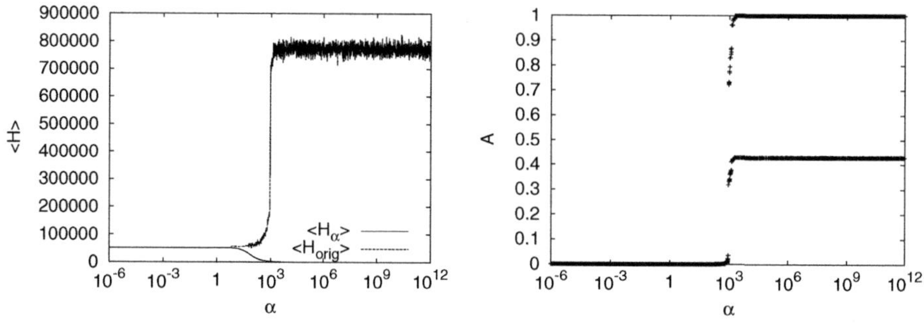

Fig. 11.8. Computational results for the mean energy $\langle \mathcal{H} \rangle$ (*left*) and the total acceptance rate A (*right*) of the PCB442 instance using the sigmoidal–concave formula (11.4) for SSS. *Left*: both the mean energy $\langle \mathcal{H}_\alpha \rangle$ according to the smoothed distance values $D_{\max} \times d_\alpha(i,j)$ and the mean energy $\langle \mathcal{H}_{\text{orig}} \rangle$ calculated with the original distances $D(i,j)$ are shown. The acceptance rate shown on *right* jumps between two values for large α

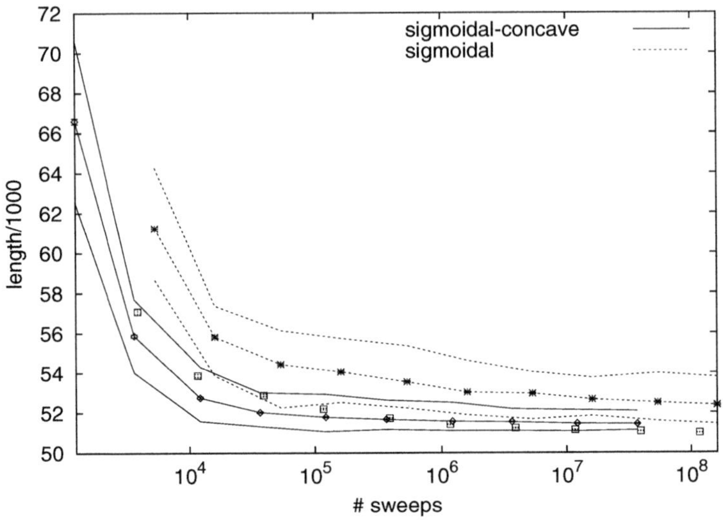

Fig. 11.9. Quality of the results achieved with SSS using the sigmoidal–concave formula (11.4) (*straight lines*) or the original sigmoidal smoothing formula (10.8) (*dotted lines*) for the PCB442 instance vs. overall number of sweeps: the three *lines* show minimum lengths, mean lengths (*error bars* are the size of the *symbols*), and maximum lengths

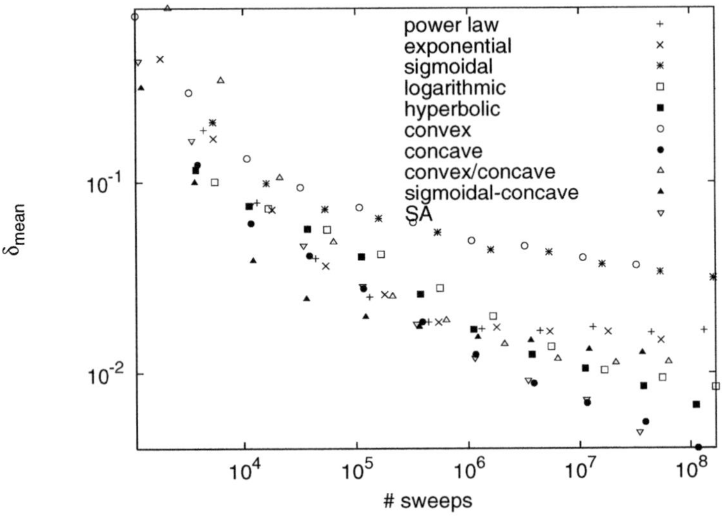

Fig. 11.10. Comparison of the qualities of results achieved for the PCB442 instance with various smoothing formulas introduced in the previous and in this chapter. δ_{mean}: mean relative deviation to the optimum as defined in formula (8.1). For comparison, the results for SA are also shown

concave approach by Coy et al. leads to an improvement over the original deviation approach of Gu and Huang.

Summarizing, Fig. 11.10 shows the mean deviation of the results achieved for the PCB442 instance for all the smoothing formulas introduced in this and in the last chapter: we clearly find that the concave smoothing formula leads to the best results, followed by hyperbolic, logarithmic, and convex/concave smoothing. The worst results are achieved with the convex and the sigmoidal smoothing formulas. The results for simulated annealing (SA), which are shown for comparison and redrawn from Fig. 7.9, show that SSS can lead to results similar to those for SA if an appropriate smoothing function is used.

11.2 Noising the System

Besides these and related techniques for smoothing the energy landscape, additional algorithms for changing the energy landscape have been developed. Based on the finding that similar problem instances usually exhibit similar ground states, the noising technique, which is also called the permutation technique, was developed. The basic idea is to change the original problem instance that is to be solved slightly again and again into other instances that look quite similar, get into a good local minimum of these, and then switch to the next altered problem instance. One hopes thus to cross barriers in the energy landscape of the original problem instance even when using the greedy algorithm for optimization.

For the traveling salesman problem (TSP), an instance can be altered by changing the entries in the distance matrix. This can be done in two ways: either one addresses the distances directly and changes them in a random way or one changes the locations of the nodes randomly, thus changing the distances. We prefer to change the locations of the nodes in order to stay with geometric Euclidean TSP instances. Furthermore, we found in our tests that it is advantageous to use some variable control parameter R that governs the size of the displacement of the nodes and that is decreased during the optimization run. Thus, the outline of our approach is as follows:

1. Create a random initial configuration σ for the proposed TSP instance.
2. Choose an appropriate starting value R of how much a node can be displaced at a maximum.
3. Derive from the original TSP instance a new instance: each node of the original TSP instance is displaced randomly within a squere of radius R around its original location.
4. Perform a greedy optimization run, starting at σ and ending at a configuration τ.
5. Set $\sigma := \tau$ and decrease R slightly.
6. If R has not yet reached some final small value, return to step 3.

7. Finally, perform a greedy optimization run starting at σ on the original problem instance (i. e., at $R = 0$) and print out the resulting configuration.

According to our tests on various TSP instances, it is advantageous to choose the initial and final value of R depending on the minimum distance $D_{\text{min}} = \min\{D(i,j)|D(i,j) > 0\}$ occurring in the original problem instance. We found empirically that it is a good approach to decrease R from $10^3 \times D_{\text{min}}$ to $10^{-2} \times D_{\text{min}}$ exponentially before finally setting $R = 0$.

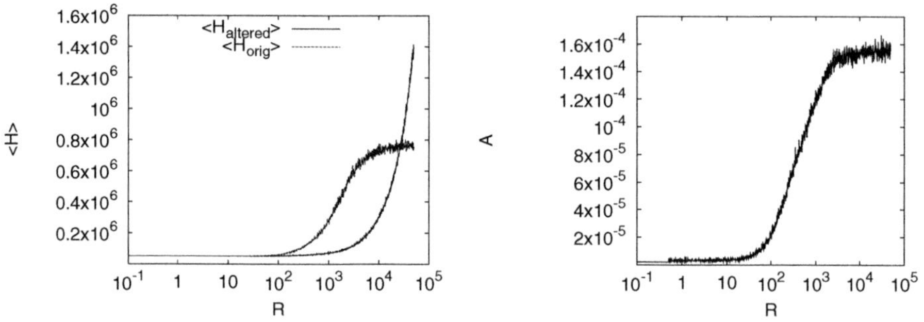

Fig. 11.11. Computational results for the mean energy $\langle \mathcal{H} \rangle$ (*left*) and the total acceptance rate A (*right*) of the PCB442 instance using the noising approach. *Left*: both the mean energy $\langle \mathcal{H}_{\text{altered}} \rangle$ of the altered instances and the mean energy $\langle \mathcal{H}_{\text{orig}} \rangle$ calculated with the original distances $D(i,j)$ are shown

The minimum distance in the PCB442 instance is 50, so that we decrease R from 5×10^4 to 0.5 exponentially with a factor of 0.99. At the end, we add a greedy step for the original instance, corresponding to $\alpha = 0$, such that we have overall 1147 steps. Figure 11.11 shows results for the mean energy and the total acceptance rate of the moves for the PCB442 instance. We have again two energies of interest: first, the energy of the altered instances that are optimized within an R-step is of interest. The curve for this $\langle \mathcal{H}_{\text{altered}} \rangle$ decreases linearly at large R. This is due to the fact that the individual nodes are also displaced far outside the boundaries of the curcuit board at these large R. In fact, the random displacements dominate the locations of the nodes at these large values. With decreasing R, the area where the nodes are located becomes smaller, such that the lengths found for these altered TSP instances decrease linearly with R. In addition, we calculate what length the sequence of the configuration would have if this sequence were the configuration of the original problem. We find that $\langle \mathcal{H}_{\text{orig}} \rangle$ takes values like those for random instances at first; thus, the system is in a kind of RW mode at these large R, when considered from the point of view of the original problem, although the acceptance rate is very small, of course, as the greedy criterion is applied. With decreasing R, $\langle \mathcal{H}_{\text{orig}} \rangle$ and the total acceptance rate A decrease sigmoidally. In this range, the noising approach tackles a series of TSP instances that become increasingly identical with the original instance. Finally, the deviations from the original instance are so small that basically the original instance is tackled all the time, so that both $\langle\mathcal{H}_{\mathrm{orig}}\rangle$ and $\langle\mathcal{H}_{\mathrm{altered}}\rangle$ freeze and finally coincide at $R = 0$.

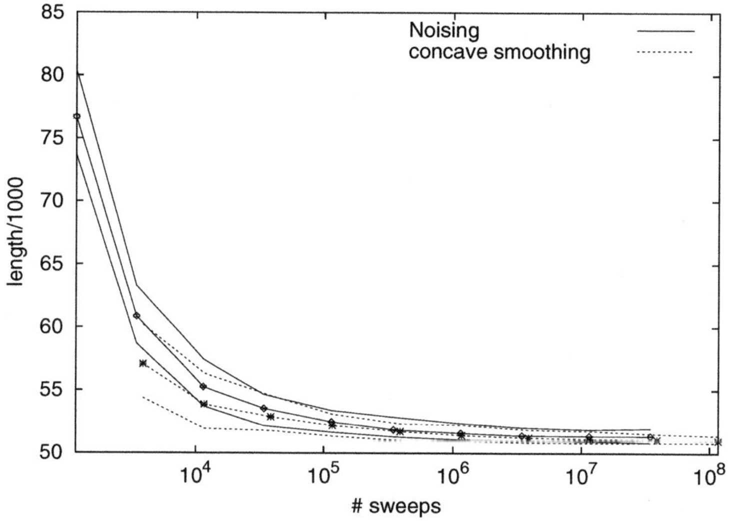

Fig. 11.12. Quality of the results achieved with the noising approach (*straight lines*) and with SSS using the concave formula (11.3) (*dotted lines*) for the PCB442 instance vs. overall number of sweeps: the three *lines* show minimum lengths, mean lengths (*error bars* are mostly the size of the *symbols*), and maximum lengths

Next we want to consider the quality of the results provided by this noising approach. In Fig. 11.12, these results are compared to the results of the concave smoothing formula (11.3), which provides the best results among all smoothing formulas for SSS introduced here. We find that the results achieved with our noising approach are only slightly worse than those achieved with SSS using the convex smoothing formula.

11.3 Weight Annealing

The weight annealing (WA) technique is closely related to the noising approach. Here weights are assigned to the individual parts of the problem in order to change the sizes of their corresponding addends in the cost function and thus their importance in the optimization run. In the application to the TSP, the individual parts are the nodes. We first rewrite the Hamiltonian of the TSP as follows:

$$\begin{aligned}\mathcal{H}(\sigma) &= \sum_{i=1}^{N} D(\sigma(i), \sigma(i+1)) \\ &= \frac{1}{2}\sum_{i=1}^{N} \left(D(\sigma(i), \sigma(i+1)) + D(\sigma(i), \sigma(i-1))\right) \\ &= \sum_{i=1}^{N} J_\sigma(\sigma(i)) ,\end{aligned} \tag{11.5}$$

with $J_\sigma(\sigma(i)) = (D(\sigma(i), \sigma(i+1)) + D(\sigma(i), \sigma(i-1)))/2$, $\sigma(0) \equiv \sigma(N)$ and $\sigma(N+1) \equiv \sigma(1)$.

Thus, each node i contributes the value

$$J_\sigma(i) = (D(i, \sigma(\sigma^{-1}(i)+1)) + D(i, \sigma(\sigma^{-1}(i)-1)))/2 \tag{11.6}$$

to the Hamiltonian $\mathcal{H}(\sigma) = \sum_i J_\sigma(i)$. If we want to strengthen the importance of a certain node i, we will thus generally enlarge its addend to the Hamiltonian by some weight factor $W(i)$. Thus, we get a weighted Hamiltonian

$$\mathcal{H}_W(\sigma) = \sum_{i=1}^{N} W(i) \times J_\sigma(i) . \tag{11.7}$$

Note that here the sum runs over the nodes and not over the tour positions as above.

This approach opens a whole slew of new possibilities for optimization algorithms, especially if a control parameter like the temperature T in SA is used: when working with SA, the individual parts of the system could be treated at different local temperatures. Local bouncing strategies could be developed. These and other techniques could look during or after the optimization run at how well the individual parts of the problem are solved and then readjust the individual weights in order to give more emphasis to obviously harder to solve parts of the problem.

We want to restrict ourselves to the combination of this WA approach with the greedy algorithm. As with the noising approach, it is our aim to overcome barriers in the energy landscape by changing the TSP instance with these weights. As one can easily see by rewriting the Hamiltonian to

$$\begin{aligned}\mathcal{H}_W(\sigma) &= \sum_{i=1}^{N} \frac{1}{2}\left(W(\sigma(i)) + W(\sigma(i+1))\right) \times D(\sigma(i), \sigma(i+1)) \\ &= \sum_{i=1}^{N} D_W(\sigma(i), \sigma(i+1)) ,\end{aligned} \tag{11.8}$$

the distances of the TSP instance are "weighted" by incorporating the weights of the nodes, such that this approach is analogous to the noising approach.

We want to start out by choosing the weights of the nodes randomly. Based on the results of the noising approach in the last section, it is advantageous to decrease the amount of noise gradually, such that one starts out with instances that are rather random and ends at the original instance. Thus, the outline of this approach is as follows:

1. Initialize the system and create some random configuration σ.
2. Choose some large initial value for the parameter α by which the degree to which the weights can deviate from 1 is governed.
3. Choose the weights $W(i)$ randomly but in dependence on the parameter α.
4. Calculate the weighted distances $D_W(i,j)$.
5. Perform a greedy step on the Hamiltonian $\mathcal{H}_W$, starting from configuration σ and ending at configuration τ.
6. Set $\sigma := \tau$.
7. Decrease parameter α.
8. If α has not yet reached its final value, return to step 3 of this approach.
9. Finally, set all weights $W(i) = 1$ and perform a greedy step on the original instance, starting at σ.

Now we must develop some function with which we calculate these weights $W(i)$. Of course, this function must depend on α, but it also must incorporate some random element. We will try the following two "weighting functions" here:

- In the first approach, which is a power law approach, we choose random numbers $r(i)$ that are uniformly distributed in the interval $[-1;1]$ and set
$$W(i) = \alpha^{r(i)} \,. \tag{11.9}$$
Thus, the weights have values in the interval $[1/\alpha;\alpha]$ and are uniformly distributed within this interval when plotting this interval using a logarithmic scale.
We start out with a large initial α (using a value of 10^5 for the PCB442 instance) and decrease α exponentially toward 1.
- In the second linear approach, we increase α linearly from 0 to 1 in steps of, e. g., 10^{-3} and choose random weight values $W(i)$ that are uniformly distributed in the interval $[\alpha; 2-\alpha]$.

Figure 11.13 shows the development of the mean energy and of the total acceptance rate for both weighting approaches. The curve for the reweighted Hamiltonian fluctuates around and decreases with the curve for the original Hamiltonian for the power law weighting technique. We find that mostly $\langle\mathcal{H}_W\rangle \leq \langle\mathcal{H}\rangle$ for small α. The acceptance rate shows a sigmoidal decrease as in the noising approach. The linear weighting approach leads to other results: here α increases linearly from 0 to 1, such that we must read the graphic from left to right here. $\langle\mathcal{H}_W\rangle$ increases with increasing α, while $\langle\mathcal{H}\rangle$ decreases. Both curves coincide for $\alpha \to 1$.

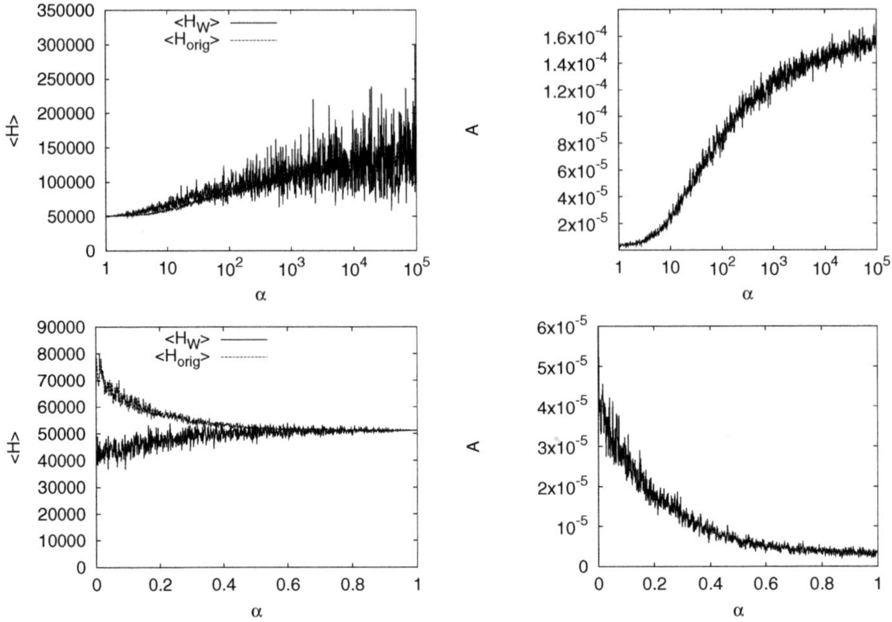

Fig. 11.13. Computational results for the mean energy $\langle \mathcal{H} \rangle$ (*left*) and the total acceptance rate A (*right*) of the PCB442 instance using the WA approach. *Left*: both the mean energy $\langle \mathcal{H}_W \rangle$ of instances with the reweighted distances and the mean energy $\langle \mathcal{H}_{\text{orig}} \rangle$ calculated with the original distances $D(i, j)$ are shown. *Top*: results for the power law weighting technique; *bottom*: results for the linear weighting technique

Finally, we want to compare the quality of the results achieved with these weighting approaches. We used 1147 α-steps when working with the power law weighting approach and 1001 α-steps using the linear weighting approach. Figure 11.14 shows the quality of the results of these approaches. We find that the linear approach leads to significantly better results than the power law approach for short computing times and remains still better for long computing times.

Of course, these two approaches are only a start in the struggle to apply this WA technique to the TSP, as they change the weights of the nodes at random. One might think of more elaborate techniques according to the philosophy of this approach: if a part of the system is badly solved, then the corresponding weight of this part will be enlarged. Thus, the question arises as to how to measure how well or how poorly a part is solved. We tried to introduce a measure with a rather simple approach: as introduced in Sect. 1.8, the misfit parameter measures how large the frustration in a TSP instance is. One can also use this parameter here: a TSP instance is solved locally optimally at some node if this node is connected to its two nearest neighbors. Let $n_1(i)$ and $n_2(i)$ be the nearest and the next nearest neighbors of node i. Then one must calculate

$$D(\sigma(i), \sigma(i+1)) + D(\sigma(i), \sigma(i-1)) - D(\sigma(i), n_1(\sigma(i))) - D(\sigma(i), n_2(\sigma(i))) \tag{11.10}$$

in order to measure how large the deviation from local optimality is for node $\sigma(i)$. This measure could then be introduced in finding new weight values $W(\sigma(i))$. We tried several approaches but found none better than our linear approach with the randomly chosen weights above. We think that this is due to the fact that usually some local part of the system should not be solved optimally in order to get to the optimum of the overall system. This is just the point when dealing with complex problems, that the overall global optimality is not given by the sum of the local optimalities.

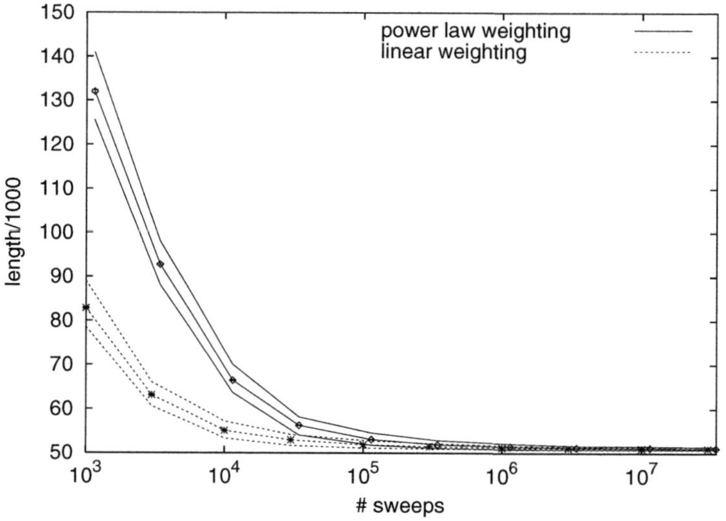

Fig. 11.14. Quality of the results achieved with WA using the power law approach (*straight lines*) and the linear approach (*dotted lines*) for the PCB442 instance vs. overall number of sweeps: the three *lines* show minimum lengths, mean lengths (*error bars* are mostly the size of the *symbols*), and maximum lengths

11.4 Final Remarks on Application of Changing Techniques

In our implementations of SSS, noising, and WA, we generally worked with a control parameter, usually called α, introduced by Gu and Huang [76]. This control parameter was gradually decreased over many steps. One sometimes also finds implementations in the literature in which the schedule for

α is decreased to only a few steps but for which still moderately or even very good results are achieved. The exact values the smoothness control parameter takes in these applications usually depend not only on, e. g., the smoothing function but also on the problem instance. Thus, in these approaches, first, these important smoothing values must be found either by trial and error or by some assumptions or within the kind of optimization runs we performed here. Of course, if only these important smoothness values are used, the calculation time in the production runs can then be decreased to a large extent. However, we prefer to present results here for straightforward implementations for which it is easier to work with a schedule with many α-values as the important ones are usually unknown a priori and are sometimes hard to find.

Secondly, we would like to comment on the basic heuristics to be used in combination with these approaches that change the energy landscape: it is not necessary to work strictly with the greedy algorithm, which does not accept any deterioration. But one cannot start a complete run with, e. g., SA at each α-step, as the local valley will be left at large temperatures, such that the guidance effect of SSS and of the other techniques is lost. As shown in [186], the greedy algorithm can be replaced by the great deluge algorithm (GDA): in that work an optimization run was performed using the GDA at each α-step. The initial value for the water level was chosen as the energy of the current configuration such that the system was forced to stay in the local valley. The GDA is thus very well suited to be combined with these approaches. But the problem with this combination is that the GDA converges very slowly, such that there is no great advantage from using the GDA instead of the greedy and if one does not want to spend too much additional calculation time. However, threshold accepting (TA) may also be used in combination with, e. g., SSS: in the threshold search space smoothing (TSSS) approach, a small constant threshold is introduced and the TA criterion is applied instead of the greedy acceptance criterion. With decreasing α, the energy differences occurring in the smoothed energy landscape become larger, such that more and more moves are rejected. Thus, the system performs both a SSS and a TA transition. However, when α approaches its final value, the system might not freeze completely. Thus, one must memorize the best solution found in this late stage of the optimization algorithm and print this solution as the result of the optimization run.

The third comment we want to make is that, although there are differences in the philosophies and the behaviors of the algorithms between SA and its relatives on the one side and SSS and its relatives on the other side, the algorithms have much in common: both types of algorithms construct a Markov chain of configurations that hopefully ends at a (quasi) optimum configuration. Both types inherit some control parameter that is changed during the optimization such that the system is transferred gradually from a quasi-RW mode into a quasigreedy mode. Now we will move on to completely different approaches. We will, however, find similar concepts in these approaches, such as the usage of a variable control parameter.

12 Application of Neural Networks to TSP

12.1 Application of a Hopfield Network

Neural networks (NNs) can be applied to the traveling salesman problem (TSP) in various ways. In this chapter, we start with the application of the Hopfield network, which was proposed by Hopfield and Tank in 1985 [88].

In order to be able to apply this NN, the TSP must be coded in a special way by using an edge matrix: let

$$S(i,a) = \begin{cases} 1 & \text{if node No. } i \text{ is} \\ & \text{the } a\text{th node in the tour}, \\ 0 & \text{otherwise}. \end{cases} \tag{12.1}$$

This coding was already mentioned in Sect. 1.4, together with the constraints that have to be considered. However, here we code these in another way as the entries of the edge matrix S will serve as the spins or neurons of the Hopfield model, such that a Hamiltonian of the form

$$\mathcal{H}(S) = -\frac{1}{2} \sum_{i,a,j,b} J_{(i,a),(j,b)} \times S(i,a) \times S(j,b) - H \times \sum_{i,a} S(i,a), \tag{12.2}$$

with J being the connectivity matrix to be discussed later, can be derived.

- The constraint that each node must be visited only once in the tour can be expressed by the penalty function

$$\mathcal{H}_1(S) = \sum_i \sum_a \sum_{b \neq a} S(i,a) \times S(i,b). \tag{12.3}$$

  This penalty function vanishes only in the case where this constraint is fulfillled; otherwise at least one of the products would add a 1.
- Analogously, there can only be one node at each tour stop:

$$\mathcal{H}_2(S) = \sum_a \sum_i \sum_{j \neq i} S(i,a) \times S(j,a). \tag{12.4}$$

- Furthermore, the matrix S must contain exactly N entries with a value of 1. As the constraints above do not consider this fact, a further penalty function is needed:

$$\mathcal{H}_3(S) = \left(\left(\sum_i \sum_a S(i,a)\right) - N\right)^2 . \tag{12.5}$$

This term can be rewritten as

$$\mathcal{H}_3(S) = \sum_{i,a} \sum_{j,b} S(i,a) \times S(j,b) - 2N \sum_{i,a} S(i,a) + N^2 . \tag{12.6}$$

As N is constant, the last addend of this term can be omitted.

- Finally, the main objective function should minimize the tour length. One could therefore record it as

$$\mathcal{H}_0(S) = \sum_i \sum_j \sum_a D(i,j) \times S(i,a) \times S(j,a+1) . \tag{12.7}$$

However, the Hopfield rule updates every neuron according to the interactions with all other neurons. As each TSP node interacts not only with its successor but also with its predecessor in the tour, it is better to use a symmetric Hamiltonian:

$$\mathcal{H}_0(S) = \frac{1}{2} \sum_i \sum_j \sum_a D(i,j) \times S(i,a) \times (S(j,a+1) + S(j,a-1)) . \tag{12.8}$$

Node j must be either the predecessor or the successor of node i in the tour such that the corresponding edge length is added. As all edge lengths are added twice, this partial objective function is divided by 2.

The complete Hamiltonian is therefore given as

$$\mathcal{H} = \mathcal{H}_0 + \lambda_1 \mathcal{H}_1 + \lambda_2 \mathcal{H}_2 + \lambda_3 \mathcal{H}_3 . \tag{12.9}$$

The main problem consists in choosing appropriate values for the Lagrange multipliers λ_1, λ_2, and λ_3. If they are chosen too large, then the Hopfield approach mostly converges to feasible solutions, which are, however, of a bad quality. If they are too small, then the constraints might be violated. Usually, all distances $D(i,j)$ are normalized between 0 and 1, such that one can transfer the chosen Lagrange multipliers from one TSP instance to another more easily.

The entries $S(i,a)$ of the edge matrix serve as the spins or neurons $S_{k=(i-1)N+a} = S(i,a)$ of the Hopfield network. By comparing the coefficients of the partial Hamiltonians with the Hamiltonian (12.2) one gets a description for the entries of matrix J, namely,

$$\begin{aligned} J_{(i,a),(j,b)} &= -2\left(\lambda_1 \delta_{i,j}(1-\delta_{a,b}) + \lambda_2 \delta_{a,b}(1-\delta_{i,j}) + \lambda_3\right) \\ &\quad - D(i,j)(\delta_{a+1,b} + \delta_{a-1,b}) , \end{aligned} \tag{12.10}$$

with the Kronecker symbol

$$\delta_{i,j} = \begin{cases} 1 & \text{if } i = j\,, \\ 0 & \text{otherwise}\,. \end{cases} \tag{12.11}$$

These weight values $J_{(i,a),(j,b)}$ serve as the interactions that can also be rewritten such that only two indices remain: $J_{k=(i-1)N+a,l=(j-1)N+b} = J_{(i,a),(j,b)}$. From the addend $\lambda 3 \mathcal{H}_3$ we get the magnetic field

$$H = 2N\lambda_3\,, \tag{12.12}$$

which is equal for all spins. As the neurons do not take the values -1 and $+1$ as in the description of the Hopfield model in Sect. 18.3 but the values 0 and 1, one must change the update rule for the neurons. Again the task is to minimize the energy function (12.2). This can be done in an analogous way to the other Hopfield rule, namely,

$$S_i(t+1) = \Theta\left(H + \sum_{j=1}^{N} J_{ij} S_j(t)\right)\,, \tag{12.13}$$

with the slightly changed Heaviside function

$$\Theta(x) = \begin{cases} 1 & \text{if } x > 0\,, \\ \text{rnd}(0,1) & \text{if } x = 0\,, \\ 0 & \text{if } x < 0\,, \end{cases} \tag{12.14}$$

in which in the marginal case $x = 0$ the new spin value is randomly set to either 0 or 1.

12.2 Computational Results for the Hopfield Network

As the number of neurons of the Hopfield network is given by N^2, with N being the number of nodes of the TSP, the number of interactions between the neurons explodes with N^4 such that only rather small TSP instances can be solved with this Hopfield network due to the finite amount of memory on a computer. For example, the smallest of our benchmark instances, the BEER127 instance, with its 127 nodes, would already require roughly 1.9 GB of memory if each of the interactions is stored in a variable with 8 bytes. This amount of required memory limits the application of this Hopfield network to small system sizes. Of course, instead of storing all these values, one could calculate them for every operation, but then the calculation time would increase strongly.

Furthermore, in applying this approach to small TSP instances, we found that the parameters λ_1, λ_2, and λ_3 must be adjusted for every TSP instance.

The most crucial parameter seems to be λ_3: if it is chosen too large in relation to the other parameters, then the attractors of the Hopfield network are the two states in which either every edge is used (i. e., $S_i = 1$ for all i) or in which no edge is used (i. e., $S_i = 0$ for all i). The Hopfield network always jumps between these two states. On the other hand, if λ_3 is chosen too small, then the initial configuration, which, e. g., consists of N spins with $S_i = 1$ and $N^2 - N$ spins with $S_i = 0$, is not changed.

For intermediate values, this approach often leads to configurations that are nearly feasible, but they have either too many or too few edges. Thus, one has to optimally derive feasible roundtrips from these configurations by either omitting or adding some edges.

For a few instances, we found nice parameters to work with, such that we obtained quite good or even optimum results. But the limitation to small system sizes and the recalibration of the Lagrange multipliers λ_i for every new instance for getting a feasible solution at all makes this approach rather difficult. Thus, we move on to another NN for solving optimization problems.

12.3 Application of a Kohonen Network

The Kohonen network can also be applied to the TSP. The basic idea of this application is that the topology of a good roundtrip by the traveling salesman is that of a closed circle. Thus, the neurons $\boldsymbol{S}_i$ are initially placed on a closed circle and gradually learn the positions of the nodes, such that the circle is iteratively deformed to a roundtrip touching each of the nodes. This approach, which is also called the elastic net approach [55], is implemented as follows:

- First, M neurons are placed on a circle lying in the area in which the nodes of the TSP lie. Thus, the neurons are given by

$$\boldsymbol{S}_i = \begin{pmatrix} C_x + R_x \cos(2\pi i/M) \\ C_y + R_y \sin(2\pi i/M) \end{pmatrix} . \tag{12.15}$$

  The coordinates of the center $\boldsymbol{C}$ of the circle and the radii of the ellipse in the x- and y-direction are determined from the borders of the area in which the nodes lie. Furthermore, the interaction matrix J, which is also called the activation profile, between the neurons is given by

$$J_{ij} = \exp\left(-\frac{D(i,j)^2}{2\lambda^2 D_{\max}^2}\right) \tag{12.16}$$

  with the distance $D(i,j)$ between the neurons according to some metric (usually the Manhattan metric on the given topology is used), the maximum $D_{\max}$ of these distances, and the stiffness parameter λ. Then the learning rate η is set to some initial value, e. g., $\eta = 1$.

- A sweep of learning steps is performed. Each learning step consists of three parts:
  - Some TSP node i at position

$$\boldsymbol{r}_i = \begin{pmatrix} x_i \\ y_i \end{pmatrix} \tag{12.17}$$

    is chosen at random.
  - The neuron j nearest to node i is chosen, i. e., the one whre the distance between the neuron and the node is minimal. Usually, the Euclidean distance $|\boldsymbol{S}_j - \boldsymbol{r}_i|$ is used as the distance metric.
  - All neurons are updated according to the Kohonen learning rule:

$$\boldsymbol{S}_k^{\text{new}} = \boldsymbol{S}_k^{\text{old}} + \eta \times J_{kj} \times \left(\boldsymbol{r}_i - \boldsymbol{S}_k^{\text{old}}\right) . \tag{12.18}$$

- After each sweep, the learning rate η and the stiffness parameter λ are decreased. A mathematical derivation of this choice can be found in [172] and a condition for practical use is given in [164].
- If the neurons have significantly changed in the last sweep, a further sweep is performed.
- Usually, the number M of neurons used exceeds the number N of nodes of the TSP. Furthermore, several or even all nodes might not be placed exactly at the positions of the nodes. Therefore, a correct TSP tour must be unhinged from the elastic band formed by the neurons. Thus, for each TSP node, one neuron that is nearest to it is marked and is located at exactly the position of the TSP node. All other neurons are deleted. If everything goes well, then N neurons remain. The sequence of these neurons determines the sequence of the TSP nodes and, thus, the solution of this problem. But if by this procedure fewer than N neurons remain, then at least one TSP node was not represented in the elastic band of neurons, so the solution is not feasible.

12.4 Computational Results for a Kohonen Network

The Kohonen network contains three crucial parameters:

- The stiffness parameter λ, which controls the extent to which neurons near neuron j also learn about the input vector $\boldsymbol{r}_i$,
- The learning rate η, which controls how strong an input vector is stored in the network, and, above all,
- The number M of neurons, i. e., the size of the NN.

Furthermore, the question is which distance metric to use for the distances $D(i,j)$ between the neurons. One could either use the Euclidean distance $D(i,j)$ between the neurons, which changes after each learning step, or, since the neurons are organized on a closed ring, one could also define a metric according to this topology, e. g., $D(i, j) = 1$ if neurons i and j are nearest neighbors on the ring and $D(i, j) = z$ if there are $z - 1$ neurons between neurons i and j. Note that these distance values are restricted to the range 0 (for the distance of a neuron to itself) to $[M/2]$ if the ring is closed. (The Gaussian brackets for $[x]$ denote the integer component of x.) These distances are constant throughout the optimization run. Thus, choosing λ and η correctly becomes less crucial for obtaining a feasible solution in the end. Thus, we use this metric type and redefine J_{ij} as $J_{ij} = \exp(-D(i, j)^2/(2\lambda^2 M^2))$. Furthermore, we start with $\eta = 1$ and $\lambda = 0.01$ and decrease them both exponentially by a factor of 0.999 after each sweep. All in all, 2000 sweeps, i. e., $2000N$ learning steps, are performed. These parameters are independent of the specific instance. We found these parameters empirically to be quite good parameters for the BEER127 and LIN318 instances. For each instance, we determined the borders of the area $([x_{\min}; x_{\max}], [y_{\min}; y_{\max}])$ in which the nodes lie. The coordinates of the center $\boldsymbol{C}$ of the circle are simply given as the mean values of the minimum and maximum x- and y-values, respectively. For the radius R_x of the ellipse in the x-direction, we set $R_x = 0.8 \times (x_{\max} - x_{\min})/2$ and analogously $R_y = 0.8 \times (y_{\max} - y_{\min})/2$ for the radius in the y-direction. Here we are mainly interested in the dependency of the number M of neurons on the quality of the results, using the parameters mentioned above simply as constants.

First, we want to study the behavior of this Kohonen algorithm. Figure 12.1 shows the initialization, various intermediate configurations that represent the roundtrip of the traveling salesman with increasing quality, and the final deletion of neurons that are not needed. In this run, $M = 3 \times N$ neurons were used. We clearly find that one must use more neurons than there are nodes in the TSP instance, as some neurons are placed on longer edges such that they do not represent a node. Thus, for larger TSP instances it is quite impossible to get a feasible solution with only N neurons, as some nodes are simply not represented by a neuron. These results for the ATT532 and for the NRW1379 instance may change if we use a larger initial value of λ (ours was rather small) and use more sweeps.

Furthermore, we are interested in the question of how many neurons are needed to get a feasible solution that contains all nodes, i. e., in which at least one neuron is associated to each node. Figure 12.2 clearly shows that the number M of neurons should be much larger than the number N of TSP nodes. For the BEER127 instance, $M = 3 \times N$ neurons are sufficient in order to achieve nearly always a feasible solution; for $M/N \geq 4$, we always get feasible solutions. This fraction M/N must be increased with increasing N: M/N must be ≥ 11 in order to always get a feasible solution for the LIN318 instance. In the case of the PCB442 instance, we get 100 feasible solutions out of an overall 100 configurations only for $M/N = 45, 60, 85, 90$, and 95. The results are even worse for the ATT532 problem: here we never reach 100% of feasible solutions; we must increase the ratio M/N even further to get

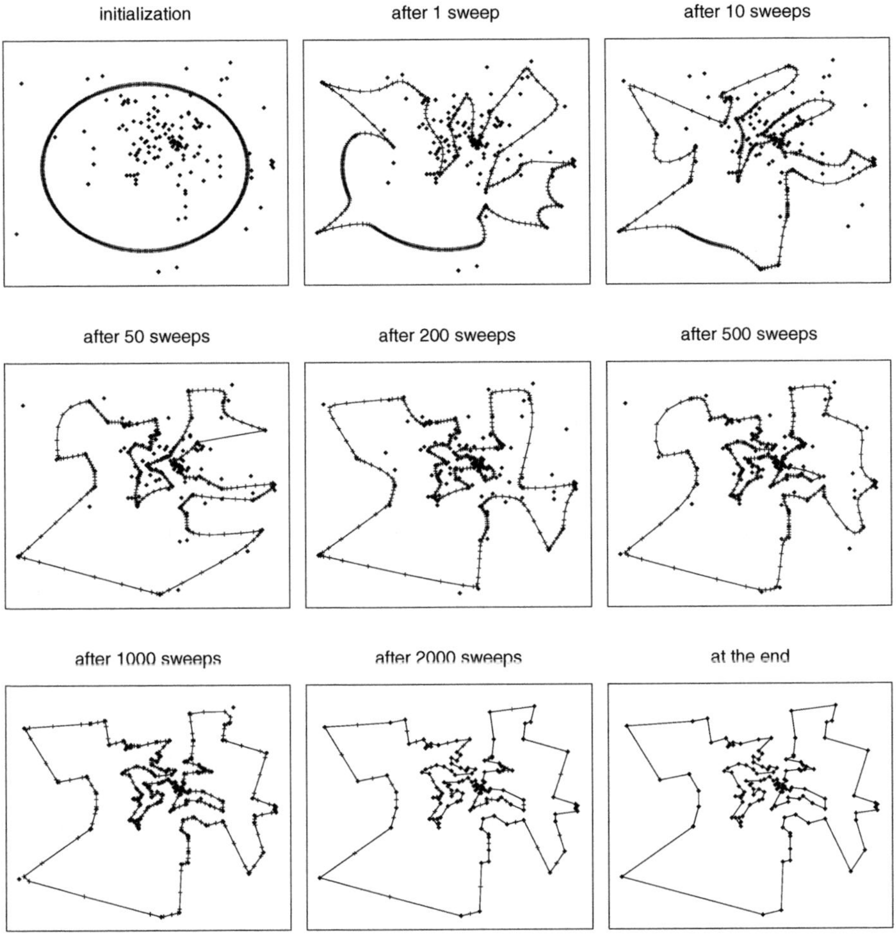

Fig. 12.1. Solution of the BEER127 instance with a Kohonen network with $M = 3 \times 127$ neurons: the neurons of the Kohonen network are initially placed on an ellipse. Then they are iteratively shifted, such that the elastic band increasingly approaches the form of a roundtrip of the traveling salesman. Finally, the unneeded neurons are deleted

a feasible solution with a reasonable probability. The curve for the ATT532 instance shows an even worse effect: the probability for getting a feasible solution with a Kohonen network is not necessarily a roughly monotonously increasing function of the ratio M/N. Instead, we find for large ratios M/N that we do not get any feasible solution at all—at least with our parameter set. We also tried the Kohonen network on the NRW1379 instance. We did not get any feasible solution for this large TSP instance with the Kohonen network.

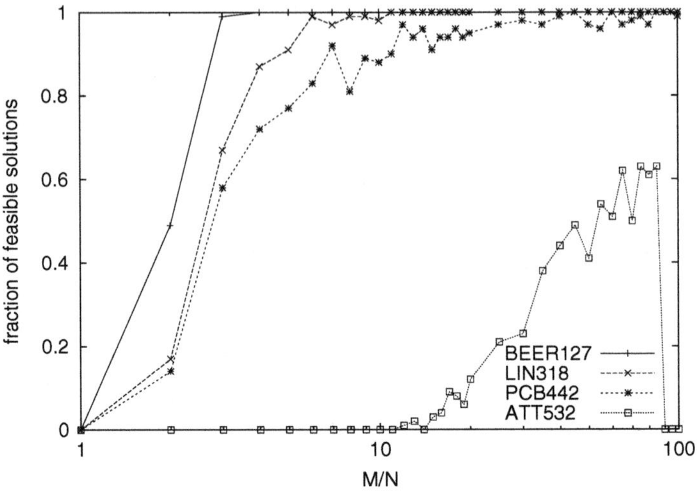

Fig. 12.2. Fraction of feasible solutions achieved with a Kohonen network for various ratios of neurons vs. number of nodes: for each of the four TSP instances (BEER127, LIN318, PCB442, and ATT532) and for each M/N ratio, 100 runs were performed

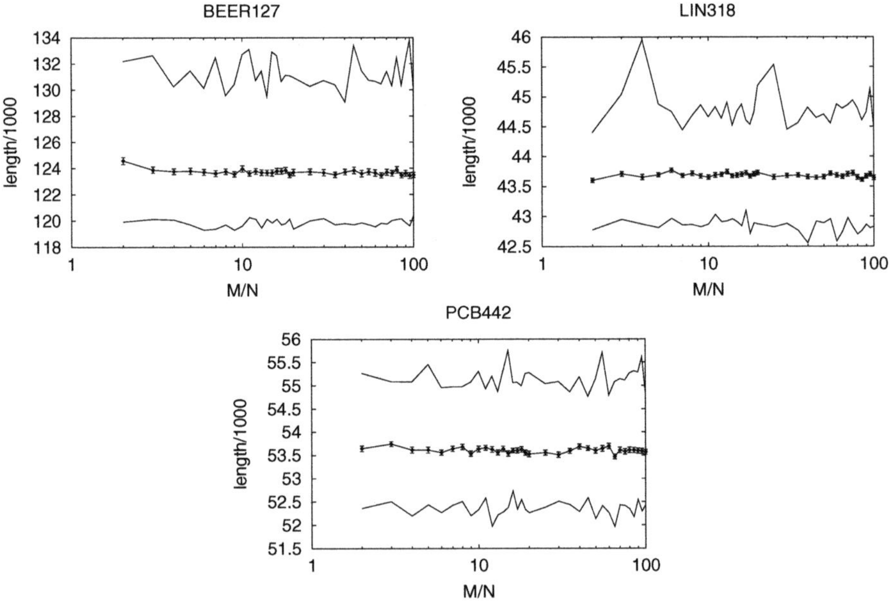

Fig. 12.3. Lengths of the solutions of TSP instances achieved with a Kohonen network for various ratios of neurons vs. number of nodes: more than 100 runs were performed for each TSP instance and for each ratio in order to get 100 feasible solutions from which the results shown are taken. The graphics show the minimum, mean (with *error bars*), and maximum length achieved

The next question is whether the results, i. e., the lengths of the roundtrips achieved with the Kohonen network, depend on the ratio M/N. Figure 12.3 shows the results for the minimum, mean, and maximum lengths achieved with various fractions M/N for the three TSP instances BEER127, LIN318, and PCB442. One finds that the mean length is virtually constant; it does not depend strongly on the ratio M/N. The mean length decreases slightly with increasing M/N for the BEER127 and PCB442 instances. Thus, increasing the number of neurons does not always help. A further disadvantage of large M values is that the calculation time increases linearly with the number of neurons M. As this number is proportional to the number of nodes N and as the number of improvement steps also increases at least linearly with N, the calculation time of the algorithm increases at least quadratically with the number of TSP nodes.

Summarizing, the Kohonen network shows that NNs can be used for optimization purposes. However, this approach can be used only for small and medium-sized TSP instances. Furthermore, the results achieved are not really good. Thus, although it is a nice and interesting idea to use NNs, the computational counterparts of brains, for optimization purposes, we cannot recommend this approach.

13 Application of Genetic Algorithms to TSP

The number of imaginable applications of genetic algorithms (GAs) to the traveling salesman problem (TSP) is huge. Many parameters can be introduced, and tuning them can strongly increase the quality of the results one can achieve with GAs. Once we implemented a parallel GA, with which we obtained the optimum results for the PCB442 and the ATT532 TSP instances. Here, however, we want to restrict ourselves to a rather simple implementation of a GA such that the interested reader can easily implement a GA for the TSP on his own after having read this chapter.

13.1 Mutations

There are a number of ways to define mutations and crossovers for the TSP, for all the various codings of the configuration. However, the one-point and two-point crossover operators, which are commonly used when working with GAs in bit-representations of the configurations, often lead to offspring configurations violating the standard constraint of the TSP, namely, that there must be exactly one closed roundtrip touching each node exactly once. Similarly, the standard mutations like bit shifting or bit inversion applied on the edge matrices (1.7) and (1.11) usually result in nonfeasible configurations. However, we prefer those mutations and crossovers that automatically lead to configurations fulfilling the TSP constraints as otherwise one would have to work with barrier or penalty functions, which increase the complexity for the algorithm and makes it usually harder to find a good solution.

The question is now what appropriate mutation and crossover operators look like for the TSP. In Chaps. 4 and 5, we already introduced the moves exchange (EXC), node insertion move (NIM), Lin-2-opt (L2O), and the four Lin-3-opts (L3Os), which we used later on and all of which lead to feasible configurations. These moves can thus serve as mutations in GAs. Indeed, these moves can even be interpreted from a genetic point of view, considering the configuration written as a permutation of numbers, as follows:

- The EXC, which exchanges two nodes in the tour, is somehow an analogon to the bit exchange.
- The NIM, which shifts one node to another position in the tour, can be seen as an analogon to the bit shift.

- The L2O, which reverses the direction of a part of the tour, is related to the bit reversal.
- The L3Os can be interpreted as combinations of bit operators shifting and reversing strings of bits.

Instead of the bits, which can take only two values, one has nodes with numbers from 1 to N.

13.2 Crossovers

We will thus use these seven moves as the mutations that are used to change single configurations. But the essential point of GAs is that there are also crossover operators that allow the configurations to mate and to have some offspring together. There are also many ways to define a crossover operator that leads to feasible offspring configurations. Usually, two child configurations K_1 and K_2 are produced from two parental configurations P_1 and P_2 that mate with each other as desribed below. In the descriptions below, only the production of K_1 will be given in detail. For K_2, the role of the two parents is simply reversed.

Some standard crossover operators for the TSP are as follows:

- Random crossover

Table 13.1. Example of the random crossover

P_1	8	1	3	0	2	7	6	4	5	9
P_2	7	4	8	1	0	2	6	5	3	9
Step 1: Choose $R(i)$ randomly.										
R	1	0	0	1	0	1	1	0	0	1
Step 2: Set $K_x(i) = P_x(i)$ if $R(i) = 1$.										
$\tilde{K}_1$	8	–	–	0	–	7	6	–	–	9
$\tilde{K}_2$	7	–	–	1	–	2	6	–	–	9
Step 3: Fill empty $K_x(i)$ in the order in which they occur in P_{3-x}.										
K_1	8	4	1	0	2	7	6	5	3	9
K_2	7	8	3	1	0	2	6	4	5	9

The simplest crossover operator for the TSP that always leads to feasible configurations is probably the random crossover (RX). The RX first produces a random bitstring R. The N bits $R(i)$ are set to 0 or 1 with equal probability. Then the RX fills $K_1(i)$ with the content of $P_1(i)$ if $R(i) = 1$. If the bit $R(i)$ is not set, then the position $K_1(i)$ remains empty at first. In the next step, the remaining empty positions of K_1 are filled with those nodes that are not yet part of K_1 in the order in which they occur in P_2. A small example in Table 13.1 illustrates this RX.

The random element in this move is rather strong. Good partial sequences that are perhaps already part of the parent configurations are usually not handed down to the child configurations. Thus, one can expect that this crossover operator might generally not lead to very good children.

- Partially matched crossover

Table 13.2. Example of the partially matched crossover

Step 1: Select positions j and k. P_1 and P_2 are cut there ($j = 3$ and $k = 7$).

P_1	8	1	3	0	2	7	6	4	5	9
P_2	7	4	8	1	0	2	6	5	3	9

Step 2: Set $R(i) = 1$ if $i \leq j$ and $i > k$.

R	1	1	1	0	0	0	0	1	1	1

Step 3: Set $K_x(i) = P_x(i)$ if $R(i) = 1$.

K_1	8	1	3	–	–	–	–	4	5	9
K_2	7	4	8	–	–	–	–	5	3	9

Step 4: Else set $K_x(i) = P_{3-x}(i)$.

K_1	8	1	3	1	0	2	6	4	5	9
K_2	7	4	8	0	2	7	6	5	3	9

Step 5: Dissolve double occurrences: Let n be such a double node. Let l_1 and l_2 be its positions with $j < l_1 \leq k$. Set $K_x(l_2) = P_x(l_1)$.

K_1	8	0	3	1	0	2	6	4	5	9
K_2	2	4	8	0	2	7	6	5	3	9

Repeat step 5 if new errors occur.

K_1	8	2	3	1	0	2	6	4	5	9
K_2	0	4	8	0	2	7	6	5	3	9

and again ...

K_1	8	7	3	1	0	2	6	4	5	9
K_2	1	4	8	0	2	7	6	5	3	9

The partially matched crossover (PMX) works in a completely different way. Here one chooses randomly two positions j and k with $1 \leq j < k \leq N$ after which the two parental configurations P_1 and P_2 are supposed to be cut. Let j_+ and k_+ be the successive positions to j and k, respectively. Thus, usually P_1 is split into the three partial sequences $P_1(1), \ldots, P_1(j)$, $P_1(j_+), \ldots, P_1(k)$, and $P_1(k_+), \ldots, P_1(N)$, and analogously P_2. Then one makes the first step in constructing the kid configuration K_1 by setting $K_1(i) = P_1(i)$ for $i \leq j$ and $i > k$. Then one sets $K_1(i) = P_2(i)$ for $j < i \leq k$. But here one faces the problem that some nodes might occur twice in K_1 while other nodes are forgotten entirely. Let, e. g., n be such a node that would occur twice, i.e., once in the partial sequence $K_1(j_+), \ldots, K_1(k)$, which K_1 inherited from P_2, and once outside this partial sequence, which K_1 inherited from P_1. Let l_1 and l_2 be the positions at which n occurs with $j_+ \leq l_1 \leq k$. To dissolve this double occurrence, set $K_1(l_2) = P_1(l_1)$. After this replacement has been done for all nodes found to occur twice in K_1, one might find that other nodes occur twice in K_1. Thus, one must iterate this search-and-repair procedure until all double occurrences are dissolved.

- Order crossover

Table 13.3. Example of the order crossover

Step 1: Select positions j and k.
P_1 and P_2 are cut there.
($j = 2$ and $k = 5$)

P_1	8	1	3	0	2	7	6	4	5	9
P_2	7	4	8	1	0	2	6	5	3	9

Step 2: Set $K_x(i) = P_x(i)$

$\tilde{K}_1$	8	1	3	0	2	7	6	4	5	9
$\tilde{K}_2$	7	4	8	1	0	2	6	5	3	9

Step 3: Delete $P_1(j_+), \ldots, P_1(k)$
in K_2 and vice versa.

$\tilde{K}_1$	–	–	3	–	2	7	6	4	5	9
$\tilde{K}_2$	7	4	8	1	–	–	6	5	–	9

Step 4: Reorder the indices:
Write "–" in the interval,
the other indices outside.
Start with index j_+.

$\tilde{K}_1$	3	2	–	–	–	7	6	4	5	9
$\tilde{K}_2$	8	1	–	–	–	6	5	9	7	4

Step 5: Set $K_x(i) = P_{3-x}(i)$
for $j < i < k$.

K_1	3	2	8	1	0	7	6	4	5	9
K_2	8	1	3	0	2	6	5	9	7	4

A further crossover operator called the order crossover (OX) is strongly related to the PMX. Again one selects two positions j and k after which the parental configurations are supposed to be cut and initially sets the kid configurations as above. However, the errors with the double occurrences of the nodes are dissolved in another way. Here one preliminarily sets $K_1(i) = P_1(i)$ for all i. Then one checks for all values in $P_2(j_+), \ldots, P_2(k)$ to see whether they occur in $K_1(1), \ldots, K_1(j)$ or in $K_1(k_+), \ldots, K_1(N)$. If at least one of them does, one marks such positions i with $K_1(i) = -$. Then one reorders the marked and unmarked indices in such a way that the marked indices are written in the exchange interval and the other indices outside, starting with the index j_+. Finally, one sets $K_1(i) = P_2(i)$ for $j_+ \leq i \leq k$.

- Cycle crossover

Table 13.4. Example of the cycle crossover

P_1	8	1	3	0	2	7	6	4	5	9
P_2	2	8	7	4	6	3	1	5	0	9

Step 1: Choose a position i_0 randomly; set $i = i_0$. Let, e. g., $i_0 = 2$.

Step 2: Iterate:
Set $K_x(i) = P_x(i)$.
Set $i = P_x^{-1}(P_{3-x}(i))$.
Repeat until $i = i_0$.

$\tilde{K}_1$	8	1	–	–	2	–	6	–	–	–
$\tilde{K}_2$	2	8	–	–	6	–	1	–	–	–

Step 3: Set empty $K_x(i) = P_{3-x}(i)$.

K_1	8	1	7	4	2	3	6	5	0	9
K_2	2	8	3	0	6	7	1	4	5	9

The cycle crossover (CX) works differently from the crossover operators mentioned above: here one randomly selects a position i_0 and sets $K_1(i_0) = P_1(i_0)$. Then one determines the position number i_1 of the node $P_2(i_0)$ in P_1 and analogously sets $K_1(i_1) = P_1(i_1)$. This ansatz of iteratively determining the tour position number i_k of the node $P_2(i_{k-1})$ in P_1 and setting $K_1(i_k) = P_1(i_k)$ is repeated until $P_2(i_{k-1}) = P_1(i_0)$. One can imagine this approach as if the numbers $P_1(i_0), \ldots, P_1(i_{k-1})$ formed a cycle with their counterparts $P_2(i_0), \ldots, P_2(i_{k-1})$. In the example in Table 13.4, the cycle is created by the numbers $1, 8, 2, 6$. For all remaining tour positions, one sets $K_1(i) = P_2(i)$.

Of course, there are many more such crossover operators, but we wish to restrict ourselves to this list of a few widely used and well-known operators [192].

13.3 Natural Selection

As already mentioned, genetic algorithms work with a large population of configurations. Thus, one starts out with a large set of either randomly initialized or already preoptimized configurations. In each evolution step, mutations can change the individuals of the population. Here one can either alter the individuals themselves or produce copies of the individuals that have been changed by the application of the mutation operators. These copies or clones correspond to an asexual reproduction. The mutations can be performed either at random or only if they lead to an improvement.

Offspring is usually produced by applying the crossover operators introduced above, which corresponds to a sexual reproduction, as two parental configurations are needed to generate a child configuration.

Due to the generation of children, the number of individuals in the population increases. But usually one wants to stay with a constant population size during the optimization run. Thus, some individuals of the population must be removed such that some children can be added to the population. Here one can think of various approaches: for example, each time a child is generated, the least fit member of the population is replaced by a child. One can alter this approach and only perform this exchange if the child has a greater fitness than this individual. But one can alternatively produce a large number of children and apply the survival-of-the-fittest approach in such a way that the populations of the parental and the children configurations are first merged and then those individuals with the least fitness are removed. Here one finds a really rich and entertaining variety of possibilities on how to simulate an evolutionary process by means of natural selection. One can also think of a maximum lifetime even for very good individuals.

13.4 Computational Results

We implemented the GA as follows:

- First, an initial population of 100 individuals is generated, each of which gets a randomly created roundtrip through the 442 drilling holes of the PCB442 TSP instance.
- Then we determine the rank of the individuals; the best individual gets the No. 1 rank, the second-best the No. 2 rank, and so on.
- Then we perform a loop over 1000 evolution steps:
  - In each evolution step, we perform ten sweeps of mutations for each individual. The seven mutations mentioned above are used with equal probability. Mutations are only accepted if they do not worsen the configuration. They are directly applied to the individuals themselves.
  - After this polishing of the individuals, we give them the opportunity to have some sexual relationships: here we assume that each individual is a hermaphrodite, i. e., each individual acts once like a male and once like a female in nature. The female selects a male individual different from herself, giving priority to individuals with a high rank. (The rank is selected according to $1+\text{int}(\text{random}()^2 \times M)$ with M being the number of individuals.) Then a crossover operator between these two individuals is performed, in which one child is generated. Thus, each individual is able to play exactly once the role of the mother, who selects a strong partner in order to produce a child with a large fitness. In the world of the males, the better individuals serve more often as paternal configurations.

  - Thus, we have created as many children configurations as there are parental configurations. Now we let the weaker half of our population die in order to return to the original population size. Of course, this approach induces a rather large evolutionary pressure.
  - Finally, we update the ranks of the individuals.
- After 1000 evolution steps, the individuals have become identical for most crossover operators. We thus return the length of the (best) configuration after this number of evolution steps as the output of the GA.

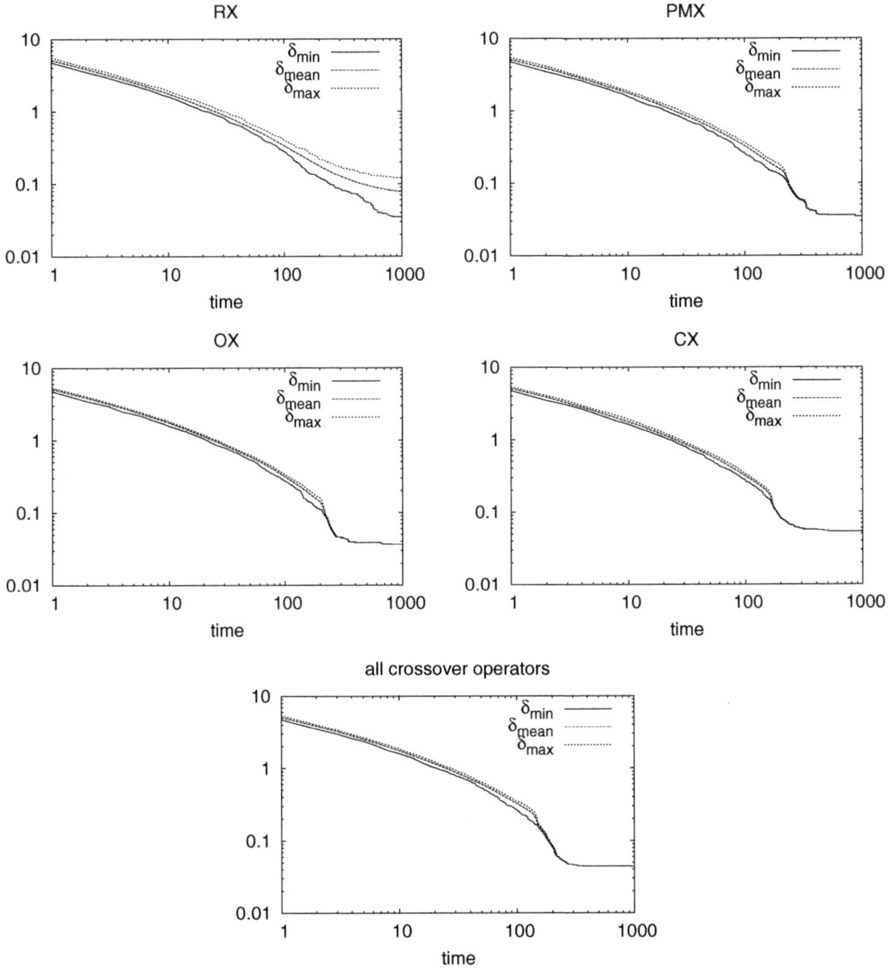

Fig. 13.1. Deviations of the minimum, mean, and maximum lengths of the individuals from the optimum of the PCB442 TSP instance vs. time measured in evolution steps

The question that arises first is which crossover operator leads to the best results. Figure 13.1 shows for five simulation runs each using a different crossover operator or using all of them with equal probability (bottom), the development of the minimum, average, and maximum lengths of the roundtrips of the individuals over time. To enlarge the effects, we plot the deviation of these observables from the optimum instead of the observables themselves. We find that the simulation run is not at all frozen after 1000 evolution steps when using the RX. Using the CX, one gets a slightly faster convergence than with the other crossover operators. In our optimization runs, the best and the worst individuals do not differ too much from each other in their fitness values.

Table 13.5. Quality of the solutions of the best individuals in a population (averaged over 100 optimization runs each) using different crossover operators for the PCB442 TSP instance

Operator	Minimum	Maximum	Mean value	± error
RX	52,366.2734	53,576.5547	53,021.3112	± 27.7
PMX	51,798.9414	53,754.0391	52,791.1574	± 37.6
OX	51,893.4336	54,245.0859	52,728.5821	± 46.9
CX	51,981.7891	54,108.1094	52,976.8450	± 46.1
all	51,687.1914	54,139.543	52,772.1975	± 51.4

Table 13.5 shows a comparison of the results that can be achieved with the various crossover operators. The PMX and the OX lead to the best results on average, the RX to the worst results. But note that the simulation runs using the RX did not converge within 1000 evolution steps. We get the overall best result by using all four crossover operators and selecting one of them at random for each reproduction process.

Of course, these results could be improved by spending more calculation time, using a larger population size, tuning various parameters, and changing the scenarios for the application of the mutations and of the crossover operators. However, there are so many possibilities for tuning GAs that we could easily fill a book the same size as this book with results for these variations. Thus, we leave it here and turn to other optimization algorithms.

14 Social Animal Algorithms Applied to TSP

14.1 Application of Ant Colony Optimization

Watching the efficiency of colonies of social animals with poor individual capabilities but complex collective behaviors, one is inspired to find out how these complex collective behaviors are created, how they work, and how they can be simulated for the usage of mankind, especially as an algorithm for optimization purposes.

As already introduced in Sect. 20.2, Part I, the capability of ants to find the shortest way around an obstacle put in their path can be translated into an optimization algorithm for problems like the traveling salesman problem (TSP) [40]. Each ant then corresponds to a traveling salesman performing a closed roundtrip through the given set of nodes, touching each node exactly once. In this ant colony optimization (ACO) heuristic, each ant is assumed to make some compromise between walking over short edges in order to minimize the total tour length locally, which corresponds to the individual intelligence, and using edges other ants already used and left their pheromones on, which corresponds to the group intelligence. These in turn lead to a global optimization scheme.

At the beginning, the system is initialized by placing M ants randomly on the nodes. Usually M is chosen to be larger than N, the number of nodes, such that at each node at least one ant starts its roundtrip. The ants do not choose the edges of their roundtrips at random. Only the starting point of the roundtrip, which has no special meaning for the standard TSP, is chosen randomly. As the ants leave some pheromones on the edges they use, the amount of pheromones on each edge has to be stored. This amount $\phi(i,j,t)$ on edge (i,j) at time t evolves in time as ants using the edge for moving either from node i to node j or from j to i add some pheromones but also as the amount of pheromones evaporates in time. Thus, the time development of this amount is given by

$$\phi(i,j,t+1) = \eta \times \phi(i,j,t) + \sum_{k=1}^{M} \Delta\phi_k(i,j,t \to t+1) \qquad (14.1)$$

with $\eta < 1$ being the evaporation coefficient and $\Delta\phi_k(i,j,t \to t+1)$ the additional amount of pheromones ant k leaves on edge (i,j) after having chosen it at time t. (Note that in this model, every edge can be traversed in the same time interval of 1, regardless of their actual lengths.)

If the ant is then at a node i at some time t, the next node j to go to is chosen with the probability

$$p(i,j,t) = \frac{\phi(i,j,t)^{\alpha}/D(i,j)^{\beta}}{\sum\limits_{\substack{k=1 \\ k\notin \text{TABULIST}}}^{N} \phi(i,k,t)^{\alpha}/D(i,k)^{\beta}} \tag{14.2}$$

containing the parameters α and β that can be adjusted to reweight the individual local search via the collective global search. The inverse distance $1/D(i,j)$ is called the visibility of edge (i,j): the shorter the distance $D(i,j)$ between nodes i and j is, the more visible node j is from node i and thus the larger the probability is that the ant will choose edge (i,j). The second term determining the probability for choosing an edge is the amount ϕ of pheromones. Thus, the ant must make some compromise between the local search visibility and the global collective approach.

Of course, to avoid initialization problems, some small amounts of pheromones must be put on the edges at the beginning. Furthermore, the ants cannot touch a node twice in the roundtrip, so this must be forbidden by introducing a tabu list for each ant in which the nodes already visited by the ant are stored. Thus, the sum in the denominator of formula (14.2), which normalizes the sum of the probabilities to be 1, runs only over the nodes not yet visited. The nodes in the tabu list get a transition probability of zero.

Summarizing, the outline of this ACO technique is as follows:

1. First, the system is initialized with small amounts of pheromones on all edges (which are either identical or random) and by putting ants on each node of the TSP.
2. M roundtrips are created in N steps:
   a) In each step, the following commands are executed for each ant:
      i. A new node is chosen randomly according to the transition probability given above.
      ii. The new node to which the ant has moved is put on the tabu list of the ant such that it cannot be selected again.
   b) The amounts of pheromones on each edge and the probabilities for choosing an edge are updated.
3. After the completion of the roundtrip, the tabu lists are emptied.
4. The heuristic returns to step 2 if some final criterion is not met.

The individual implementations of the ACO heuristic differ mainly in the amounts of pheromones added by an ant to the edge:

- In the Ant-density model, the amount added by ant k on edge (i,j) is given by

$$\Delta\phi_k(i,j,t \to t+1) = \begin{cases} Q_d & \text{if the } k\text{th ant goes from node } i \\ & \text{to node } j \text{ or vice versa between} \\ & t \text{ and } t+1\,, \\ 0 & \text{otherwise}\,. \end{cases} \tag{14.3}$$

Thus, the amount is given by the number of ants using the edge. This says nothing about how good the idea is of using this edge in the first place. The amount only refers to how many other ants have already chosen the edge. The quality of the edge is seen in its visibility in the probability function.

- In the ant-quantity model, the amount is given by

$$\Delta\phi_k(i,j,t \to t+1) = \begin{cases} Q_q/D(i,j) & \text{if the } k\text{th ant goes} \\ & \text{from node } i \text{ to node } j \\ & \text{or vice versa} \\ & \text{between } t \text{ and } t+1\,, \\ 0 & \text{otherwise}\,. \end{cases} \tag{14.4}$$

  Thus, this model also considers the length of a specific edge and thus reinforces the visibility of an edge here.
  However, these two models can be mapped on each other simply by changing the exponent β in the probability formula. Furthermore, although the impact of short edge lengths is enlarged in this ant-quantity model, the basic task of the traveling salesman to find an overall short roundtrip is not considered. The traveling salesman might have to choose a worse edge locally in order to get an overall shorter roundtrip. But these ant-density and ant-quality models only consider a locally optimum way in the probability function when using their descriptions of the amounts of pheromones to add.
- Thus, the ant-cycle model moves point 2b) in the outline above to point 3 and updates the amounts of the pheromones after having created the complete roundtrip according to

$$\Delta\phi_k(i,j,t \to t+N) = \begin{cases} Q_c/L_k & \text{if the } k\text{th ant goes} \\ & \text{from node } i \text{ to node } j \\ & \text{or vice versa} \\ & \text{in its roundtrip}\,, \\ 0 & \text{otherwise}\,, \end{cases} \tag{14.5}$$

  with L_k being the overall length of the roundtrip the kth ant performs. Thus, the shorter the overall tour length is, the more pheromones are added to the edges that are part of the tour.

Of course, at first this ACO heuristic recalls the tour construction heuristics introduced in Chap. 3. There the nodes were taken one by one from a bag. The nearest neighbor heuristic adds the best node at the end of the partially already created roundtrip. The ACO approach does basically the same initially, except that it introduces a random element, namely, the probability function by which the next node is chosen. However, due to the outcome of many ant steps performed in parallel, this probability function is changed. Furthermore, the algorithm does not end after having created a complete roundtrip. Instead, this construction approach is iterated again and again in order to improve the roundtrips created. Thus, one can consider the ACO heuristic an iterative construction heuristic. But one can also include this heuristic among the tabu search heuristics, as the amount of pheromones left is some kind of adaptive memory.

14.2 Computational Results

The three models of the ACO heuristic contain several parameters on which the results achieved depend, namely,

- The fraction M/N, i. e., the number of ants divided by the number of nodes, which should be ≥ 1, as already shown in [40];
- The evaporation coefficient η, which should be chosen somewhere in the range $0.5 \leq \eta < 1$;
- The basic amounts Q_i of pheromones to be added on an edge, which is, depending on the model, divided by the length of the edge or the overall tour length;
- The two exponents α and β, which weigh the two quantities pheromones and visibility, in their influence on the probability of which edge to choose.

In [40], various values were given for these parameters: $M/N \in \{1, 3, 10, 30\}$, $\eta \in \{0.5, 0.7, 0.9\}$, $Q \in \{1, 10, 100, 1000, 10000\}$, and $(\alpha, \beta) \in \{0.5, 1, 2, 5, 10\}^2$. But no direction was given as to which values for these parameters were optimal or how the ratio between them should be. We performed 100 simulation runs for each quintuple $(M/N, \eta, Q, \alpha, \beta)$ applying the ant-density, the ant-quantity, and the ant-cycle model to the BEER127 instance, which took months of calculation time. We were unable to find a good configuration for the BEER127 instance. None of the quintuples seemed to be significantly better than all other quintuples.

However, it is still worthwile to have a closer look at the algorithm and see how the ants gradually reduce the lengths of their roundtrips. To study the behavior of the algorithm, we randomly selected the values $M/N = 3$, $\eta = 0.7$, $Q = 100$, $\alpha = 1$, and $\beta = 1$ and performed one simulation run each for the ant-density, ant-quantity, and ant-cycle models. The decrease in the minimum, mean, and maximum lengths of the roundtrips the ants perform between the 127 beergardens of Augsburg is shown in Fig. 14.1. In each time step, all ants performed a complete roundtrip. After they all finished their roundtrip, the amount of chemical substances on an edge, and thus the probability that a specific edge would be chosen by an ant, was updated for the next time step.

We find that there is a strong decrease in the lengths already at the beginning with the ant-density and for the ant-quantity models, whereas the

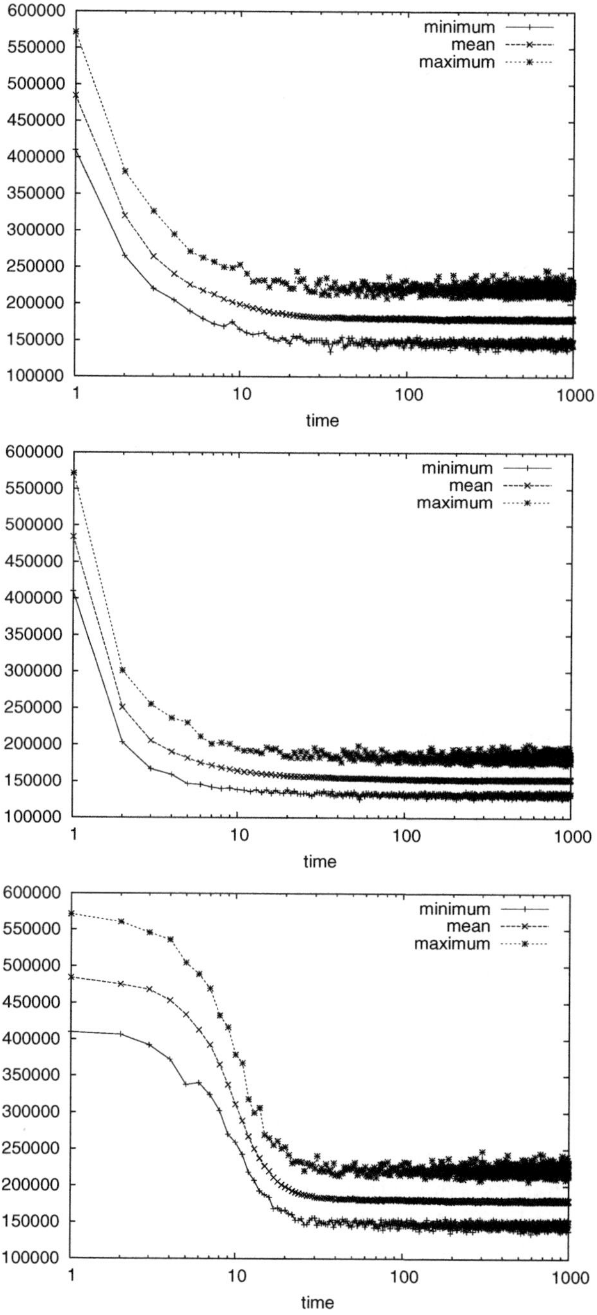

Fig. 14.1. Decrease in the minimum, mean, and maximum lengths of the roundtrips performed by 3 × 127 ants between the 127 beer gardens of Augsburg, using the ant-density (*top*), ant-quantity (*middle*), and ant-cycle (*bottom*) models vs. time

lengths decrease sigmoidally for the ant-cycle model. But for all three models we find that there is no longer any decrease after 300 time steps: the minimum length in the ant-density model fluctuates finally around 145,000, in the ant-quantity model around 130,000, and in the ant-cycle model around 150,000. These values are much worse than the optimum value of 118,293.52, so that we cannot recommend the ACO method, at least not in these variants.

14.3 Application of Bird Flock Model

Particle swarm optimization (PSO) is an exciting new methodology in evolutionary computation that is somewhat similar to a genetic algorithm in that the system is initialized with a population of random solutions. Unlike other algorithms, however, each potential solution (called a particle) is also assigned a randomized velocity and then flown through the problem hyperspace. We apply this PSO algorithm to the TSP as follows. First, we generate an initial population consisting of a number M of randomly created configurations. Then a random walk (RW) is performed for ten sweeps, in which the node insertion move (NIM), the exchange (EXC), and the Lin-2-opt (L2O) are applied to the various individuals. For each individual i, we store the current configuration $\sigma(i)$ and its cost function value $\mathcal{H}(\sigma(i))$ and the best configuration $\sigma_{\text{best}}(i)$ found by the individual during the RW and its cost function value $\mathcal{H}(\sigma_{\text{best}}(i))$. Furthermore, we store the overall best configuration σ_{super} (called the superconfiguration) found during the RW and its cost function value $\mathcal{H}(\sigma_{\text{super}})$.

After the RW, we perform several sweeps containing $N \times M$ moves. In each move, an individual i is randomly selected. Then this individual can act in three different ways:

- With some probability p, it performs a random move, i. e., one of the three moves NIM, EXC, and L2O is chosen at random, and the move is then applied as in the RW.
- But with some probability q, individual i tries to do something similar to the best configuration $\sigma_{\text{best}}(i)$ it has found so far. Again one of the three moves NIM, EXC, and L2O is selected at random:
  - If the NIM is selected, one edge of this configuration $\sigma_{\text{best}}(i)$ is chosen at random. If this edge is also part of the current configuration $\sigma(i)$, then another edge is chosen at random. Then one searches for the two nodes a and b that are connected via an edge in $\sigma_{\text{best}}(i)$ in the configuration $\sigma(i)$. Then the NIM is applied in such a way that b is inserted directly behind a. Thus, edge (a, b) becomes part of the current configuration $\sigma(i)$.
  - If EXC is selected, again one edge (a, b) of $\sigma_{\text{best}}(i)$ is chosen at random. If this edge is also part of $\sigma(i)$, then another edge is chosen at random. Let c be the node that succeeds node a in $\sigma(i)$. Then c and b are replaced by the EXC, such that edge (a, b) becomes part of $\sigma(i)$.

- If the L2O is selected, again one edge (a, b) of $\sigma_{\text{best}}(i)$ is chosen at random. Again one chooses another randomly chosen edge if this edge is also part of $\sigma(i)$. Let us consider here only the case in which node a is approached by the traveling salesman before node b is approached in $\sigma(i)$. Again let c be the node succeeding node a in $\sigma(i)$. Then a L2O is performed that turns around the partial sequence containing all nodes from node c to node b, thus introducing edge (a, b) in $\sigma(i)$.

These three cases are simply three different ways to insert edge (a, b), which is part of the best configuration $\sigma_{\text{best}}(i)$, into the current configuration $\sigma(i)$.
- With the remaining probability $1 - p - q$, the individual i tries to do something similar to the overall best configuration σ_{super} found so far. An edge (a, b) of this configuration is chosen at random, which is not part of $\sigma(i)$. Like above, this edge is then inserted with either the NIM or the EXC or the L2O.

  Thus, the same approach as just described is used, but with the overall best configuration σ_{super} instead of the best configuration $\sigma_{\text{best}}(i)$ individual i has found itself.

After each such move, it is checked whether $\sigma_{\text{best}}(i)$ and σ_{super} must be updated.

At the end of the optimization run, the best solution found σ_{super} and its cost function value $\mathcal{H}(\sigma_{\text{super}})$ are returned.

14.4 Computational Results

This algorithm contains three basic parameters, namely, the ratio M/N between the number M of individuals and the number N of nodes, the probability p with which a random movement is performed, and the probability q with which the individual introduces an edge of its best-so-far configuration. We applied this algorithm to the BEER127 TSP instance.

We varied p and q independently in the range between 0 and 1 in steps of 0.01, of course considering the fact that $p + q \leq 1$. We used $M/N = 1$, 3, 10, and 30. The RW consisted of ten sweeps for each individual. Then we performed $3000N^2$ moves. At the end of the optimization run, the cost function value of σ_{super} was returned.

Figure 14.2 shows the results for this super configuration averaged over 100 optimization runs, for various values of p and q and for $M = N$. We see at first glance that we obviously achieve the best results for very small p, i. e., if the individuals do not make random moves. Furthermore, it seems that q should also be chosen rather small. Therefore, it is obviously optimal if the individuals consider the best performance of the best individual, which is precisely the philosophy behind this algorithm.

Figure 14.3 shows the results again, but now only for small values of p and q and for various values of M/N. To emphasize the differences, the graphic

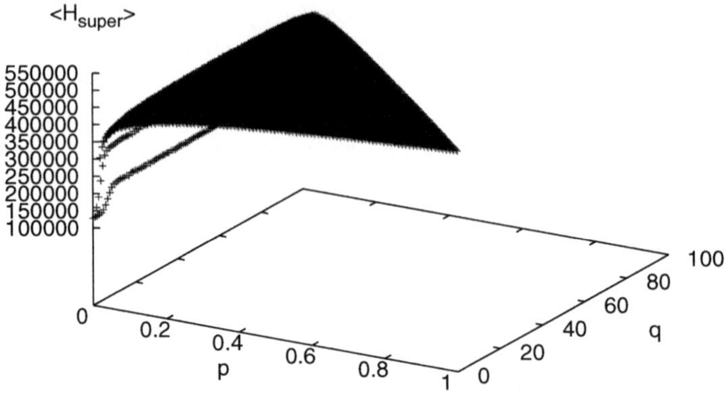

Fig. 14.2. Best results found for the BEER127 TSP instance for various values of p and q and for $M/N = 1$: the results are averaged over 100 optimization runs

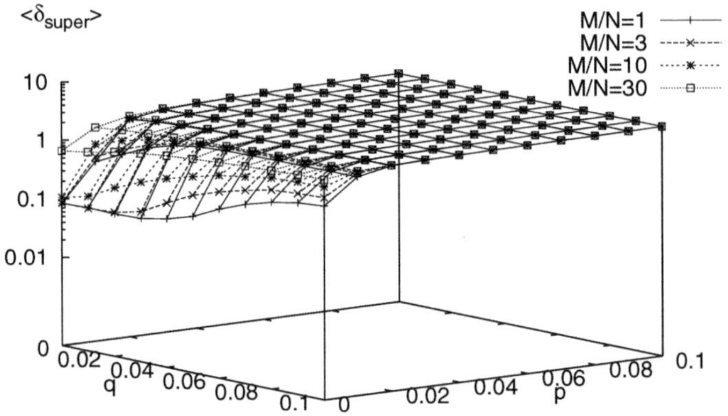

Fig. 14.3. Best results found for the BEER127 TSP instance for small values of p and q and for various values of M/N: the results are averaged over 100 optimization runs

shows the mean relative deviation of the super configuration instead of the cost function value itself. We find that $p = 0$ is optimal and that q should be chosen in the range $0 \leq q \leq 0.03$. One should also set $M = N$. We obtained the best result for $M/N = 1$, $p = 0$, and $q = 0.01$, namely, a value of 119441.664. This is rather close to the optimum value of the BEER127 TSP instance.

Summarizing, the PSO algorithm, which is motivated by biological systems like flocks of birds, provides an interesting approach to optimization, leading to quite good results.

15 Simulated Trading Applied to TSP

15.1 Application of Simulated Trading to the TSP

Simulated trading (ST) was originally developed for the vehicle routing problem (VRP), which is an extension of the traveling salesman problem (TSP), as already mentioned in Chap. 2. In their original publication [13], Bachem et al. considered each truck driver as an agent who buys and sells goods to his customers.

In contrast, there is only one traveling salesman in a TSP. However, the ST algorithm can also be applied to the TSP by splitting the roundtrip such that several agents $1, \ldots, M$ each get a partial sequence $\sigma(k_i), \sigma(k_i+1), \ldots, \sigma(k_i+N_i-1)$, containing N_i nodes, of the whole tour. Of course, all these sequences put together must result in the closed roundtrip of the traveling salesman, i. e., $\sum_{i=1}^{M} N_i = N$.

Now, the ST algorithm requires that there be items to be bought or sold and that there be agents willing to give a buy or a sell order. In this section, we want to stick closely to the original approach as described in [13]. The individual items in this case are the nodes of the tour, which are sold and bought by agents, thus removing or adding one node from or to their partial sequences.

The outline of such a "trading move" is as follows:

- First, it is determined whether an agent wants to buy or to sell something. For simplicity, only the case where each agent buys or sells one node will be considered. Thus, optimally there will be as many buying as selling agents.
- Then each selling agent i must submit a sell order:
  - He/She calculates for each of the nodes $\sigma(k_i+1), \ldots, \sigma(k_i+N_i-2)$ the savings $S(j,i)$ that would occur if the jth node in the tour was removed, i. e.,

$$S(j,i) = \Delta_j^- , \tag{15.1}$$

with

$$\Delta_j^- = D(\sigma(j-1), \sigma(j)) + D(\sigma(j), \sigma(j+1)) - D(\sigma(j-1), \sigma(j+1)) . \tag{15.2}$$

Note that we do not consider the first and last nodes of such a partial sequence as they are connected with the outer nodes of the partial se-

quences of the neighboring agents and as the agents are considered to act independently of each other.
- Now the question of the behavior or manners of each agent arises. If they are rather greedy, they would want to offer that node $\sigma(j)$ for which $S(j,i)$ is maximal. However, as was already mentioned in [13], it is advantageous to work with a nongreedy behavior. Therefore, probabilities $p(j)$ are assigned to the nodes $\sigma(j)$

$$p(j) = \frac{S(j,i)}{\sum\limits_{l=k_i+1}^{k_i+N_i-2} S(l,i)} \tag{15.3}$$

for all nodes $\sigma(j)$ with $k_i+1 \le j \le k_i+N_i-2$. Note that the probability of a particular j is large if the corresponding savings $S(j,i)$ is large compared to the other savings values.
- Then a uniformly distributed random number r is chosen by which a particular $\tilde{j}$ is determined according to the probabilities $p(j)$. Thus, agent i offers the node $\sigma(\tilde{j})$, for which he/she gets the savings $S(\tilde{j},i)$, for sale. Let us introduce the abbreviation $S(\sigma(\tilde{j})) := S(\tilde{j},i)$ here, thus assigning the savings value to node $\sigma(\tilde{j})$.

- All selling agents send their sell orders with the associated savings values to the central stock market, which summarizes them in a public selling list. This selling list, containing Z offers, is sent to all buying agents.
- Each buying agent must then go through each individual offer:
  - For each offered node z, an agent i has to find out where to best insert z into his/her partial sequence. So here a greedy approach is used. Let $C(z,i)$ denote the minimum costs for inserting node z into the partial sequence of agent i:

$$C(z,i) = \min_{j=k_i,\dots,k_i+N_i-1} \{\Delta^+_{z,j}\}, \tag{15.4}$$

with

$$\Delta^+_{z,j} = D(\sigma(j),z) + D(z,\sigma(j+1)) - D(\sigma(j),\sigma(j+1)). \tag{15.5}$$

  - Then agent i must determine which node $\tilde{z}$ he/she actually wants to buy. If only the costs for insertion are considered, then nearly every agent would refuse to buy, as the costs for insertion are nonnegative and only zero if node z lies on a straight line between $\sigma(j)$ and $\sigma(j+1)$. Therefore, the gain for removing a node in the old sequence and inserting it into the new one, i. e.,

$$G(z,i) = S(z) - C(z,i), \tag{15.6}$$

must be considered. If this gain is positive, then the shifting of the node to the new agent decreases the tour length. Now agent i determines the minimum gain m_i,

$$m_i = \min_{z \in \text{ selling list}} G(z, i) \,. \tag{15.7}$$

If $m_i < 0$, then there is at least one offer with negative gain. Of course, the single agent i would not want to accept such an offer. Whats more, if all gains are negative, he would not want to buy anything at all. On the other hand, it might be advantageous for the system as a whole if he accepted an offer with negative gain so that the overall system would not get stuck in local minima. Therefore, the algorithm distinguishes between the cases $m_i < 0$ and $m_i \geq 0$:

- If $m_i \geq 0$, then all of the gains of an agent are nonnegative, so that he can be happy with any offer. Analogously to the behavior of the selling agents, the buying agent i calculates the probabilities

$$p(z, i) = \frac{G(z, i) - m_i}{\sum\limits_{x=1}^{Z} (G(x, i) - m_i)} \tag{15.8}$$

  for the various offers z. However, checking this formula closely, one finds that the least attractive offer has a probability of 0. To overcome this problem, the value m_i is slightly decreased, e. g., set to $0.97 \times m_i$, before the probabilities are calculated. Then a random number is calculated by which an offer $\tilde{z}$ that agent i would accept is determined. Thus, agent i tells the stock market that he wants to submit a buy order for node $\tilde{z}$ with a gain $G(\tilde{z}, i)$.
- Otherwise, in the case $m_i < 0$, a trick is used: agent i adds the virtual $(Z+1)$th offer $\hat{z}$ with the supposed gain $G(\hat{z}, i) = 0$ to the list of offers. Analogously to the case above, m_i is decreased by 3% by multiplying it by 1.03, such that the least attractive offer also has some probability of being chosen. The probabilities $p(z_1, i)$, $p(z_2, i)$, …, $p(z_Z, i)$, and $p(\hat{z}, i)$ for the individual offers are calculated as follows:

$$p(z, i) = \frac{G(z, i) - m_i}{\sum\limits_{x=1}^{Z+1} (G(x, i) - m_i)} \,. \tag{15.9}$$

  By means of a uniformly distributed random number, a particular offer $\tilde{z}$ is chosen from the list of offers. If $\tilde{z} \neq \hat{z}$, then agent i tells the stock market that he/she wants to buy node $\tilde{z}$ with the determined gain $G(\tilde{z}) = G(\tilde{z}, i)$. If, however, $\tilde{z} = \hat{z}$, then he/she tells the stock market that he/she does not want to buy anything.

- The stock market collects all these buy orders from the buying agents in a list. Comparing this list with the selling list, there are three possibilities for each node on the selling list:

  - There might be no buy order for a node on the selling list. Then the agent who made this sell order cannot sell his/her node. Therefore, this sell order is removed from the selling list.
  - There is exactly one buy order for a node on the selling list. In this case, the stock market creates a link between the buy order and the corresponding sell order. The link is weighted with the gain that would be received for shifting the node from the selling agent to the buying agent.
  - There might be more than one buy order for a node on the selling list. In this case, the stock market goes through all of these buy orders, determines that buy order with the maximum gain, and creates a link between this buy order and the corresponding sell order. This link is weighted with the gain.
- Finally, the stock market sums up the gains over all existing links. If the sum over all gains is nonnegative, then the trading move is accepted, i. e., for all existing links between a sell order from a selling agent and a buy order from a buying agent, the node is shifted from the selling to the buying agent. Note that some of the partial gains might be negative, producing local deteriorations. However, if the sum of all gains is positive, this trading move leads to an overall improvement.

If the trading move is always performed in this way, then it always exhibits in some sense the same size, as each selling agent gives a sell order. In order to add even more randomness and to improve the results, the authors introduced a probability q with which a selling agent actually places a sell order [13].

This trading move is the key ingredient of the ST algorithm. Thus, the outline of the ST algorithm is as follows:

- Read the data, determine the distance matrix, and initialize the system with a random tour.
- Do several times:
  - Split the tour into an even number of partial sequences. These sequences should be roughly the same length. Each of these sequences must contain at least two nodes. Furthermore, do this splitting in such a way that a tour position is selected randomly that is the starting position of the partial sequence of the first agent.
  - Do several times:
    - Determine which agents want to buy and which ones want to sell. The number of buying and selling agents should be roughly the same.
    - The individual selling agents determine by means of a random number whether they actually want to sell. This random number must be smaller than the probability q mentioned above. If it is, then the agents check whether there are at least three nodes in their partial sequences of the tour and determine the node they want to sell together with the savings for removing this node from the partial sequence.
    - The stock market collects all sell orders from the selling agents and summarizes them in a public selling list.

    - If the selling list contains at least one item, then it is sent to the buying agents.
    - The buying agents go through all offers and determine for each offered node where to best insert it into their partial sequences. Then they choose one of the offers randomly as described above and send a buy order for this offer together with its gain to the central stock market.
    - The stock market goes through all buy orders and creates links between the sell orders and those buy orders that provide the maximum gain for the corresponding sell orders. The links are weighted with the determined gains.
    - If the sum over all these gains is nonnegative, then this trading-move is accepted, i. e., for each link between a selling or a buying agent, the offered node is removed from the partial sequence of the selling agent and inserted at the best possible position into the partial sequence of the buying agent. Otherwise, if the move is not accepted, nothing happens.
  - After several such trading moves, the partial sequences are merged to one closed tour such that it can next be split into other parts.
- Print out the final result.

15.2 Computational Results

We performed several simulations with the algorithm sketched above. We set the number of buying agents equal to the number M_s of selling agents such that M_s is given by $M/2$, with M being the overall number of agents. We give each agent roughly the same number of nodes when splitting the tour into several partial sequences, namely, $[N/M]$ or $[N/M] + 1$, with $[x]$ being the integer part of x, such that the sequences merged together result in a feasible configuration containing each node once. Furthermore, we add the first node of each partial sequence to the end of the previous sequence such that there are no connections of the tour that must not be cut. Thus, $\sum_{i=1}^{M} N_i = N + M$.

For us, the two main variables of interest are the number $M_s = M/2$ of selling agents and the probability q with which a selling agent actually wants to sell something. Figure 15.1 shows the mean results (with error bars) for three TSP instances achieved with the ST heuristic. In each simulation, 10000 sweeps were performed, which is according to our empirical tests enough to "freeze" the system within this heuristic. In each sweep, first the tour was split among the agents, giving the individual agents partial sequences as described above. Then N trading moves were tried. At the end of each sweep, the partial sequences were merged together to form a closed roundtrip through all nodes again. Note that the results shown generally in this chapter, and thus also in Fig. 15.1, are averaged over only 50 optimization runs each.

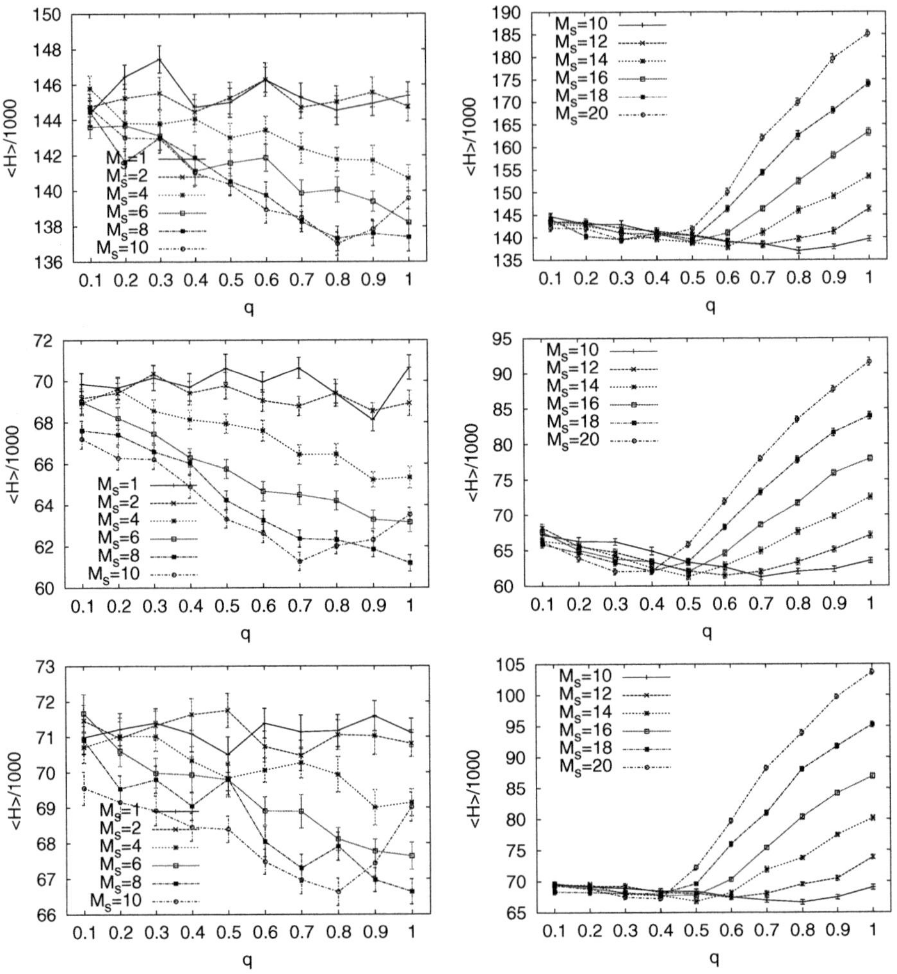

Fig. 15.1. Quality of the results achieved with the ST heuristic for three TSP instances [BEER127 (*top*), LIN318 (*middle*), and PCB442 (*bottom*)] depending on the propability q and on the number M_s of selling agents (*left*: results for $M_s \leq 10$, *right*: results for $M_s \geq 10$)

We find that the results must be differentiated between a small number of agents and a large number of agents. This splitting can be done at $M_s = 10$ for all three instances: for $M_s < 10$, the results improve with increasing M_s and increasing q. For all three instances, the curve for $M_s = 10$ starts to rise again at $q \to 1$. This effect becomes dramatic for $M_s > 10$: with increasing q, the results become worse. The more agents are used, the worse the results become. Obviously, this is the reason for introducing the parameter q into the algorithm, namely, to obtain much better results for larger numbers of agents. It seems that there is an optimum number of agents taking part in such a trading move: if the number of agents is too small, then the number of sell offers is small, so that there is no real stock market. The trading move simply performs one or a few node insertion moves (NIMs) at the same time. Thus, the trading move is identical to either the NIM or to some special cases of the Lin-n-opts with small n. These moves are accepted if they either improve the configuration or do not worsen it too much. If the number of sell offers is large, the trading move, which creates a large amount of links between sell and buy offers and thus changes the system overall in many ways, is no longer a move of the local search type. As already discussed in Chap. 5, the more cuts are introduced in a Lin-n-opt, the less probable the move will lead to an improvement if the current configuration is already rather good. This finding for the Lin-n-Opts that $n \leq 3$ is optimal is found here again for the trading move that leads to the best results if the number of selling agents taking part in the trading move is roughly 7 or 8. The shift in the optimum number of cuts is given by the various acceptance rules, which are rather elaborate for the trading move.

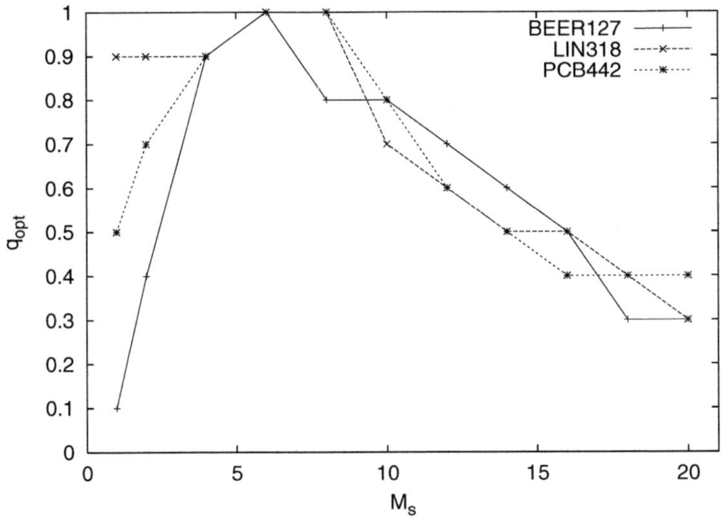

Fig. 15.2. Optimum probability leading to the best results as a function of M_s, the number of selling nodes, for three TSP instances

Figure 15.2 shows the optimum value q_{opt} found for the probability q for various numbers of selling agents. Please note that the results at small M_s are not significant as there the error bars overlap with each other. But basically we again obtain the picture that an introduction of the probability q with values $q \leq 0.4$ is necessary for a large number of agents in order to decrease the number of agents taking part in a trading move and thus to increase the probability that the trading move will lead to a good result.

15.3 Discussion of Simulated Trading

The ST heuristic is at first glance a very elaborate technique for getting solutions to combinatorial optimization problems. The approach of ST seems to be well balanced between a strict set of rules and some kind of randomness, which is necessary to arrive at quite good results with a stochastic optimization algorithm.

However, as the results in the last section show, this heuristic is not well suited for the TSP. (Thus, we did not consider even applying it to larger instances.) In fact, the results are at least more than 10% worse than the optimum value. Of course, we must admit that this algorithm was originally not developed for the TSP and that the artificial splitting of the tour into partial sequences leads to a localization of the problem. A complex problem like the TSP, however, is defined by long-range interactions of its individual parts. One must look at the complete problem in a global way. Thus, ST can surely not compete with global optimization algorithms like simulated annealing for the TSP. It may be better suited for VRP and depot localization problems, for which it was designed and for which such a splitting into the subsets of customers served by one truck or by one depot even makes sense.

But if the ST approach wants to make use of the local search concept, then the number of agents taking part in a trading move must not be too large. For a small number of them, the trading move is obviously applied in a greedylike way, i. e., it mostly leads to improvements, until no further improvements can be found due to the limitations in the conception of the trading move. With an increasing number of orders, good moves can be found even more seldom.

15.4 Simulated Trading and Working

According to Benedict of Nursia's *Ora et labora*, a man should pray and work. At least getting agents to work is easy. In the original ST approach, the agents only trade but do not work, e. g., to improve their parts of the tour. Thus, we extend this original approach now, so that after each trading move all agents whose tours contain at least $N_i = 4$ nodes perform N_i moves in the greedy mode. The moves are chosen randomly to be either an exchange or a node insertion move or a Lin-2-opt with equal probability. Larger moves were not used in our experiments as the partial sequences are rather short for a larger number of agents.

Figure 15.3 shows the results achieved with this simulated trading and working (STW) approach. We find that the results for a small number of agents improve to a large extent: the best results are achieved for $M_\mathrm{s} = 1$, followed by $M_\mathrm{s} = 2$ and $M_\mathrm{s} = 4$. Here the number of agents is minimal, meaning their partial sequences are rather long. Furthermore, we find that the results for $M_\mathrm{s} = 1$ and $M_\mathrm{s} = 2$ are rather independent of q, in contrast

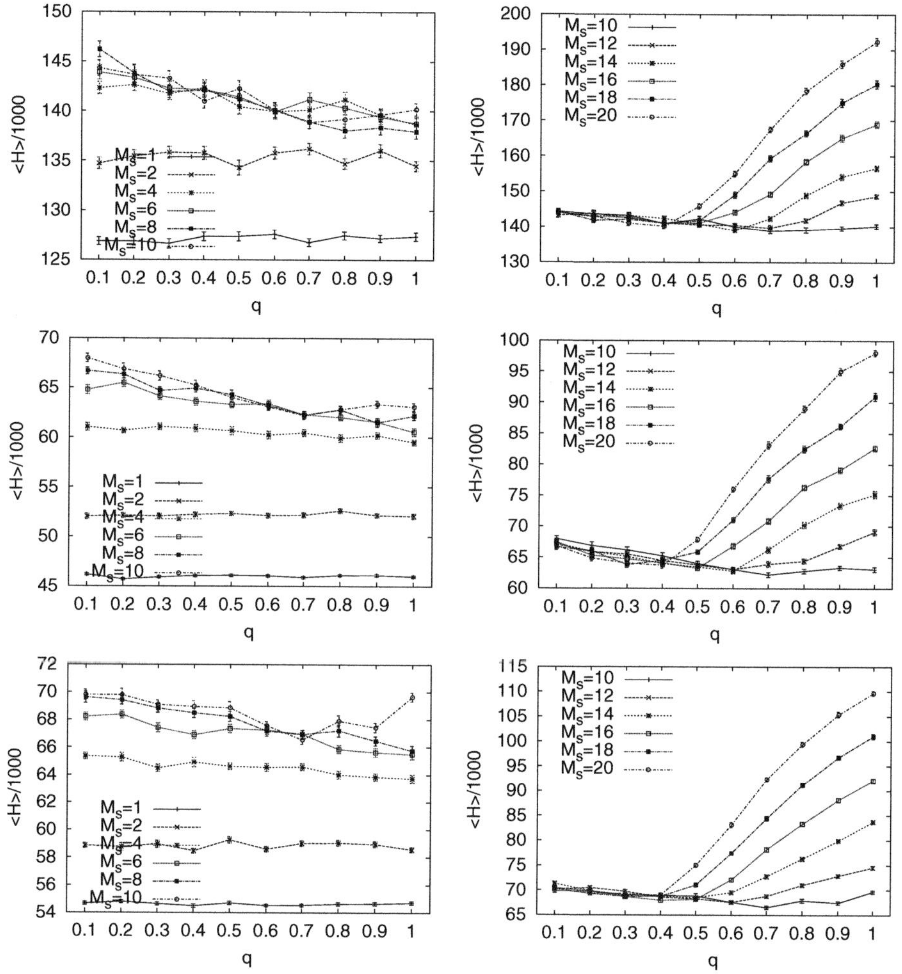

Fig. 15.3. Quality of the results achieved with the STW heuristic for three TSP instances [BEER127 (*top*), LIN318 (*middle*), and PCB442 (*bottom*)] depending on the propability q and on the number M_s of selling agents (*left*: results for $M_s \leq 10$, *right*: results for $M_s \geq 10$)

to the results for larger M_s. As for these small M_s, the largest improvements were found in comparison with the pure ST approach. We conclude that these improvements are mainly due to the small moves performed in the greedy mode. Of course, here the greedy algorithm must be most successful as the partial sequences are rather long for small M_s. For $M_s \geq 10$, we again find that the results become worse with increasing probability q. Thus, we have the same dependency on q and M_s here as before for large M_s. However, the results are worse than those for the original ST approach for large M_s and large q, although the greedy moves should be able to optimize short partial sequences very well. But obviously the trading moves not only destroy the local order generated by the greedy moves but are even less able than before to lead to reasonable configurations. This is due to the fact that with preoptimized partial sequences, the gain for removing a node is mostly much smaller than the costs for inserting it into another partial sequence. Thus, a good trading move can hardly be found anymore. One could even say that the local greedy moves work against the trading moves.

Summarizing, we get in this STW approach better results for small M_s, where we basically benefit from the greedy algorithm when it is applied with the small moves but worse results for large M_s than for the original ST approach. In all cases, however, the quality of the results is not satisfactory. Thus, we recommend that the agents start to pray.

16 Tabu Search Applied to TSP

The very general tabu search paradigm allows for a very wide variety of explicit implementations of tabu search to optimization problems. Within the covers of this book, we surely cannot present implementations and computational results for all tabu search algorithms we might be able to think of.

In searching for popular implementations of tabu search to the traveling salesman problem (TSP) a few years ago, we repeatedly encountered the remark that the well-known Lin–Kernighan heuristic for the TSP [130] exhibited all features of tabu search such that it could be said to be the standard tabu search application to the TSP. In the meantime, however, several applications following the guidelines of tabu search have been published, e. g., by Misevičius [141], who presented results both for a basic and an advanced tabu search implementation.

However, we are not aware of a tabu search implementation to the TSP which is now (March 2006) widely considered to be optimal or at least to be setting a standard. Thus, we decided to create a tabu search implementation of our own here (partially based on the work in [141]) in four steps, by which the interested reader will obtain a good understanding of why all ingredients of tabu search are needed in order to get good solutions.

16.1 Definition of a Tabu List

The basic strategy of tabu search is to search the whole neighborhood of a configuration for the best move that is not tabu. Thus, the first task when working with tabu search is to define a tabu list. But in this context, one must also reconsider the neighborhood of a TSP configuration. In previous chapters, we applied various moves, like the Lin-2-opt (L2O), the node insertion move (NIM), the exchange (EXC), and the four variants of the Lin-3-opt (L3O) to the TSP, such that the neighborhood of a configuration was rather large. To decrease this size as a first step toward decreasing the size of the tabu list, we restrict ourselves to using the L2O only. Furthermore, to save calculation time, we will not go through the entire neighborhood of a configuration generated by this L2O but randomly select $100N$ configurations that can be reached from the current configuration by applying a L2O.

As already mentioned in Chap. 22 in Part I of this book, the main problem when working with tabu search is to prevent the tabu list size, i. e., the number of items declared to be tabu, from exploding. Therefore, we must think of an efficient way to define a tabu list with the properties that it cannot explode in size and that it does not take too much computing time to determine whether a move should be forbidden because the resulting configuration is declared as tabu. But why should a tentative new configuration be tabu? According to the philosophy behind tabu search, the answer is: "This configuration contains properties that were part of many former configurations; but as one intends to investigate other configurations as well without these properties, these properties, and therefore also other configurations containing them, have been declared tabu." In the case of the TSP, such properties are the edges used in the configurations. Thus, we introduce an edge matrix η between pairs (i, j) of nodes with

$$\eta(i,j) = \begin{cases} 0 & \text{if the edge between node } i \text{ and node } j \text{ is tabu}\,, \\ 1 & \text{otherwise}\,. \end{cases} \tag{16.1}$$

At the beginning, we set all $\eta(i,j) = 1$ and create a random configuration. Then the best L2O must be found among $100N$ randomly selected possibilities for the L2O. Let k and l be the tour position numbers of the nodes after which the tour is cut by the best possible L2O, and let k_+ and l_+ be the position numbers succeeding k and l, respectively. The L2O replaces then the edges from tour position k to k_+ and from l to l_+ by the edges from k to l and from k_+ to l_+. The simplest approach would now be to store the old configuration in a tabu list. This approach, however, would lead to a linear increase of the tabu list in time, slowing down the velocity of the tabu search algorithm, because more and more configurations would have to be checked for being tabu.

But we can now make use of the edge matrix and set $\eta(k,l) = 0$, $\eta(l,k) = 0$, $\eta(k_+,l_+) = 0$, and $\eta(l_+,k_+) = 0$. A new tabu-search-L2O that again selects the tour position numbers k and l would try to return to the old configuration. But all moves must check whether both of the two edges they intend to introduce are tabu. In this way, the inverse L2O leading back to the previous configuration is forbidden.

Please note that by using the L2O only instead of all seven small move variants, we can keep the tabu list small, as otherwise one would also have to store the corresponding move that had been applied to get to the current configuration in the tabu list.

Furthermore note that the definition of a move within this tabu search context is very different from that in, e. g., the simulated annealing (SA) context: within tabu search, a tabu-L2O move consists of first randomly selecting $100N$ applications of the L2O move, by which two edges are removed and two new edges are added to the system, then evaluating them for which of these variants is the best among those that are not tabu, and finally performing this variant. In contrast, a SA-L2O checks only for one randomly chosen variant and decides whether to accept or reject it according to the Metropolis criterion. Thus, it makes no sense to compare tabu search and SA in times given by the numbers of applied moves.

A tabu search move is always accepted. At the beginning, the tabu search approach will usually lead downhill, as each tabu search move intends to perform the maximum improvement found if that one is not tabu. But then, when the system has, e. g., reached a local minimum in the energy landscape, it will search for that configuration in the neighborhood that is not tabu and that leads to the smallest deterioration possible. Thus, a tabu search does not get stuck in local minima. But one must store the best-so-far solution and return this solution as the output of the tabu search algorithm, as the tabu search algorithm might never return to this configuration and might then work on configurations that are significantly worse than this best-so-far solution.

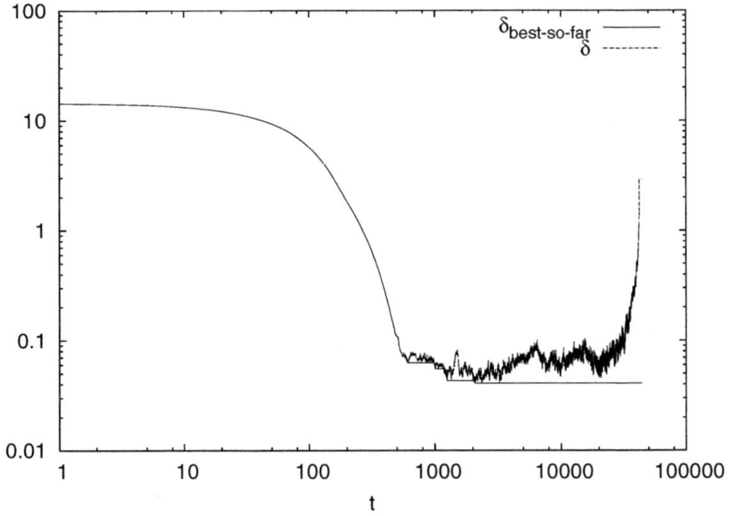

Fig. 16.1. Development of the deviation δ and of the best-so-far deviation $\delta_{\text{best-so-far}}$ vs. time t for the tabu search application as described in Sect. 16.1 for the PCB442 TSP instance

Figure 16.1 shows the development of the energy of this best-so-far solution and of the energy itself with time when applying tabu search moves to the PCB442 TSP instance after starting out with a randomly generated configuration. The time is measured in tabu search moves. In order to magnify small effects, we again show the relative deviation from the optimum $\delta = (\mathcal{H} - \mathcal{H}_{\text{opt}})/\mathcal{H}_{\text{opt}}$ instead of the energies themselves. We find that the deviations decrease first together. But then the best-so-far deviation starts to freeze and can only seldom be improved by the system. Finally, the deviation δ increases again and freezes at some value.

This last behavior can be easily explained: more and more edges are declared tabu. Thus, the system must introduce longer and longer edges as many of the shortest edges are already tabu. Finally, the system can no longer perform any move at all as all moves are declared tabu.

Thus, in our first rather simple tabu search implementation the system worsens and finally freezes because all edges are memorized as being tabu. Nevertheless, the average results for this approach (Table 16.1) are quite promising for such a simple approach but of course not very good. (Note that all results in this table were achieved after a maximum of $100N$ tabu search moves.) Therefore, we carry on with a first improvement of the tabu search algorithm in order to overcome the problem that the system must introduce long edges as all the short ones are declared tabu.

Table 16.1. Quality of the solutions (averaged over 100 optimization runs each) created by our increasingly enhanced variants of tabu search (as described in Sects. 16.1–16.4) for the PCB442 TSP instance

Section	Minimum	Maximum	Mean value ± error
16.1	51,855.922	54,667.465	53,338.336 ± 49.9
16.2	51,798.410	53,790.672	52,602.840 ± 36.0
16.3	51,222.250	53,218.211	52,069.723 ± 38.5
16.4	51,245.555	52,954.289	51,967.270 ± 38.8

16.2 Introduction of Short-Term Memory

So far, we have used a so-called long-term memory, i. e., if some property got the tabu seal, then it was tabu forever. But as we saw, this approach is only successful for a rather limited time, after which it leads to deteriorations. Therefore, the system would be better off forgetting after some time that some property has been declared tabu. This approach is usually called short-term memory. Of course, we now have an additional tuning parameter τ for the tabu search algorithm that marks the number of time steps after which a property that has been declared tabu at some time t is no longer tabu. In our implementation, we simply set $\tau = 10N$, without much testing.

Figure 16.2 shows how the results for the deviation and the best-so-far deviation change when using a short-term memory. The system no longer freezes but is able to arrive at better and better solutions. The comparison of the results in Table 16.1 yields that a short-term memory is superior to a long-term memory, at least for the problem we study here. Now we want to improve our results even further by adding further ingredients of the tabu search algorithm.

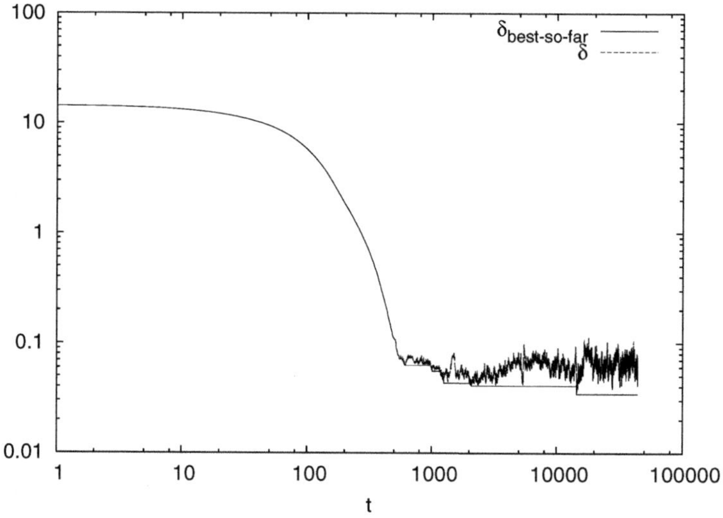

Fig. 16.2. Development of the deviation δ and of the best-so-far deviation $\delta_{\text{best-so-far}}$ vs. time t for the tabu search application as described in Sect. 16.2 for the PCB442 TSP instance

16.3 Adding some Aspiration

One of the key ingredients in tabu search is the aspiration criterion. This criterion defines whether a tabu seal may be overlooked such that a property that is marked as tabu can become a part of the new configuration. A widely used aspiration criterion is the test of whether the introduction of properties that are tabu would lead to a configuration that is better than all other configurations visited so far. We added this type of an aspiration criterion to our program and got the results shown in Fig. 16.3 and Table 16.1. We find that the introduction of this aspiration criterion leads indeed to a further improvement in the results. However, we find that $\delta_{\text{best-so-far}}$ stays constant for rather long times, such that we want to add further ingredients of tabu search in order to get an improvement here, too.

16.4 Adding Intensification and Diversification

Two key ingredients of tabu search are still missing in our approach: intensification and diversification. The concept of intensification requires that the system be somehow forced to reinvestigate the neighborhood of a very good configuration that had already been visited by the system before. This method is mostly applied when the optimization run has not found a new best-so-far solution for several sweeps. Then the system is either forced to perform a series of moves leading the system back into the neighborhood of

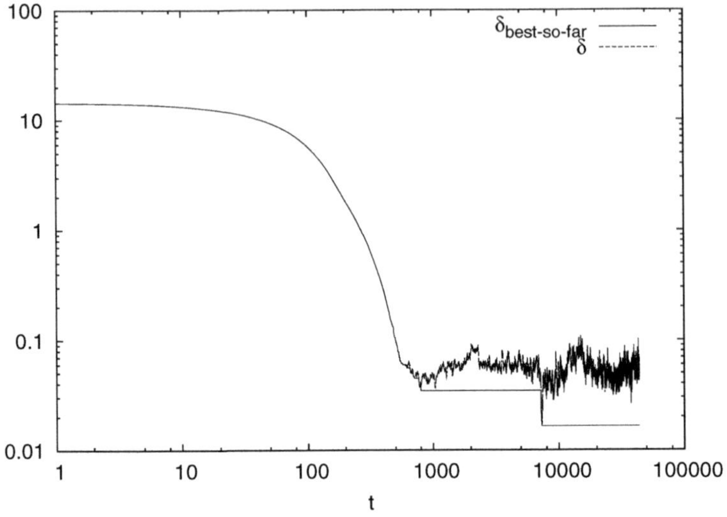

Fig. 16.3. Development of the deviation δ and of the best-so-far deviation $\delta_{\text{best-so-far}}$ vs. time t for the tabu search application as described in Sect. 16.3 for the PCB442 TSP instance

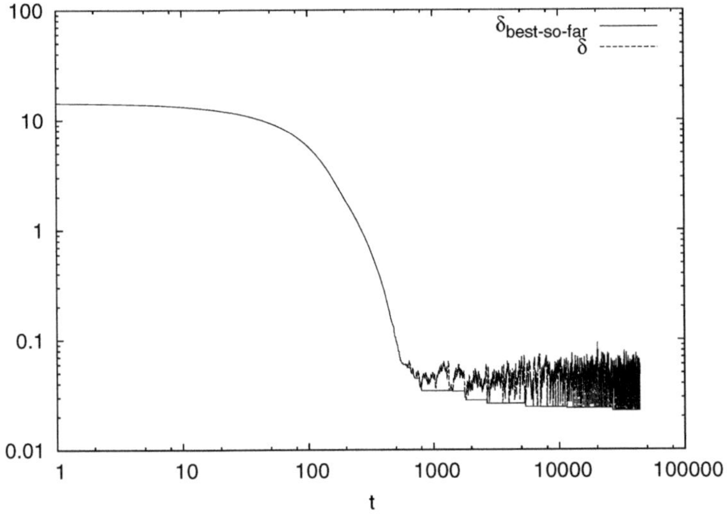

Fig. 16.4. Development of the deviation δ and of the best-so-far deviation $\delta_{\text{best-so-far}}$ vs. time t for the tabu search application as described in Sect. 16.4 for the PCB442 TSP instance

one of the best solutions or to jump immediately to one of these solutions. This intensification approach is often combined with a diversification strategy, according to which the system does not jump to, e. g., the desired, stored good configuration but to, e. g., a configuration that is rather similar to this good configuration. Usually, this configuration was created by making some changes in the stored configuration.

In our simple approach, we check after each sweep whether a new best-so-far solution was found in this sweep. If not, then the system jumps to a configuration created by the application of a Ruin & Recreate move to the best-so-far solution. Here we use the random ruin strategy, which removes 10% of the nodes from the system. The nodes to be removed are chosen at random. They are then reinserted by the bestinsertion heuristic.

Figure 16.4 shows that the quality of the best solution increases gradually with time and that the system is not at all frozen after 100 sweeps. But taking the results after this time, as in the other three variants, we find that we obtain here the best results on average, as can be seen in Table 16.1.

Of course, the results can be further improved by spending more calculation time here, by using better memory, aspiration, intensification, and diversification strategies, and by tuning the various parameters of tabu search. As with genetic algorithms, an entire book the size of this one could be filled with elaborate strategies for tabu search, but we end here.

17 Application of History Algorithms to TSP

In this chapter, we study the application of some exemplary algorithms to the traveling salesman problem (TSP) in which the information about all or many previously visited configurations influences the choice of whether a new move is accepted or rejected. These algorithms are either physically motivated like the multicanonical algorithm (MUCA), multicanonical annealing (MUCANN), and the simulated annealing (SA) variant acceptance simulated annealing (ASA) or motivated by the desire to change the energy landscape, as is the case for guided local search (GLS). All of these algorithms memorize in some way previous energies, properties, former configurations, or moves performed. As a result of such memorization, one might want to count these algorithms as variants of tabu search, although they do not exhibit all typical properties of a tabu search implementation such as working both with intensification and diversification strategies.

17.1 The Multicanonical Algorithm

There are various ways of applying the MUCA. We want to study here probably the simplest application by working directly with a histogram $n(\mathcal{H})$ and applying the transition probability

$$p(\sigma \to \tau) = \begin{cases} 1 & \text{if } n(\mathcal{H}(\sigma)) \geq n(\mathcal{H}(\tau)) \,, \\ \dfrac{n(\mathcal{H}(\sigma))}{n(\mathcal{H}(\tau))} & \text{otherwise} \end{cases} \tag{17.1}$$

to the move $\sigma \to \tau$. We initialize all histogram bins with 1 at the very beginning. If a move $\sigma \to \tau$ is accepted, then $n(\mathcal{H}(\tau))$ is incremented by 1; if it is rejected, then $n(\mathcal{H}(\sigma))$ is incremented by 1. In order not to get some overflows, we check after each 10^6 sweeps whether at least one histogram entry has become larger than 10^9. If so, then all histogram entries are halved, i. e., $n \to n/2 + 1$.

We start each simulation with a random configuration. At the beginning, the system will therefore be in the range of the completely unordered and high-energy configurations. By increasing the histogram entries for these energy values, some pressure is put on the system to leave this energy range and to move to configurations with either smaller or larger energy values until finally the probability for the system to be at a configuration σ with some energy value $\mathcal{H}(\sigma)$ is the same for all energy values. Of course, this pressure depends strongly on the width $\Delta\mathcal{H}$ of the histogram bins: if the histogram bins are very narrow, then it will take a long time before each of the bins within some wider energy interval is incremented often enough to induce a general pressure to leave this wider energy interval. If the bins are too wide, then some random walks (RWs) might be performed within the energy intervals of the bins, until finally some move leads to a configuration in the energy interval of a new bin. Thus, we may expect that there is an optimum intermediate energy width $\Delta\mathcal{H}$ of a bin.

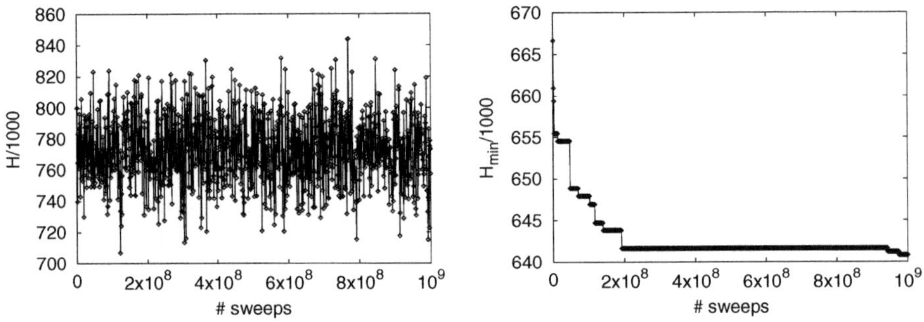

Fig. 17.1. Development of the energy (*left*) and the minimum energy found (*right*) in one simulation run applying MUCA to the PCB442 instance vs. time measured in number of sweeps with a bin width of $\Delta\mathcal{H} = 1$. The energy values (*left*) were always recorded after 10^6 sweeps; the minimum energy was recorded continuously. The energy fluctuates in the range of the energies of random configurations, and the minimum energy found decreases very slowly over time

We applied this algorithm to the PCB442 instance and monitored both the best energy found and the current energy of the system. We found that the best energy found decreased very slowly over time. The simulation run for the results shown in Fig. 17.1 took a few days. Here we used a narrow bin width of $\Delta\mathcal{H} = 1$. One might wonder whether a wider bin width would lead to a significant speedup of this algorithm, leading to much better solutions in shorter times.

As these simulation runs take such a long time, we only performed ten simulation runs for each of the bin widths $\Delta\mathcal{H} = 1$, 10, 10^2, 10^3, and 10^4. Furthermore, we stopped these runs after an overall number of 10^9 sweeps each. We clearly find in Fig. 17.2 a dependency of the best result achieved on the bin width $\Delta\mathcal{H}$. Furthermore, the minimum energy value found decreases linearly if plotted vs. a logarithmic time scale, but only for the time scales shown here, which seem to be rather short for the algorithm, as the results obtained are not very good but are more typical of random configurations.

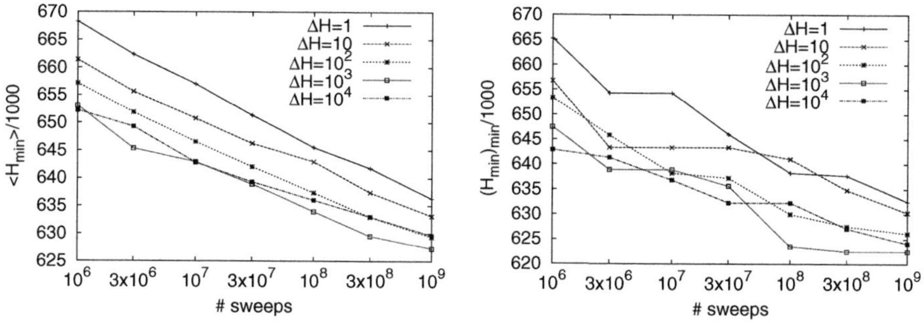

Fig. 17.2. Development of the minimum energy found in time (measured in sweeps) for 10 simulation runs applying MUCA to the PCB442 instance with various bin widths $\Delta\mathcal{H} = 1,\ 10,\ 10^2,\ 10^3$, or 10^4. *Left*: averages over minimum energies; *right*: minima of minima. The best results are achieved for the intermediate value $\Delta\mathcal{H} = 10^3$. Note that the optimum value for this TSP instance is $\approx 51 \times 10^3$ and that the energy range of its random configurations lies around 773×10^3

We believe that this decrease will not stay linear but will slow down even further, as the number of states grows exponentially with increasing energy for small energy values, such that it is quite unlikely to choose a move leading to a better configuration if already being at a quite good configuration. But, assuming that the decrease will stay linear on a logarithmic time scale, we need roughly 10^{76} sweeps on average in order to reach the optimum using a bin width $\Delta\mathcal{H} = 10^3$ such that it is useless to search for the optimum with this implementation of MUCA.

We find that increasing $\Delta\mathcal{H}$ from 1 to 10^3 yields improved minimum results. Obviously, a widening of the histrogram bins leads to an increased pressure on the system to leave the area of the random configurations. This can be easily explained as many more bins must be filled up in order to generate a hill in the histogram from which the system can roll downwards, if more but narrower bins are used. However, we get the best results for $\Delta\mathcal{H} = 10^3$; the results are already worse for $\Delta\mathcal{H} = 10^4$. As mentioned before, this worsening is due to the fact that many moves lead to a tentative new configuration that lies in the same energy bin as the current configuration if using such wide energy intervals such that the system performs a RW within this energy bin.

Summarizing, the MUCA is not suitable for application to the TSP in this version. Generally, when working with the MUCA, the histogram or the microcanonical inverse temperature $\beta(\mathcal{H})$ and the fugacity $\alpha(\mathcal{H})$ must be determined. For our purposes here we assumed that they are not known a priori, so that we had to sum up the histogram during the optimization run. Usually, for new optimization problems occurring in real life, one must choose the approach introduced here if one wants to work with the MUCA. Sometimes other algorithms like parallel tempering in REMUCA are used to determine $\alpha(\mathcal{H})$ and $\beta(\mathcal{H})$ with these before starting the MUCA run. Often some windowing methods are also used that determine $\alpha(\mathcal{H})$ and $\beta(\mathcal{H})$ for several "windows" of energy intervals and then merge the results at the interval boundaries. There exist also some applications that try to guess overall functions $\alpha(\mathcal{H})$ and $\beta(\mathcal{H})$ from the knowledge achieved so far in the MUCA simulation by summing up the histogram values and iterate this guessing procedure several times. The windowing and the iterative methods require much calculation time before the functions are derived, so they are often unsuitable for optimization purposes. If, however, the functions $\alpha(\mathcal{H})$ and $\beta(\mathcal{H})$, or at least good guesses for these functions, are known beforehand, then one might be able to perform optimization runs based on this knowledge with a reasonable amount of calculation time.

17.2 Multicanonical Annealing

Multicanonical annealing (MUCANN) is a combination of the MUCA and the great deluge algorithm (GDA). Basically, the same acceptance criterion as in the MUCA is used, except that, if the energy of the tentative new configuration is larger than a certain value $\mathcal{T}$, then the move is always rejected. Thus, we have an acceptance criterion as follows:

$$p(\sigma \to \tau) = \begin{cases} 1 & \text{if } n(\mathcal{H}(\sigma)) \geq n(\mathcal{H}(\tau)) \text{ and } \mathcal{H}(\tau) \leq \mathcal{T}\,, \\ \dfrac{n(\mathcal{H}(\sigma))}{n(\mathcal{H}(\tau))} & \text{if } n(\mathcal{H}(\sigma)) < n(\mathcal{H}(\tau)) \text{ and } \mathcal{H}(\tau) \leq \mathcal{T}\,, \\ 0 & \text{if } \mathcal{H}(\tau) > \mathcal{T}\,. \end{cases} \tag{17.2}$$

One can consider the GDA to be a special case of MUCANN with only one bin in the histogram covering the whole energy range between the best and the worst configuration. We will thus use "$\Delta\mathcal{H} = \infty$" here as a synonym for the GDA.

We use the same approach for multicanonical annealing as we used for the MUCA earlier, but we additionally add this water level $\mathcal{T}$: the initial value of $\mathcal{T}$ is set to the energy of the initial random configuration. After every 10^6 sweeps, $\mathcal{T}$ is lowered by a factor of 0.99. If the new value of $\mathcal{T}$ is smaller than the energy of the current configuration, then $\mathcal{T}$ is increased to this energy value. Thus, we use the same approach for decreasing $\mathcal{T}$ as we use in the GDA. As already mentioned in Sect. 12.4 in Part I and Sect. 9.3 in Part II for the GDA, using this cooling method the water level $\mathcal{T}$ is decreased very slowly.

Just as with the MUCA, we start out our investigation of multicanonical annealing with the development of the current energy and of the minimum energy found in a simulation run. Figure 17.3 shows a behavior of these two values that is very different from the behavior of these observables when working with the original MUCA: here we find a nice decrease in these observables over time. Finally, they freeze at some very good values; the minimum energy

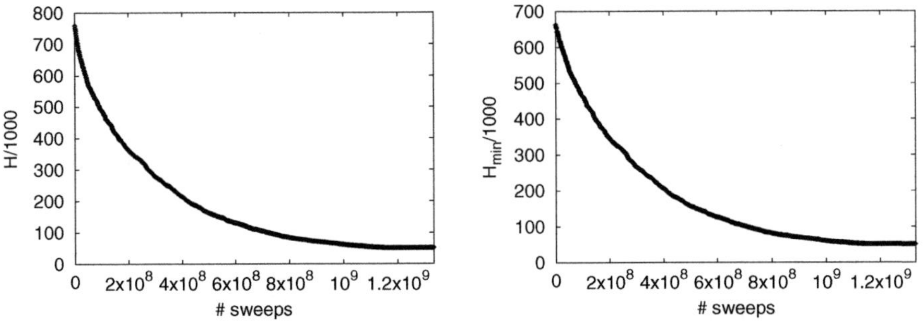

Fig. 17.3. Development of the energy (*left*) and the minimum energy found (*right*) in one simulation run applying MUCANN to the PCB442 instance vs. time measured in number of sweeps with a bin width of $\Delta\mathcal{H} = 1$. The energy values (*left*) were recorded always after 10^6 sweeps; the minimum energy was recorded continuously. Both the current energy and the minimum energy decrease and finally freeze at some very low energy values

found in this simulation run is 50,839.132 and the final value of the energy is 50,845.4896. Thus, it makes sense also to save the minimum energy found so far. This is not much of an additional effort as one must work with total energies anyway when using this algorithm. Despite the fact that this simulation run ended at a good solution, we must keep in mind that it took $1.16 \cdot 10^9$ sweeps in this run to freeze into a local minimum; such a result could be easily achieved with SA with less calculation time.

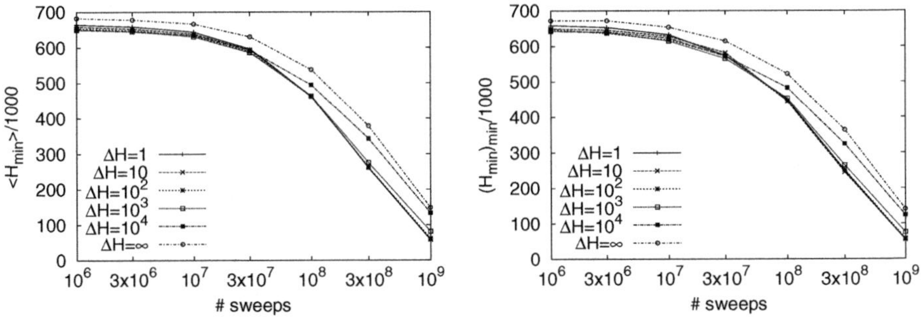

Fig. 17.4. Development of the minimum energy found in time (measured in sweeps) for 10 simulation runs applying MUCANN to the PCB442 instance with various bin widths $\Delta\mathcal{H} = 1$, 10, 10^2, 10^3, 10^4, or ∞ ($\Delta\mathcal{H} = \infty$ corresponds to the GDA). *Left*: averages over minimum energies; *right*: minima of minima

Next we want to find out whether a faster convergence can be achieved by using wider energy intervals for the bins in the histogram. The results for some given calculation times and for various bin widths are shown in Fig. 17.4. We find again a dependency on the bin width: for relatively short calculation times of "only" 10^6 sweeps, we get the best results with a bin width of $\Delta\mathcal{H} = 10^3$, followed by the bin width $\Delta\mathcal{H} = 10^4$. But at these short calculation times, the results are hardly better than those of the MUCA. Spending 10^9 sweeps, however, we get the best results for the narrow bin widths $\Delta\mathcal{H} = 1$ and $\Delta\mathcal{H} = 10$. The worst results are achieved for $\Delta\mathcal{H} = \infty$, i. e., for the original GDA.

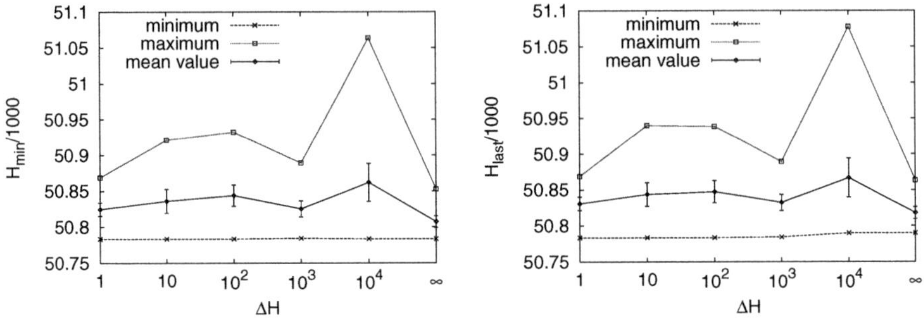

Fig. 17.5. Minimum energies found (*left*) and energies of the last configurations, when the system is already frozen (*right*), in 10 simulation runs applying MUCANN or the GDA to the PCB442 instance. For MUCANN, the bin widths $\Delta\mathcal{H} = 1$, 10, 10^2, 10^3, or 10^4 were used. $\Delta\mathcal{H} = \infty$ corresponds to the GDA

However, after these 10^9 sweeps, the system is not yet frozen; the system needs "slightly" more than 10^9 sweeps to freeze into a local minimum. We consider the system to be frozen when the water level is identical with the current energy of the system and when there is no change in the energy in the following 10^7 sweeps. Figure 17.5 shows the energies of the last configurations of the simulation runs. We find that the energies of the configurations in which the simulation runs freeze do not differ strongly from the best energies found during the optimization runs, and sometimes they are exactly the same. The GDA leads on average both to the best minimum and to the best final results, followed by MUCANN with $\Delta\mathcal{H} = 1$ and $\Delta\mathcal{H} = 10^3$. However, only when using a bin width of $\Delta\mathcal{H} \leq 10^2$ does the system freeze in at least one of ten simulation runs in an optimum configuration. Some simulations applying the GDA or MUCANN with larger bin widths were able to find an optimum configuration, but they froze in slightly worse configurations.

Finally, we consider the number of sweeps the system actually needs in order to freeze. As Fig. 17.6 shows, this time strongly increases with an increasing bin width and is maximal for the original GDA, for which it takes nearly twice as much time as for multicanonical annealing with $\Delta\mathcal{H} = 1$.

Summarizing, we recommend using rather narrow bins in the histogram if applying MUCANN as the results are nearly as good as for the GDA but only half of the calculation time is needed for convergence. MUCANN can also be

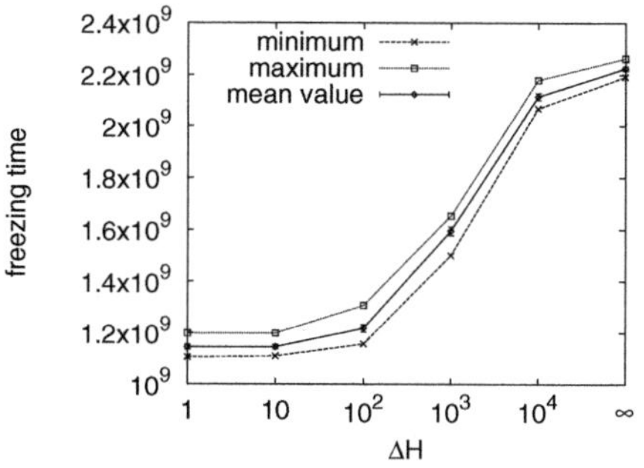

Fig. 17.6. Freezing time in number of sweeps for ten simulation runs applying MUCANN or the GDA to the PCB442 instance. For MUCANN, the bin widths $\Delta\mathcal{H} = 1$, 10, 10^2, 10^3, or 10^4 were used. $\Delta\mathcal{H} = \infty$ corresponds to the GDA

considered from the point of view of the GDA: in the GDA, the decreasing water level presses the system down to better and better configurations. Although the system is allowed to perform a restricted RW below the water level in the GDA, it always likes to stay slightly below the water level. In MUCANN, an additional pressure is induced to go downwards, as the histogram entries of the energies of the configurations slightly below the water level grow. This is the reason for why we get the shortest freezing times for the narrowest histogram bins $\Delta\mathcal{H} = 1$ and $\Delta\mathcal{H} = 10$, as there the pressure is naturally strongest: hardly any nontrivial move will lead to a configuration in the same energy interval. Nevertheless, the speedup is only a factor of ≈ 2 by this additional effect such that MUCANN is still unable to compete with algorithms like SA.

17.3 Acceptance Simulated Annealing

ASA is a SA variant introduced by Puchta [167]. As in SA, a move is accepted with the Metropolis acceptance probability, which depends both on the energy difference between the current and the tentative new configuration and on the control parameter, the temperature T of the system. In contrast to standard SA, this control parameter is not simply given from outside but is determined via the last m improvements the system has found: here the control parameter is the probability p with which a move leading to a worse configuration shall be accepted. Instead of a cooling schedule for the temperature T, one has here a decreasing schedule for the probability p, i. e., p is decreased exponentially by a factor of, e. g., 0.99 from an initial value of 1 to some small final value, e. g., to 10^{-5}. Let the average over the energy differences of the last m improving moves be $\overline{\Delta\mathcal{H}}$. Then p will be the probability with which a move will be accepted that leads to a new configuration whose energy is $|\overline{\Delta\mathcal{H}}|$ worse than the energy of the current configuration. For this purpose, the temperature T is set to the value

$$T = \frac{\overline{\Delta\mathcal{H}}}{\ln(p)}\,. \tag{17.3}$$

In the implementation of this algorithm, first, a short RW must be performed that must be long enough to memorize the energy differences of m improving moves. With the average of these energy differences an initial temperature can be calculated. Then, at each value of p, several sweeps are performed. If a move leads to an improvement, then its energy difference is stored and replaces the energy difference of the oldest move still in the memory. Thus, the mean value of these stored energy differences changes, leading to a new value of the temperature T, which is immediately updated. At the end, some greedy sweeps are performed to guide the system to a local minimum if it is not yet frozen. The number of greedy sweeps is identical with the number of sweeps per p-step. For small m, we use the same amount of sweeps in the initial RW. If, however, m is so large that the memory cannot be filled within this time, then $3m$ moves are performed in the RW.

Thus, ASA determines the temperature in an adaptive way: initially, the average energy difference of moves in the RW determines the temperature. With decreasing p, the temperature T decreases, leading the system to a range of more ordered configurations. The energy differences of improving moves in this regime might be different from those in the regime of the RW. Of course, this might not be true for any system, i. e., there might be systems in which the mean value of these energy differences stays constant. However, if these energy differences change according to the amount of ordering of the configurations taking part in the moves, then ASA will react to that: if the energy differences become absolutely smaller in size, then the temperature will be lowered additionally according to this effect. If they become absolutely much larger, then the temperature is increased intermediately despite the constant or just decreased control parameter p.

First, we want to investigate the behavior of the algorithm. Figure 17.7 shows for the PCB442 instance what temperatures are chosen by ASA with various values of m at some probability value p if only a small amount of calculation time is used. In order not to burden the reader with too many fluctuations, only the final temperature value T at each p is plotted. We find that the behavior of the $T(p)$ curve is different for various m: for $m = 1$, one can clearly distinguish three scenarios: at large values of p, T decreases and fluctuates strongly such that this decrease can be fitted either with $\approx 10^4 p^3$ or with $\approx 200\exp(4p)$. At very small p, no more moves are accepted, such that the average energy difference remains constant. Thus, we get here $\approx -3 \times 10^{-14}/\ln(p)$. Between these large- and small-p scenarios we get a wide transition in which the curve jumps up and down. These fluctuations occur due to the fact that here only the energy difference of the last move that led to an improvement is used. Already for $m = 3$, we find that the fluctuations are mostly suppressed. At

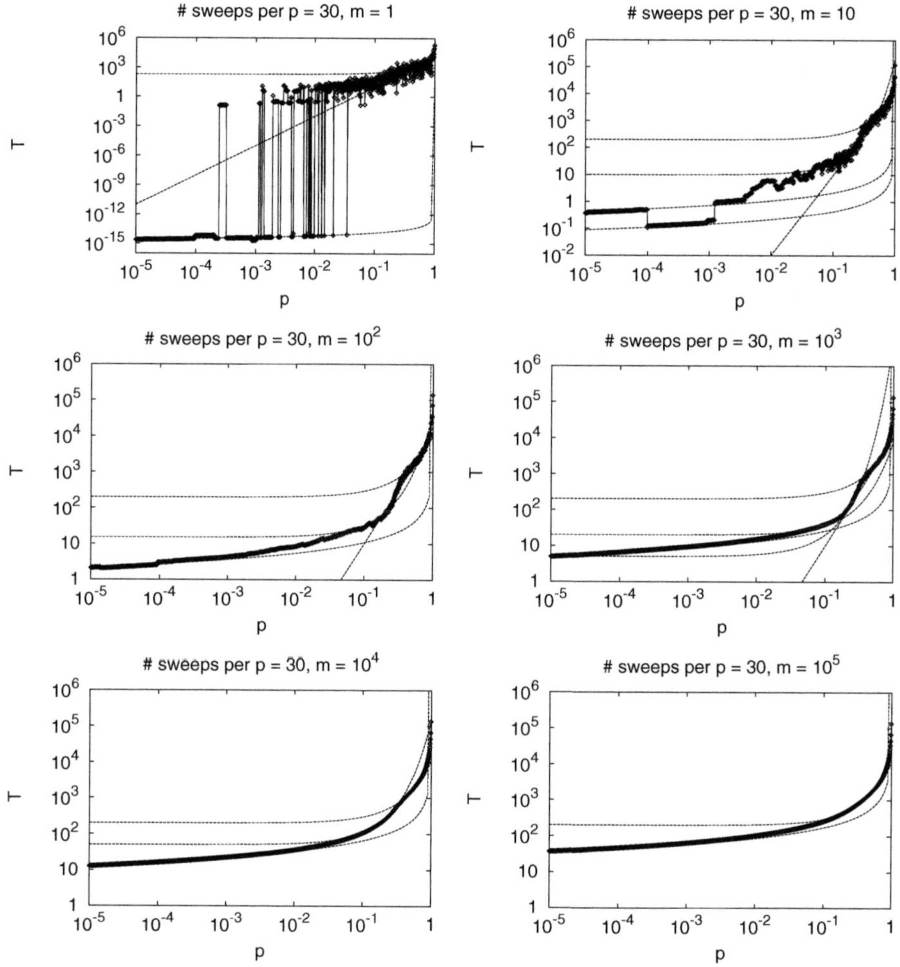

Fig. 17.7. Final temperature T chosen by the ASA run, which was applied to the PCB442 instance, for each probability value p. The behavior of ASA for a short calculation time (30 sweeps per probability step) and various numbers m of stored energy differences can be approximated stepwise with fit functions as described in the text

$m = 3$, we can again fit the large-p behavior with the fit function $\approx 10^4 p^3$. But here we already find that this is an overall fit for this area, not taking more detailed structures into account. We can also fit the curve with $\approx 200 \exp(4p)$ for $0.4 \leq p \leq 0.85$ and $\approx 10 \exp(10p)$ for $0.05 \leq p \leq 0.25$. For very small p, when the system is frozen, we get $\approx -1/\ln(p)$ for $10^{-4} \leq p \leq 10^{-3}$ and $\approx -5/\ln(p)$ otherwise. For $m = 100$, one can still describe the large-p behavior with the fit function $\approx 10^4 p^3$. But one would clearly neglect the more detailed structure visible here, as there are hardly any fluctuations left.

We get again $\approx 200\exp(4p)$ for $0.4 \leq p \leq 0.85$ and additionally $\approx 15\exp(8p)$ for $0.1 \leq p \leq 0.25$. There is a smooth transition between these fitting curves. For small p, we get $\approx -25/\ln(p)$, i. e., again the const$/\ln(p)$ behavior. Considering next $m = 10^3$, we get again the curve $\approx 10^4 p^3$ as a rough estimate for large p. Again we can fit the curve better stepwise with exponential functions: we get $\approx 200\exp(4p)$ for $0.4 \leq p \leq 0.75$, $\approx 5\exp(14p)$ for $0.2 \leq p \leq 0.35$, and $\approx 20\exp(6p)$. For small p, the curve converges to $-60/\ln(p)$. For $m = 10^4$, we do not see any possibility for a p^3 fit anymore. Here we get $\approx 200\exp(4p)$ for $0.4 \leq p \leq 0.7$, $\approx 50\exp(8p)$ for $0.1 \leq p \leq 0.3$, and $\approx -150/\ln(p)$ for $10^{-5} \leq p \leq 3\times 10^{-3}$. Finally, we get for $m = 10^5$ the fit curves $\approx 200\exp(4p)$ for $0.25 \leq p \leq 0.75$ and $\approx -400/\ln(p)$ for $10^{-5} \leq p \leq 3 \times 10^{-3}$.

Summarizing these results, we can distinguish between the special case $m = 1$, in which there occurs a breakdown to very small values of T, and the case for $m \geq 10$, in which we do not see such a breakdown. In all cases, the initial value of the temperature is much larger than the freezing and the critical temperature of the PCB442 instance. The breakdown for $m = 1$ to $T(p) \approx -3 \times 10^{-14}/\ln(p)$ leads to much too small temperatures at the end, and the system is quenched down in the case $m = 1$. This small value of -3×10^{-14} is due to the fact that also trivial moves with an energy difference of $\Delta\mathcal{H} = 0$ can occur. Due to rounding errors (as usual, we calculate with REAL*8 here), we get instead of $\Delta\mathcal{H} = 0$ a slightly negative energy difference, which leads to this minuscule temperature. On the other hand, the simulations with large m led to a final temperature, which was much too large. One must choose a final temperature smaller than 1 when working with SA on the PCB442 instance. The final 30 sweeps in the greedy mode are otherwise not sufficient to lead the system to a good local minimum within the valley of the energy landscape in which the system currently stays.

Now let us consider the corresponding curves for long calculation times, shown in Fig. 17.8. Here we use again 1145 steps, over which the probability is reduced from 1 to 10^{-5} by a factor of 0.99, and an additional greedy step at the end. Here 30000 sweeps per probability step are performed. For $m = 1$, we get the same fit functions as above; we can again fit the data with either $\approx 200\exp(4p)$ in the range $0.2 \leq p \leq 0.9$ or, even better, with $\approx 10^4 p^3$ even for $0.1 \leq p \leq 1$. Then we again get a fluctuation regime, which is located here at $0.05 \leq p \leq 0.1$. After that the temperature breaks down and can be fitted with $\approx -3\times 10^{-14}/\ln(p)$. Thus, the fit functions stay the same in the case of $m = 1$ when increasing the calculation time by a factor of 1000. But we get a different picture for $m = 10$. Here we can again fit the large-p behavior with either $\approx 10^4 p^3$ for $0.15 \leq p \leq 0.95$ or with $\approx 200\exp(4p)$ for $0.3 \leq p \leq 0.9$.

For small p, we see a breakdown of the $T(p)$ curve but without the fluctuations as with $m = 1$. However, the data points can be fitted again with $\approx -3 \times 10^{-14}\ln(p)$ for $10^{-5} \leq p \leq 0.1$. For $m = 10^2$, we get exactly the same fit functions as for $m = 10$. For $m = 10^3$, we get again roughly the same behavior, with the only exception being that the breakdown is slightly

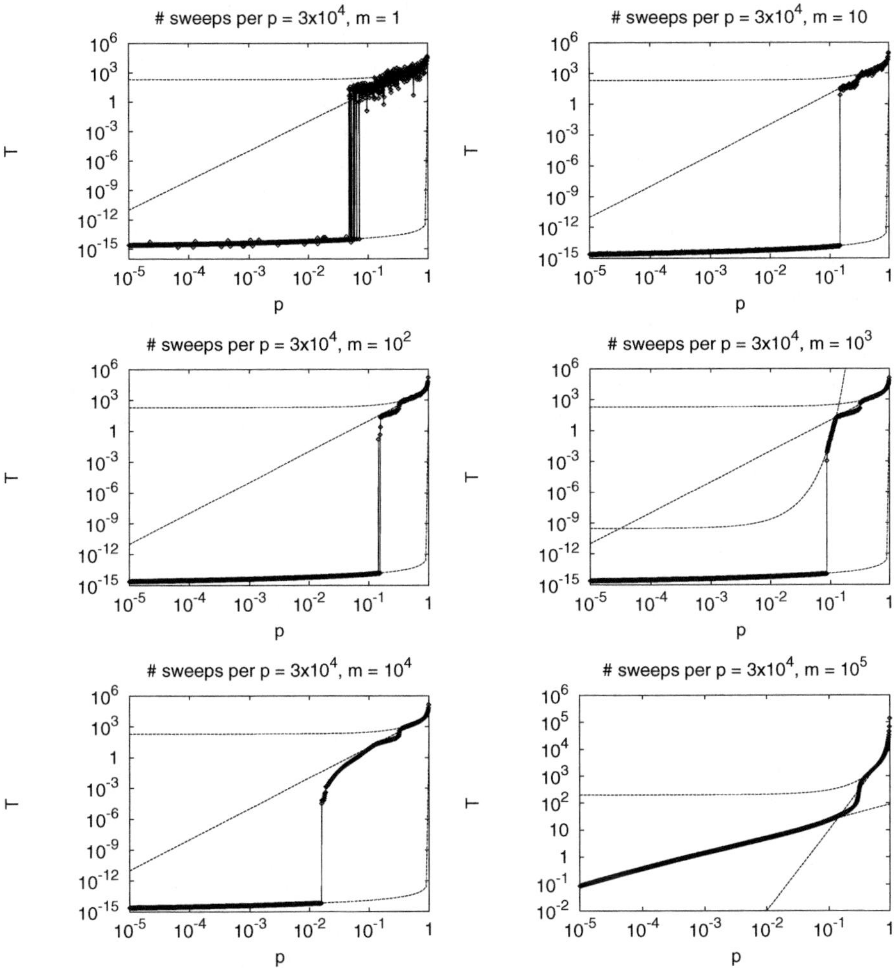

Fig. 17.8. Final temperature T chosen by the ASA run, which was applied to the PCB442 instance, for each probability value p. The behavior of ASA for a long calculation time (30000 sweeps per probability step) and various numbers m of stored energy differences can be approximated stepwise with fit functions as described in the text

delayed here—it happens now after a short steep decrease of the curve that can be fitted to $\approx 3 \times 10^{-10} \exp(200p)$. For $m = 10^4$, we find again all of these fit functions, except for the additional steep decrease, which was found only for $m = 10^3$. Instead of this steep decrease, we find that the breakdown is delayed even more and is much smoother, such that it can be described here as well with the p^3 behavior. For $m = 10^5$, we can describe the behavior for $0.35 \leq p \leq 0.75$ with again either $\approx 10^4 p^3$ or $\approx 200 \exp(4p)$. But then there is no breakdown afterward but a sigmoidal transition to a $\approx 90p^{0.6}$ curve, which describes the behavior for $10^{-5} \leq p \leq 0.1$.

Summarizing these results as well, we find the breakdown that we saw only for $m = 1$ as we used 30 sweeps now for $m \leq 10^4$. Thus, the transition of the behavior of the algorithm is shifted with increasing calculation time. Especially at small temperatures, when the system is frozen such that no new improving moves can be found anymore, one finds a $1/\ln(p)$ behavior as expected. However, we find an interesting power law decrease for $m = 10^5$ at small p.

As p is decreased exponentially in time and as all graphics in Figs. 17.7 and 17.8 are plotted with logarithmic p-axis, we can read all the curves also as functions of time as $t \propto |\ln(p)|$. We get only in the case $m = 10$ and 30,000 sweeps per step the exponential cooling schedule with appropriate initial and final values for the temperature, which is what we roughly use as well. The question now is: To what results can this standard and the other chosen cooling schedules $T(t)$ lead?

Thus, now we investigate the quality of the results achieved with ASA. Figure 17.9 shows the results for the PCB442 instance, for various values of m, and for various calculation times. Additionally, the results for SA from Sect. 7.4, which were achieved in the same calculation times and also by using a factor of 0.99, are redrawn with the label $m = 0$. To visualize the differences for very good results, we do not show the results themselves but the deviation of these results from the optimum given by $\delta_{\text{mean}} = (\langle\mathcal{H}\rangle - \mathcal{H}_{\text{opt}})/\mathcal{H}_{\text{opt}}$, as defined in formula (8.1) in Sect. 8.1. For all values of m, we generally find that the results for δ_{mean} improve with an increasing number of sweeps, i. e., with increasing calculation time. (The overall calculation time in sweeps is given as the product of the number of sweeps per p-step in ASA times the number of these steps, which is 1146.) The global optimum of the PCB442 instance is reached with $m = 3 \times 10^3$ for 10^3 sweeps, $m = 3 \times 10^3$ and $m = 3 \times 10^5$ for 10^4 sweeps, and $m = 10^4$, 3×10^4, 10^5, and 10^6 for 3×10^4 sweeps,

However, we also clearly see that there is no general best value for the number m of stored energy differences: obviously, one must use small m for short calculations, as the best results are on average achieved with $m \leq 10^2$. This boundary might depend on the system size. The worst results at these short calculation times are achieved with the largest values of m, $m = 10^6$ and $m = 10^7$, whose mean values coincide at the beginning. On the other hand, we find for intermediate calculation times, like 30 or 100 sweeps per step, that one should also use an intermediate number of stored energy differences: here $m = 10^4$ leads to the best results, after $m = 10^3$ has led to the best results for 10 sweeps per step. Increasing the calculation time further, the optimum value for m is increased as well: $m = 10^5$ leads to the best results for 10^3 sweeps per step, $m = 10^6$ for 10^4 sweeps, and $m = 10^7$ for 3×10^4 sweeps. Obviously, we get here a general law: $m = 10^x$ leads to the best results for 10^{x-2} sweeps per step. These optimum values also lead to better results than standard SA at the same calculation time in sweeps. Furthermore, we see that

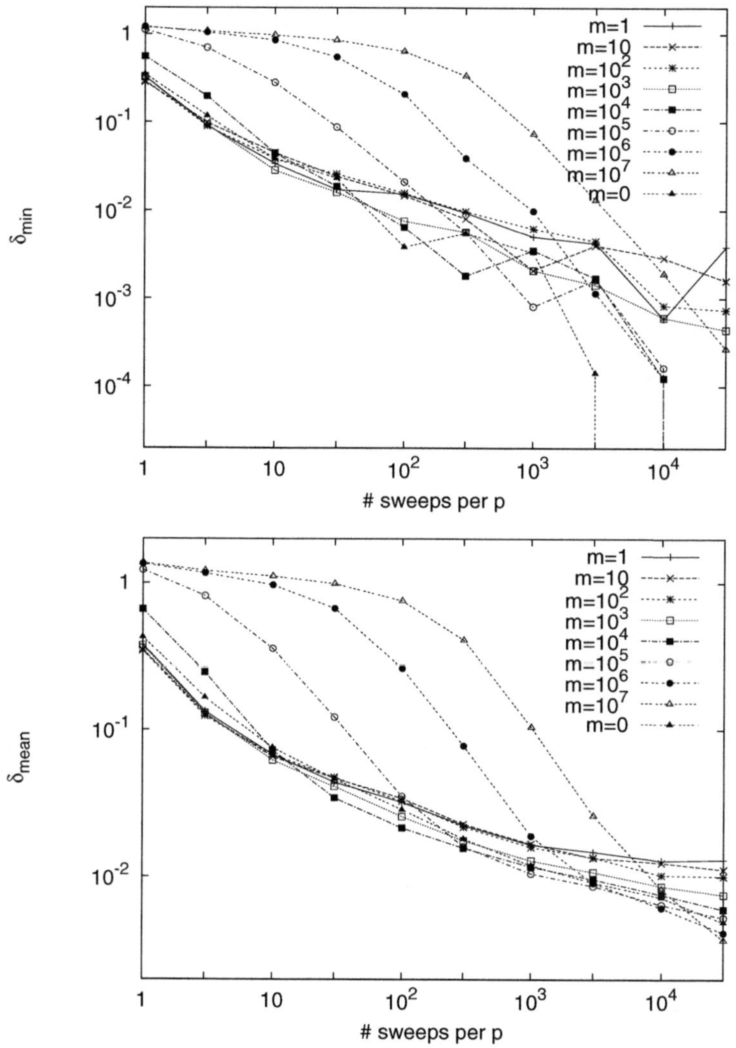

Fig. 17.9. Quality of the results achieved with ASA for the PCB442 instance vs. number of sweeps per probability step (i. e., vs. calculation time) for various numbers m of stored energy differences. *Top*: deviation of the best final configuration out of 100 optimization runs to the global optimum; *bottom*: average deviation of the energies of these 100 final configurations to the global optimum. For comparison, the results for the original SA algorithm ($m = 0$) are shown. *Error bars* are smaller than the *symbols*

one should on no account use $m > 10^x$ for only 10^{x-3} sweeps, as the results are horribly bad there. This is due to the fact that there are larger energy differences still stored from the start of the optimization run such that the final temperature of the system is overestimated. In turn, the system is not cooled down completely, which leads to the worsening of the results. Usually, the results for $m = 10^x$ are only slightly worse than the results for smaller m such that we want to set the boundary between good and bad results at that value. The results described here can be seen even better in Fig. 17.10. Here each curve stands for one proposed calculation time and the question is which m is optimal to use.

Finally, the question arises as to whether these conclusions about the best value and the maximum good value of m for a given number of sweeps depend on the system size N. Therefore, we performed the same set of optimization runs also for the BEER127 instance. Figure 17.11 shows the results for the BEER127 instance, which are plotted in the same way as in Figs. 17.9 and 17.10. We find the same dependencies here, namely, that the optimum value of m is given by the number of sweeps multiplied by a factor of 100 and that one should on no account use a value of m that is larger than 1000 times the number of sweeps. Thus, at least for the range of the system size N we use, this law seems to hold generally.

Summarizing, we studied in this section a highly interesting SA variant that makes use of a memory in which the energy differences of former improving moves are stored in order to determine an appropriate temperature. We found that the size of the memory is crucial for the average quality of the solutions in our application.

As a memory is used in ASA, one might also want to include this algorithm in the wider field of tabu search or to speak of a hybrid algorithm, although

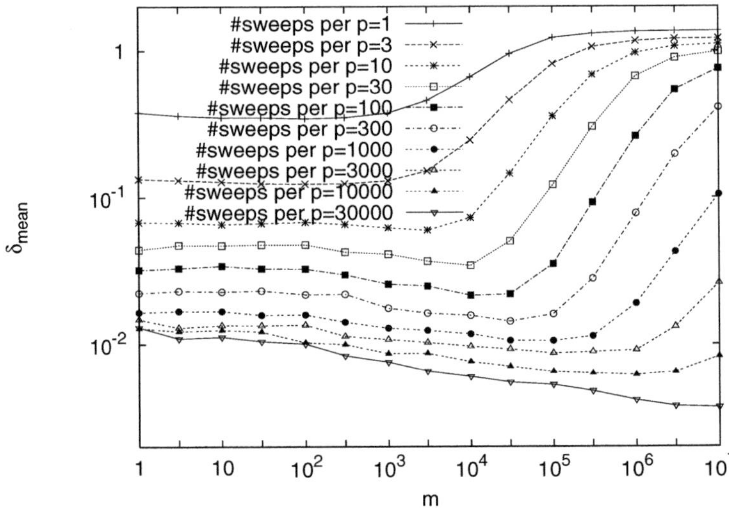

Fig. 17.10. Results redrawn from Fig. 17.9, but now plotted vs. the number m of memorized energy differences

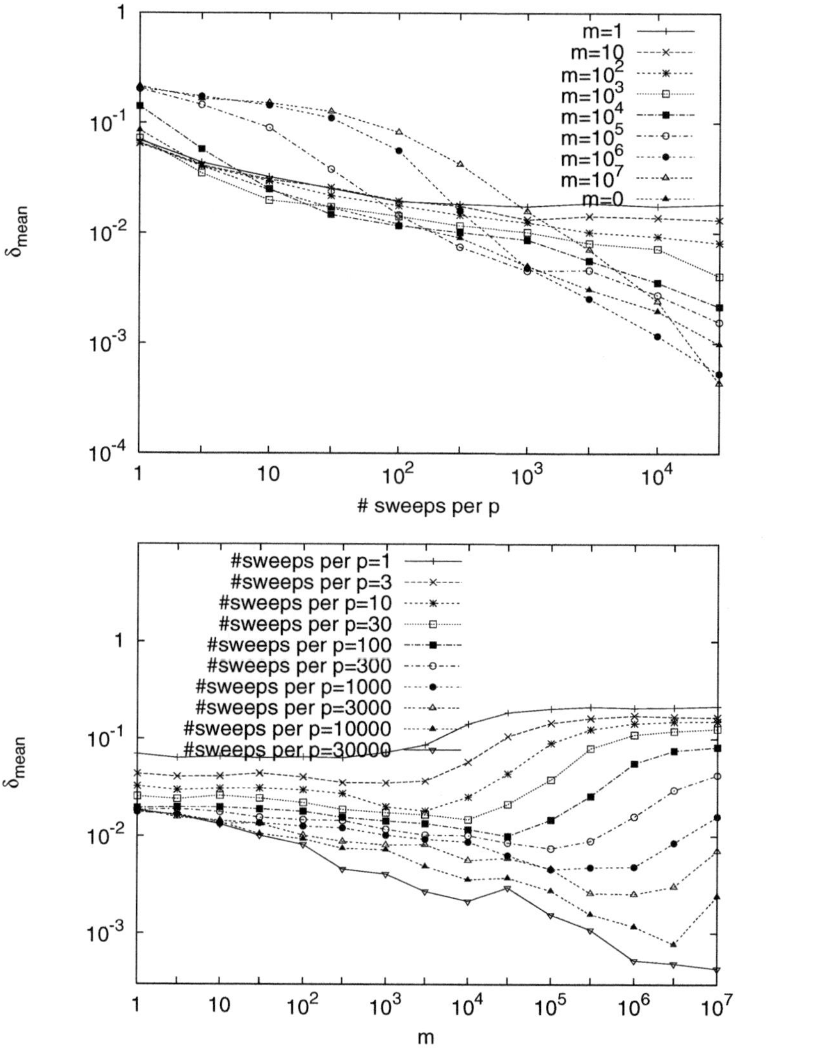

Fig. 17.11. Quality of the results achieved with ASA for the BEER127 instance for various numbers of sweeps per probability step and for various numbers m of stored energy differences. *Top*: the results for δ_{mean} are plotted as in Fig. 17.9; *bottom*: as in Fig. 17.10

basic tabu search strategies like intensification and diversification are missing here. Thus, for us, it is mainly an implementation of a SA variant in which the temperature is determined adaptively.

17.4 Guided Local Search

The application of guided local search (GLS), which was introduced in Sect. 23.1 in Part I, to the TSP is rather straightforward: the features a solution σ does or does not contain are the edges that can occur. The costs assigned to these edges (i, j) are the distance values $D(i, j)$.

We use as a basic algorithm the greedy algorithm, in which a neighboring configuration is selected at random and is accepted if it is either equally good as or better than the current configuration.

Indeed, we can use our original greedy algorithm program and have only to extend it by a further loop over the iterations of the GLS algorithm and by assigning penalty values $p(i, j)$ and utility values $u(i, j)$ to the edges. The distances we use in the greedy algorithm are no longer the original distances $D(i, j)$ but the distances

$$\tilde{D}(i, j) = D(i, j) + \lambda p(i, j) , \tag{17.4}$$

with λ being the Lagrange multiplier by which the penalties $p(i, j)$, which take integer values, are related to the original costs.

Thus, the outline of the GLS algorithm is as follows. First, a random configuration σ is created, the penalty values for all edges are initialized with $p(i, j) = 0$, and the altered distances are set to the original distances, i. e., $\tilde{D}(i, j) = D(i, j)$. As reported in [212], a good value for λ is

$$\lambda = a \times \mathcal{H}_{2-\text{optimum}}/N , \tag{17.5}$$

with $\mathcal{H}_{2-\text{optimum}}$ being the length of a 2-optimum tour, i. e., a solution that cannot be further improved with the Lin-2-opt, N being the number of nodes as usual, and a being a parameter that must be changed to tune the algorithm. Voudouris and Tsang report that the best performance was achieved for $a \approx 0.3$: when a was too small, the algorithm processed too slowly; when it was too large, it converged rather fast to only a handful of good local minima. If a was chosen correctly, then the algorithm generated a sequence of gradually improving solutions. We work with $a = 0.3$ and set $\mathcal{H}_{2-\text{optimum}} = 51{,}000$ for the PCB442 instance for which we present results here, such that we get $\lambda \approx 34.6$.

After this initialization, we start out with a conventional greedy run starting from the configuration σ and using the distance matrix $\tilde{D}$ (which is identical to D in the first iteration) and store the final configuration as τ. Then we derive a symmetric edge matrix η from τ with

$$\eta(i, j) = \begin{cases} 1 & \text{if node } j \text{ is either predecessor or} \\ & \text{successor of node } i \text{ in} \\ & \text{the configuration } \tau , \\ 0 & \text{otherwise} . \end{cases} \tag{17.6}$$

Then we calculate the utility values $u(i,j)$ for all edges (i,j) according to

$$u(i,j) = \frac{\eta(i,j) \times D(i,j)}{1 + p(i,j)}\,. \tag{17.7}$$

We go through all (i,j) and determine the maximum of all $u(i,j)$. That edge (i,j) for which $u(i,j)$ is maximum (we generally restrict to one edge, as it is usually one; if there are more than one, the first edge found with a maximum utility value is punished) is punished by incrementing its penalty value, i. e., $p(i,j) = p(j,i) := 1 + p(i,j)$. Then we set

$$\tilde{D}(i,j) = \tilde{D}(j,i) := D(i,j) + \lambda \times p(i,j)\,. \tag{17.8}$$

After that we set $\sigma := \tau$ and perform the next greedy optimization run, which hopefully removes the punished bad edge (i,j). We end up at a new final configuration τ and proceed as above.

This algorithm is iterated over and over. In each iteration, the cost function $\mathcal{H}$ is to be minimized according to a new distance matrix $\tilde{D}$,

$$\begin{aligned} \mathcal{H}(\sigma) &= \sum_{i=1}^{N} \tilde{D}(\sigma(i), \sigma(i+1)) \\ &= \sum_{i=1}^{N} D(\sigma(i), \sigma(i+1)) + \lambda \times \sum_{i=1}^{N} p(\sigma(i), \sigma(i+1)) \\ &= \mathcal{H}_0(\sigma) + \lambda \times \sum_{i=1}^{N} p(\sigma(i), \sigma(i+1))\,, \end{aligned} \tag{17.9}$$

with $\sigma(N+1) \equiv \sigma(1)$ and the original cost function $\mathcal{H}_0$. It might be that the algorithm finds on its way a rather good solution according to the original cost function $\mathcal{H}_0$ but gets to worse solutions in the next iterations. Thus, we always store the best final configuration τ found so far and return this τ_{best} with its cost value $\mathcal{H}_{0,bsf} = \mathcal{H}_0(\tau_{\text{best}})$ as the result of the optimization run.

In our implementation, first a random configuration is created. This configuration is improved in 10^5 greedy sweeps in the original energy landscape. Then a loop over $100 \times N^2$ GLS iterations is performed in which first an edge is selected for which the penalty value is incremented according to the description above. Second, 30 greedy sweeps with the newly changed distance matrix $\tilde{D}$ are performed in order to get the system into the perhaps new local minimum. Finally, 30 greedy sweeps with the original distance matrix D are performed.

Figure 17.12 shows the development of the observables $\mathcal{H}$, $\mathcal{H}_0$, and $\mathcal{H}_{0,\text{bsf}}$ over time. The time is measured in the number of GLS iterations. From this one can get the time in sweeps by multiplying the iteration number by 30 (as 30 greedy sweeps were performed in each iteration $i \geq 1$) and adding 10^5 (as 10^5 greedy sweeps were performed in the initial iteration

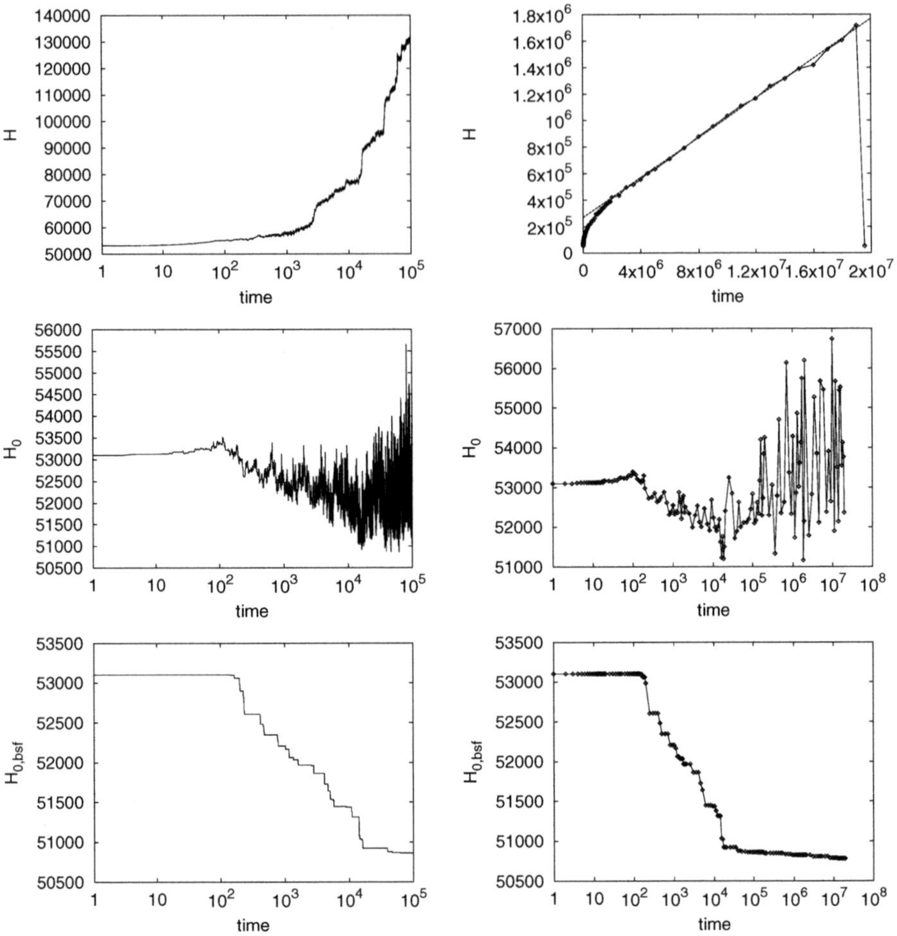

Fig. 17.12. Development of the total cost function $\mathcal{H}$ (*top*), of the original cost function $\mathcal{H}_0$ (*middle*), and of the best cost function value found so far $\mathcal{H}_{0,\mathrm{bsf}}$ according to the original cost function (*bottom*). *Left*: development for short time scales. The time is measured in iterations of the GLS algorithm. *Right*: the long-time developments of the observables are shown by plotting 147 exemplary data points each and connecting them with *lines*. Of course, this way one does not see the full extent of the fluctuations of $\mathcal{H}_0$ in the long time range, but if plotting all 19,536,400 data points, one only sees noise for $\mathcal{H}_0$ and gets no additional information for $\mathcal{H}$ and $\mathcal{H}_{0,\mathrm{bsf}}$. The final value 52,373.9531 for $\mathcal{H}$ and $\mathcal{H}_0$ is considerably larger than the optimum value 50,783.5469 of the PCB442 instance, which was found in at least one iteration

$i = 0$, in order to get into or at least near a local minimum). Of course, 30 sweeps in one greedy iteration might not be sufficient to get from the formerly locally minimum configuration to a local minimum nearby in the new energy landscape. However, we did not want to spend even more calculation time.

First, we want to consider the development of the total energy $\mathcal{H}$. We find that $\mathcal{H}$ increases over time, which is quite obvious, as the penalty values are incremented over and over. Despite this increase, sometimes a local minimum configuration will stay locally minimal such that the next greedy iteration cannot lead to an improvement. In the long time scale, $\mathcal{H}$ increases linearly according to $\mathcal{H} \approx 0.075 \times i + 2.7 \times 10^5$, with i being the iteration number. From the value 0.075 we can derive that in the long time limit, the system usually jumps to a new local minimum in most iterations. Otherwise the curve would increase with one $\lambda \approx 34.615$ per iteration. Finally, in the last iteration, the system is optimized according to the original cost function $\mathcal{H}_0$, such that the value of $\mathcal{H}$ drops to a value of $\mathcal{H}_0$.

Secondly, we consider the original cost function $\mathcal{H}_0$. We find that $\mathcal{H}_0$ increases slightly in the first few iterations. Thus, in order to remove the punished long edges, the system is forced into overall slightly longer roundtrips at first. But already in the 18th iteration, the system is for the first time able to get into a local minimum according to $\mathcal{H}$, which is better according to $\mathcal{H}_0$ than the previous one. After roughly 150 iterations, $\mathcal{H}_0$ starts to decrease below the value of the first iteration, such that this iterative algorithm starts to pay off. With the further decrease, more and more fluctuations in $\mathcal{H}_0$ occur. On average, $\mathcal{H}_0$ increases again, but the fluctuations also grow larger, such that one finds better and better intermediate solutions.

Thus, we must always store the best solution found so far with its energy value $\mathcal{H}_{0,\text{bsf}}$, especially as the optimization run will usually not find its way back to this solution in the end. The development of $\mathcal{H}_{0,\text{bsf}}$ over time is shown at the bottom of Fig. 17.12. We find that $\mathcal{H}_{0,\text{bsf}}$ stays constant at the beginning and then decreases gradually in a sigmoidal way, i. e., the last improvements are the smallest. After roughly 1.3×10^7 iterations, the system finds a globally optimum solution for the first time.

Finally, we consider the quality of the results that can be achieved with GLS. Of course, we use for this examination the final values for $\mathcal{H}_{0,\text{bsf}}$. We stay with overall 19,536,400 GLS iterations but change the number of sweeps within them to between 1 and 30. Figure 17.13 shows that the quality of the results is improved with an increasing number of sweeps. We get the optimum value in 6% of the runs with 10 sweeps and in 54% of the runs with 30 sweeps. The results in the other runs are nearly as good as the optimum value.

For the results shown up to now, we always used for a the value 0.3, as was recommended by the inventors of the algorithm. However, the question arises as to whether this value is generally an optimum value or whether this value of 0.3 is simply a good compromise for arriving at rather good solutions rather fast. Thus, we vary this control parameter of the GLS technique now. Figure 17.14 shows for the two shorter calculation times of one sweep and three sweeps per iteration both the mean deviation $\delta_{\text{mean}} = (\langle\mathcal{H}\rangle - \mathcal{H}_{\text{opt}})/\mathcal{H}_{\text{opt}}$ from the optimum and the probability p_{opt} of reaching the global optimum in an optimization run. Of course, the curve for p_{opt} is not too smooth,

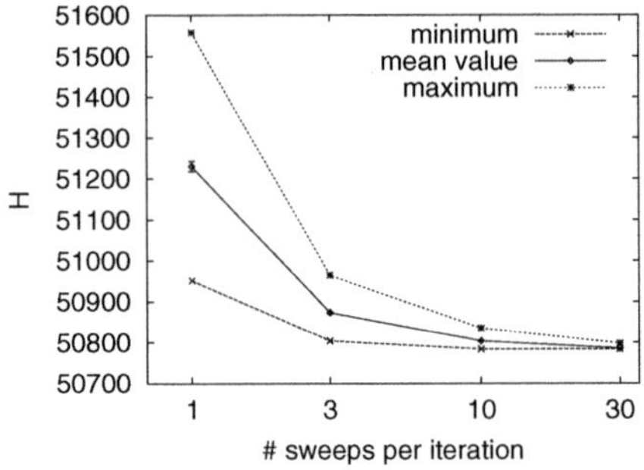

Fig. 17.13. Quality of the best results achieved with 100 simulation runs applying GLS to the PCB442 instance vs. number of sweeps in each iteration. As already mentioned, in the zeroth iteration 10^5 greedy sweeps were performed. In the succeeding 19,536,400 GLS iterations, 1, 3, 10, or 30 greedy sweeps were performed using the altered distance matrix

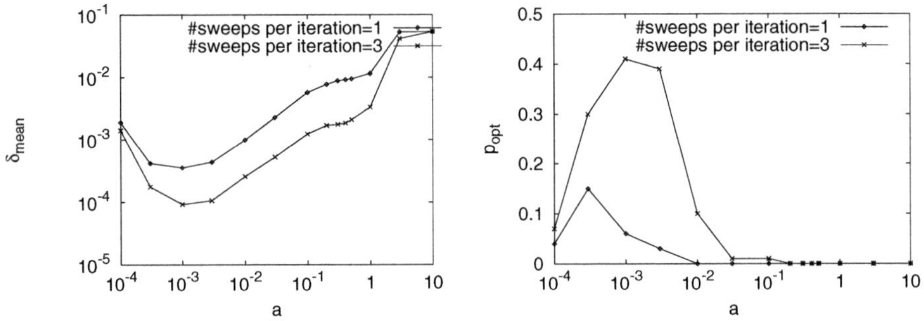

Fig. 17.14. Quality of the best results achieved with 100 simulation runs applying GLS to the PCB442 instance vs. algorithm parameter a for 1 or 3 sweeps per iteration. As already mentioned, in the zeroth iteration 10^5 greedy sweeps were performed. In the succeeding 19,536,400 GLS iterations, 1 or 3 greedy sweeps were performed using the altered distance matrix. *Left*: mean deviation from the optimum $\delta_{\text{mean}} = (\langle\mathcal{H}\rangle - \mathcal{H}_{\text{opt}})/\mathcal{H}_{\text{opt}}$; *right*: probability p_{opt} of reaching the global optimum in an optimization run

as only 100 optimization runs were performed for each value of a. We find immediately that the optimum value for a is much smaller than the recommended value of 0.3. Indeed, the optimum value for optimizing the PCB442 instance with these calculation times is $a \approx 10^{-3}$. Please note that we generally assume that we do not know the optimum value, so that we cannot break the optimization run early if the optimization run has already found the optimum or at least a quasioptimum solution. Furthermore, note that our calculation times are probably much larger than the inventors of the GLS algorithm had intended. Although we find for the PCB442 instance a much smaller optimum value for the parameter a, they might be quite right in recommending a larger value of a for the sake of reaching good solutions in very short times. Indeed, our two curves show jumps to much worse solutions when crossing the value $a = 1$ such that surely a value $a < 1$ must be chosen..

Furthermore, Fig. 17.14 shows on its right-hand side that a globally optimum configuration is reached in 41 out of 100 optimization runs when using 3 sweeps per iteration and the value $a = 10^{-3}$. Thus, GLS must be considered an excellent optimization algorithm.

Comparing GLS with SA (we take the results in Sect. 7.4 for comparison), we can roughly say that the results for one sweep here correspond to the results for 10,000 sweeps in roughly 1800 temperature steps and the results for 3 sweeps to the results for 30,000 sweeps. Of course, one must find an appropriate value of a for each problem instance when using GLS. But taking small values for a, we generally get better results on average for GLS than for SA. Furthermore, we must state that it depends on the problem whether SA or GLS is faster: GLS with an underlying greedy algorithm accepts much fewer moves than SA. If the update of the configuration plays the dominant role in the calculation time, then GLS is surely faster, except if one is dealing with a very complex problem in which one first must build up the complete tentative new configuration in order to calculate its energy. On the other hand, if determining which of the properties of a configuration will be punished takes much time, then SA is superior. An advantage of GLS over SA is that it is easier to stop immediately if one is already content with the achieved result quality so far or if a solution is suddenly needed urgently. This advantage arises from the fact that SA cools the system gradually from high-energy configurations and reaches very good solutions only at the end of the optimization run, whereas GLS, with its underlying greedy dynamics, produces in each iteration a configuration that is up to the order of 10% worse than the optimum. One might wonder whether this GLS algorithm can be as widely applied to various optimization problems as general algorithms like SA. We think that the approach shown here for the TSP can be applied to any problem with some interaction matrix like the distance matrix here. One might also easily find appropriate properties for other problems, for which some penalties can be introduced, such that GLS might be an algorithm that can be as generally applied as SA.

18 Application of Searching for Backbones to TSP

18.1 Definition of a Backbone

Now the searching for backbones (SfB) algorithm will be applied to the traveling salesman problem (TSP). As was already mentioned in Chap. 24 in Part I, one starts by creating a set of solutions and compares these for common parts. Figure 18.1 shows two quite good solutions for the PCB442 problem. At first glance, one finds the differences between these solutions. On the other hand, there are also common parts, namely, sequences of nodes, that are identical in both solutions.

Therefore, it is quite clear that this type of similarity should be used for the TSP for defining a backbone: as the TSP is a sequencing problem, this definition corresponds to the nature of the problem. Let us here now only consider the case of a symmetric TSP, i. e., $D(i, j) = D(j, i)$ for all pairs (i, j)

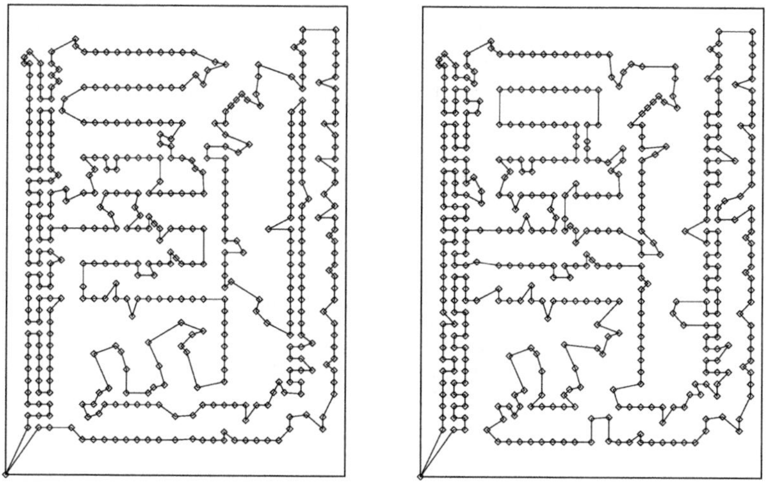

Fig. 18.1. Comparison of two quite good but not optimal solutions for the PCB442 problem. One finds many differences between these two solutions. However, there are also many sequences of nodes identical in both solutions

of nodes. Traversing each configuration either clockwise or counterclockwise leads to the same overall tour length.

The basic parts of the system are the individual nodes. The simplest relation between them is to ask whether they are neighbors of each other in one of the solutions. Let σ^ν be the solution of the run with number ν and let p be the number of compared solutions, which shall be kept constant here. Then the overlap matrix η_S for the symmetric TSP is defined as follows: the overlap between nodes i and j is given as

$$\eta_S(i,j) = \sum_{\nu=1}^{p} \sum_{k=1}^{N} \delta_{i,\sigma^\nu(k)} \cdot \left(\delta_{j,\sigma^\nu(k-1)} + \delta_{j,\sigma^\nu(k+1)}\right) , \tag{18.1}$$

with $\sigma^\nu(0) \equiv \sigma^\nu(N)$ and $\sigma^\nu(1) \equiv \sigma^\nu(N+1)$. Therefore, one gets an overlap between two nodes i and j in the solution σ^ν if j is either the predecessor or the successor of i in the solution σ^ν. If $\eta_S(i,j) = p$, then j is a neighbor of i in all solutions. In this case, they belong in one backbone. Using this overlap matrix, one can determine the backbones easily. A backbone is a sequence of nodes $i_1, i_2, \ldots, i_n$ with $\eta_S(i_1,i_2) = \eta_S(i_2,i_3) = \ldots = \eta_S(i_{n-1},i_n) = p$. There is no node i_0 with $i_0 \neq i_2$ and $\eta_S(i_0,i_1) = p$. Analogously, there is no such node i_{n+1} with $i_{n+1} \neq i_{n-1}$ and $\eta_S(i_n,i_{n+1}) = p$.

After the backbones have been determined, they are supposed to be used for further optimization runs, in which they must not be destroyed. The simplest approach for achieving this is to ask in every move whether a chosen connection may be cut or not. This question can simply be answered by, e. g., looking at the corresponding entry in the overlap matrix η_S and checking whether it is maximal. However, this costs a large amount of calculation time. The more backbones one finds, the worse this situation is. Therefore, one must simplify this situation for subsequent optimization runs.

An obvious way to achieve this is to represent each backbone by a pair of nodes. Let us visualize this approach with a small example of a symmetric TSP instance with $N = 10$ nodes, for which the following four solutions were produced:

0	9	1	2	3	4	5	6	7	8
0	3	2	1	4	5	6	7	8	9
0	9	7	6	5	4	8	1	2	3
0	4	5	6	7	8	3	2	1	9

From these solutions one derives the overlap matrix

$$\eta_S = \begin{pmatrix} 0 & 0 & 0 & 2 & 1 & 0 & 0 & 0 & 1 & 4 \\ 0 & 0 & 4 & 0 & 1 & 0 & 0 & 0 & 1 & 2 \\ 0 & 4 & 0 & 4 & 0 & 0 & 0 & 0 & 0 & 0 \\ 2 & 0 & 4 & 0 & 1 & 0 & 0 & 0 & 1 & 0 \\ 1 & 1 & 0 & 1 & 0 & 4 & 0 & 0 & 1 & 0 \\ 0 & 0 & 0 & 0 & 4 & 0 & 4 & 0 & 0 & 0 \\ 0 & 0 & 0 & 0 & 0 & 4 & 0 & 4 & 0 & 0 \\ 0 & 0 & 0 & 0 & 0 & 0 & 4 & 0 & 3 & 1 \\ 1 & 1 & 0 & 1 & 1 & 0 & 0 & 3 & 0 & 1 \\ 4 & 2 & 0 & 0 & 0 & 0 & 0 & 1 & 1 & 0 \end{pmatrix}.$$

(Of course, the maximum entries here are identical to the number of solutions $p = 4$) and from that (using the first solution for splitting it into the backbones) the following set of backbones:

$$\boxed{0\quad 9}$$
$$\boxed{1\quad 2\quad 3}$$
$$\boxed{4\quad 5\quad 6\quad 7}$$
$$\boxed{8}$$

These backbones, which must not be destroyed in subsequent optimization runs, are now coded for the next iteration in such a way that the work for the processors is as easy as possible. One obvious way to do this is to code every backbone as a pair of nodes. In the next optimization runs, the tour may only be cut after pairs of nodes, i. e., after the second, fourth, sixth, etc. node. This can easily be achieved by producing only even integer random numbers.

Using this approach, all edge points of the backbones get a new number, and the nodes inside the backbones are removed. Therefore, the backbone set of the example above is coded in the following way:

$$\boxed{0\quad 9} \Longrightarrow \boxed{0\quad 1}$$
$$\boxed{1\quad 2\quad 3} \Longrightarrow \boxed{2\quad 3}$$
$$\boxed{4\quad 5\quad 6\quad 7} \Longrightarrow \boxed{4\quad 5}$$
$$\boxed{8} \Longrightarrow \boxed{6}$$

One gets the following tour:

0–1 2–3 4–5 6–6

This tour is identical with the first solution of the previous SfB iteration. The "–" signs indicate that these connections must not be destroyed. Note that backbones containing only one node are doubled in the tour, such that this approach of choosing only even edges in the moves can be used.

Furthermore, one derives a smaller 7×7 distance matrix $\tilde{D}$ from the original 10×10 distance matrix D, which contains for example the following entries:

$$\widetilde{D}(0,1) = D(0,9), \widetilde{D}(2,3) = D(1,2) + D(2,3), \widetilde{D}(0,6) = D(0,8) .$$

This distance matrix and this tour are used for the next optimization runs. Of course, this coding never leads to a distance matrix $\tilde{D}$ with more entries than D. However, the tour could contain twice as many nodes as the original tour if the solutions are so different that no backbone containing at least two nodes can be found. If there are many backbones containing only one node, this coding costs additional time; on the other hand, it pays off if the solutions are rather similar to each other. Using either this or a related coding or the overlap matrix is necessary so as not to destroy the backbones.

In the second iteration of the algorithm, again a number of solutions is generated, but with a slightly altered program in which only edges after an even tour position number are allowed to be cut such that the backbones are not destroyed. Again the solutions shall be generated independently of each other. These new solutions must be decoded with the old backbones. Extending the example above, the new solutions could be

0–1 2–3 6–6 4–5
1–0 4–5 2–3 6–6
0–1 6–6 3–2 5–4
1–0 4–5 6–6 3–2

These solutions are decoded to:

0 9 1 2 3 8 4 5 6 7
9 0 4 5 6 7 1 2 3 8
0 9 8 3 2 1 7 6 5 4
9 0 4 5 6 7 8 3 2 1

As already described in Chap. 24 in Part I, these new solutions are supposed to be better than the previous ones and furthermore supposed to give a better representation of those parts of the problem instance that are already solved optimally. Therefore, the old solutions are discarded, but their inheritance consists of the backbones that were not allowed to be destroyed, such that they are also part of the new solutions. The new set of backbones is only constructed with the new solutions:

0	9		
1	2	3	8
4	5	6	7

With an increasing number of iterations, one gets fewer but longer backbones. The system to be optimized becomes smaller, and the calculation time per iteration therefore decreases. In this example, the tour now contains only six nodes and the distance matrix is of size 6×6.

18.2 Application to the Completely Asymmetric TSP

The asymmetric TSP (ATSP) exhibits a nonsymmetric distance matrix, i. e., there is at least one entry $D(i,j)$ with $D(i,j) \neq D(j,i)$. However, one must distinguish two cases for the ATSP: a completely asymmetric TSP has the property that for all pairs (i,j) of nodes $D(i,j) \neq D(j,i)$ and furthermore there is no sequence of nodes whose length stays the same if turned around. However, one usually finds the partially ATSP in practical applications in which there are some one-way streets in the problem instance by which only some percentage of distances becomes asymmetric.

The outline for the SfB algorithm for an ATSP is identical to that for the symmetric TSP; however, the definition of the overlap matrix, the determination and coding of the backbones, and the decoding of the coded solutions are different.

In the case of the completely asymmetric TSP, for which there is a preferable direction for any subsequence of nodes and for which there is no symmetry between clockwise and anticlockwise traversing the tour, the backbones must be determined in the following way: again the various solutions generated in the first iteration are compared with each other. However, only the traversion direction that is actually used in the individual solutions is used, i. e.,

$$\eta_{\mathrm{A}}(i,j) = \sum_{\nu=1}^{p} \sum_{k=1}^{N} \delta_{i,\sigma^\nu(k)} \cdot \delta_{j,\sigma^\nu(k+1)} \,, \tag{18.2}$$

with $\sigma^\nu(N+1) \equiv \sigma^\nu(1)$. This overlap matrix for the completely asymmetric TSP is therefore asymmetric. Then the backbones are determined according to the maximum entries, which again must have the value p. As these backbones are traversed in only one direction, it is sufficient to code them by only one node as they are not allowed to be either cut or turned around. This will be demonstrated in the following example. Let us consider a completely asymmetric TSP with $N = 10$ nodes, for which $p = 4$ solutions of the following form have been produced:

0	9	1	2	3	4	5	6	7	8
0	9	3	2	1	4	5	6	7	8
0	9	4	5	6	7	1	2	3	8
0	9	8	4	5	6	7	1	2	3

Then the overlap matrix is given by

$$\eta_{\mathrm{A}} = \begin{pmatrix} 0 & 0 & 0 & 0 & 0 & 0 & 0 & 0 & 0 & 4 \\ 0 & 0 & 3 & 0 & 1 & 0 & 0 & 0 & 0 & 0 \\ 0 & 1 & 0 & 3 & 0 & 0 & 0 & 0 & 0 & 0 \\ 1 & 0 & 1 & 0 & 1 & 0 & 0 & 0 & 1 & 0 \\ 0 & 0 & 0 & 0 & 0 & 4 & 0 & 0 & 0 & 0 \\ 0 & 0 & 0 & 0 & 0 & 0 & 4 & 0 & 0 & 0 \\ 0 & 0 & 0 & 0 & 0 & 0 & 0 & 4 & 0 & 0 \\ 0 & 2 & 0 & 0 & 0 & 0 & 0 & 0 & 2 & 0 \\ 3 & 0 & 0 & 0 & 1 & 0 & 0 & 0 & 0 & 0 \\ 0 & 1 & 0 & 1 & 1 & 0 & 0 & 0 & 1 & 0 \end{pmatrix}.$$

From this matrix the following backbone set is created:

0 9
1
2
3
4 5 6 7
8

These backbones are coded as follows:

0
1
2
3
4
5

From this one creates the new tour

$$0 \quad 1 \quad 2 \quad 3 \quad 4 \quad 5\,,$$

which may be cut after each position, and a smaller 6×6 distance matrix $\tilde{D}$ with asymmetric distances between the various backbones, e. g.,

$$\tilde{D}(0,1) = D(9,1), \tilde{D}(1,0) = D(1,0), \tilde{D}(0,4) = D(9,4), \tilde{D}(4,0) = D(7,0)\,,$$

and furthermore distances inside the one-node backbones that have to be considered, e. g.,

$$\tilde{D}(0,0) = D(0,9), \tilde{D}(4,4) = D(4,5) + D(5,6) + D(6,7)\,.$$

In contrast to the symmetric TSP, the number of nodes in the tour, the number of nodes in the distance matrix, and the number of backbones are identical. Here also the diagonal elements of the distance matrix are used in order to calculate the length of a configuration. In conclusion, this coding already pays off if at least two nodes can be combined to one backbone when using a basic serial optimization algorithm that does not require a calculation of the total energy of the configuration when performing a change.

18.3 Application to Partially Asymmetric TSP

The coding is most difficult for the partially asymmetric TSP, as both symmetric and asymmetric backbones must be determined. Analogously to the two marginal cases of the completely symmetric and the completely asymmetric TSP above, first, the overlap matrices η_{S} and η_{A} are created. Then, again one of the solutions is chosen according to which the backbones are built. If $\eta_{\mathrm{S}}(i,j) = p$ for a pair (i,j) of nodes, then these nodes belong to one backbone. After the construction of the backbone set by using the symmetric overlap matrix η_{S} one must check, for all backbones containing more than one node, whether $\eta_{\mathrm{A}}(i_1,i_2) = p$ for the first two nodes in it. If this is the case, then automatically $\eta_{\mathrm{A}}(i_2,i_3) = \ldots = \eta_{\mathrm{A}}(i_{n-1},i_n) = p$. In this case, the backbone must be marked as an asymmetric backbone; otherwise it is a symmetric backbone. Incidentally, if all backbones containing at least two nodes are marked as asymmetric, then the asymmetries in the distance matrix obviously dominate the system, such that one can proceed as in Sect. 18.2. Otherwise, one must represent each backbone by two nodes in the tour, as in the case of the symmetric TSP. However, the symmetric backbones are then coded with two different nodes, whereas the asymmetric backbones and the one-node backbones are represented by only one node, which is doubled in the tour. Again the diagonal elements of the distance matrix $\tilde{D}$ also must be added up when calculating the length of the solution.

At the end of the algorithm, all solutions will be identical, so that only one backbone containing all nodes is left. If the ground state of the problem is degenerate, a small number of backbones will remain at the end.

Generally, this algorithm is based on the idea that the backbones are considered to be optimal and, even more, to be part of the globally optimum solution. Of course, it is impossible to determine a priori which part belongs to an optimum solution. The assumption in this algorithm is that the statistical averaging over many good solutions will reproduce these backbones. This requires comparing an appropriate number of solutions. If there are too few solutions, then the backbones are too long already at the very beginning. Often they consist of nodes not connected in an optimum way. On the other hand, if there are too many solutions, then there are so many differences between them that pieces cannot be connected with each other as there is at least one solution voting for another possibility. The algorithm cannot converge in such a case. Therefore, the question of whether there is an optimum number of used solutions is of central importance.

We will present results for symmetric TSP instances. The calculations were performed on the massive parallel computer jump of the John von Neumann Institute for Computing at the Research Center Jülich, Germany. The IBM Fortran compiler mpxlf for the programming language Fortran 77 and the parallelization library MPI were used, working on $2^n, 1 \leq n \leq 7$ (2, 4, 8, 16, 32, 64, and 128) processors. Each processor created one solution in each SfB iteration.

Note that in all applications of the SfB algorithm, only the nearest-neighbor interaction between the nodes is considered, as described above: if two nodes are neighbors of each other in all solutions, then the link between them is added to the backbone list. Of course, this nearest-neighbor interaction is only a marginal case of the vast variety of correlations between various parts of a problem instance. However, due to the success this application has, it is sufficient to work with this nearest neighbor interaction only.

18.4 Computational Results

Now we investigate the results for an implementation of the SfB algorithm. We will concentrate here on the PCB442 instance, as this instance might be rather hard to solve for this algorithm due to the degeneracy of the ground state and of higher energy states of this instance. As the basic serial optimization algorithm used for providing solutions, we use simulated annealing (SA), as SA has the property that there are no restrictions in the search for good solutions due to the absence of constructive elements in SA. We always start with an initial temperature of 10^4 and cool the system down to a final temperature of 0.1 exponentially with a cooling factor of 0.99 and add a greedy step at the end. Thus, we used the same parameters for SA as in Sect. 7.4. Furthermore, each run starts with a random configuration, which is created by putting the nodes in a random order in the first iteration and the backbones in a random order in the other iterations. We use the node insertion move (NIM), the Lin-2-opt (L2O), and the four variants of the Lin-3-opt (L3O) with equal probability and leave out the exchange here. From the second iteration on, in which two-node-backbones instead of single nodes are used, the node insertion move becomes an edge insertion move: this move, which is also widely used for the standard TSP, shifts a pair of neighboring nodes to a new position. Please note again that also the moves used do not contain any constructive elements. The SfB algorithm ends either after a maximum of 1000 iterations or if less than three backbones are left, as at least three backbones are needed for performing a L3O or a NIM.

In [187] and [183], results for several observables describing the behavior of the SfB algorithm were already studied for several numbers p of compared solutions, but only for one fixed amount of calculation time. Thanks to the generous grant of computing time by the John von Neumann Institute, here we can study the quality of the algorithm both for various values of p ($p = 4$, 8, 16, 32, 64, and 128) and for various numbers of sweeps per temperature step in the SA algorithm (1, 3, 10, 30, 100, 300, 1000, and 3000 sweeps). If looking again at the graphics for the PCB442 instance in Fig. 7.9, we find that the quality of the results depends very strongly on the calculation time. For 10,000 and 30,000 sweeps per temperature step, we already obtained with some probability a ground state of the PCB442 instance with such a serial run such that we restrict ourselves now to shorter computing times for each SfB iteration. Furthermore, we must notice from Fig. 7.9 that rather different solutions are compared for these different computing times: for very short computing times, one compares rather bad solutions, which surely have much less in common with each other and with the global optimum, as there was no time to freeze these systems in locally minimum configurations, than those produced with larger amounts of computing time.

Thus, the question generally arises as to whether also for short computing times backbones can be found. To investigate this question we additionally performed some test SfB runs in which random configurations were compared for common parts. In the first iteration, the random configurations were created as usual; in the next iterations, the backbones were placed in a random order. If comparing only two random configurations in each iteration of the SfB algorithm for common parts, the algorithm converges to one random configuration within 1000 iterations. A small number of short backbones containing more than one node is even found in this time when using $p = 3$. For $p \geq 4$, no common structures in p different random configurations can be found within this time limit of 1000 iterations. Now we return to comparing solutions generated with SA.

Figure 18.2 shows the maximum number of nodes within a backbone for various computing times and numbers p of compared solutions. When working with $p = 128$, $p = 64$, $p = 32$, and $p = 16$, we find that the maximum number of nodes in a backbone remains 1 or close to 1 for short computing

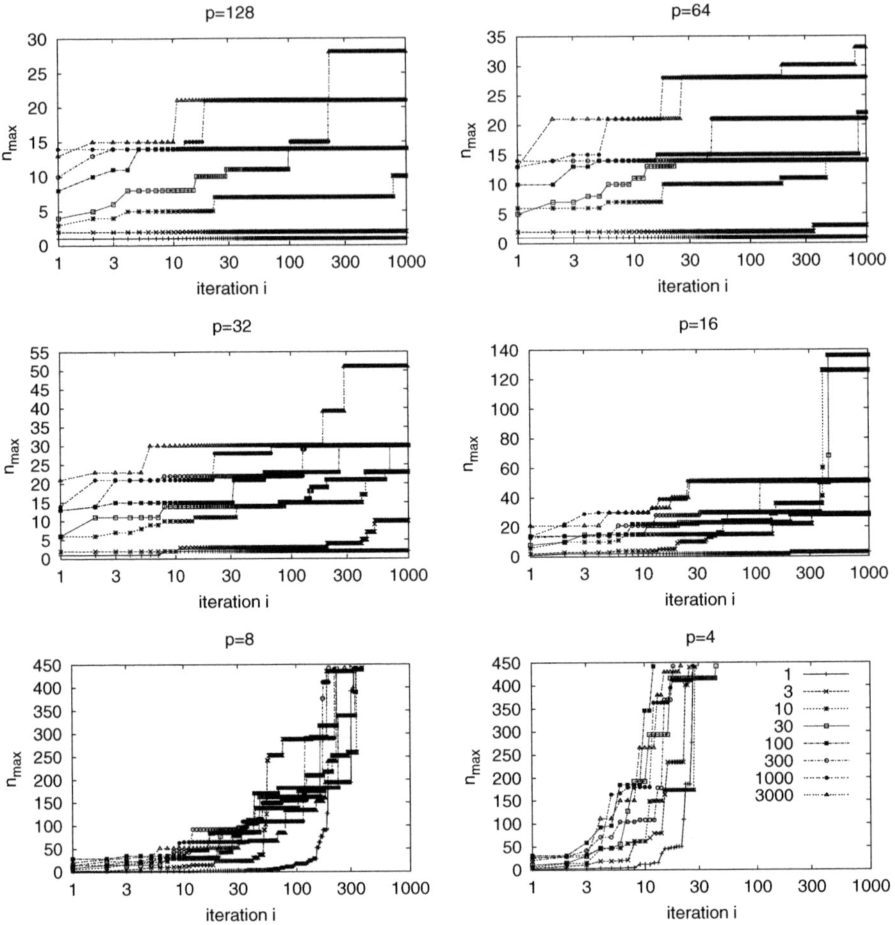

Fig. 18.2. Results for the application of SfB to the PCB442 instance for various numbers p ($p = 128, 64, 32, 16, 8, 4$) of compared solutions and for various calculation times. For the basic SA optimization algorithm, we use 1, 3, 10, 30, 100, 300, 1000, or 3000 sweeps per temperature step. These graphics show the maximum number of nodes $n_{\max}$ within a backbone vs. iteration i

times. Thus, either no backbone can be created or only very short ones. The solutions exhibit too many differences as the optimization processes hardly had any time to push the systems into local minima. If more computing time is invested, the maximum backbone size increases with an increasing number of iterations, but it never reaches the overall number of nodes in the system. Thus, ultimately the processors do not agree on a common solution. Contrarily, when working with $p = 8$ or $p = 4$, the maximum number of nodes in a backbone increases toward the system size within 1000 iterations for all calculation times. However, one cannot deduce from this behavior that the SfB algorithm would then always lead to the optimum solution. This behavior only shows that one can often find common structures if the number of solutions is appropriately small.

The relative mean deviation of the quality of the results achieved with the SfB algorithm to the optimum value of 50,783.5... of the PCB442 instance is shown in Fig. 18.3. For large numbers of processors, we get that this mean deviation, and therefore the average quality of the solutions, decreases only slightly with an increasing number of iterations. For short computing times,

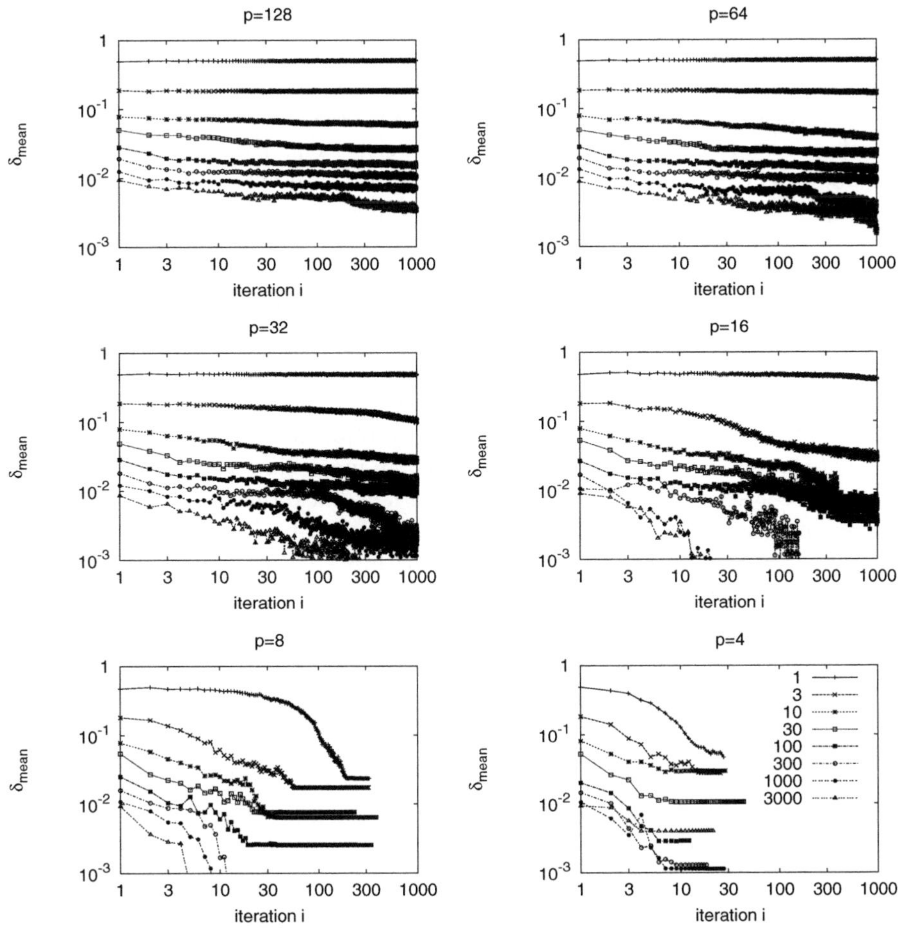

Fig. 18.3. Results of applying SfB to the PCB442 instance for various numbers p ($p = 128, 64, 32, 16, 8, 4$) of compared solutions and for various calculation times: for the basic SA optimization algorithm, we use 1, 3, 10, 30, 100, 300, 1000, or 3000 sweeps per temperature step. These graphics show the mean deviation of the results achieved in iteration i from the optimum

the average quality of the results is basically determined by the amount of calculation time invested. Given more time, some longer backbones could be created, such that obviously the SfB algorithm could lead to a decrease in the mean deviation with an increasing number of iterations. For $p = 32$ and $p = 16$, we find that the SfB algorithm does a really good job improving the average quality of the results if the serial calculation time is sufficiently long. For a small number of solutions, we also find strong improvements in the average quality of the solutions. However, the final average quality is also determined by the serial calculation time.

Figure 18.4 shows analogously the deviation of the best result found in an iteration from the optimum value for the PCB442 instance. Again we see that both the serial calculation time in each iteration and the number of processors have an influence on the results. Generally, one finds that if already a very good solution is found in some iteration, this does not necessarily mean that the best result of the next iteration will be either equally good or even better, as one might think because the backbones of the previous best solution are of course part of the new solutions. Thus, one should also always store the best solution found so far with the SfB algorithm, as the final solution might be worse.

Following the discussion about the improvement of the quality of the results by the SfB algorithm, we consider the convergence behavior of this algorithm. As the SfB algorithm unites several nodes to backbones, which will finally unite to only one backbone containing all nodes, it is useful to have a look at the number of backbones in the system, shown in Fig. 18.5. We find that this number gradually decreases with an increasing number of iterations. Thus, the system is always able to create new backbones consisting of a few former one-node backbones or to unite some longer backbones. However, only in the cases $p = 8$ and $p = 4$ is the algorithm able to converge to only one backbone within 1000 iterations.

Besides the number of backbones, which our coding routine makes equal to half the number of nodes in the tour, one can also investigate the convergence behavior by looking at the number of nodes in the distance matrix, which of course also decreases due to our coding as the system converges. This number is shown in Fig. 18.6. Like the number of backbones, the number of nodes in the distance matrix decreases gradually with an increasing number of iterations. Thus, the SfB algorithm also saves memory space, such that even for originally large TSP instances the distance matrix might finally fit in the cache, speeding up the calculation considerably.

This convergence behavior can best be described by the introduction of some order parameters for the algorithm. First, we want to define a parameter

$$\varphi(\eta_{\mathrm{S}}) = \frac{\sum_{i,j} \delta_{\eta_{\mathrm{S}}(i,j),p}}{\alpha} \tag{18.3}$$

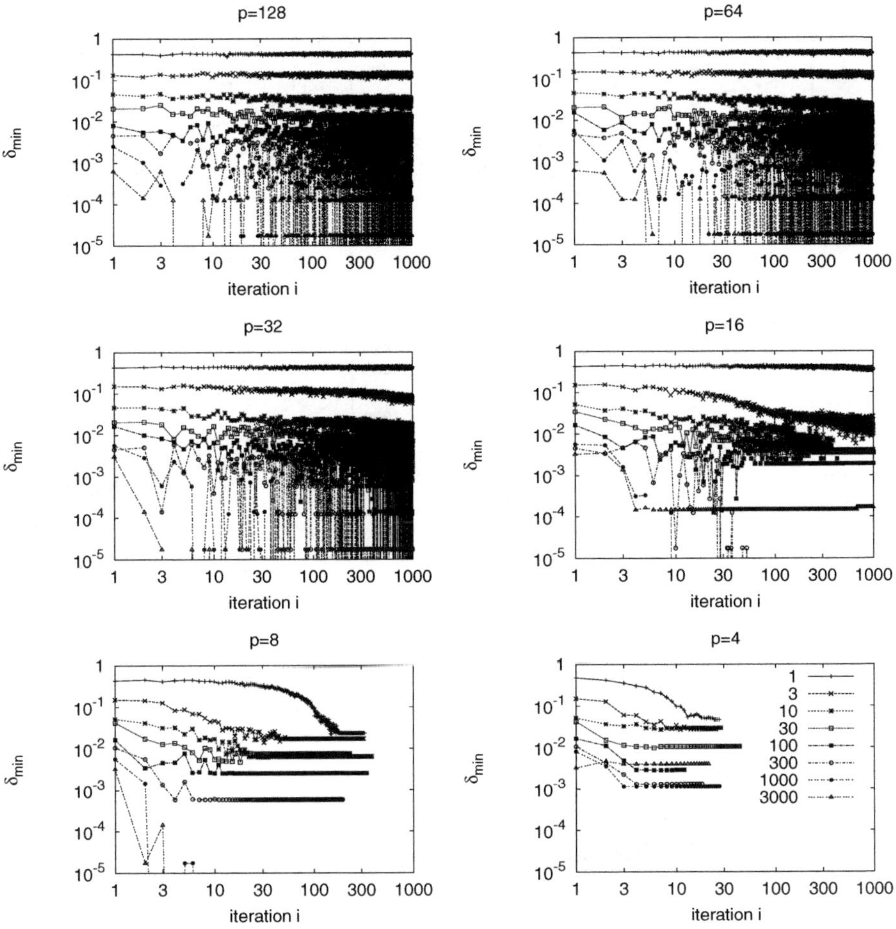

Fig. 18.4. Results of applying SfB to the PCB442 instance for various numbers p ($p = 128, 64, 32, 16, 8, 4$) of compared solutions and for various calculation times. For the basic SA optimization algorithm, we use 1, 3, 10, 30, 100, 300, 1000, or 3000 sweeps per temperature step. These graphics show the deviation of the best results in iteration i from the optimum

with $\alpha = 2N$ for the symmetric TSP. In the case of the completely asymmetric TSP, one sets $\alpha = N$ and uses η_A instead of η_S. If all solutions coincide, then two arbitrarily chosen nodes i and j are connected with each other in every solution, if they are connected in at least one solution, such that $\varphi = 1$. Contrarily, if there are so many differences between the solutions that no backbone consisting of at least two nodes can be created, then $\varphi = 0$. Figure 18.7 shows the results for this order parameter. We find that these graphics are like mirror images of the graphics for the number of nodes in the distance matrix shown in Fig. 18.6.

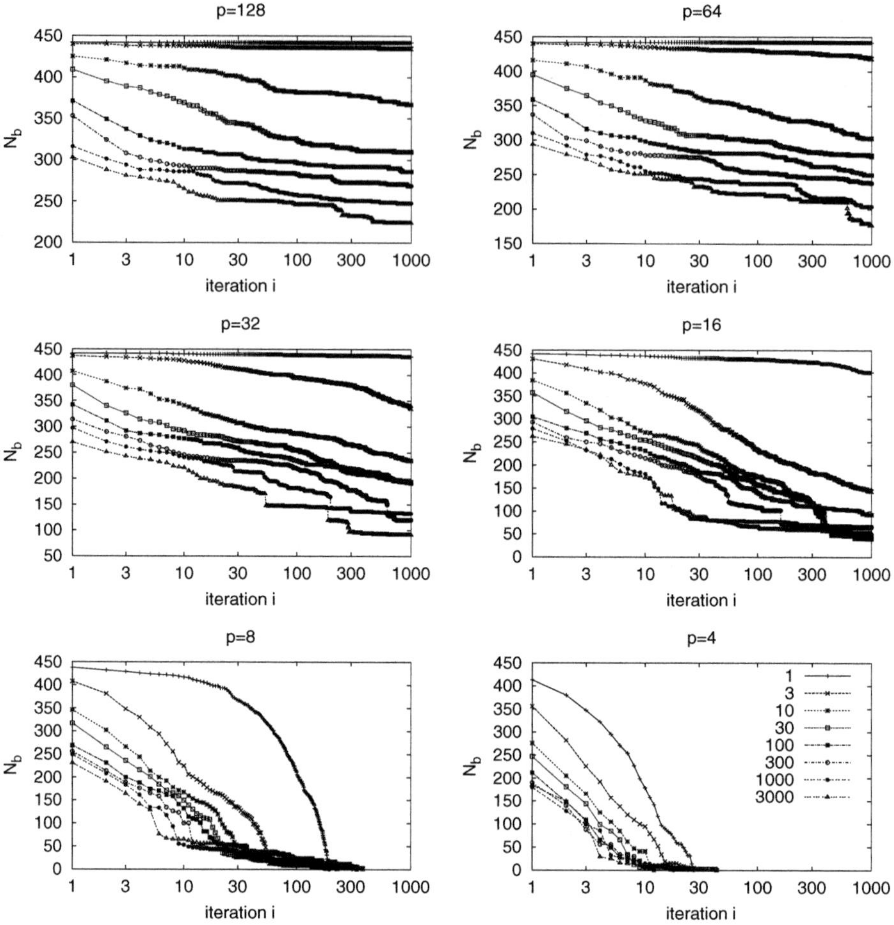

Fig. 18.5. Results of applying SfB to the PCB442 instance for various numbers p ($p = 128, 64, 32, 16, 8, 4$) of compared solutions and for various calculation times. For the basic SA optimization algorithm, we use 1, 3, 10, 30, 100, 300, 1000, or 3000 sweeps per temperature step. These graphics show the number of backbones, which is identical with half the number of nodes in the tour

A further order parameter that depends not on the maximum value of the entries in the edge matrix but on the number of zeroes in it can be defined as

$$\psi(\eta_{\mathrm{S}}) = 1 - \frac{-\alpha + \sum_{i,j}(1 - \delta_{\eta_{\mathrm{S}}(i,j),0})}{\alpha(p-1)} . \qquad (18.4)$$

If all solutions are the same, then there are only $2N$ nonvanishing entries in the edge matrix of a symmetric TSP or N nonvanishing entries in the edge

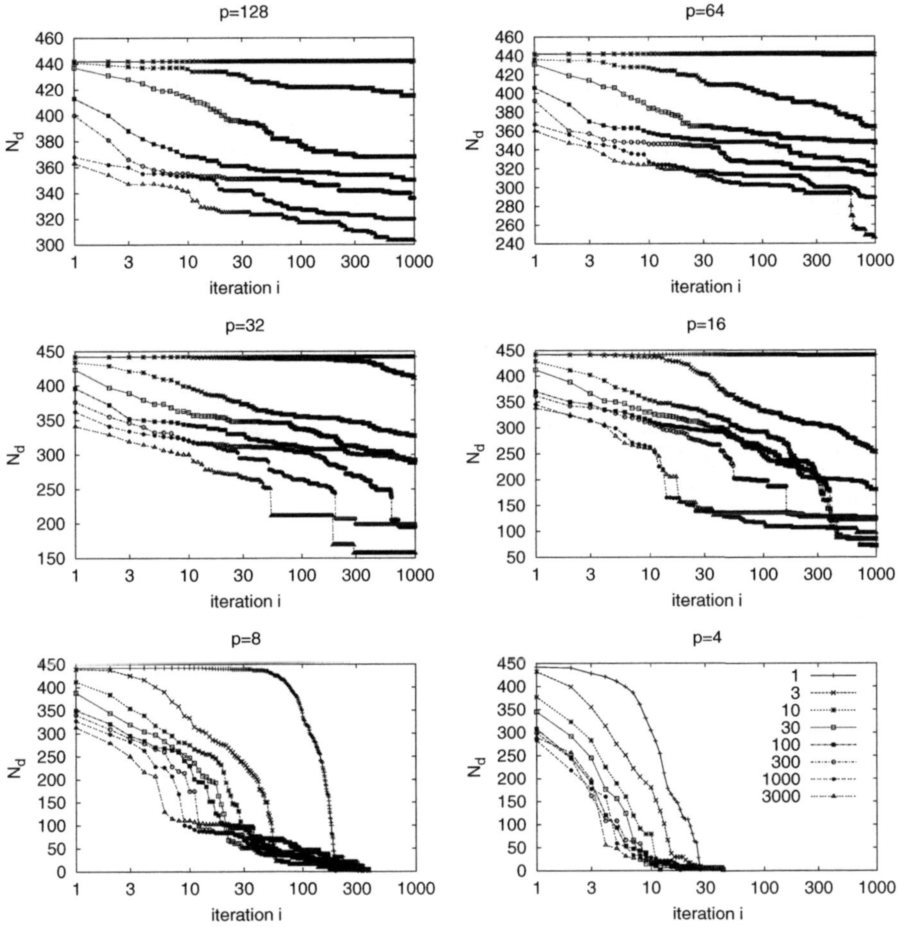

Fig. 18.6. Results of applying SfB to the PCB442 instance for various numbers p ($p = 128, 64, 32, 16, 8, 4$) of compared solutions and for various calculation times. For the basic SA optimization algorithm, we use 1, 3, 10, 30, 100, 300, 1000, or 3000 sweeps per temperature step. These graphics show the number of nodes in the distance matrix

matrix of an ATSP that are nonzero, thus $\psi = 1$. If all solutions are totally different (i. e., if node i is connected to node j in one solution, then these two nodes are not connected in any other solution), then there is a maximum number of nonzeroes in the edge matrix, namely, $2Np$ in the symmetric case and Np in the case of the ATSP, thus $\psi = 0$.

The results for the order parameter ψ are shown in Fig. 18.8. One sees at first glance that there are significant differences from the results for the order parameter φ in Fig. 18.7 for large values of p. In all cases, ψ is significantly

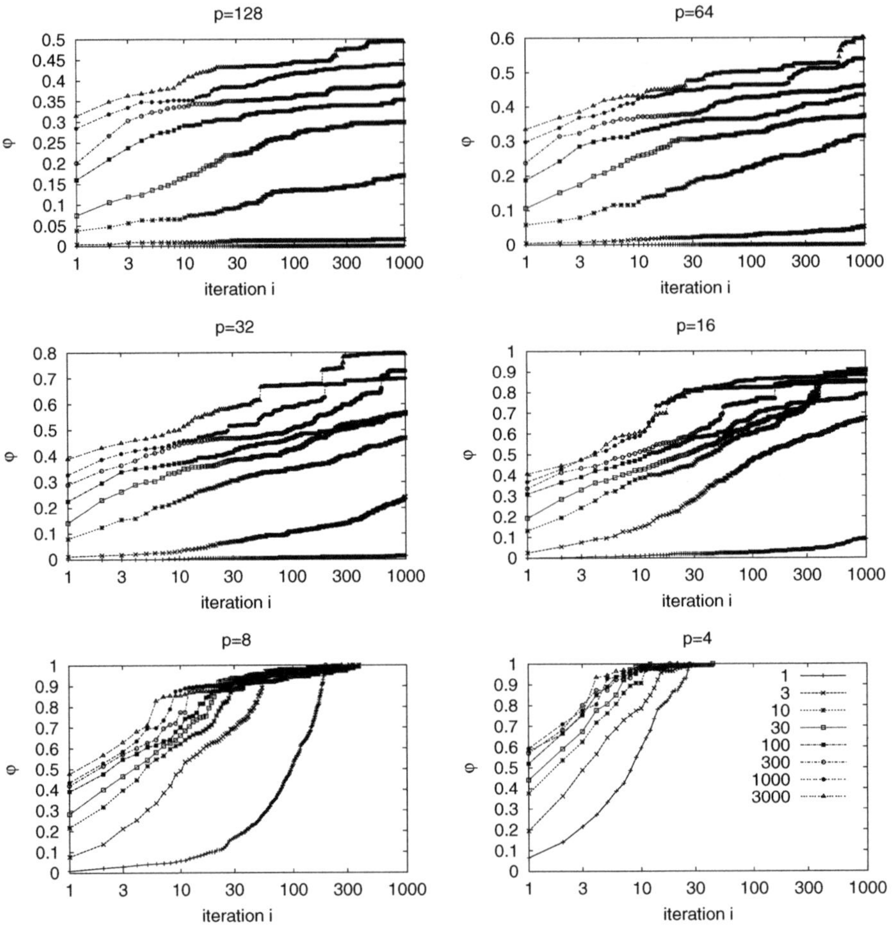

Fig. 18.7. Results of applying SfB to the PCB442 instance for various numbers p ($p = 128, 64, 32, 16, 8, 4$) of compared solutions and for various calculation times. For the basic SA optimization algorithm, we use 1, 3, 10, 30, 100, 300, 1000, or 3000 sweeps per temperature step. These graphics show the order parameter φ as defined in the text

larger than 0 already at the beginning. We get the smallest values for small p. For $p = 128$ and $p = 64$, ψ stays virtually constant, whereas ψ reaches a value of 1 for $p = 8$ and $p = 4$.

In the case $p = 2$, which is not shown here, φ and ψ generally coincide. For $p > 2$, one always gets $\psi > \varphi$. The difference between these parameters increases with an increasing number p. One can also provide a mathematical proof for this relation, which was done by Froschhammer and which is published in [187, 181, 182].

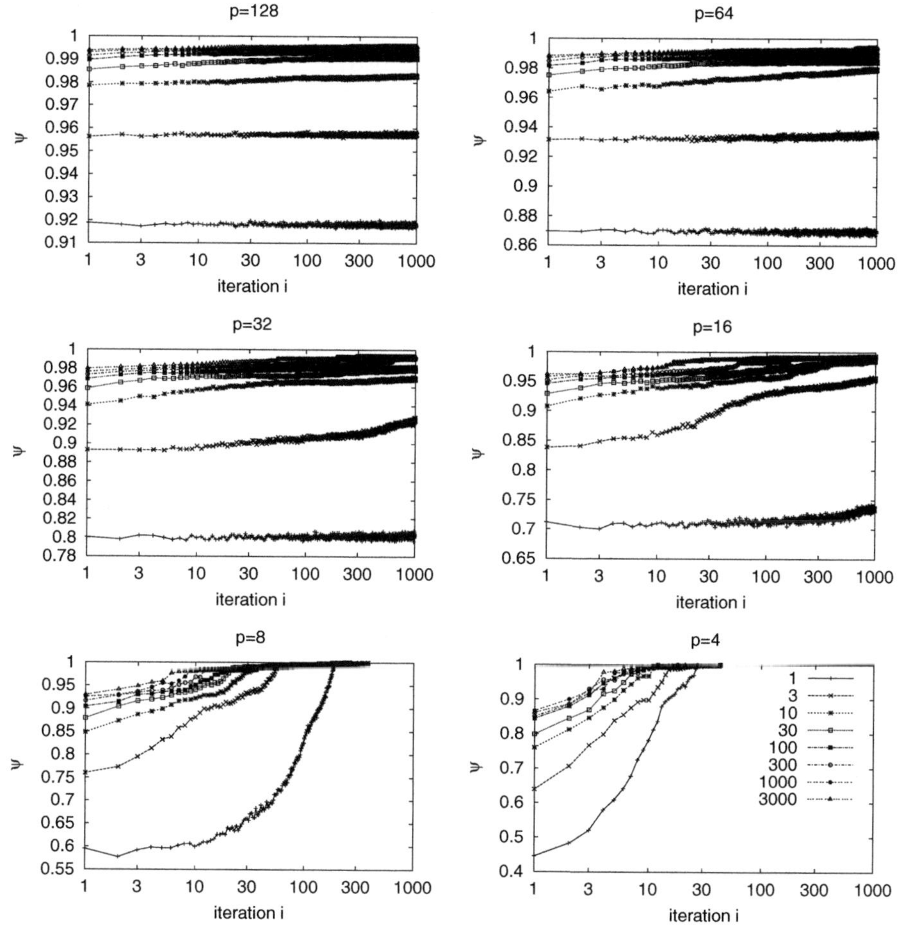

Fig. 18.8. Results of applying SfB to the PCB442 instance for various numbers p ($p = 128, 64, 32, 16, 8, 4$) of compared solutions and for various calculation times. For the basic SA optimization algorithm, we use 1, 3, 10, 30, 100, 300, 1000, or 3000 sweeps per temperature step. These graphics show the order parameter ψ as defined in the text

Finally, we want to define a parameter for the overlap that measures the percentage of those nodes that have another specific node as predecessor or successor in all solutions. One can write this order parameter as

$$\xi = \frac{N - N_{\mathrm{b}}}{N}\,. \tag{18.5}$$

ξ vanishes if each backbone consists of one node only. This parameter is shown in Fig. 18.9. Of course, these curves are related to the curves in Fig. 18.5, which shows the number of backbones in the system.

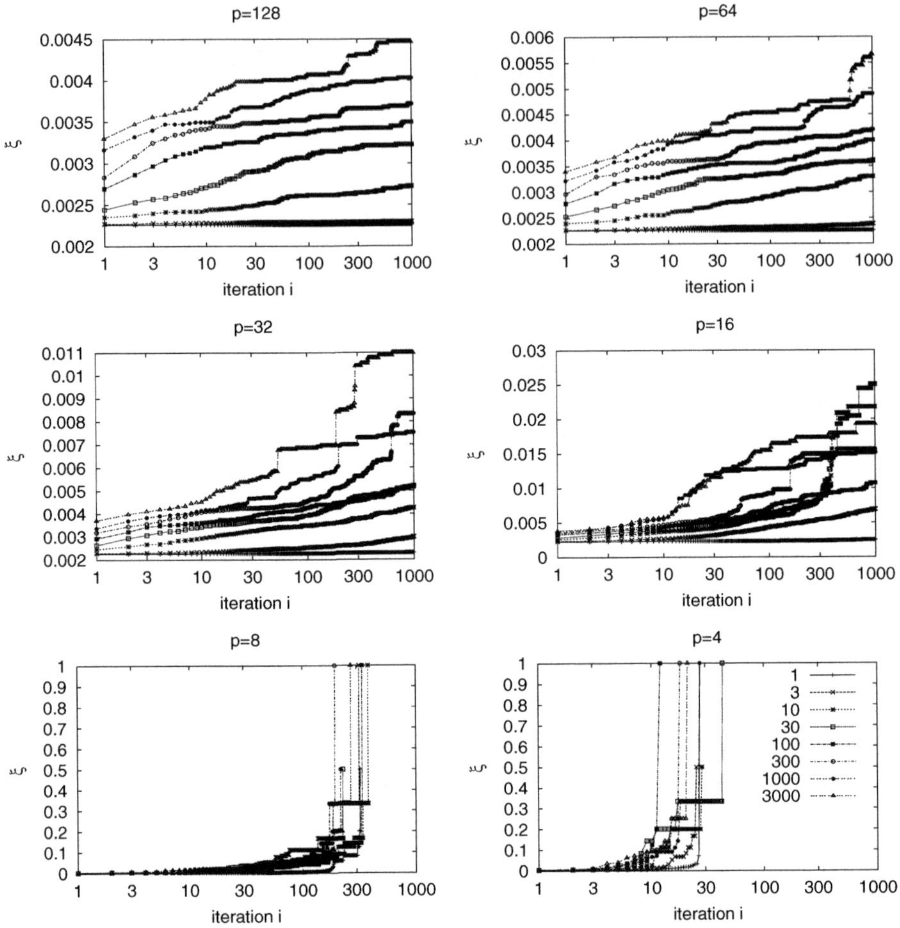

Fig. 18.9. Results of applying SfB to the PCB442 instance for various numbers p ($p = 128, 64, 32, 16, 8, 4$) of compared solutions and for various calculation times. For the basic SA optimization algorithm, we use 1, 3, 10, 30, 100, 300, 1000, or 3000 sweeps per temperature step. These graphics show the order parameter ξ as defined in the text

19 Simulating Various Types of Government with Searching for Backbones

Till now, all solutions created have been compared for the process of finding the backbones. As shown above, this searching for backbones (SfB) algorithm needs an appropriate amount of solutions to be compared in order to construct a suitable set of backbones. If the number of solutions is too small, then backbones are created that are nonoptimal. On the other hand, if the number of solutions is too large, convergence problems occur [187].

It could also be shown that the better the solutions are, the faster the convergence to even better solutions takes place. We still want to use a large number of processes in order to create a large amount of solutions. But we want to overcome the convergence problem for these large numbers of solutions.

19.1 An Aristocratic Approach

Thus, we introduce an "aristocratic" approach to SfB [187, 183]: instead of comparing all p configurations, we only compare the best q out of p configurations for constructing the backbones. Thus, we overcome the convergence problem in two ways: first, the number of configurations that are compared decreases. Secondly, if comparing the best q configurations only, we may expect that the convergence of this aristocratic approach will be not only faster than the original approach using all p configurations but also faster than using q configurations randomly selected form the p configurations. (This would be identical to the original approach using q processors that generate one configuration each.) The reason for this is that the best q out of $p \gg q$ configurations are on average much better than the average for the overall p configurations.

As the standard SfB algorithm reaches the optimum value of the PCB442 instance if some larger amount of calculation time is invested in each iteration, we work here only with 100 sweeps per temperature step. Figure 19.1 shows that the achieved quality of the solutions with this aristocratic approach is much better than what is achieved with the standard approach if the fraction of solutions taken for determining the backbones is chosen appropriately. We recommend using a moderate-size number, such as in this case 16 or 32 out of 128 solutions: If the number of solutions to be compared is chosen too small, then again the convergence takes place rather fast, but the quality of

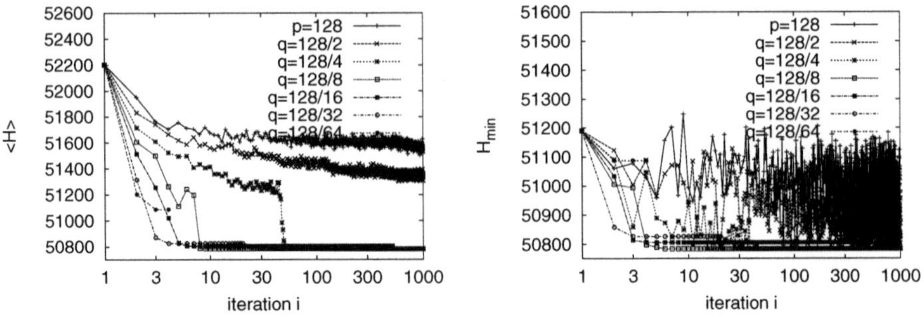

Fig. 19.1. Results of applying the aristocratic SfB approach to the PCB442 instance for 100 sweeps per temperature step. In each iteration $p = 128$ solutions were generated, but only the $q = p/\omega$ best solutions were used to determine the backbones. The graphics show the average quality $\langle \mathcal{H} \rangle$ of the solutions in each iteration (*left*) and the length $\mathcal{H}_{\min}$ of the best solution (*right*). For comparison, the result of the standard SfB algorithm with $p = 128$ is also shown

the solutions is not optimal. If, for example, 64 solutions are used, then the algorithm runs into convergence problems again.

Figure 19.2 shows the convergence behavior at the number of backbones in the system and at the number of nodes in the distance matrix. For 32 out of 128 solutions, we see a sharp drop both in the number of backbones and in the number of nodes in the distance matrix. Thus, suddenly the system size is reduced considerably. Summarizing, the aristocratic approach is superior to the original SfB algorithm and leads to very good results if the fraction $1/\omega$ of used solutions is appropriately chosen.

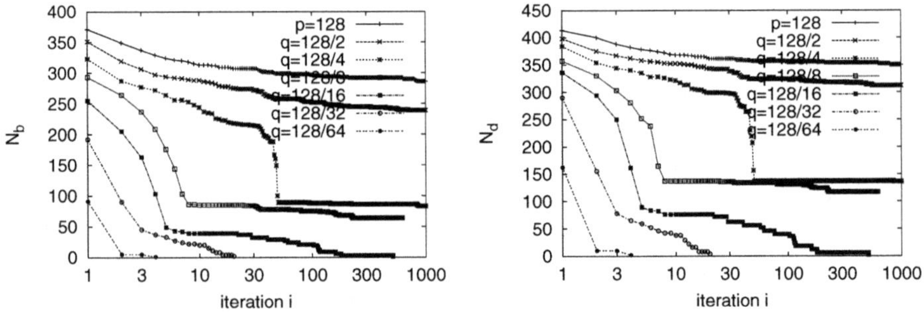

Fig. 19.2. Results of applying the aristocratic SfB approach to the PCB442 instance for 100 sweeps per temperature step. In each iteration $p = 128$ solutions were generated, but only the $q = p/\omega$ best solutions were used for determining the backbones. The graphics show the numbers N_{b} of backbones (*left*) and N_{d} of nodes in the distance matrix (*right*). For comparison, also the result of the standard SfB algorithm with $p = 128$ is shown

19.2 A Democratic Approach

Reconsidering the aristocratic approach in the last section, the question arises as to whether it is really necessary to use the best solutions. Thus, we introduce now a democatic approach to SfB. In a democracy, the majority rules, both in the majority of voters at the poll who decide what party/parties can form the new government and in the majority of the members of parliament or congress who decide for or against a new law. Such majority decisions have already been implemented in various ways in computer algorithms. In the field of neural networks (NNs), the committee machine that decides according to the majority of the neurons is widely used. Here we want to use not the strict majority/committee approach but a slightly different approach.

In modern democracies, there is usually some head of state called president, prime minister, or chancellor. We identify the best solution achieved in each iteration with this head of state. Now the head of state wants to bring new laws through the parliament. In our case, he/she wants his/her edges to become part of the set of backbones. For this purpose, he/she needs the majority in the parliament. Thus, an edge will become part of a backbone if more than half of the solutions also contain this edge.

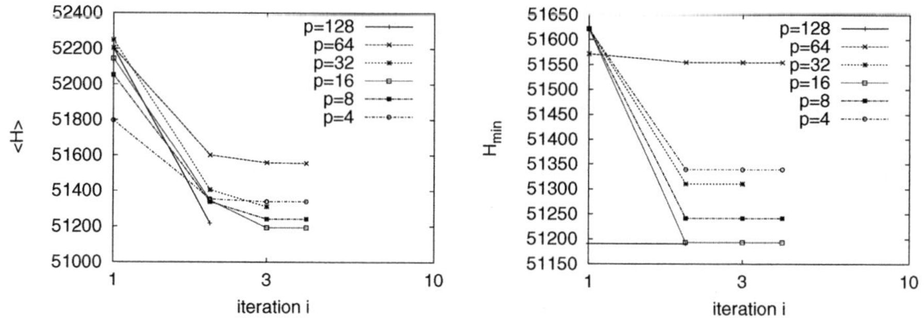

Fig. 19.3. Results of applying the democratic SfB approach to the PCB442 instance for 100 sweeps per temperature step. In each iteration p solutions were generated. The graphics show the average quality $\langle \mathcal{H} \rangle$ of the solutions in each iteration (*left*) and the length $\mathcal{H}_{\min}$ of the best solution (*right*)

In our simulations of this democratic approach, we stay with the 100 sweeps per temperature step that we used in our aristocratic approach. Figure 19.3 shows the mean and minimum quality of the solutions in each iteration. We see at first glance that this approach converges very rapidly to one common solution in only a few iteration steps. However, due to this rapid convergence, this democratic variant is not always able to improve the quality of the solutions with an increasing number of iterations, once, indeed, the quality of the best solution does not change at all. Generally, the final quality of the solutions is much worse than in the aristocratic approach.

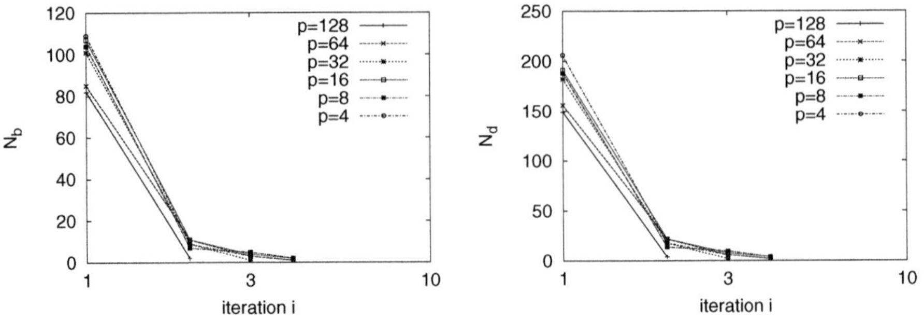

Fig. 19.4. Results of applying the democratic SfB approach to the PCB442 instance for 100 sweeps per temperature step. In each iteration p solutions were generated. The graphics show the number of backbones in the system (*left*) and the number of nodes in the distance matrix (*right*)

Figure 19.4 allows a better look at the convergence behavior of this democratic variant. We see that the number of backbones decreases to less than 20 in only two iterations of the algorithm, and the number of nodes in the distance matrix is reduced in an analogous way. Thus, this democratic variant exhibits a very rapid convergence, but it cannot be recommended for practical use as the quality of the resulting configurations is not good enough.

19.3 Solution of the PCB442 Problem

More than two million optimum solutions of minimum length were found by the aristocratic variant of the SfB algorithm for the PCB442 instance, which has a highly degenerate ground state. Thus, we would like to use these solutions for a discussion of the ground state of this problem. Comparing all these solutions with each other, one gets backbones for these optimum solutions. There are a total of 89 backbones. A few are rather long, but 41 of them consist of one node only:

# nodes	1	2	3	4	6	10	17	25	27	32	33	39	51
# backbones	41	30	4	1	1	3	1	1	1	2	2	1	1

These 89 backbones are shown in Fig. 19.5. Taking a closer look at this figure, one finds that these 89 pieces are not independent of each other. On the contrary, the pieces are only split from each other in some areas in which there are a few possibilities to connect them optimally. Fixing the direction of a backbone consisting of at least three nodes, one finds that the directions of most other backbones are then also fixed. As shown in Fig. 19.6, these backbones belong to three groups: there are 41 backbones consisting of only one node. Then there are 16 "blinkers", which are backbones consisting of two nodes each, for which there exist both possibilities for traversing them

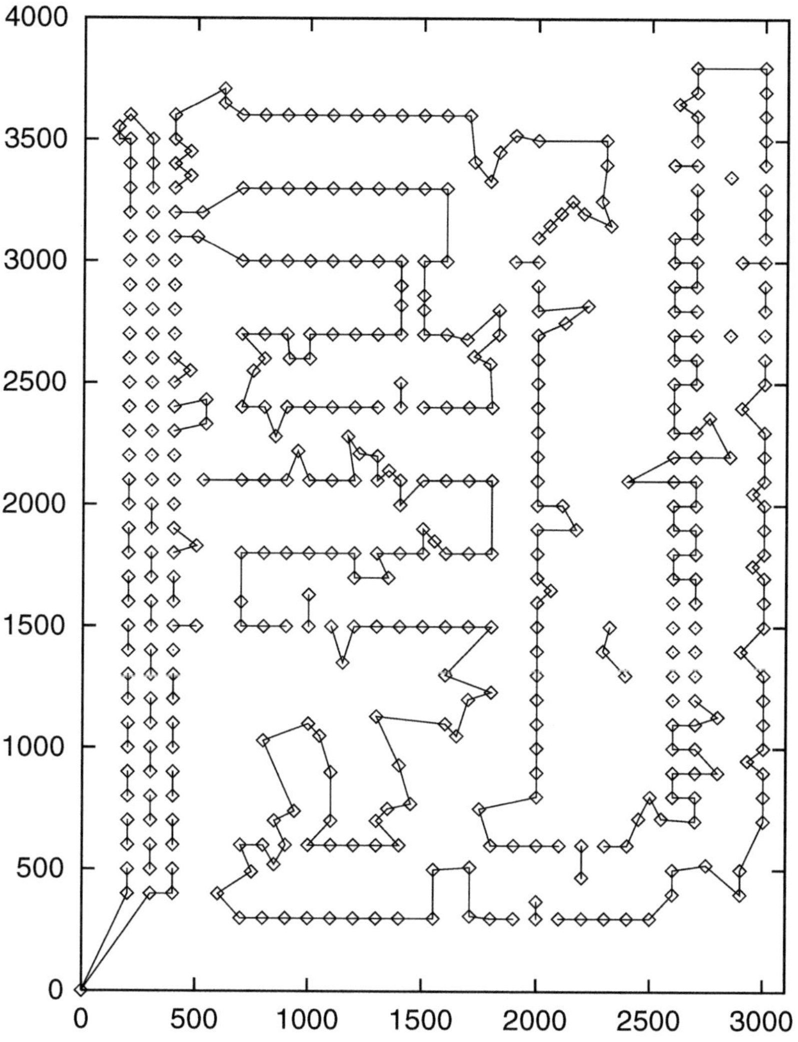

Fig. 19.5. Comparing over two million optimum solutions for the PCB442 TSP instance that were found with the SfB algorithm, 89 backbones common to all solutions can be identified

in an optimum solution if the directions of the other backbones are fixed. Finally, there is a "superbackbone" consisting of 369 nodes.

Note that this local determinedness in the fixation of the direction of all parts of the superbackbone if one part is fixed is a distinguishing feature of the PCB442 instance. One can easily construct TSP instances that do not show such a local determinedness. Examples are given in Fig. 19.7 [182]. The Möbius strip is a standard example in mathematics for areas with boundaries

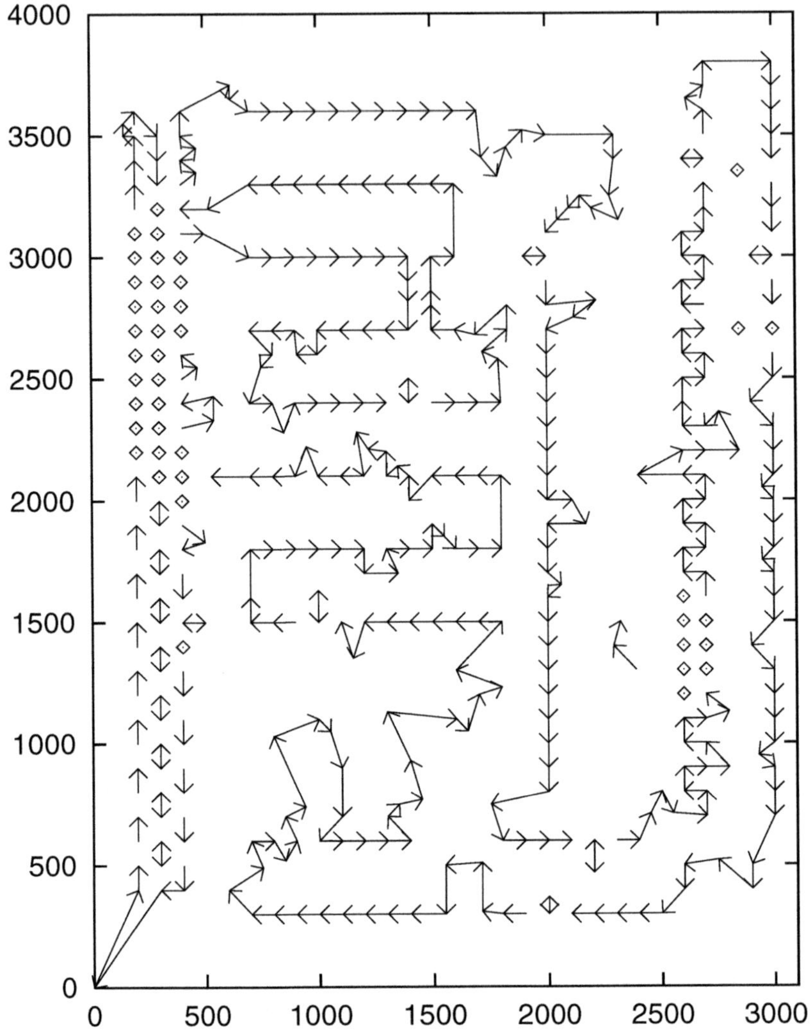

Fig. 19.6. If the direction of one of the longer backbones (consisting of at least three nodes) shown in Fig. 19.5 is fixed, the directions of all other backbones, with the exception of 16 "blinkers" consisting of two nodes each, are also fixed

that cannot be orientated [29]. It is created by a rectangle with the extensions $[0; 1] \times [-1; 1]$ for which the points $(0, t)$ and $(1, -t)$ are identified for all $t \in [-1; 1]$. We place a regular square lattice with two columns on this Möbius strip. Due to the Möbius geometry, the successor of the upper left point is the lower right point and vice versa. Additionally, we place two point sets on this strip that are connected to one backbone each in every optimum solution. This example at the top of Fig. 19.7 shows that there is no local determinedness on nonorientable areas.

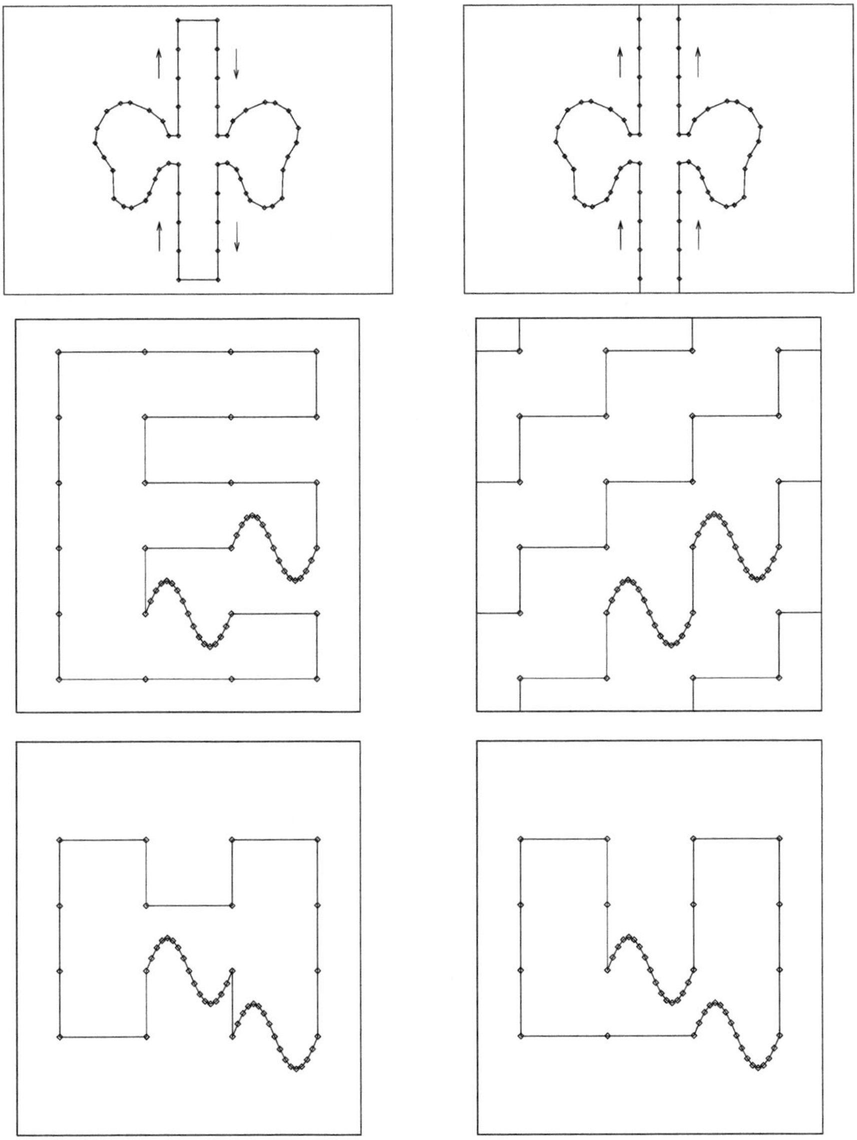

Fig. 19.7. No general local determinedness of the relative directions of backbones. These graphics show three examples of how there is no general local determinedness of backbones in TSP instances as in the PCB442 instance. Each of these three instances contains two partial sequences that will be registered as backbones by the SfB algorithm and some structures that allow for various connections. *Top*: Möbius strip with ears; *middle*: outline of an *E* on a torus: *bottom*: structure on a plain that can be connected to, e. g., the outlines of an *H* and a *U*. Note that the configurations on the *left* are the same length as the corresponding configurations on the *right*

The second example in Fig. 19.7 shows points that form the outline of the large letter E. Additionally there are two point sets connected to two backbones each. The rectangular area $[-1;1] \times [-1;1]$ will be formed into a torus by identifying $(-1, t)$ with $(1, t)$ and $(t, -1)$ with $(t, 1)$ for all $t \in [-1;1]$. The right graphic shows another possible optimum solution for this TSP instance on a torus. Again we find that when the direction of one backbone is fixed, then the direction of the other backbone is not fixed. Finally, the bottom of Fig. 19.7 shows a TSP instance in which the point sets form the outlines of the large letters H and U. Additionally, there are again two point sets that will be connected in two backbones in every optimum solution. In the case of the H, both backbones are crossed in the same direction, in the case of the U, in the opposite direction. Thus, the PCB442 instance is rather special to show this local determinedness in the fixation of the directions of the backbones.

Still, the superbackbone and the blinkers are not a complete description of the ground state of the PCB442 instance, which was introduced by Grötschel. The parts of the superbackbone can only be connected with the other parts in an optimum way by using certain edges. At some places, it is easy to see that such an edge is contained in an optimum solution with a probability of 50%. However, there are other, more complicated, local structures as well in which the probability of an edge that can be part of an optimum solution actually being within the given optimum solution must be calculated carefully and can be, e. g., 25%. These "connecting links" must also be considered when speaking of the full ground state of the problem and when projecting the current solution in an optimization run on the ground state in order to calculate the order parameter. Furthermore, these connecting edges are correlated with each other. If a certain edge is part of an optimum configuration, then this often requires that at least one additional edge be part of this optimum configuration.

19.4 Can Humans Do This, Too?

In our introduction to the SfB algorithm, we started out with some assumptions: we proposed that if several processors that try to optimize a given problem arive at the same solution parts independently of each other, then we can assume them to be optimally solved. Of course, JJS had such assumptions in mind when he considered real-world outcomes: if some person A tells you strange things, you probably would not believe him/her. If a second person tells you independently of the first one the same strange stories, then you are already tempted to believe them. Now if many people tell you the same thing seemingly independently of each other, then you are basically forced to believe these stories. Now in the real world, it might be that all of these people are simply repeating some widespread prejudice. But this cannot happen on a parallel computer, in which the nodes are really independent of each other.

Now we want to go in the reverse direction after having developed this SfB algorithm: can we use humans as the nodes of a parallel processor, let them generate a number of solutions, then determine the backbones, then tell them to generate new solutions using these backbones, and so on? Does the convergence behavior of this approach differ from the convergence behavior on a computer system?

At a conference on the emergence of identity and self-organization in socIal systems organized by Sorin Solomon in Jerusalem in December 2003, JJS handed out to the attendees, most of whom were physicists and computer scientists, graphics with the locations of the 48 nodes of a TSP benchmark instance. JJS asked the participants to draw a closed tour by hand. Looking at the resulting solutions, it became immediately clear that most people had done a very poor job generating the solution. Indeed, the solutions looked rather similar to those that are achieved with construction heuristics like the nearest neighbor heuristic, in which the last nodes are attached via very long edges to the tour. Now perhaps the conference participants were caught by surprise or were thinking about buying Christmas presents. However, a comparison of all solutions returned to JJS revealed that not a single backbone could be found. A random selection of half the solutions showed these remaining solutions had exactly one edge in common.

Thus as the convergence would be rather slow and as it would be tedious and also slightly crazy to ask people at every upcoming conference to draw a solution, we decided to transfer this task back to computers. They should simulate this human behavior by generating solutions according to a slightly modified bestinsertion technique: starting out from a randomly chosen backbone, one iteratively selects a backbone at random and determines where to best insert it into the tour. In addition to the standard bestinsertion technique, we have to take both directions of the backbone into account. Here the proceeding is the same as with the original SfB algorithm, except that the solutions are not generated with SA but with this construction heuristic.

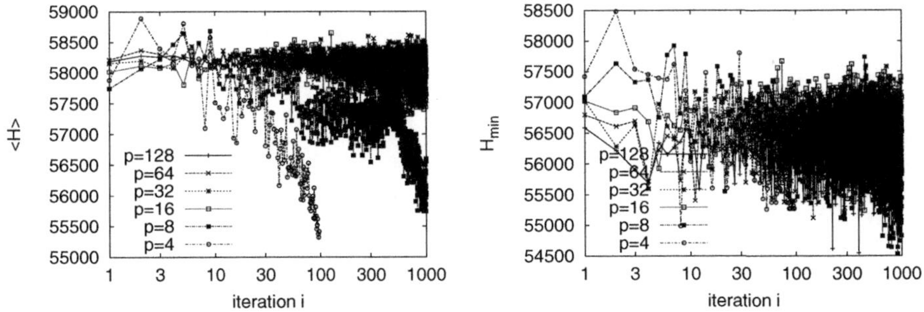

Fig. 19.8. Results of applying the human SfB approach to the PCB442 instance for various numbers of solutions p. The graphics show the average quality $\langle \mathcal{H} \rangle$ of the solutions in each iteration (*left*) and the length $\mathcal{H}_{\min}$ of the best solution (*right*)

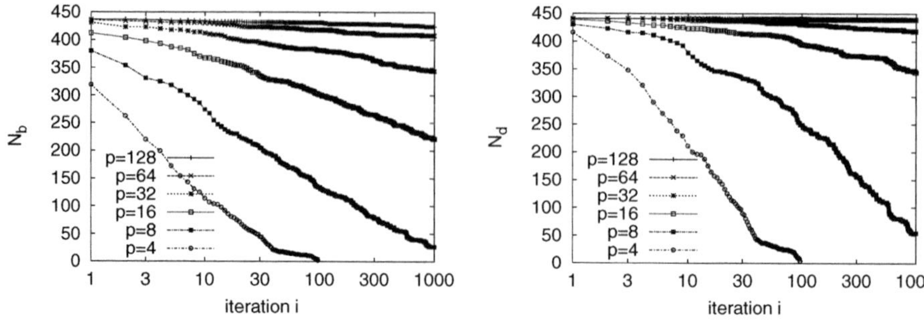

Fig. 19.9. Results of applying the human SfB approach to the PCB442 instance for various numbers of solutions p. The graphics show the numbers N_{b} of backbones (*left*) and N_{d} of nodes in the distance matrix (*right*)

Figures 19.8 and 19.9 show the results of this approach. We find that this approach converges after 96 iterations for $p = 4$ and after 1783 iterations for $p = 8$. Although the curves for N_{b} and N_{d} decrease nicely at the beginning, no convergence was achieved for $p \geq 16$ with more than 38,000 iterations. The final quality of the solutions was so bad that this approach cannot be recommended.

But here we used the standard SfB technique, so what happened? We simply replaced the SA routine by the bestinsertion routine. The SfB algorithm gradually reduced the system size and improved the quality of the solutions. However, obviously the bestinsertion heuristic only generates a certain type of solution, which is not even very good. These solutions are only part of a specific subset of the configuration space. Thus, in contrast to the results achieved with SA, the configuration space is not appropriately sampled by the bestinsertion technique. Thus, we strongly recommend using a basic serial optimization scheme that is able to sample the configuration space properly and to let the systems optimize to their full extent such that real locally minimum configurations are compared by the SfB algorithm.

Applications B
The Constraint Satisfaction Problem

20 The Constraint Satisfaction Problem

20.1 Sources of Constraint Satisfaction Problems

A large class of tasks that are important in practice can be given a common expression as constraint satisfaction problems (CSPs). Such a problem has the following form. A set of variables, x_i, must each be assigned values drawn from an appropriate domain, D_i, so that all or as many as possible of a set of constraints $C_j(x_1, x_2, \ldots, x_N)$ are satisfied. Tasks that fit in the class are widely encountered. For example, scheduling times for work activities, times for events, use of classrooms, or even frequencies to be employed for wireless transmission of data packets can all be expressed in this form. Less obvious examples would be obtaining a legal layout of all the circuits in a VLSI Si chip or verifying the correctness of the logic designed for such a chip.

These problems are large. Modern hardware designs often involve millions of elements, each of which may require a number of variables to be assigned values before the element is completely specified. Problems in classical areas of logistics, such as scheduling airline flights, equipment, and their crews, easily reach this scale. The rapidly growing field of bioinformatics also involves analyzing genetic sequences involving hundreds of thousands to a few million coding steps, or attempting to predict the structures of biologically active chain molecules (proteins) with similarly large numbers of basic molecular building blocks.

At the core of this large class of problems are simpler randomly generated classes of CSPs, not arising in industrial practice but created for purposes such as testing solution methods for constraint satisfaction by providing reliably difficult problems to try. Exploring the nature of the solutions to these problems is providing clues to why and under what conditions the more irregular realistic problems are either easy or difficult. The randomly generated problems typically draw each variable from the same finite domain, typically a restricted set of the integers, such as between 1 and k. Variables could even be restricted to be Boolean, with only two values, “0” or “1”. In any of these examples, the difficulty in the solution comes from the fact that it may be necessary to explore a finite fraction of the entire state space of the independent variables, which is at least of size 2^N for N variables, in order to find an acceptable solution.

The results we will discuss in the next few chapters focus on ensembles of randomly generated problems since the results obtained on these prove very illuminating with respect to understanding methods of dealing with any CSP. Some experts in constraint satisfaction believe that the "real problems" that are suggested by or extracted from actual applications are different from random problems because of the structure embedded in them and, as a result, are harder to solve. We are skeptical that the difference is that profound.

As we shall see, the search for local iterative improvements form the basis for one class of successful heuristics in CSPs. In these, all variables are assigned values within their domains, and the values are changed one at a time until an acceptable solution is found. A second class of methods with a structure like the construction heuristic also play an important role. In the latter methods, the variables are assigned values one after another. Sophisticated versions of this keep sufficient records to make it possible to "backtrack", undoing some of the assignments when an insoluble conflict is reached, and starting off again in a new search direction. The hope is that such a sequence of assignments can be successfully completed, leading to a solution of the problem in which all constraints are satisfied.

The sequential assignment approaches, when supported by the possibility of backtracking, can provide exact results. Suppose the variables are Boolean. Then each time a contradiction is found, the search backs up one step and resumes after assuming the opposite value of the variable just before the contradiction. If a search for a satisfying solution ends by backtracking to the origin, we have proved that the problem does not have a satisfying solution, since no configuration of the subset of variables that have been searched is allowed. This may be a valuable conclusion in itself, and reaching this conclusion without having to consider all of the N variables can provide a considerable savings of computation over exploring all 2^N possibilities.

Constraint satisfaction has entered the public eye with the popular game of Sudoku, published in newspapers and on Web sites around the world. The name of this game, which was invented more than 25 years ago, was artificially created from the Japanese "Suji wa dokushin ni kagiru", which means that a number must stay alone. An example is shown in Fig. 20.1. It is a fancy version of Latin squares. In a Latin square, each square is to be filled in with an integer from a restricted range so that no row or column contains the same number more than once. Sudoku goes one step further by identifying additional neighborhoods (the 3×3 squares shown bounded by double lines in Fig. 20.1) in which no number can appear more than once. The constraint comes from the fact that some of the squares are already filled in. If 30 or more are filled in, the puzzle is easy, but when only 25 or fewer of the normal 81 squares are filled in, the puzzles are considered "evil" or "diabolical" and provide a very enjoyable challenge for human appliers of logic. In the practical problems that we study, there are many solutions, but none of them is easy to find. Here, in contrast, the puzzle solutions are claimed to be unique, and an effort is made to have the solutions discoverable

	7		2			3	9	
						2		
		9					1	
9		1	4		7		3	
			1		5			
	4		6		9	1		7
	8					4		
		7						
	6	3			4		5	

Fig. 20.1. The popular logic game Sudoku is a handcrafted CSP. Each empty *square* is to be filled with a number between 1 and 9. Each row, column, or 3×3 array of squares bounded by the heavier *lines* must use each of the nine numbers exactly once. We are told that the problem can be solved using only logical reasoning, not trial and error, and that the solution is unique. The uniqueness of the solution makes Sudoku problems somewhat different from naturally arising CSPs

by applying complex chains of logical inferences, without requiring extensive depth-first search and backtracking.

20.2 Benchmarks and Competitions

Naturally, any good problem, such as constraint satisfaction, creates an opportunity for intense competition. In constraint satisfaction, annual contests are run to determine the best all-around constraint satisfaction solver [94, 95]. Baskets of 1000 or more challenging problems, left unsolved in previous contests or never seen before, are compiled. The problems typically include pathological examples constructed by hand, problems drawn from or suggested by industrial practice, and randomly generated problems. CSP solvers are run automatically in a test environment against all of the problems in the basket and compared on both the total running time they require to solve them and on the number of problems in the basket that they actually succeed in solving before reaching an agreed-upon upper time limit for each particular problem.

In a typical year, the winning programs solve about 75% of the challenge problems, and the CPU time required varies by a factor of as much as 10, even among just the top five or six programs entered. This contest structure prevents programs tuned to solve a particular type of program from getting very far, but it has not to date answered the question of exactly how the best techniques for solving actual problems might differ from the simplest methods that have been tested extensively on randomly generated problems.

In our discussion, we will explore the properties of CSP solutions and solution methods for a particularly simple class of randomly generated problems, because of the more general insights that these suggest. We will not discuss hand-generated pathological examples.

20.3 Randomly Generated Models and Their Complexity

Two classes of random constraint satisfaction models have attracted considerable interest, K-SAT and K-XOR-SAT. As a result we now understand the characteristics of their solutions in considerable detail and can hope to extrapolate the insights gained to other more structured problems. In each problem, we have N Boolean variables, x_i, and generate a formula, F, at random. We then ask to exhibit a configuration of the Boolean variables for which the formula evaluates to TRUE. The formula is most easily expressed as the AND of a collection of M clauses, each involving K of the Boolean variables, selected at random. If all of the clauses are individually true, then their logical AND, and the formula as a whole, will also be true.

The parameter that characterizes the overall difficulty of satisfying a typical formula is the ratio M/N, which we will call α. In addition to considering methods of solving large formulas, we shall be interested in the asymptotic behavior of this problem, the results of which will almost certainly be seen when N and M are allowed to become very large while their ratio, α, remains constant. When α is small, the formulas are almost always satisfiable. When α is very large, the formulas will almost surely not be satisfiable. Somewhere in between, physical reasoning suggests, there will be a phase boundary between the satisfiable phase (SAT) and the unsatisfiable (UNSAT), and the crossover between these two will become sharp as $N \to \infty$. Experiments show that this indeed happens, and both experiment and theory now provide considerable insight into the details of this surprisingly rich transition. Whether or not the transition also occurs in realistic CSPs and occurs with some of the same phenomena is currently an active area of research.

K-SAT and K-XOR-SAT differ in the definition of their clauses. In words, each clause in K-XOR-SAT takes K distinct randomly selected variables and demands that their parity (the XOR of the K variables) be either 0 or 1, chosen at random with equal probability. In K-SAT, again K variables are chosen at random for each clause. Each variable is assigned a desired value, chosen at random to be either "0" or "1" with equal probability. The clause is deemed to be satisfied as long as one or more of the variables takes its desired value. In other words, each clause is the OR of K comparisons between the variables and their desired values. This logical structure, the AND of OR clauses, is known as conjunctive normal form, or CNF, so another name for random K-SAT is random K-CNF.

Both problems are easily solved when $K = 2$ but become difficult for $K \geq 3$. Values of K that are not integers can also be studied by considering formulas that are mixtures of two values of K. Thus $K = 2.5$ can be considered a mixture of equal numbers of two-variable and three-variable clauses. The clauses in K-XOR-SAT are satisfied by half of the assignments of variables, regardless of the value of K, while the clauses in K-SAT are satisfied unless all K of the variables differ from their desired values; thus they fail to be satisfied for only 2^{-K} of the assignments. As a result, the value of α at which the K-XOR-SAT problem becomes unsatisfiable is much smaller than the value of α at which the K-SAT problems become unsatisfiable for the same value of K.

There is a connection between the $K = 2$ problem and stochastic branching processes, which is at the heart of why $K = 2$ is an easy problem. Consider one clause in a 2-SAT formula, for example "X OR Y". For the formula to be true, at least one of the two clauses in it must be true. So if X is not true, Y must be, and if Y is not true, X must be. We can represent this situation as a directed graph, in which an arrow leads from $\neg$X to Y and a second arrow leads from $\neg$Y to X. If we construct this graph of implications for an entire formula, the arrows will form chains whose length is finite when $\alpha < 1$ and diverges when $\alpha > 1$. This is because the expected number of outgoing links from each clause is α. The formula is satisfiable iff there is no path leading from some variable to imply its negation. If the paths formed by the arrows of implication that represent the 2-clauses are all short in length, it is possible that no such path is present. But if the paths connect and span a significant fraction of the variables, then almost certainly there will be many such contradictions. Thus for 2-SAT, the runaway seen in the branching process is also the phase boundary between a satisfiable and an unsatisfiable phase. Determining whether the formula connects a finite fraction of the variables requires a simple breadth-first search, whose cost is linear in the number of clauses present.

When 3-clauses are present, there is no compact way of searching for the logical implications. Searches for satisfying assignments to the variables or for logical contradictions are inherently capable of branching and thus require a cost exponentially increasing with the size of the problem.

20.4 Randomly Generated Models and Their Phase Diagrams

Early interest in the satisfiability problem arose from reports that it was in fact easy when posed against an ensemble of randomly generated problems. For example, Goldberg [68] described a class of random SAT problems for which a construction heuristic, augmented by backtracking [47], can almost always find a satisfying solution. In Goldberg's model, each clause is generated by selecting variables with a fixed probability. This leads to clauses of varying length. It was soon realized that in this model, the cost of solution is polynomial for almost all instances [60]. It is difficult to generate computationally hard problem instances. This does not mean that they do not exist, only that they are rare, and may in fact not be observable in practice.

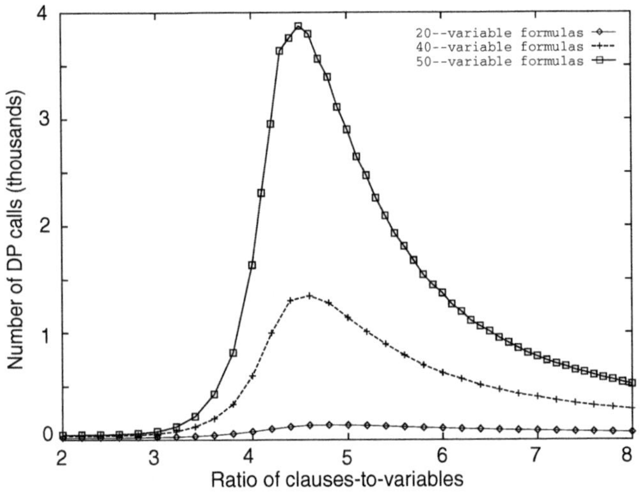

Fig. 20.2. The cost (in number of constructive forward search steps) of solving 3-SAT problems displays a peak at the SAT-UNSAT boundary, which sharpens dramatically when $N \to \infty$ (from [113])

However, once the length of each clause is fixed, the K-SAT problem does reliably generate problems that are hard to solve by any known algorithm, at least for certain ranges of α. In Fig. 20.2, we see that the cost of finding a solution peaks at $\alpha \approx 4.3$ for 3-SAT. In Fig. 20.3, we see that in this range of α, the chance that the randomly generated 3-SAT formula is simply not satisfiable first becomes significant. The width of the peak in the solution cost roughly corresponds to the range of α over which this probability change occurs. In this regime, the cost of finding a solution is observed to grow

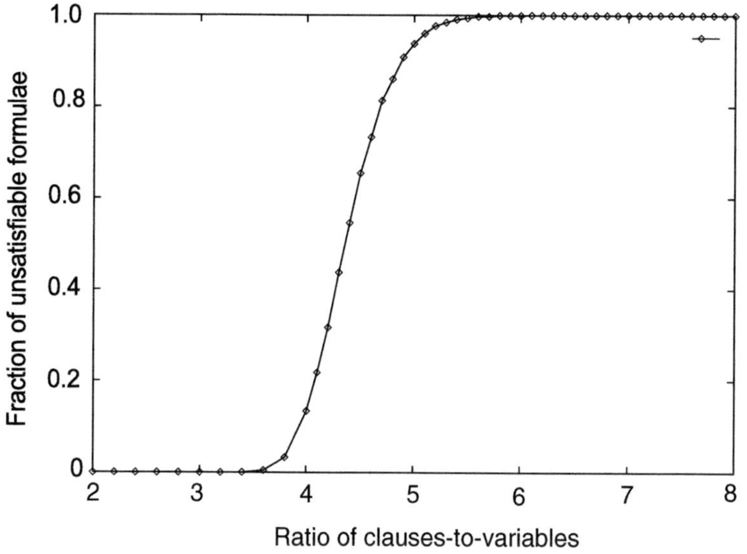

Fig. 20.3. Fraction of 3-SAT problems found to be unsatisfiable when formulas are generated with 50 variables, $M = \alpha N$ clauses (from [113])

exponentially with N, the number of variables. The narrowing and apparent divergence of the peak in the cost of solving 3-SAT problems suggests that some sort of critical phenomena are taking place.

A survey of modest-sized problems in K-SAT for values of K ranging from 2 to 6 is summarized in Fig. 20.4. The threshold curves, showing the fraction of unsatisfiable formulas in the ensemble increasing from near 0 to near 1, appear to sharpen up and pivot around a common point for each value of K, implicitly suggesting that this is the asymptotic position of the phase transition between the satisfiable phase and the unsatisfiable phase. "Finite-size scaling" is a useful tool in the study of phase transitions. The basic idea is to find a simple and general transformation of the scale over which the parameters (here, α) vary when passing through the critical regime, so that with this transformation data obtained from samples of various sizes can be reduced to a common form. The transformation should depend only on N and the distance from the critical point. A standard form that we have employed is to plot data for the fraction of formulas that are unsatisfiable against, not α, but a rescaled coordinate, $\alpha' = N^{1/\nu}(\alpha - \alpha_c)/\alpha_c$. This mapping has the effect of "opening up" the transition region as it narrows with increasing N.

In Fig. 20.5, we see that the data for smaller samples collapse into a single crossover form and coalesces into a single curve at larger values of N. Here the values of the parameters used are $\alpha_c = 4.2$ and $\nu = 1.5$. These values are not very precisely determined; other nearby values of α_c and ν will also collapse

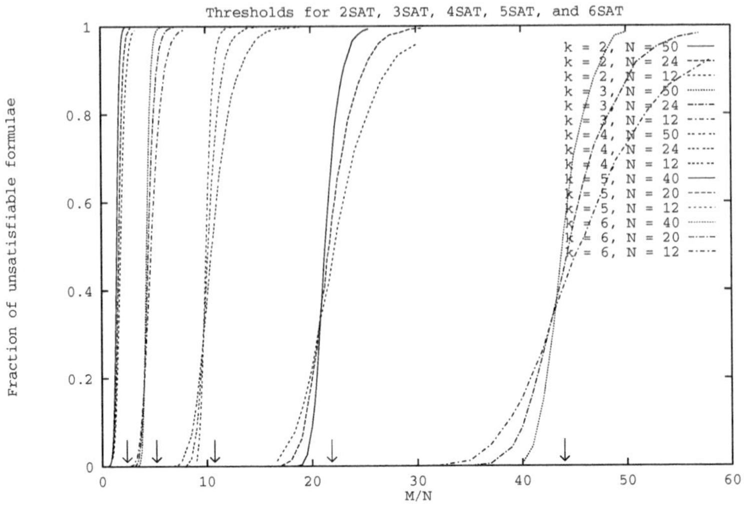

Fig. 20.4. Phase boundary between the SAT and the UNSAT phase seen for K-SAT with $K = 2, 3, 4, 5$, and 6 (from [198])

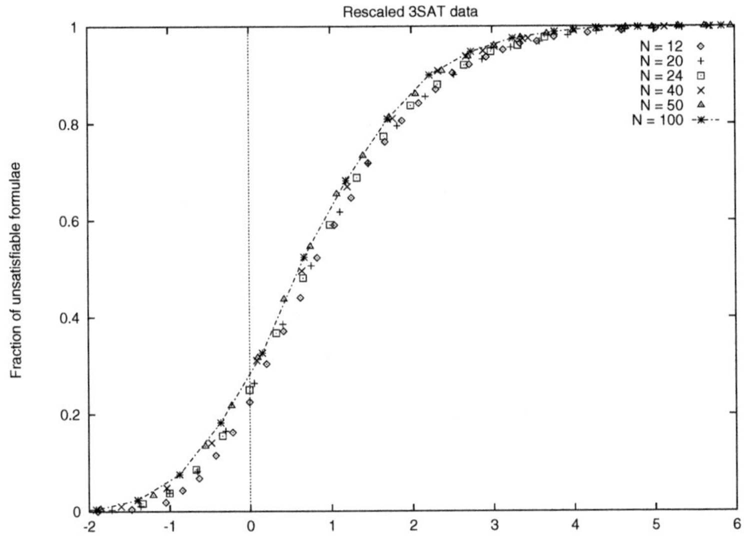

Fig. 20.5. Fraction of unsatisfiable formulas for $K = 3$, various N, rescaled to lie on a single transition *curve* (from [198])

the data and look almost as good to the eye. Still, the evidence of a behavior that becomes simple in the limit of very large problem size is quite compelling. This type of "finite-size scaling" analysis can be applied to many problems in combinatorics in which one has a threshold between two characteristics. Such "phase transitions" are becoming a well-understood aspect of large complex systems.

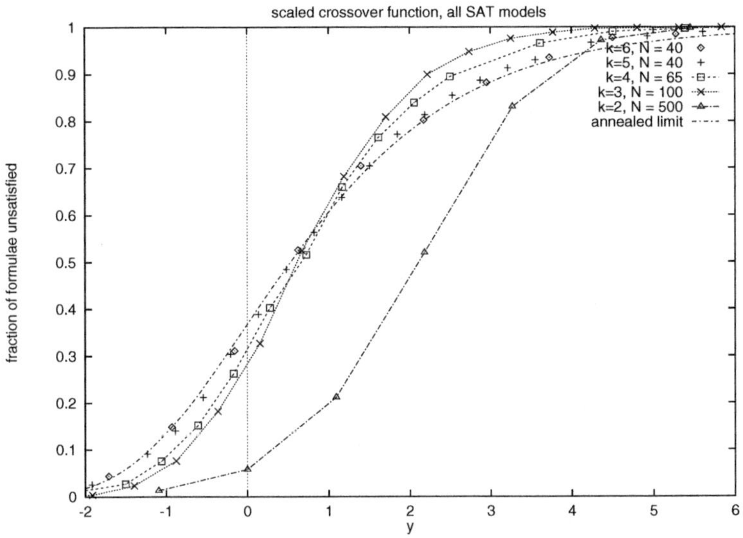

Fig. 20.6. Crossover *curves* connected by finite-size scaling for $K = 2, 3, 4, 5$, and 6. The data *points* for $K = 6$ are plotted against the crossover *curve* predicted in the annealed limit (from [198])

Figure 20.6 shows that additional information can be inferred from the shapes of the rescaled crossover curves seen in the various cases with $K = 2, 3, 4, 5$, and 6. For $K = 2$, the shape of the crossover curve is quite different from that of the other cases. As K reaches 6, we also find that the curve takes a shape with $\nu = 1$ and a simple form that can be predicted by arguments based on averages in which we neglect all correlations between the values of the variables in different clauses. This is often called an "annealed" average in the study of disordered materials. If we define

$$y_{\mathrm{ann}} = N(\alpha - \alpha_{\mathrm{ann}})/\alpha_{\mathrm{ann}} \, , \tag{20.1}$$

then the probability that a formula will remain unsatisfied in all 2^N configurations can be estimated as $e^{-2^{-y_{\mathrm{ann}}}}$. The data for $K = 5$ and 6 tend to this limit in Fig. 20.6.

20.5 Mixtures of easy and hard CSPs

In contrast to the models with a mixture of all sizes of clauses, shown in [60] to almost always be satisfiable, if we study mixtures that interpolate between two adjacent values of K, the results are more interesting. The case $K = 2$ we know can be solved by a search whose cost is linear in N, while the case $K = 3$ is a classic NP-complete problem. What happens in between has now been carefully studied [145, 146, 147] by experiments and theory on formulas with random mixtures of 2- and 3-clauses. Consider a formula with M clauses, pM of which are 3-clauses and $(1-p)M$ of which are 2-clauses, with $0 \leq p \leq 1$. This "$(2+p)$-SAT" model smoothly interpolates between 2-SAT ($p = 0$) and 3-SAT ($p = 1$). For a given value of N the number of 2-clauses cannot exceed N, since then even in the absence of any 3-clauses the formula would likely be unsatisfiable. This sets an upper limit of $\alpha_c(2+p) \leq \alpha_c(2)/(1-p) = 1/(1-p)$. Using rigorous arguments, Achlioptas et al. [3] showed that $1/(1-p)$ is also a lower bound for $p \leq 0.4$. Above $p = 0.4$ the upper and lower bounds separate, suggesting that only above $p = 0.4$ do the 3-clauses begin to play a role in inhibiting satisfiability of the constraints in the problem. In Fig. 20.7, we compare experimental results obtained by depth-first search with backtracking with the upper and lower bounds of [3] and a theoretical prediction obtained by methods from statistical mechanics that are outside the scope of this book. All three agree that from $p = 0$ to $p = 0.4$

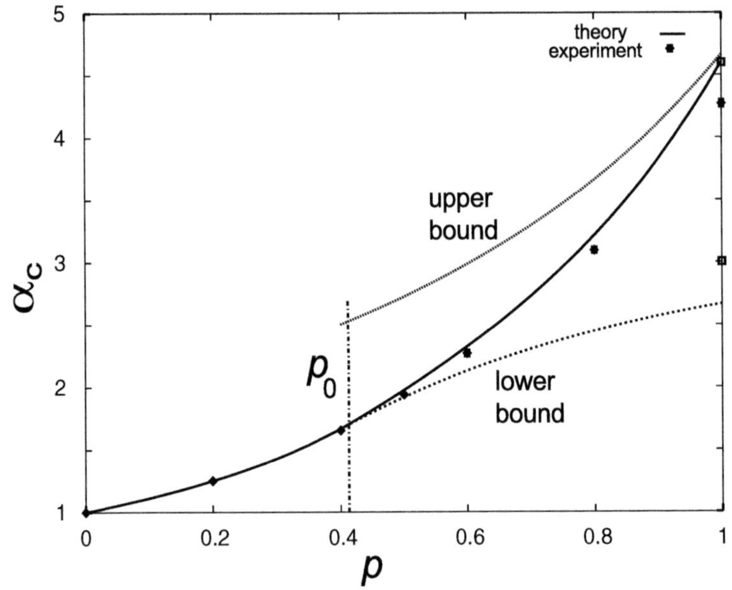

Fig. 20.7. Theoretical and experimental results for the SAT/UNSAT transition in the $2+p$-SAT model (from [145, 146])

the typical behavior of these mixture models is that of its 2-SAT subproblem, with the 3-clauses appearing to play no role. From $0.4 \leq p \leq 1.0$, both sets of clauses are apparently important in providing constraints. The "replica symmetric" theory of Monasson et al. [145] gives a pretty good account of the transition point, making at worst a 10% overestimate at $p = 1$, while both sets of results lie within the rigorous bounds on the transition that are derived in [3].

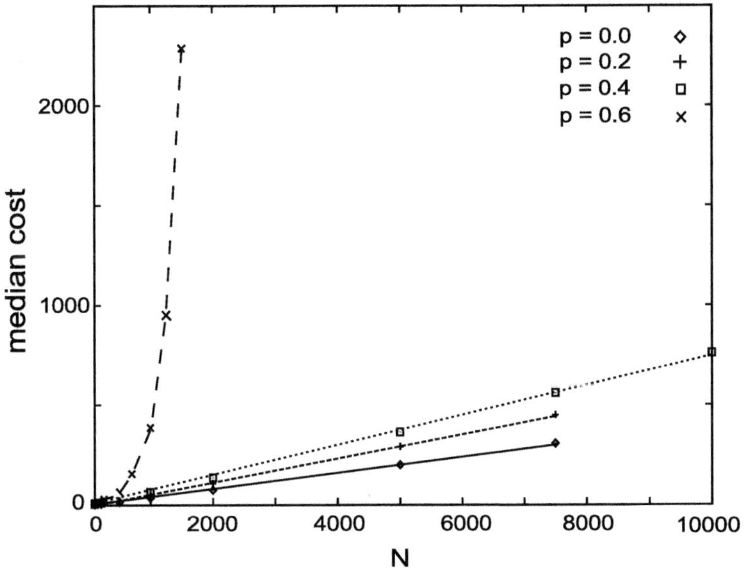

Fig. 20.8. Median computational cost of proving a formula SAT or UNSAT using a depth-first search with backtracking, for p ranging from 0 to 1. Note the extreme difference in the behavior of this cost as seen for $p = 0.6$ and as seen for $p \leq 0.4$

Another observation, which tends to confirm the interpretation we have given of the role of the 2-clauses and 3-clauses as functions of p, is the cost of solving these problems as N increases. Figure 20.8 provides data from studies with $p = 0, 0.2, 0.4$, and 0.6. The cost increases linearly with N until $p = 0.4$, and more steeply for $p > 0.4$. The figure shows data from $p = 0.6$, but the effect is evident even earlier than that.

21 Construction Heuristics for CSP

21.1 Application of the Bestinsertion Heuristic to the 3-SAT Problem

As a first approach to solving a constraint satisfaction problem (CSP) with a construction heuristic, one might think of simply applying a standard construction heuristic like the bestinsertion heuristic. Thus, one would start out with a tabula rasa, in which each binary variable x_i is unset. Then one selects one of the variables x_i at random and asks whether one should set x_i to true or to false according to the bestinsertion paradigm, i. e., x_i is set to true if a larger number of clauses are fulfilled than if setting it to false and vice versa. If exactly the same number of clauses is fulfilled when set to either true or false, then it is randomly set to either true or false with equal probability. This method of randomly selecting a variable that is not yet set to either true or false and then setting it to the better value is repeated until all variables have been set.

Of course, such an approach might be sufficient for very simple problems. However, we tested them at a few benchmark sets of the 3-SAT problem called uf$N$-M, consisting of several benchmark instances, with $20 \leq N \leq 250$ and $91 \leq M \leq 1065$, all of which lie in the phase-transition region, are satisfiable, and can be downloaded from [91]. We performed 100 optimization runs for each benchmark instance. Figure 21.1 shows that the bestinsertion heuristic only rarely found the solution for small systems; it failed completely for $N \geq 100$.

Here the question arises of the extent to which the bestinsertion heuristic was unable to solve these problem instances. In order to investigate this problem, we measured for every instance the average and the maximum fraction of unsatisfied clauses in the series of our 100 applications of the bestinsertion heuristic to this problem. Taking the average of these numbers over all the instances with a specific value of M, we find that the average of the average fraction of unsatisfied clauses stays rather constant, whereas the maximum fractions decrease, as shown in Fig. 21.2.

Of course, one can argue now that we have applied the simple bestinsertion heuristic to these specific benchmark instances that lie in the transition range and can thus be considered rather hard. Therefore, we also generated our

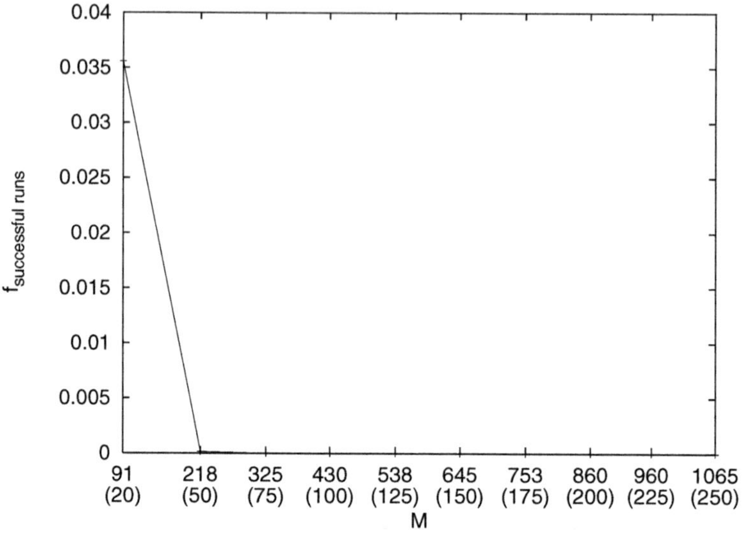

Fig. 21.1. Fractions of successful optimization runs using the bestinsertion heuristic, which is applied to the benchmark instances mentioned in the text. For each benchmark instance, we performed 100 optimization runs. We plot these fractions vs. the number M of clauses in the benchmark instances. Below, we give the number N of variables in brackets. For most values of N, the benchmark library contains 100 benchmark instances, except for $N = 20, 50$, and 100, for which there are 1000 benchmark instances. We find that only 3.5% of the runs are able to find a feasible solution for $N = 20$ and $M = 91$. For $N = 50$ and $M = 218$, 0.015% of the runs were successful, for $N = 75$ and $M = 325$, only one instance was solved once. For all larger instances, the bestinsertion heuristic always failed to provide a feasible solution

own instances, with $N = 1000$ variables each and with various numbers M of clauses, in the range between $M = 100$ and $M = 2000$. For each clause, three different binary variables were selected that were inserted as themselves into the clause with 50% probability or as their negations with 50% probability.

Figures 21.3 and 21.4 show the results averaged over these instances. We find that the fraction of runs ending in a feasible configuration decreases sigmoidally from 1 to 0 and finally vanishes for $M/N = 1.3$. For large fractions M/N, we get a linear increase of the fraction of unsatisfied clauses.

Now one could also try to vary the bestinsertion heuristic. For example, one might want to put more emphasis on the unsatisfied clauses. Thus, one would select a randomly chosen unsatisfied clause in which at least one of the three variables is not yet set to either true or false. Then one would randomly select one of these variables and set it to either true or false according to the bestinsertion paradigm. Please note that this does not mean here that

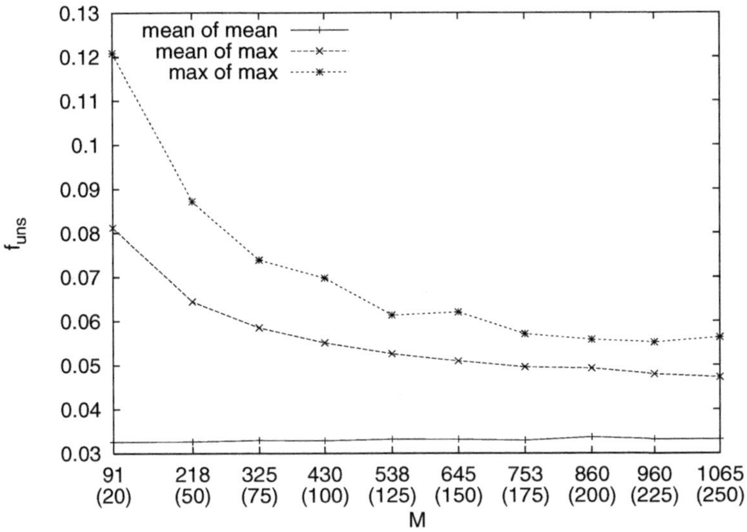

Fig. 21.2. Fraction f_{uns} of clauses remaining unsatisfied after applying the bestinsertion heuristic vs. overall number M of clauses. (Again we give the corresponding values of N in *brackets*.) For each benchmark instance, the average and maximum fraction of unsatisfied clauses was determined. The graphic shows the average of this average fraction ("mean of mean"), the average of the maximum fraction ("mean of max"), and the maximum of all maxima ("max of max"). We find that the *curve* "mean of mean" stays rather constant and that the maximum fraction of violated clauses decreases with increasing system size

the clause that was selected originally is then fulfilled, as the bestinsertion heuristic prefers to introduce the variable in such a way that a maximum number of clauses are satisfied and not that a specific clause is satisfied. We also implemented this variant of the bestinsertion heuristic. The results for this variant are also shown in Figs. 21.3 and 21.4. We find that this variant leads to worse results than the original bestinsertion heuristic. This result is in accordance with corresponding results for the traveling salesman problem: there most of the variants of the bestinsertion heuristic led to worse results than the bestinsertion heuristic itself, although their rule sets were more elaborate.

Summarizing, the application of a simple standard construction heuristic to a 3-SAT problem fails nearly completely when the fraction M/N reaches values in which the 3-SAT problem becomes interesting. For hard CSP problems we will need to develop much more clever and fast approximate strategies. But for easy problems (which do also arise in practice), only slight improvements in this brute force construction heuristic are required.

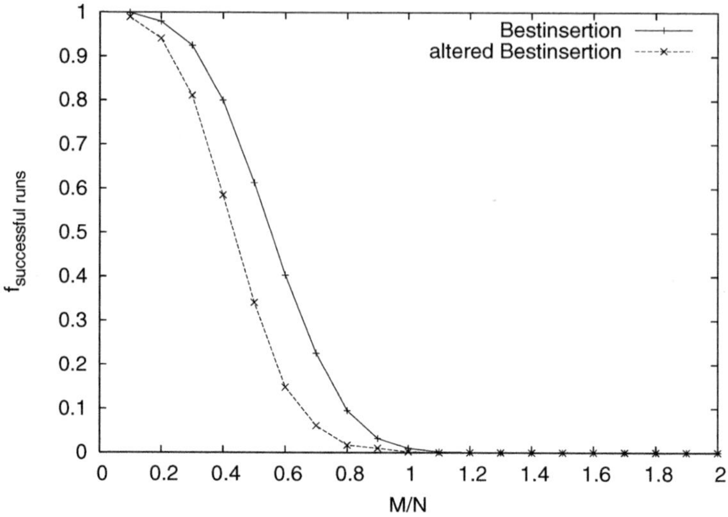

Fig. 21.3. Fractions of successful optimization runs using the bestinsertion heuristic or an altered variant of the bestinsertion heuristic, which is applied to instances generated by us. We generated 100 instances and performed 10 optimization runs on each of these instances. We plot these fractions vs. the fraction M/N in the benchmark instances

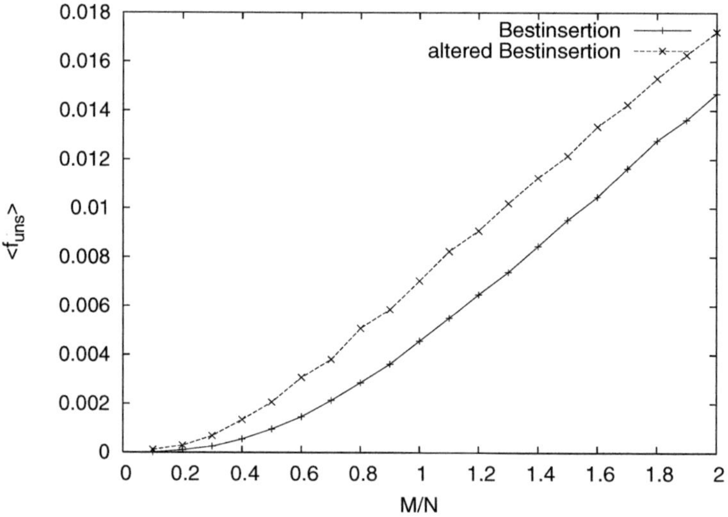

Fig. 21.4. Fraction f_{uns} of clauses remaining unsatisfied after applying the bestinsertion heuristic or an altered variant of the bestinsertion heuristic vs. fraction M/N. These results were achieved in the same optimization runs as were used for Fig. 21.3

21.2 Assertion, Decimation, and Resolution

The construction process described in the preceeding section has been used and studied under several names. The oldest is "assertion". This name arises in the study of logical expressions and systems for determining their capabilities of expression. If we think of our variables as logical variables and our formulas as logical expressions, then it is natural to speak of "asserting" the truth (or falsity) of a "variable" and to use the term "literal" for the form in which a variable or its negation appears in a logical expression. Thus our 3-SAT formulas consist of clauses, each of which is the logical OR of three literals. Asserting a variable to be either true or false, whichever makes the resulting literal in the clause evaluate to true will make the clause true. The term "decimation" arises from the fact that each assertion that satisfies one or more clauses reduces the size of the part of the formula remaining unsatisfied. If we find enough useful assertions, the formula will disappear entirely.

For 3-SAT, there is a clever series of assertions that has been shown to decimate away simpler formulas entirely. Suppose we list, for each variable, the number of clauses that require the variable to be true and the number of clauses that require it to be false. A variable that needs to be always true or always false, and will find no conflicting clause, is called a "pure literal". Asserting such a "pure literal" to be true removes all of its clauses from the formula. This will require correcting the lists for the remaining variables, some of which were in the clauses just removed. Some of these may become "pure literals" as well. Broder et al. [30] showed that decimating pure literals (the "pure literal rule") will, with high probability, solve 3-SAT for all values of $\alpha < 1.63$. Dealing with pure literals first, rather than decimating at random, as in the bestinsertion heuristic, seems a simple price to pay for such a large improvement.

Alekhnovich and Ben-Sasson [7] were able to take the pure literal result a step further and show that this result implies that there exists an upper bound to the cost of running local improvement heuristics, to be discussed in the next chapter, for all 3-SAT instances with $\alpha < 1.63\ldots$.

21.3 Analyzable Assertion Protocols

To get even stronger results, and to learn some intriguing things about the structure of the solution space for SAT problems, we next restrict our attention to the XOR-SAT problem. 3-XOR-SAT can also be expressed as a model in which the variables are Ising spins (taking the values +1 or −1), and the "clauses" are replaced by products of three of the spins, multiplied by either +1 or −1, with equal probability. We shall use the XOR-SAT language in this discussion, considering the variables to be Boolean and the clauses to be the truth or negation (chosen at random) of three distinct variables, selected at random. There is an equivalent of the pure literal rule for this problem. Consider a variable that participates in only one clause. Such "1-variables" can be asserted to do whatever the clause and the remaining variables require, and as a result the clause can always be satisfied. So for this problem, the process of decimation consists of finding all 1-variables, removing them and their clauses from consideration, finding any variables that are now exposed as 1-variables, removing them, and continuing until no 1-variables remain. This process has been studied by several groups [171, 37].

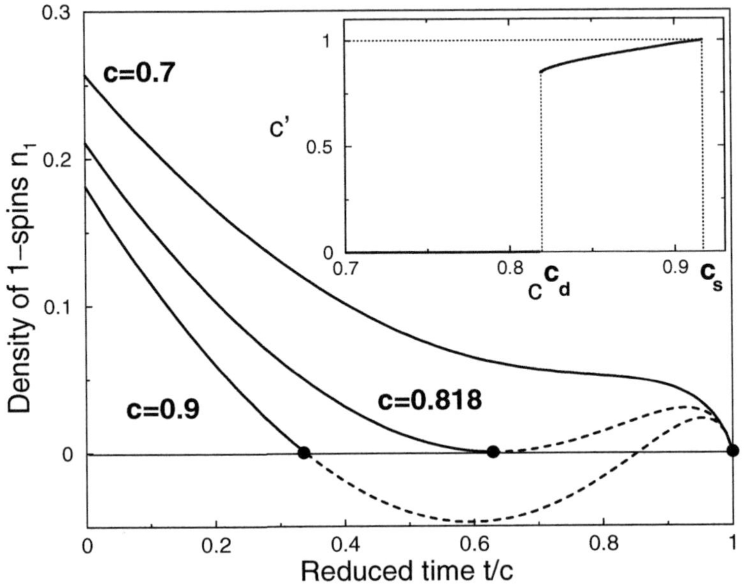

Fig. 21.5. Density of 1-spins (1-variables) generated by the decimation procedure in 3-XOR-SAT as the decimation progresses to completion. For concentrations of less than $c = 0.818$, the decimation proceeds until all clauses have been eliminated. Above $c = 0.918$, no 1-spin decimation is possible, and formulas are unsatisfiable (from [37])

In Fig. 21.5 we see the result of this decimation process. For concentrations c of clauses (equivalent to our parameter α) less than 0.818, the decimation process is almost always successful. Between $c = 0.818$ and $c = 0.918$ the decimation process reduces the size of the formula somewhat but stops after removing a relatively small fraction of the variables and clauses, as shown in the inset of Fig. 21.5. However, an analysis based on tracing the propagation of inferred signs of variables, similar to the construction of inference chains described for 2-SAT, can be carried out here. This shows that the chains are short, and thus satisfiability is likely for $0.818 \leq c \leq 0.918$. In Fig. 21.6, we see that the threshold sharpens up in the canonical fashion discussed in

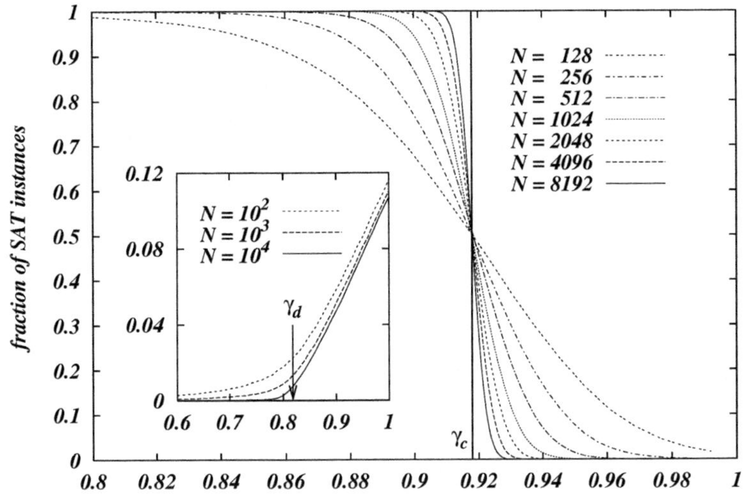

Fig. 21.6. Finite-size sharpening of the SAT-UNSAT threshold in the 3-XOR-SAT model as a function of c and of the number of variables, N. *Inset*: energy reached by a deterministic rule that is not strong enough to identify satisfying configurations in the hard-SAT region (from [171])

Chap. 20, indicating a phase transition into an UNSAT phase taking place at precisely 0.918. In the intermediate region, one is likely to be forced to accept a good, but not optimal, solution. The inset of Fig. 21.6 shows the minimum cost reached by a simple deterministic strategy for finding such ground states. We call this intermediate regime a hard-SAT region, and its occurrence is a common feature of these combinatoric problems defined over ensembles of random logic graphs.

21.4 Solution Space Structure of XOR-SAT

The XOR-SAT model is unusual in that we know the number of ground states and many things about their distribution over the configuration space of the model as functions of the concentration of clauses. By the methods developed in [37] we find the entropy s (the log to base 2 of the number of ground states) is given by

$$s = 1 - c\,, \tag{21.1}$$

as shown in Fig. 21.5. The entropy is continuous at the SAT/UNSAT threshold, $c_{\mathrm{s}} = 0.918\ldots$, but for the larger values of c describes low-lying states in which there are inevitably some unsatisfied clauses. The interesting region lies between c_{d}, defined as the "dynamical threshold", and c_{s} because of the difficulty to reach ground states that was shown in Fig. 21.2. Below c_{d}, the

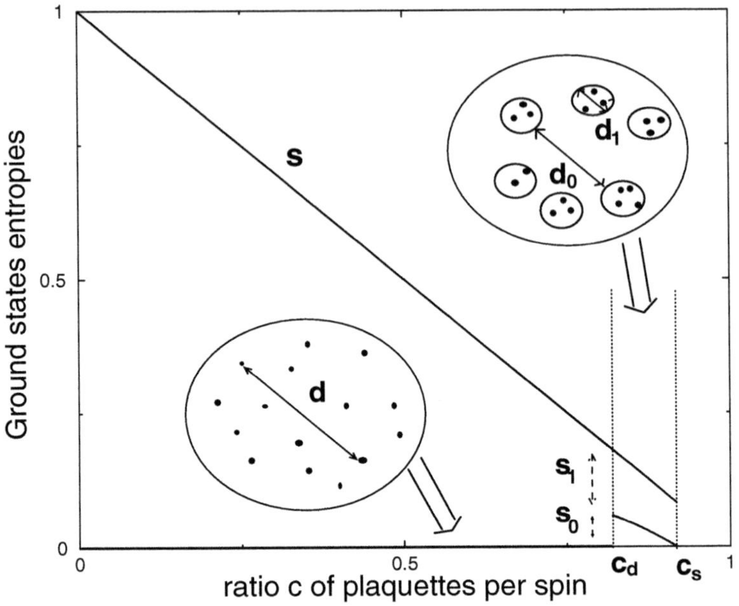

Fig. 21.7. Nature of the ground states in the 3-XOR-SAT model as a function of the concentration c. The total entropy (logarithm of the number of solutions) decreases as $s = 1 - c$ and is nonvanishing at the threshold, $c = 0.918\ldots$. In the hard-SAT region, the entropy is the sum of two terms, the logartithm of the number of clusters (s_0) and an intra cluster entropy (s_1)

ground states are uniformly distributed over the entire configuration space, since they constitute all possible configurations of the variables remaining when the decimation of 1-variables is completed. Above c_{d} the decimation completes leaving most of the variables still interacting. The entropy now separates into two components. The number of ground states of the reduced, decimated system is still of order $\exp(s_0 N)$, and each such ground state can be considered the seed of a large number of possible additional states in which the variables removed during the decimation process take on all possible values. On average, this adds s_1 in entropy to each of the reduced ground states.

There is also a geometric interpretation to this more complex set of ground state configurations. The ground states of the reduced formula can be thought of as the centers of $\exp(s_0 N)$ clusters. The clusters are separated on average by a Hamming distance of $N/2$, as were all ground states for $c < c_{\mathrm{d}}$. The additional states of each cluster are grouped more tightly, as suggested by the sketch in Fig. 21.7. The average separation between states in the same cluster, measured in Hamming distance, is ≤ 0.15, since that is the fraction of variables removed in the 1-variable decimation, and these take essentially random values.

This suggests a general hypothesis that the extra hardness associated with the transition from a solvable to an unsolvable problem in combinatorics or optimization is a consequence (or at least is accompanied by) a change in the geometry and statistics of the configurations that are sought. The presence of a satisfiable but still much harder-to-solve region before the transition is also a feature that was subsequently confirmed in other models. However, XOR-SAT is an exactly solvable model, at least in its satisfiable phase, and may be somewhat simplistic. K-SAT, and perhaps other models that have not been solved, appear to display this pattern but exhibit additional finer structure in the approach to their phase boundaries from the SAT side.

22 Random Local Iterative Search Heuristics

22.1 RWalkSAT

In this chapter we consider a fairly wide range of algorithms for K-SAT and other constraint satisfaction problems (CSPs), all of which are in the family of local rearrangement rather than construction heuristics. Many of them are somewhat carelessly called "walksat", although there are several different heuristics lumped under the general phrase. We shall separate them by their histories.

The first proposal for a local improvement heuristic for random CSPs was made by Papadimitriou in 1991 [160]. In this method, one starts with a random configuration of the variables of a SAT formula and focuses on the unsatisfied clauses. In each step, one chooses an unsatisfied clause, selected at random, and chooses one variable found within that clause, also selected at random. Reversing the value of that variable has the primary effect of satisfying the clause. It may also satisfy other clauses that were unsatisfied or cause other clauses to become unsatisfied. These secondary effects should cancel out initially, but as the density of unsatisfied clauses decreases, one would expect that the clauses that are made unsatisfied would increase in number and limit the effectiveness of this heuristic. However, Papadimitriou was able to prove that this heuristic will solve 2-SAT formulas with $\alpha \leq 1$ (in the SAT phase) in at worst N^2 steps. We will refer to this simple rule as RWalksat, since it is essentially a random walk with one clever trick.

Actually, the heuristic works much better than that. Alekhnovich and Ben-Sasson [7] proved that it will reach a solution of 3-SAT random formulas with $\alpha < 1.63$ at a cost in steps that is linear in N. They optained an upper bound for the length of the search by use of the pure literal rule. In fact, it works in linear time over an even wider range than this analysis could prove. Two recent studies [200, 16] discovered that RWalkSAT finds satisfying configurations for 3-SAT in time linear in N as long as $\alpha \leq 2.7$. Although the search cost increases only linearly with increasing N, the proportionality constant diverges as α approaches its "dynamical transition", α_d. When $\alpha > \alpha_\mathrm{d}$, the cost of solving the problem using RWalkSAT increases exponentially with N. There is a simple explanation for this. In the easy region, $\alpha \leq \alpha_\mathrm{d}$, the random walk reaches a ground state directly, but at larger values of α, the initial decrease in energy reaches a nonzero average value, but with large fluctuations about that average value. Above α_d, the time to reach a satisfying configuration using RWalkSAT is dominated by the waiting time for a large enough fluctuation about this average to occur that a ground state is reached.

Desroulers and Monasson [48] have shown that the characteristics of the endpoint of the linear region and onset of exponential cost are common to several algorithms. They propose, based on an analysis of decimation rules, that the probability of finding a solution with these simple heuristics has a universal form, decreasing as $\exp(-N^{1/6})$. In addition, the width of the crossover region scales as $N^{-1/3}$. They argue that this form should apply to all heuristics that work for sufficiently easy problems, those outside the Hard-SAT region, so this could include both construction heuristics and methods based on local moves, such as RWalkSAT.

22.2 WalkSAT

Selman, Kautz, and various coworkers have extended Papadimitriou's idea into a widely used local rearrangement search method for solving many sorts of CSP problems, both model problems and those arising in real-world contexts [196, 197]. These methods are all called WSAT, but at least six variants of the actual moves employed have been proposed. They are reviewed in [135]. The most widely used procedure is still their original version, which adds two tricks to the random variable selection of [160]. After choosing an unsatisfied clause to satisfy, they define a "greedy" move as choosing to reverse the vari-

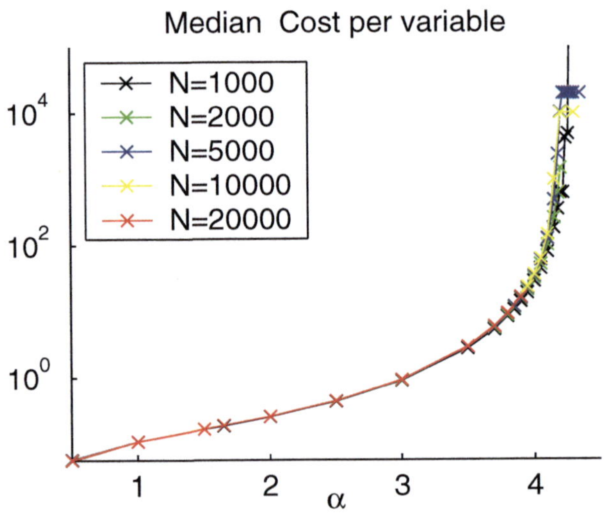

Fig. 22.1. Median cost of WSAT random walk steps per variable taken to solve 3-SAT formulas with α ranging from 0.5 to 4.3 (from [12])

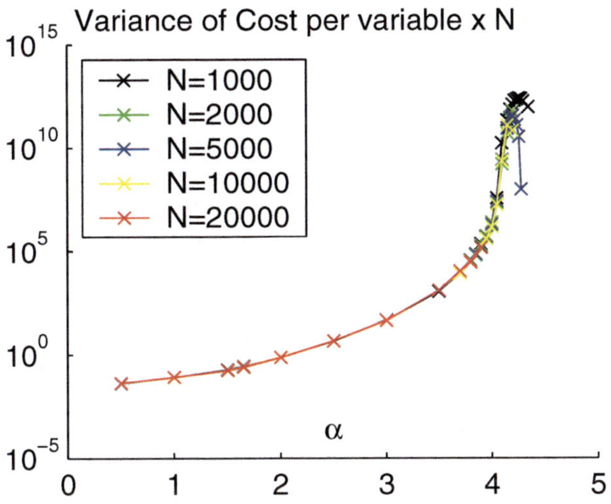

Fig. 22.2. Variance of the WSAT median cost, multiplied by N (from [12])

able that breaks the fewest previously satisfied clauses by its reversal. (One could also choose to reverse the variable for which the number of clauses that become satisfied minus the number that become unsatisfied is maximized, but this seems not to do as well.) Next, for robustness they mix moves using the random selection with moves using the greedy selection of a variable. Their recommendation is an equal proportion of the two moves, selected at random, but the fraction of each type of move used is an obvious tuning parameter. Finally, Kautz and coworkers test each variable in the unsatisfied clause to see if it is a "pure literal". If any variable proves to be a pure literal, they reverse it without considering any of the other moves.

Using the "greedy" move but without testing for pure literals, Barthel et al. [16] found little improvement in power over pure RWalkSAT. However, by testing first for pure literal moves, Aurell et al. [12] found that the cost of solving 3-SAT formulas with the full WSAT remained linear in N up to roughly $\alpha = 4.15$, as shown in Fig. 22.1. The cost per variable of the WSAT solution increases by about six orders of magnitude over this range of α, and the distribution of times observed in [12] is broad. Aurell et al. also evaluated the first four moments of this distribution to ensure that it became concentrated with increasing N on the linear dependence reported. Their results for the variance of the solution cost, scaled up by N, are shown in Fig. 22.2. The third and fourth moments of the distribution narrow in proportion to N and to N^2, respectively. Thus the cost of WSAT behaves in exactly the way that a process governed by the usual laws of large numbers is expected to behave, in spite of the very large increase in the magnitude of the effects observed.

22.3 Simulated Annealing

After this discussion of the application of algorithms specifically designed for the satisfiability problem, we would like to mention that this problem can also be solved by a standard approach like simulated annealing (SA). For the application of SA, a cost function $\mathcal{H}(\sigma)$ for a configuration $\sigma = (x_1, \ldots, x_N)$ must be defined that is then minimized with SA. We chose $\mathcal{H}$ to simply be the number of unsatisfied clauses. Thus, when the Hamiltonian $\mathcal{H}$ reaches a value of 0 some time in the optimization run, we know that all clauses are fulfilled and thus a feasible configuration has been reached.

After initializing all binary variables x_i randomly with either true or false, we simply perform a standard SA run, starting at an initial temperature given by the overall number of clauses and then reducing the temperature exponentially by a factor of 0.9 to a final temperature of 10^{-2}. In each temperature step, we performed 1000 sweeps, i. e., $1000N$ moves. Each move simply selects one variable x_i at random and intends to perform the move $x_i \rightarrow \neg x_i$. This move usually changes the number of satisfied clauses. The move is then accepted or rejected according to the Metropolis acceptance criterion. A next higher move (but there was no need to implement it) would be to try to change two randomly selected variables at the same time.

We downloaded two large libraries of benchmark instances provided by SATLIB [94, 91], namely, the satisfiable uniform random 3-SAT instances that lie in the phase-transition region and all random 3-SAT instances with backbone-minimal subinstances contributed by Singer. With our simple SA approach, we were able to solve all of these instances with system sizes (N, M) between $(20, 91)$ and $(250, 1065)$. For the smaller instances, usually only one optimization run was needed to find a configuration in which all clauses were fulfilled. For the larger instances, it was sometimes necessary to run the SA program with up to ten different random seeds in order to get a feasible configuration.

Figure 22.3 shows the results of applying SA to one of the benchmark instances, using the parameters given above. Just as for the traveling salesman problem, we find that the mean energy decreases sigmoidally with decreasing temperature and finally freezes in the optimum value of 0. The curve of the specific heat is of course rather rough, due to the shortness of the optimization run. The specific heat exhibits its peak in the temperature range between 0.1 and 1. There might be a multipeak structure because of clustering and ordering effects. But, due to the small amount of calculation time, it could also be that the occurrence of a few peaks is only a nonequilibrium effect. The acceptance rate also decreases sigmoidally. However, it does not vanish for the smallest energies, although the energy no longer changes there. Thus, trivial moves can be performed that do not change the energy at all. Therefore, this problem instance has a degenerate ground state.

As these benchmark instances could so easily be solved with a simple SA approach, there was no need to think of more complex optimization schemes,

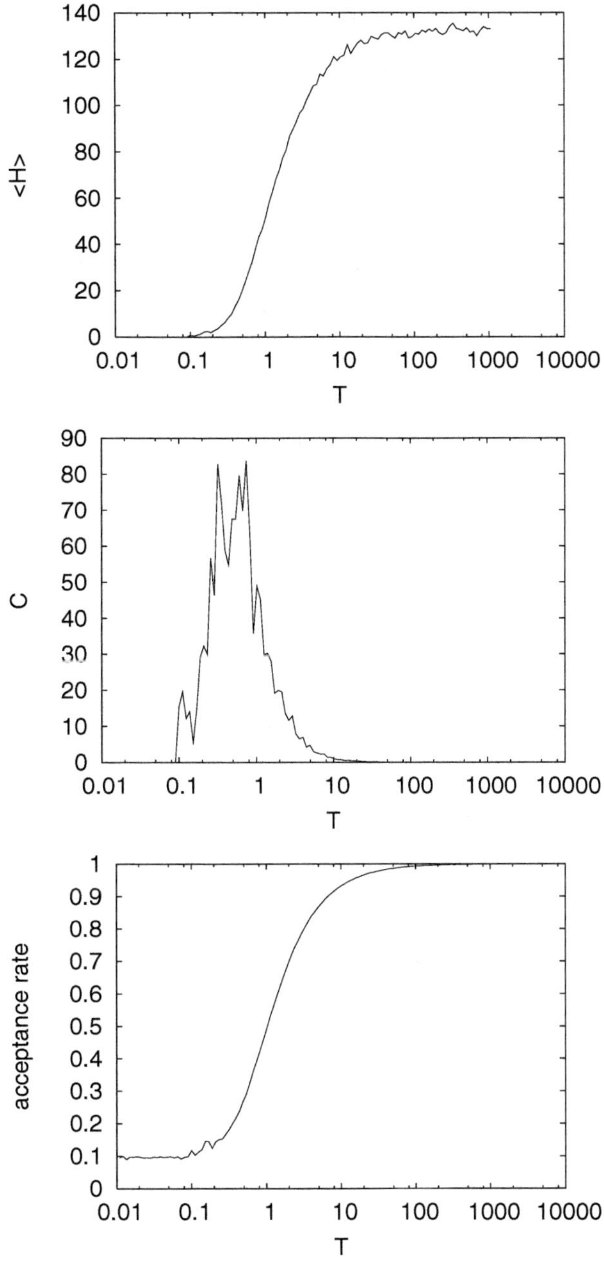

Fig. 22.3. Results of applying SA to the uf250-01.cnf 3-SAT benchmark instance [91]. The graphics show the change of the mean energy $\langle \mathcal{H} \rangle$ (*top*), of the specific heat C (*middle*), and of the rate at which the move was accepted (*bottom*) with decreasing temperature T

like, for example, ruin & recreate (R & R) moves or the searching for backbones (SfB) algorithm. But of course for more difficult instances, one might use these approaches: when working with R & R, one would randomly select some variables, remove them from the system in such a way that neither true nor false values were assigned to them, and reinsert them in a random order while trying to maximize the number of satisfied clauses by setting them to either true or false.

In the SfB algorithm, one would compare the solutions for equal parts, i. e., find out whether a variable x_i is set to either true or false in all solutions. If so, then it is again assumed that this setting also applies to the optimum solution. Therefore, the variable is removed from the system by replacing it with the value of true or false in all clauses of which it is a part. If it is replaced by true, then the clauses are automatically fulfilled, such that they can be removed from the system. If it is false, then the clause containing three variables that are connected by the OR operator are reduced to two variables with an OR operator in them. Thus, in both cases, the complexity of the system is reduced.

23 Belief Propagation and Survey Propagation

23.1 Belief Propagation, Message Passing, and Cavities

Belief propagation [162] is a fairly old strategy, popular in artificial intelligence, for using the calculus of probabilities to estimate where solutions to complicated discrete problems such as constraint satisfaction are most likely to be found. The mathematics behind it is usually not rigorous, but it offers the promise of replacing a difficult integer program with a more tractable linear or quadratic evaluation, which may often give a solution in which all the variables are in fact integers as desired, or may be accurately rounded off to the nearest integers. Belief propagation is generally performed as an iterative algorithm in which each probability, of "belief", is updated in the light of the information currently available for the beliefs of the other variables with which the variable on which the belief is based directly interact. To physicists, this sort of iterative update procedure is reminiscent of the "cavity models", of mean field theory used in early treatments of magnetic ordering. Finally, because the iterative evaluation of beliefs is carried out by passing information in messages that flow from one variable to another along the graph that is defined by the interactions in a problem, it is natural to think of distributing the calculation so that it can proceed asynchronously and in parallel without central coordination.

To make these ideas concrete, we shall show how they can be applied to constraint satisfaction, in particular to K-SAT. A recent series of papers drawing upon some general ideas from the statistical mechanics of disordered materials have given deep insight into the nature of the SAT-UNSAT phase transition [138, 139, 27, 161]. This work relies on the concept of "replicas" of the random system being studied, identical copies of that system that are studied together. The basic insight, going back to Edwards and Anderson [56], is that ordering in random systems is a matter of stability, rather than regular structure evident by its symmetry. Thus an ordered structure is observed because it forms repeatedly in the different replicas of the system. The simplest forms of order in random systems are called "replica symmetric".

The ground states of the XOR-SAT problem discussed in the previous chapter were first understood using the replica analysis. More complicated structures with additional hierarchy are now known. The clustered solutions found in the XOR-SAT problem in the hard-SAT region are one example of these. Most of the formal methods that follow this line of investigation are beyond the scope of this book, but practical methods of obtaining some of the results using message passing and iterative "cavity" evaluation of quantities called "surveys" that are similar in spirit to the standard beliefs have become available, and will be explored next, following the derivation in [12].

An iterative "belief propagation" (BP) [162] algorithm for K-SAT can be derived to evaluate the probability, or "belief", that a variable will take the value TRUE in the set of configurations that satisfy the formula considered. To calculate this, we first define a message ("transport") sent from a variable to a clause:

- $t_{i\to a}$ is *the probability that variable* x_i *satisfies clause* a.

In the other direction, we define a message ("influence") sent from a clause to a variable:

- $i_{a\to i}$ is *the probability that clause* a *is satisfied by another variable than* x_i.

In 3-SAT, where clause a depends on variables x_i, x_j, and x_k, BP gives the following iterative update equation for its influence:

$$i^{(l)}_{a\to i} = t^{(l)}_{j\to a} + t^{(l)}_{k\to a} - t^{(l)}_{j\to a} t^{(l)}_{k\to a} \,. \tag{23.1}$$

The BP update equations for the transport $t_{i\to a}$ involve the products of influences acting on a variable from the clauses that surround x_i, forming its "cavity", V_i, sorted by which literal (x_i or $\neg x_i$) appears in the clause:

$$A^0_i = \prod_{b\in V_i,\, y_{i,b}=\neg x_i} i_{b\to i} \quad \text{and} \quad A^1_i = \prod_{b\in V_i,\, y_{i,b}=x_i} i_{b\to i} \,, \tag{23.2}$$

with

$$y_{i,b} = \begin{cases} x_i & \text{if } x_i \text{ is part of clause } b \\ \neg x_i & \text{if } \neg x_i \text{ is part of clause } b \end{cases} \,. \tag{23.3}$$

The update equations are then

$$t^{(l)}_{i\to a} = \begin{cases} \dfrac{i^{(l-1)}_{a\to i} A^1_i}{i^{(l-1)}_{a\to i} A^1_i + A^0_i} & \text{if } y_{i,a} = \neg x_i \,, \\[2ex] \dfrac{i^{(l-1)}_{a\to i} A^0_i}{i^{(l-1)}_{a\to i} A^0_i + A^1_i} & \text{if } y_{i,a} = x_i \,. \end{cases} \tag{23.4}$$

The superscripts (l) and $(l-1)$ denote iteration. The probabilistic interpretation is as follows: suppose we have $i^{(l)}_{b\to i}$ for all clauses b connected to variable i. Each of these clauses can either be satisfied by another variable (with probability $i^{(l)}_{b\to i}$) or not be satisfied by another variable (with probability $\left(1-i^{(l)}_{b\to i}\right)$) and also be satisfied by variable i itself. If we set variable x_i to 0, then some clauses are satisfied by x_i, and some must be satisfied by other variables. The probability that they will all be *satisfied* is $\prod_{b\neq a, y_{i,b}=x_i} i^{(l)}_{b\to i}$. Similarly, if x_i is set to 1, then all these clauses b are satisfied with probability $\prod_{b\neq a, y_{i,b}=\neg x_i} i^{(l)}_{b\to i}$. The products in Eq. (23.4) can therefore be interpreted as joint probabilities of independent events. Variable x_i can be 0 or 1 in a solution if the clauses in which x_i appears are satisfied either directly by x_i itself or by other variables. Hence

$$\mathrm{Prob}(x_i) = \frac{A_i^0}{A_i^0 + A_i^1} \quad \text{and} \quad \mathrm{Prob}(\neg x_i) = \frac{A_i^1}{A_i^0 + A_i^1} \,. \tag{23.5}$$

23.2 Message Passing as Side Information for Decimation

To use BP for decimation, we select the variables with the largest probability to be either true or false. We then assign them their likely value, recalculate the beliefs for the reduced formula, and repeat. As with removal of 1-variables in K-XOR-SAT, decimation using BP proceeds until there are no clauses left and leaves the remaining variables untouched. The entropy is therefore given by the number of remaining variables, or 1—the "depth of decimation" plotted with the dashed line in Fig. 23.1. This method succeeds in finding satisfying configurations (ground states) of large 3-SAT formulas up to nearly $\alpha = 3.9$, which is believed to be the beginning of a Hard-SAT regime for this problem. Since we studied only large formulas, and the SAT-UNSAT transition is now understood to occur at $\alpha = 4.267\ldots$ [139], the formulas studied were almost certainly all satisfiable. On the rising part of the BP depth of decimation curve in Fig. 23.1, the decimation stops when the BP equations fail to converge. At this point, use of the full WalkSAT algorithm in the form described in the previous chapter finds a satisfying configuration in every case studied.

In the hard-SAT region, a hierarchical decomposition into clusters of solutions that are more separated from each other, as occurs in XOR-SAT, is plausible. A more complicated set of messages, designed for use in a single cluster of solutions, has been called survey propagation (SP) [27] and shown to provide a viable decimation scheme in this region. To arrive at SP we introduce a modified system of beliefs: every variable falls into one of three classes: TRUE in all solutions (1), FALSE in all solutions (0), and TRUE in some and FALSE in other solutions ($free$). Thus a decimation scheme will

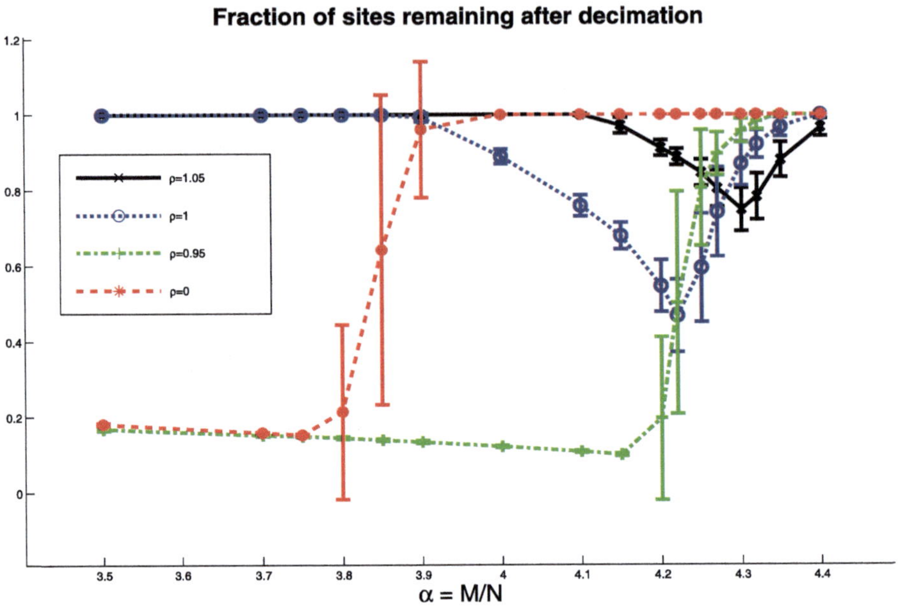

Fig. 23.1. Depth of decimation achieved by BP, SP, and two mixed cases ($\rho = 0.95$ and $\rho = 1.05$) as a function of α, the ratio of the number of clauses to the number of variables in 3-SAT (from [12])

attempt to identify the most frozen variables, those that are constant over a cluster of solutions. The message from a clause to a variable (an influence) is the the same as in BP above. Although we will again only need to keep track of one message from a variable to a clause (a transport), it is convenient to first introduce three ancillary messages:

- $\hat{T}_{i\to a}(1)$ is *the probability that variable x_i is true in clause a in all solutions*,
- $\hat{T}_{i\to a}(0)$ is *the probability that variable x_i is false in clause a in all solutions*,
- $\hat{T}_{i\to a}(free)$ is *the probability that variable x_i is true in clause a in some solutions and false in others.*

Note that there are here three transports for each directed link $i \to a$, from a variable to a clause, in the graph. As in BP, these numbers will be functions of the influences from clauses to variables in the preceeding update step. Taking again the incoming influences independently, we have

$$\begin{aligned}
\hat{T}^{(l)}_{i\to a}(free) &\propto \textstyle\prod_{b\in V_i\setminus a} \quad i^{(l-1)}_{b\to i}\,, \\
\hat{T}^{(l)}_{i\to a}(0) + \hat{T}^{(l)}_{i\to a}(free) &\propto \textstyle\prod_{b\in V_i\setminus a, y_{i,b}=x_i} \; i^{(l-1)}_{b\to i}\,, \\
\hat{T}^{(l)}_{i\to a}(1) + \hat{T}^{(l)}_{i\to a}(free) &\propto \textstyle\prod_{b\in V_i\setminus a, y_{i,b}=\neg x_i} \; i^{(l-1)}_{b\to i}\,.
\end{aligned} \tag{23.6}$$

The proportionality indicates that the probabilities are to be normalized. We see that the structure is quite similar to that in BP. But we can make it closer still by introducing $t_{i\to a}$ with the same meaning as in BP. In SP it will then, as the case might be, be equal to $T_{i\to a}(free)+T_{i\to a}(0)$ or $T_{i\to a}(free)+T_{i\to a}(1)$. That gives [cf. Eq. (23.4)]:

$$t_{i\to a}^{(l)} = \begin{cases} \dfrac{i_{a\to i}^{(l-1)} A_i^1}{i_{a\to i}^{(l-1)} A_i^1 + A_i^0 - A_i^1 A_i^0} & \text{if } y_{i,a} = \neg x_i\,, \\[2ex] \dfrac{i_{a\to i}^{(l-1)} A_i^0}{i_{a\to i}^{(l-1)} A_i^0 + A_i^1 - A_i^1 A_i^0} & \text{if } y_{i,a} = x_i\,. \end{cases} \tag{23.7}$$

The update equations for $t_{i\to a}$ are the same in SP as in BP, i. e., one uses Eq. (23.1) in SP as well. Similarly to Eq. (23.5), decimation now removes the most fixed variable, i. e., the one with the largest absolute value of $(A_i^0 - A_i^1)/(A_i^0 + A_i^1 - A_i^1 A_i^0)$. Given the complexity of the original derivation of SP [138, 139], it is remarkable that the SP scheme can be interpreted as a type of belief propagation in another belief system. And even more remarkable is the fact that the final iteration formulas differ so little.

A modification of SP that we will consider in what follows is to interpolate between BP ($\rho = 0$) and SP ($\rho = 1$)[1] by considering

$$t_{i\to a}^{(l)} \propto \begin{cases} \dfrac{i_{a\to i}^{(l-1)} A_i^1}{i_{a\to i}^{(l-1)} A_i^1 + A_i^0 - \rho A_i^1 A_i^0} & \text{if } y_{i,a} = \neg x_i\,, \\[2ex] \dfrac{i_{a\to i}^{(l-1)} A_i^0}{i_{a\to i}^{(l-1)} A_i^0 + A_i^1 - \rho A_i^1 A_i^0} & \text{if } y_{i,a} = x_i\,. \end{cases} \tag{23.8}$$

We do not have an interpretation of the intermediate cases of ρ as belief systems.

Figure 23.1 shows the depth of decimation resulting from SP as well as the variants that are shifted by use of the ρ parameter slightly in the direction of BP and slightly away from BP ("overrelaxed"). We see that the SP decimation is effective only in the hard-SAT region, since below $\alpha = 3.9$ it considers all variables as "free" to take either value in the configurations that can be constructed. In the hard-SAT region, it has the effect of removing the core variables, leaving WalkSAT with a problem to solve that is no more difficult than finding ground states at $\alpha = 3.9\ldots$, the entry to the hard-SAT

[1] This interpolation has also been considered and implemented by Zecchina and coworkers.

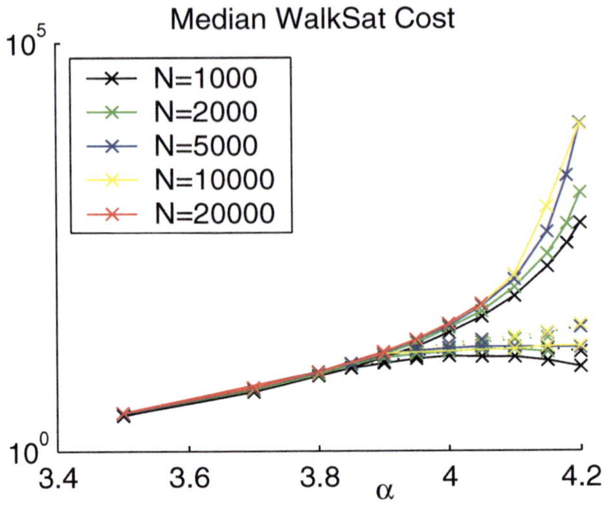

Fig. 23.2. Median cost of a WalkSAT solution close to the SAT-UNSAT transition, with and without first performing the SP-induced decimation to reduce the complexity of the formula (from [12])

region. This is suggested by the cost data contained in Fig. 23.2. Setting the parameter ρ to 0.95 produces an interesting result. The decimation proceeds to completion, just as in BP, but the method continues working in the hard-SAT region of the parameter space. Overrelaxed SP ($\rho = 1.05$) gives what may be reasonable recommendations even above the SAT–UNSAT boundary, where these methods can be used to arrive at configurations with the smallest number of unsatisfied clauses.

23.3 Belief Propagation and Sudoku

The Sudoku puzzle that we showed in Fig. 20.1 poses a difficult challenge for stochastic methods of resolving constraint problems. The puzzles are generated in apparently limitless quantities on popular Web sites. In our study, we made use of puzzles from `www.websudoku.com`. Programs that automatically solve the puzzles exist and are used in the filtering process that leads to automatic generation of good puzzles, but to the best of our knowledge all employ a long list of complex logic rules, followed by exhaustive search and backtracking to explore the hardest cases.

The puzzles are graded by level of difficulty. "Easy" puzzles typically have about 35 squares already filled in. "Medium" puzzles will have 32 clues. One can solve puzzles at these levels by repeatedly using the rule that no square can have a value that is already taken by another square in the same row, column, or 3×3 square. As soon as you discover a square that has only one possible value available to it, you insert that value and see what other discoveries the assignment may enable. A second rule is essential. If a square is the only place in a given row, column, or 3×3 square that can take on a particular value, then it must be assigned that value. These two rules can solve any "easy" or "medium" Sudoku puzzle, and many "hard" puzzles as well. "Hard" puzzles typically have 28 to 30 squares filled in initially. "Evil", sometimes called "diabolical", puzzles start with 24 to 26 squares filled in, and these two simple rules seldom provide more than one or two additional assignments.

A plausible way to develop a stochastic Sudoku solver is to use belief propagation to replace a hard discrete optimization problem with a softer problem using real numbers. For each square, we might calculate the nine probabilities that the square contains each one of the nine allowed numbers. If we let $P(i, j)$ be the probability of square j taking value i, then we can evaluate this probability as the product of the probabilities that no other square in the same row, column, or 3×3 square will take value i:

$$P(i, j) = \prod_{k \text{ in ngbhd of } j} (1 - P(i, k)) . \tag{23.9}$$

After evaluating Eq. (23.9) for each of the nine choices of i, we normalize the results so that the probabilities in a single square sum to unity. Unfortunately, this takes advantage of only one of the two basic rules. After iterating a while, we may find that the total probability of finding a "3" somewhere in a given row does not sum to unity but can take almost any value. Since we have already normalized the beliefs after iterating with Eq. (23.9), there is no convenient place left to incorporate the second rule in the evolution of the probabilities. Use of decimation, however, gets around some of this difficulty [22].

After iterating the beliefs to convergence, one selects the square and the value that are most strongly indicated and asserts that value in that square. One can augment this procedure with rules to prevent making assignments that are manifestly wrong (e. g., assigning the same number to two squares in the same row, column, or 3×3 square). The result is much stronger than simply applying the two basic rules to eliminate choices and solves nearly all "easy" to "hard" puzzles, and some "evil" puzzles as well. Probably the fact that the use of only nine numbers for the squares is now built in adds some of the effects of the second logical rule, which forces values when they cannot be assigned elsewhere in their local neighborhood.

6	7	8	2	5	1	3	9	4
3	1	4	7	9	8	2	6	5
2	5	9	3	4	6	7	1	8
9	2	1	4	8	7	5	3	6
7	3	6	1	2	5	8	4	9
8	4	5	6	3	9	1	2	7
5	8	2	9	6	3	4	7	1
4	9	7	5	1	2	6	8	3
1	6	3	8	7	4	9	5	2

Fig. 23.3. Solution to the puzzle posed in Fig. 20.1. A human puzzle solver solved this "evil" puzzle in only 12 min. It appears to be at about the limit for automatic solution (using belief propagation) today (March 2006)

The puzzle in Fig. 20.1 was solved by a practiced human Sudoku solver in 12 min (Fig. 23.3) and by belief propagation in seconds, but the belief propagation program needed several tries to find a successful solution. The program solved another "evil" puzzle on the first try, while the human solver required 14 min to solve it. This work is still in progress, but it appears likely that these extremely complicated logical inference puzzles will yield to stochastic optimization with modest further effort.

Part III

Outlook

24 Future Outlook of Optimization Business

24.1 $\mathcal{P} = \mathcal{NP}$?

At present, optimization problems can be divided into two classes. First are the problems for which exact algorithms exist, solving these problems in polynomial time, i. e., the time needed to solve a problem instance of size N is dominated by a power of N, e. g., N^3. These problems belong to the class $\mathcal{P}$. The second group includes problems for which no such algorithms have been found so far. A widely studied subclass of these problems has the property that while it may be very hard to find a solution, once we claim to have found a solution it is trivial to check that it is a solution. Thus with SAT, if we claim to have a configuration that satisfies all the clauses in the formula, one only has to check each clause to see if some Boolean variable in the proposed solution satisfies it. It thus takes time linear in the size of the problem to check the answer. The common consensus so far is that no polynomial time algorithms can exist for these problems, so that the approach of simply testing all the exponentially many possible solutions may not be so wasteful after all.

In practice, one would use exact algorithms of the branch & bound type for solving such problems, but these algorithms also require a calculation time that grows exponentially with the system size N. After this possibility is exhausted, the other choice is to use a heuristic to get a quite good solution for the problem.

However, there is a subclass of these problems, each of which can be mapped into any of the others in polynomial time. One calls problems in this subclass nondeterministic polynomial complete or $\mathcal{NPC}$. If in the future somebody was able to find some exact algorithm for solving even one of the problems in polynomial time, then there would be an algorithm for solving all of these problems in polynomial time, as one simply needs to transform any other problem to this solved problem in polynomial time first. In this case, all problems would be solved automatically.

In the last few decades, attempts have been made to provide polynomial algorithms for at least one of the problems classified as $\mathcal{NPC}$. So far none of these attempts has been successful. Even if somebody were to provide such an algorithm in the future, it is questionable whether it would be of any practical use. Some of the attempts made so far have led to polynomials

$p(N)$ of a comparatively large order. Usually in practical computations, one works with algorithms linear in N like the calculation of the scalar product, quadratic in N like the matrix–vector multiplication, and cubic in N, like the matrix–matrix multiplication. But that is somehow the end one can reach with a reasonable calculation time for larger system sizes. Thus, if somebody really were to discover an exact polynomial algorithm with a leading term of, e. g., N^6, then it might not be of any use at all for larger system sizes.

Thus, we want to conclude that the types of heuristics described in the book will always have their fields of application in practice, depending on the time one is willing to invest to create a solution.

24.2 Quantum Computing

Perhaps the challenges of optimization could be swept away by fundamental changes in the nature of computation itself. Several approaches currently under active investigation propose to accomplish just this. Truly new approaches to computing modify two standard elements of computers of the past century. They may involve new representations of the problem to be solved, representations for which there are novel mechanisms with which to compute things, not simply binary bits and switching circuits. And they will certainly involve very high degrees of parallelism in the computation, perhaps even distributing the data and the computing elements over a wide area.

Quantum computing, discussed in this section, and molecular computing, also known as "DNA computing", which we will discuss in the next section, do both of these things at the same time. Both will have to evolve considerably before we can consider them to be efficient methods for solving practical problems, although they show great potential power. A less radical step, already hinted at in the chapters on constraint satisfaction, is parallel distributed computing, perhaps using message passing and similar stochastic approaches to replace exact algorithms with much faster but approximate ones. This uses conventional representations and arithmetic but attempts to offer parallelism on the same scale as the size of the problem. It will be discussed in the final section below.

Of course, completely revising the nature of computation means giving up a lot that we now take for granted. It will take a long time to develop software as rich and usable as the compilers, development environments, and vast libraries of working modules out of which we build solutions to complicated problems and distribute the answers to those who ask the questions today.

Quantum computing provides a new representation, based on quantum bits (Qbits). Each conventional bit can only take on two states, '0' or '1'. N conventional bits can represent 2^N possible states of a system, but we can only explore one such state in each computation that we perform on these bits. Qbits are defined in terms of two possible states, called a "basis", denoted $|0\rangle$ or $|1\rangle$, but a Qbit can be any of the possible normalized linear combinations of these two, $\alpha|0\rangle + \beta|1\rangle$, where α and β are complex numbers. Arithmetic with classical bits is replaced by the quantum evolution of a linear combination of Qbits, in which the derived state is a new linear combination of the original states. When N Qbits are present, 2^N coefficients are required to describe the quantum state, since the "basis" of the N-Qbit state consists of all possible products of the N Qbits in states $|0\rangle$ or $|1\rangle$. Thus the N-Qbit system evolves through a configuration space with exponentially many degrees of freedom, and its evolution in a rough sense explores these many possibilities in parallel.

Many practical issues make this miracle extraordinarily difficult to realize as a realistic computational tool. For a quantum mechanical system to explore this very rich configuration space, it must be almost completely decoupled from the rest of the world, yet to write information in and take results out we must couple to the system. Other sources of noise also add to what is called "decoherence", which limits the extent to which the quantum system can evolve before it degrades to behaving classically. Strategies for error correction have been proposed and demonstrated, but these add considerable complexity. The result of these real-world effects is that quantum computation has been demonstrated with up to seven Qbits, but the systems in which this has been done do not appear to permit further extensions. Other systems are currently under study and may someday permit the thousandfold expansion in the number of Qbits required before quantum computing will bring presently impossible computations within reach.

24.3 DNA Computing

When defining a toy computer, Turing gave an example of a very simple device consisting of a pair of tapes and a control unit that read instructions from the input tape while moving simultaneously along the output tape reading and writing data there. This device is now called a Turing machine and is the basis of every existing computer.

In biological systems, similar events can be observed: chromatides of the desoxyribonucleic acid, or DNA for short, are read by a molecule called DNA polymerase that creates a mirror image of the chromatide, thus allowing the DNA to reproduce. The DNA is a very long molecule basically consting of a sequence of four different nuclein bases called adenine (A), thymine (T), guanine (G), and cytosine (C). A, T, G, and C are also called the four letters of the DNA alphabet. Adenine and thymine are complementary bases, i. e., in the double helix of the chromosome with its two strands, each adenine is connected to a thymine via a hydrogen bond. Analogously, guanine and cytosine form pairs of complementary bases. In a chromatide, there is only one of these two strands. If a polymerase molecule gets at a starting unit of a single chromatide, then it immediately starts reading the sequence of bases of this chromatide and producing a mirror image in the way that it attaches an A when it reads a T, a T when it reads an A, a G when it reads a C, and a C when it reads a G. If the original strand gets together with its counterpart, they form a complete DNA again.

As the computer scientist Leonard Adleman came across these biological findings, he immediately thought whether this biological type of computer, which resembles a Turing machine so well, could also be used like a real computer for solving mathematical problems [5]. The basic ingredients for such a biological computer were already at hand, as they were provided by nature that had developed them over millions of years of evolution:

- Only complementary strands of DNA can combine themselves via the Watson–Crick pairing as described above. If two strands get together that are not at least partially complementary, they cannot form a pair.
- The polymerase molecules are able to create a complementary copy of the DNA.
- Then there is the DNA ligase that binds two strands of DNA in proximity and bonds them covalently into a single molecule. In nature, this ligase is used to repair breaks in the DNA strands.
- Contrarily, nucleases search for a specific sequence of bases and cut the DNA there into two pieces.

Adleman made use of gel electrophoresis, a common technique for separating charged molecules in an electric field. He used this technique to separate DNA molecules of different lengths, which are charged negatively. The longer the molecules were, the slower they moved. Finally, he got bands in the gel, each containing DNA molecules of some specific size.

If this DNA computer is to solve real problems, then it must be possible to store information in a DNA molecule, i. e., to create a DNA molecule that contains the desired information. This can be done with modern DNA synthesis methods, which can generate as a starting point 10^{18} molecules of DNA, nearly all of which contain the desired sequence of A, T, G, and C letters. These molecules only need to be placed in a small tube, while the first electronic computers filled whole rooms.

To show the possibilities but also the difficulties of DNA computing, we briefly sketch the work of Adleman, who was able to solve by these means a small instance of the Hamiltonian path problem (HPP), which is closely related to the traveling salesman problem (TSP): in the Hamiltonian path problem, which is an NP-complete problem, a set of nodes and a set of edges between the nodes are given. The traveling salesman cannot move from some node to any other randomly chosen node but can walk only over those edges that are allowed. Generally, these edges are unidirectional, i. e., if there is an edge supporting the salesman on his way from node i to node j, there is not necessarily an edge for returning directly from j to i. The task in this HPP is to find a path from a given initial node to a proposed final node, touching each of the other nodes exactly once.

To make this problem solvable for his DNA computer, Adleman used the following simple algorithm for solving the HPP. The first step is to generate a set of random paths through the graph. For each of these paths, one must verify whether they start and end at the correct nodes. If a configuration does not fulfill this condition, it is removed from the set. Then it is checked whether the path passes through the correct number of nodes; if not, the path is also removed. After that it is checked whether every node is actually passed in the path; if not, the path is removed. Finally, if the set of paths is not empty, there is a Hamiltonian path; otherwise there is not. Of course, this final outcome is only exact if every possible path was created at the beginning.

The problem instance Adleman was able to solve with this algorithm for his DNA computer at that time consisted of 7 nodes and 14 edges. This required 7 days in the laboratory. (A human could have found the solution in about 1 min, but Adleman's experiment was intended only to demonstrate that this method worked.) He coded each node by a DNA name consisting of eight nucleotide bases. The first four bases served as a type of first name, the last four as a surname. An edge from a node i to a node j is then coded by the surname of node i and the first name of node j. Omitting the initial and the final nodes of the Hamiltonian path, the first name of the first intermediate stop, and the surname of the last intermediate stop, the solution of the Hamiltonian path could thus be coded with 32 bases. After generating 10^{14} DNA molecules, Adleman put them in a simple tube, added water and salt to reproduce biological conditions and the molecules needed such as ligase. Within a second, the DNA computer was able to solve this problem due to its massively parallel processing. However, the biological computer contained not only the correct solution but also many different random paths. The solution to this problem was first to reproduce exponentially the correct paths, while paths with both a wrong start and a wrong end node were not duplicated at all. If exactly one of these two end nodes was correct, then the growth was much slower than that of the molecules meeting both conditions. Then Adleman used gel electrophoresis to extract the molecules with the correct length.

After that Adleman made use of a procedure called affinity separation in which the complementary names of the desired nodes, which should be intermediately visited in the path, are attached to microscopic iron balls. These short molecules like to attract their counterparts in the path configurations. Then he placed a magnet against the wall of the tube and poured out those molecules that contained the corresponding node. After adding a new solvent and removing the magnet, he raised the temperature to break the connection between the molecules and the probes. Next he extracted the iron balls with a magnet. This approach was repeated for any node that should be visited along the path. After some final reheating and cooling and a final electrophoresis step, the first DNA computation was complete: the DNA computer, i. e., the tube, contained the desired Hamiltonian path solution.

In the meantime, more and larger problems have been solved with DNA computing. Currently, more and more operations can be performed automatically by machines in biochemical labs, so that this work is not as tedious as it was for Leonard Adleman in 1994. However, to our knowledge, so far no example of DNA computing has proven potentially superior to the existing algorithms in the standard computing world with silicon computers, but perhaps this will change in the future.

24.4 How Will the Problems Evolve?

We are currently studying optimization methods that scale to extremely large problems because such problems exist and are becoming more typical as well as larger with each passing year. At points in the past, it has seemed as if things had gotten as big and complex as they will ever be, but at every turn, some small improvement in technology has had a big impact, and the old limits have disappeared. Consider the telephone system. Pictures from the 1920s show city skies dominated by telephone wires, telephone poles crowding the streets, and huge rooms filled with telephone operators connecting calls. At the time, it was fashionable to say that this growth could not continue because within a few years everyone would have to become a telephone operator. The introduction of automatic dialing did, in fact, make everyone a telephone operator, and further automation of long-distance connection has managed to stabilize and slowly reduce the population of live, professional operators.

In addition, improvements in senders and receivers and the invention of multiplexing made it possible to pass more calls over a wire at faster rates. Optical communication has further increased the rates at which calls move, and thus the number of calls, and has introduced the possibility of sending multiple streams of data at different light wavelengths along the same "wire". In many parts of the world where wires and poles never got as dense as in Western cities, the introduction of wireless telephony has continued this growth without requiring the wired infrastructure at all. The result is that today there are one or more telephones for each person on the planet. Is that the limit? Probably not, since we are beginning to find reasons to use telephones to communicate with machines. Our computers talk to us, and even more often to each other, over communication lines that have grown out of the telephone system. Computing power is being inserted into ever smaller appliances, devices with a few well-understood functions, such as sending and receiving e-mail, playing music in earphones, or washing clothes with minimum waste of water and discharge of detergents. These machines need to talk to us and to each other increasingly often. So the limits of the telephone system, which has become a telephone and data communications network, don't seem to be pressing upon us.

Data themselves are udergoing a similar explosion. Observational sciences, such as biology, high-energy physics, astronomy, and sciences that exploit observations of the earth and its atmosphere, increasingly take advantage of "virtual observatories" of data that have been gathered and characterized so that multiple types of analysis can be applied to them. Experiments seeking new phenomena can combine searches for anomalies within the existing data with new measurements to explain them. Problems that produced gigabytes (1 GB $\approx 10^9$ bytes) of data a few years ago are beginning to produce terabytes (1 TB $\approx 10^{12}$ bytes) of data in current years, which are managed in single installations but must be joined in data federations around the world. The petabyte (1 PB $\approx 10^{15}$ bytes) and exabyte data accumulations (1 EB $\approx 10^{18}$ bytes) are not far away. Commercial data are also on the same growth curve. There is also an increasing tendency for these data to be available for carefully controlled public access. Banking records and catalogs of information such as Google and Amazon or similar online stores' Web sites

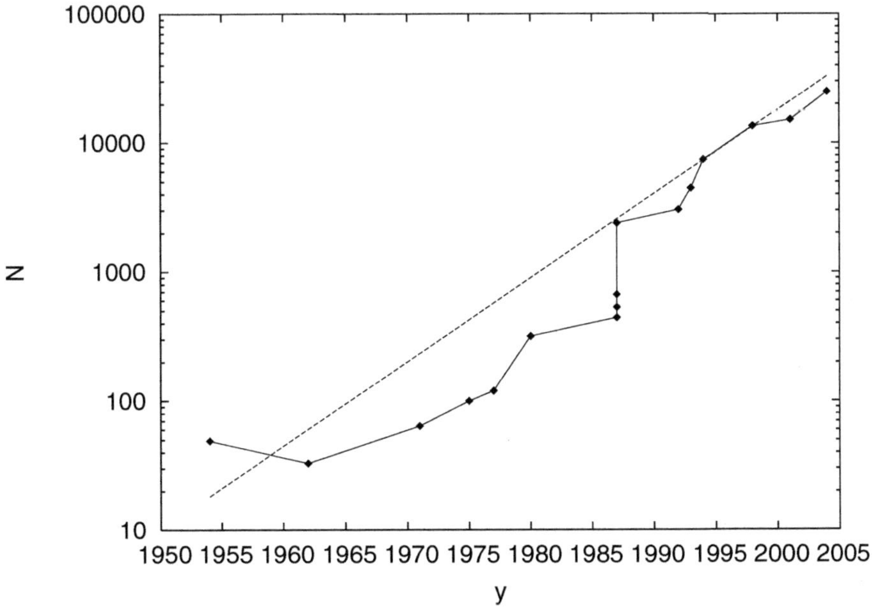

Fig. 24.1. Increase of the system size N of the TSP instances optimally solved in the year y: one clearly sees an instantaneous increase in the system sizes in the mid-1980s in which the algorithms for solving TSP instances exactly were considerably improved. Afterwards, an exponential increase in the size of this system size took place. The fitting line is given by $10 \times \exp((y - 1950) \times 0.15)$. The data used for this graphic were taken from the pictorial history site of the TSP on the Web [96], according to which the current world record (March 2006) of a TSP instance optimally solved is held by Applegate, Bixby, Chvátal, Cook, and Helsgaun, who found the optimal tour through 24978 cities in Sweden in 2004

are good examples. It is probably no accident that each time the volume of data that we are aware of existing in the world doubles, the price per bit of storing that data and making it available online is nearly halved.

Even in our special world of TSP solving, the growth in the size of systems solved exactly shows an exponential growth trend, as shown in Fig. 24.1. TSPs in excess of 10^4 sites are now solved, but not routinely, as each of the new records in Fig. 24.1 has required some months of dedicated supercomputer time.

The focus of computing strategies to deal with this growth has increasingly turned to distributed computation. Unlike the revolutionary strategies of quantum or molecular computing, distributed computing is evolutionary. The representation of events is as close as possible to whatever is currently used in centralized computing, but the form of the computation changes both because of the overall scale and because as much as possible, computation is delegated to occur in as many places as the data reside in.

As a result, prospects for the optimization business seem excellent. There are opportunities for optimizing access to data by intelligent placement of related information. There is a serious need to optimize queries against distributed data so that they can avoid aggregations of huge partial results that are not ultimately needed. The gathering of fresh data about the world needs planning to avoid intolerable communications expense. And the message-passing strategies and stochastic or randomized algorithms that we discuss in this book are increasingly common in this world which we see developing.

Acknowledgments

First, JJS would like to thank Prof. Ingo Morgenstern (University of Regensburg, Germany), from whom he learned the methods of optimization in physics as well as their application to both physical models and real-world applications. JJS is especially grateful to him as he gave him a great measure of liberty during his time as a diploma and a Ph.D. student, but also a great deal of background support as other people complained about the large amounts of computing time JJS used in those days. Secondly, JJS would like to thank Johannes Maria Singer (formerly University of Regensburg and University of Zurich, Switzerland) for the discussion of many questions about optimization and for the fruitful collaboration that resulted in several papers. For providing a very thorough introduction to numerical methods and for constructive criticisms, JJS would like to thank Prof. Hans-Georg Matuttis (University of Electro-Communications, Chofu, near Tokyo, Japan). Thirdly, JJS would like to thank Prof. Gunter Dueck (formerly Scientific Center IBM Heidelberg, now IBM Mannheim, Germany) and the members of his former TopC optimization group at the former Scientific Center IBM Heidelberg, especially Tobias Scheuer (now SAP Walldorf, Germany), Gerhard Schrimpf (now IBM Frankfurt, Germany), and Hermann Stamm-Wilbrandt (now IBM Stuttgart, Germany), from whom he learned additional methods of fine-tuning optimization algorithms and with whom he developed new algorithms for optimization. Fruitful discussions with Andreas Lang, Prof. Andreas Rudolph, Prof. Marcus Spies, Klaus Volk, Gerald Möse, Reinhard Peters, Peter Altevogt, Ute Twisselmann, and Hans-Martin Wallmeier of the same group at that time are also acknowledged. Furthermore, JJS would like to thank Prof. Uwe Krey (University of Regensburg) and Prof. Sigismund Kobe (Dresden University of Technology, Germany) for explaining the properties of spin glasses, which serve as an excellent problem for applying optimization algorithms. For providing a very interesting insight into the history of computational physics and for discussions about random number generation, JJS would like to thank Erich Stoll (University of Zurich). JJS would like to thank Herbert Maier for teaching him the basics of evolutionary and behavioral biology and Josef Graminger for teaching him the basics of statistical physics. For his useful criticism and humorous remarks, JJS would like to thank Prof. Dietrich Stauffer (University of Cologne, Germany). Furthermore, JJS would like to thank Prof. Kurt Binder (Johannes Gutenberg University of Mainz,

Germany) for his explanations of order parameters and phase transitions and for his patience during the years in which this book was written. For demonstrating algorithms and auxiliary methods from computer science for the optimization of packing problems, JJS would like to thank Prof. Elmar Schömer (Johannes Gutenberg University of Mainz).

JJS would also like to thank Mr. Hautmann and Mr. Pammer (BMW Regensburg), Mr. Wohlleben and Mr. Schömig (Siemens Semiconductor Regensburg), Peter Korevaar (formerly IBM Scientific Center Heidelberg), and Wolfgang Rudolph and Andreas Brinkmann (BMW Munich, Germany) for providing additional opportunities for applying optimization algorithms to a wide variety of optimization problems in the real world.

Furthermore, JJS would like to thank Florian Baumgartner, Günter Bauer, Sebastian Beichele, Johannes Bentner, Johann Birneder, Jürgen Britze, Markus Dankesreiter, Anja Ebersbach, Christine Froschhammer, Christian Hirtreiter, Isabel Käser, Ralf Kinder, Bernhard Niedermeier, Thomas Pfadenhauer, Thomas Pongratz, Markus Puchta, Martin Ransberger, Uta Sasse, Martin Schmid, Martin Schwendke, Günther Stattenberger, Ralf Vater, and Matthias Weiser (formerly University of Regensburg, partially still there), Peter Hüsser, Samo Pliberšek, and Yuan Shen (formerly University of Zurich), Helga Labermeier (formerly Swiss Federal Institute of Technology in Zurich, Switzerland (ETHZ)), Matan Ninio (The Hebrew University of Jerusalem, Israel), Boris Brodda, Sebastian Golke, Michael Hawlitzky, and Tobias Preis (Johannes Gutenberg University of Mainz), Elena Ramírez Barrios (University of Kiel, Germany), and Juan Diaz Ochoa (University of Bremen, Germany) for fruitful discussions on optimization.

Finally, JJS would like to thank several institutions for computing time, ranging from the permission to use a few computers when they were free (University of Berne, Switzerland, University of Zurich) via the usage of computer clusters or parallel computers (University of Regensburg, ETHZ, GMD, Swiss Center for Scientific Computing in Manno near Lugano, Switzerland) to huge amounts of computing time on massively parallel supercomputers (John von Neumann Institute for Computing at the Research Center Julich, Germany) in recent years, without which this book would not have been possible.

SK would very much like to acknowledge the many people at IBM and IBM's partners who have contributed their insights into the real issues that arise in solving practical optimization problems, but the list is even longer than that of JJS above, and he does not wish to offend those who might inadvertently be overlooked. SK's three advisors, Professors Tom Carver at Princeton, Henry Ehrenreich at Harvard, and Morrel Cohen at the University of Chicago, do, however, deserve special thanks. And SK is grateful to Dan Gelatt, then at IBM and presently at Northern Micrographics, Inc. for the many discussions that led to a clear formulation of simulated annealing and its underpinnings. JJS and SK would like to thank Glenn Corey for copyediting the book. Mr.

Corey contributed not only his copyediting skills but also relevant comments regarding some terminology use, which he considered to be his two cents worth but which the authors value at no less than three cents.

During the period that JJS and SK worked together at the Hebrew University of Jerusalem, JJS was supported by a DARPA program on "ANTS: autonomous negotiating teams". Some subsequent work reported in this book has been carried out under the EVERGROW European integrated project, No. 1935, of the Sixth Framework. Since JJS took up his position at the University of Mainz, our continued collaboration has been supported by the Leibniz Foundation and the visitor program of the Hebrew University's Interdisciplinary Center for Neural Computation.

References

1. M. Abramowitz, I.A. Stegun (eds.) (1964), *Handbook of Mathematical Functions with Formulas, Graphs, and Mathematical Tables*, Dover, New York
2. D. Achlioptas (2000), Setting two variables at a time yields a new lower bound for random 3-SAT, in: 32nd Ann. ACM Symp. on Theory of Computing, Portland, OR, ACM, New York, 28–37
3. D. Achlioptas, L. Kirousis, E. Kranakis, D. Krizanc (1997), Rigorous results for random (2+p)-SAT, in: Proc. RALCOM **97**, 1–10
4. D. Achlioptas, G. Sorkin (2000), Optimal policies for greedy 3-SAT algorithms, in: 41st Ann. Symp. on Foundations of Computer Science, IEEE Computer Society Press, Los Alamitos, CA
5. L.M. Adleman (1998), Computing with DNA, Sci. Am. August 1998, 34–41
6. F. Alcuinus, (1978) *Propositio de lupo et capra et fasciculo cauli*, ≈ 800, in: M. Folkerts (ed.), *Die älteste mathematische Aufgabensammlung in lateinischer Sprache: Die Alkuin zugeschriebenen Propositiones ad acuendos iuvenes*, Österreichische Akademie der Wissenschaften, Mathematisch-Naturwissenschaftliche Klasse, Denkschriften **116**, 6, Abhandlung, Vienna
7. M. Alekhnovich, E. Ben-Sasson (2004), Linear Upper Bounds for Random Walk on Small Density Random 3CNFs, Electronic Colloquium on Computational Complexity (ECCC), ECCC-TR04-016
8. D.J. Amit (1989) *Modelling Brain Function: The World of Attractor Neural Networks*, Cambridge University Press, New York
9. I. Andricioaei, J.E. Straub (1996), Generalized simulated annealing algorithms using Tsallis statistics: application to conformational optimization of a tetrapeptide, Phys. Rev. E **53**, R3055–R3058
10. I. Andricioaei, J. E. Straub (1997), An efficient Monte Carlo algorithm for overcoming broken ergodicity in the simulation of spin systems, Phys. A **247**, 553
11. N. Ascheuer, M. Fischetti, M. Grötschel (2001), Solving the Asymmetric Travelling Salesman Problem with time windows by branch-and-cut, Math. Prog. Ser. A **90**, 475–506
12. E. Aurell, U. Gordon, S. Kirkpatrick (2005), Comparing Beliefs, Surveys and Random Walks, in: L. K. Saul, Y. Weiss, L. Bottou (eds.), *Advances in Neural Information Processing Systems 17*, MIT Press, Cambridge, MA, 49–56
13. A. Bachem, W. Hochstättler, M. Malich (1993), The Simulated Trading Heuristic for Solving Vehicle Routing Problems, Report No. 93.139, Mathematisches Institut, University of Cologne, Germany
14. A. Bachem, W. Hochstättler, M. Malich (1994), Simulated Trading – a new parallel approach for solving Vehicle Routing Problems, Parallel Computing:

Trends and Applications, in: Proc. Int. Conf. ParCo93, Grenoble, France, 7–10 September 1993

15. G. T. Barkema, N. Mousseau (1996), Event-Based Relaxation of Continuous Disordered Systems, Phys. Rev. Lett. **77**, 4358–4361
16. W. Barthel, A. K. Hartmann, M. Weigt (2003), Solving satisfiability problems by fluctuations: the dynamics of stochastic local search algorithms, Phys. Rev. E **67**, 066104–1–21
17. J. Bentner, G. Bauer, G. M. Obermair, I. Morgenstern, J. Schneider (2001), Optimization of the time-dependent Traveling Salesman Problem with Monte Carlo Methods, Phys. Rev. E **64**, 036701–1–8
18. B. A. Berg (1999), Introduction to Multicanonical Monte Carlo Simulations, arXiv:cond-mat/9909236v1
19. B. A. Berg, T. Neuhaus (1991), Multicanonical algorithms for first order phase transitions, Phys. Lett. **B267**, 249–253
20. B. A. Berg, T. Neuhaus (1992), Multicanonical Ensemble: a new approach to simulate first-order phase transitions, Phys. Rev. Lett. **68**, 9–12
21. D. J. Bertsimas (1992), A Vehicle Routing Problem with stochastic demand, Oper. Res. **40/3**, 574–585
22. D. Bickson, HUJI (2006), pers. commun.
23. K. Binder, D. W. Heermann (1992), *Monte Carlo Simulation in Statistical Physics*, Springer, Berlin Heidelberg New York
24. B. Bollobas, C. Borgs, J. Chayes, J. H. Kim, D. B. Wilson (2001), The scaling window of the 2-SAT transition, Random Struct. Algor. **18**, 201–256
25. R. Borndörfer, M. Grötschel, A. Löbel (1998), Alcuin's transportation problems and integer programming. In: P. L. Butzer, H. Th. Jongen, W. Oberschelp (eds.), *Charlemagne and His Heritage: 1200 Years of Civilization and Science in Europe, Vol. 2: The Mathematical Arts*, Brepols, Turnhout, Belgium; preprint SC 95-27, Konrad-Zuse-Zentrum Berlin, 1995
26. A. Braunstein, M. Mezard, M. Weight, R. Zecchina (2003), Constraint satisfaction by survey propagation, arXiv:cond-mat/02124551; Volume on Computational Complexity and Statistical Physics, Oxford University Press, Oxford
27. A. Braunstein, M. Mezard, R. Zecchina (2002), Survey Propagation: an algorithm for satisfiability, arXiv:cs.CC/0212002
28. A. Braunstein, R. Zecchina (2004), Survey Propagation as local equilibrium equations, arXiv:cond-mat/0312483 v5
29. Th. Bröcker (1995), *Analysis I*, Spektrum, Heidelberg
30. A. Z. Broder, A. M. Frieze, E. Upfal (1993), On the satisfiability and maximum satisfiability of random 3-CNF formulas, in: Proc. 4th Ann. ACM-SIAM Symp. on Discrete Algorithms, Austin, TX, ACM, New York, 322–330
31. N. J. Cerf, J. Boutet de Monvel, O. Bohigas, O. C. Martin, A. G. Percus (1997), The random link approximation for the Euclidean traveling salesman problem, J. Phys. I **7**, 117–136
32. V. Cerny (1985), Thermodynamical approach to the Traveling Salesman Problem: an efficient simulation algorithm, J. Opt. Theory Appl. **45**, 41–51
33. M.-T. Chao, J. Franco (1986), Probabilistic Analysis of two heuristics for the 3-SAT problem, SIAM J. Comput. **15**, 1106–1118
34. I. Charon, O. Hudry (1993), The Noising Method: a new method for combinatorial optimization, Oper. Res. Lett. **14**, 133–137

35. V. Chvatal, B. Reed (1992), Mick gets some (the odds are on his side), in: 33rd Ann. Symp. on Foundations of Computer Science, Pittsburgh, IEEE Computer Society, Los Alamitos, CA, 620–627
36. G. Clarke, J. W. Wright (1964), Scheduling of Vehicles from a Central Depot to a Number of Delivery Points, Oper. Res. **12**, 568–581
37. S. Cocco, O. Dubois, J. Mandler, R. Monasson (2003), Rigorous Decimation-based construction of Ground Pure States for Spin-Glass Models on Random Lattices, Phys. Rev. Lett. **90**, 047205–1
38. S. Cocco, L. Ein-Dor, R. Monasson (2004), Analysis of backtracking procedures for random decision problems, in: A. K. Hartmann, H. Rieger (eds.), *New Optimization Methods in Physics*, Wiley, New York
39. B. Codenotti, G. Manzini, L. Margara, G. Resta (1996), Perturbation: an efficient technique for the solution of very large instances of the Euclidean TSP, INFORMS J. Comput. **8**, 125–133
40. A. Colorni, M. Dorigo, V. Maniezzo (1991), Distributed optimization by ant colonies, in: Proc. ECAL91 – European Conference on Artificial Life, Paris, 134–142
41. B. Coluzzi, G. Parisi (1998), On the Approach to the equilibrium and the equilibrium properties of a glass-forming model, J. Phys. A **31**, 4349–4368
42. S. P. Coy, B. L. Golden, E. A. Wasil (2000), A computational study of smoothing heuristics for the Traveling Salesman Problem, Research report, University of Maryland, Eur. J. Oper. Res. **124**, 15–27
43. G. Cybenko (1988), *Continuous Valued Neural Network with Two Hidden Layers are Sufficient*, Technical report, Department of Computer Science, Tufts University, Medford, MA
44. G. B. Dantzig, D. R. Fulkerson, S. M. Johnson (1959), On a Linear Programming-Combinatorial Approach to the Traveling Salesman Problem: Notes on Linear Programming and Extensions – Part 49, Research Memorandum RM-2321, The RAND Corporation, Santa Monica, CA; Oper. Res. **7**, 58–66
45. Ch. Darwin, *On the Origin of Species by Means of Natural Selection, or the Preservation of Favoured Races in the Struggle for Life*, J. Murray (London, 1859), can be downloaded from the Banned Books Online page: `http://digital.library.upenn.edu/books/banned-books.html`
46. M. Davis, G. Logemann, D. Loveland (1962), A machine program for theorem-proving, Commun. ACM **5**, 394–397
47. M. Davis, H. Putnam (1960), A computing procedure for quantification theory, J. ACM **7**, 201–215
48. C. Deroulers, R. Monasson (2004), Critical behaviour of combinatorial search algorithms, and the unitary-propagation universality class, Europhys. Lett. **68**, 153–159
49. M. Desrochers, J. Desrosiers, M. Solomon (1992), A new optimization algorithm for the Vehicle Routing Problem with Time Windows, Oper. Res. **40**, 342–354
50. C. Desroulers, R. Monasson (2004), Critical behaviour of combinatorial search algorithms, and the unitary-propagation universality class, Europhys. Lett. **68**, 153–159
51. G. Dueck (1993), New optimization heuristics: the Great Deluge Algorithm and the Record-to-Record Travel, J. Comput. Phys. **104**, 86

52. G. Dueck (2004), *Das Sintflutprinzip – Ein Mathematik-Roman*, Springer, Berlin Heidelberg New York
53. G. Dueck, T. Scheuer (1990), Threshold Accepting: a general purpose optimization algorithm appearing superior to Simulated Annealing, J. Comput. Phys. **90**, 161–175
54. G. Dueck, T. Scheuer, H.-M. Wallmeier (1993), Toleranzschwelle und Sintflut: Neue Ideen zur Optimierung, Spektrum der Wissenschaft **1993/3**, 42–51
55. R. Durbin, D. Willshaw (1987), An analogue approach to the travelling salesman problem using an elastic net method, Nature **326**, 689–691
56. S. F. Edwards, P. W. Anderson (1978), Theory of spin glasses, J. Phys. F Metal Phys. **5**, 965–974
57. A. Fachat (1997), *Gleichgewichtsverteilungen bei Threshold Accepting Algorithmen*, Poster **DY 19.42**, DPG-Tagung, Munster
58. K. H. Fischer, J. A. Hertz (1991), *Spin Glasses*, Cambridge University Press, Cambridge, UK
59. J. Franco (2001), Results related to threshold phenomena research in satisfiability: lower bounds, Theor. Comput. Sci. **265**, 147–157
60. J. Franco, N. M. Paull (1983), Probabilistic analysis of the Davis Putnam procedure for solving the satisfiability problem, Discrete Appl. Math. **5**, 77–87
61. E. Friedgut (1999), Necessary and sufficient conditions for sharp thresholds of graph properties and the K-SAT problem, J. Am. Math. Soc. **12**, 1017–1054
62. M. R. Garey, D. S. Johnson (1979), *Computers and Intractability. A Guide to the Theory of NP-Completeness*, Freeman, San Francisco
63. R. S. Garfinkel, G. L. Nemhauser (1972), *Integer Programming*, Wiley, New York
64. S. B. Gelfand, S. K. Mitter (1985), Analysis of Simulated Annealing for optimization, in: Proc. 24th Conf. on Decision and Control, 779–786
65. S. Geman, D. Geman (1984), Stochastic Relaxation, Gibbs Distributions, and Bayesian Restoration of Images, IEEE Trans. Pattern Anal. Mach. Intell. **6**, 721–741
66. R. J. Glauber (1963), Time-dependent statistics of the Ising model, J. Math. Phys. **4**, 294–307
67. F. Glover, M. Laguna (1997), *Tabu Search*, Kluwer, Dordrecht
68. A. Goldberg (1979), On the complexity of the satisfiability problem, Courant Comput. Sci. Rep. No. 16, New York University, New York
69. D. E. Goldberg (1989), *Genetic Algorithms in Search, Optimization and Machine Learning*, Addison-Wesley, Reading, MA
70. P. Grassberger (1993), On correlations in "good" random number generators, Phys. Lett. A **181**, 43
71. G. S. Grest, C. M. Soukoulis, K. Levin (1986), Cooling rate dependence for the spin glass ground state energy: implications for optimization by simulated annealing, Phys. Rev. Lett. **56**, 1148
72. M. Grötschel (1984), Polyedrische Kombinatorik und Schnittebenenverfahren, Preprint No. 38, University of Augsburg, Germany
73. M. Grötschel, O. Holland, Solution of Large-Scale Symmetric Travelling Salesman Problems, Math. Prog. **51**, 141–202 (1991)
74. M. Grötschel, M. Padberg (1999), Odysseus als Handlungsreisender, Spektrum der Wissenschaft **1999/4**, 76–85

75. M. Grötschel, M. Padberg (2001), The Optimized Odyssey, AIROnews **VI:2**, 1–7
76. J. Gu, X. Huang (1994), Efficient Local Search Space Smoothing: a case study of the Traveling Salesman Problem (TSP), IEEE Trans. Syst. Man Cybern. **24**, 728–735
77. J. van Haemmen, I. Morgenstern (1983), Heidelberg Colloquium on Spin Glasses, Springer, Berlin Heidelberg New York
78. B. Hajek (1985), A tutorial survey of theory and applications of Simulated Annealing, in: Proc. 24th Conf. on Decision and Control, 755–760
79. K. Hamacher (2006), Adaption in Stochastic Tunneling Global Optimization of Complex Potential Energy Landscapes, Europhys. Lett. **74**, 944–950
80. M. Hawlitzky (2001), *Untersuchungen zu dynamischen Erweiterungen an Kohonen-Karten*, Diploma thesis, University of Mainz, Germany
81. S. Haykin (1994), *Neural Networks – A Comprehensive Foundation*, MacMillan, Ontario
82. D. O. Hebb (1949), *The Organization of Behavior*, Wiley, New York
83. Foto: Stadt Heidelberg
84. M. Held, R. M. Karp (1970), The traveling-salesman problem and minimum spanning trees, Oper. Res. **18**, 1138–1162
85. M. Held, R. M. Karp (1971), The traveling-salesman problem and minimum spanning trees. II. Math. Programm. **1**, 6–25
86. J. Hertz, A. Krogh, R. G. Palmer (1991), *Introduction to the Theory of Neural Computation*, Addison-Wesley, Reading, MA
87. J. Holland (1973), Genetic algorithms and the optimal allocations of trials, SIAM J. Comput. **2**, 88–105
88. J. J. Hopfield, D. W. Tank (1985), Neural computation of decisions in optimization problems, Biol. Cybern. **52**, 141–152
89. `http://web.cba.neu.edu/~msolomon/problems.htm`
90. `http://www.informatik.uni-heidelberg.de/groups/comopt/software/TSPLIB95`
91. `http://www.intellektik.informatik.tu-darmstadt.de/SATLIB/benchm.html`
92. `http://www.klettern-im-elbsandstein.de`, with friendly permission of Peter Horntrich
93. `http://www.math.princeton.edu/tsp/history.html`
94. `http://www.satlib.org`
95. `http://www.satisfiability.org/SAT04`
96. `http://www.tsp.gatech.edu//history/pictorial/pictorial.html`
97. `http://www-apache.imag.fr/~paugerat/VRP/INSTANCES`
98. K. Hukushima, K. Nemoto (1996), Exchange Monte Carlo method and application to spin glass simulations, J. Phys. Soc. Jpn. **65**, 1604–1608
99. K. Hukushima, H. Takayama, H. Yoshino (1998), Exchange Monte Carlo Dynamics in the SK Model, J. Phys. Soc. Jpn. **67**, 12–15
100. D. A. Huse, D. S. Fisher (1986), Residual energies after slow cooling of disordered systems, Phys. Rev. Lett. **57**, 2203–2206
101. P. Jaillet (1985), Probabilistic Traveling Salesman Problem, Ph.D. thesis, Operations Research Center, Massachusetts Institute of Technology, Cambridge, MA

102. P. Jaillet, A. R. Odoni (1988), The Probabilistic Vehicle Routing Problem. In: B. L. Golden, A. A. Assad (eds.), *Vehicle Routing: Methods and Studies*, North-Holland, Amsterdam, 293–318
103. D. S. Johnson, L. A. McGeoch, E. E. Rothberg (1996), Asymptotic experimental analysis for the Held–Karp Traveling Salesman Bound, in: Proc. 7th Ann. ACM-SIAM Symp. on Discrete Algorithms, 341–350
104. D. Jungnickel (1999), *Graphs, Networks and Algorithms, Algorithms and Computation in Mathematics*, Vol. 5, Springer, Berlin Heidelberg New York
105. A. C. Kaporis, L. M. Kirousis, E. G. Lalas (2002), The Probabilistic Analysis of a Greedy Satisfiability Algorithm, ESA 2002, 574–585
106. J. Kennedy (1997), The Particle Swarm: social adaptation of knowledge, in: IEEE Int. Conf. on Evolutionary Computation, Indianapolis, IN
107. J. Kennedy, R. Eberhart (1995), Particle Swarm Optimization, in: Proc. IEEE Int. Conf. on Neural Networks, **4**, 1942–1948
108. J. Kennedy, R. Eberhart, Y. Shi (2001), *Swarm Intelligence*, Morgan Kaufmann, San Francisco
109. W. Kerler, P. Rehberg (1994), Simulated-tempering procedure for spin-glass simulations, Phys. Rev. E **50**, 4220–4225
110. S. Kirkpatrick, C. D. Gelatt Jr., M. P. Vecchi (1983), Optimization by Simulated Annealing, Science **220**, 671–680
111. S. Kirkpatrick, G. Gyorgyi (1994), The Statistical Mechanics of k-Satisfaction, in: J. D. Cowan, G. Tesauro, J. Alspector (eds.), *Advances in Neural Information Processing Systems 6*, Morgan Kaufman, San Francisco, 439
112. S. Kirkpatrick, B. Selman (1994), Critical behavior in the satisfiability of random formulas, Science **264**, 1297–1301
113. S. Kirkpatrick, B. Selman (2001), Statistical physics and computational complexity, in: N. Phuan Ong, R. N. Bhatt (eds.), *More is Different – Fifty Years of Condensed Matter Physics – A Festschrift for PW Anderson*, 331–345
114. S. Kirkpatrick, D. Sherrington (1978), Infinite-ranged models of spin-glasses, Phys. Rev. B **17**, 4384–4403
115. S. Kirkpatrick, E. Stoll (1981), A very fast shift-register sequence random number generator, J. Comput. Phys. **40**, 517–526
116. S. Kirkpatrick, G. Toulouse (1985), Configuration Space Analysis of the Travelling Salesman Problem, J. Phys. **46**, 1277–1292
117. J. Klos, S. Kobe (2001), Generalized Simulated Annealing algorithms using Tsallis statistics: application to $\pm J$ spin glass model, in: S. Abe, Y. Okamo (eds.), *Nonextensive Statistical Mechanics and Its Applications*, Lecture Notes in Physics, Vol. 560, Springer, Berlin Heidelberg New York, 253–258
118. S. Kobe, T. Klotz (1995), Frustration – how it can be measured, Phys. Rev. E **52**, 5660–5663 and references therein
119. T. Kohonen (1989), *Self-Organization and Associative Memory*, Springer, Berlin Heidelberg New York
120. U. Krey (1995), pers. commun.
121. U. Krey (1996), *Spingläser und Neuronale Netze*, script, University of Regensburg, Germany
122. P. J. M. van Laarhoven, E. H. L. Aarts (1987), *Simulated Annealing: Theory and Applications*, Reidel, Dordrecht
123. Landesbildstelle Baden

124. H. Labermeier (1998), *Simulation and Optimisation of the Storage and Delivery of Goods with Stochastic/Seasonal Depending Demand*, Ph.D. thesis, Eidgenössische Technische Hochschule, Zurich
125. E. W. Lang (1996), *Neuronale Netze*, script, University of Regensburg, Germany
126. E. L. Lawler, J. K. Lenstra, A. H. G. Rinnoy Kan, D. B. Shmoys (1985), *The Traveling Salesman Problem*, Wiley, New York
127. I. Lee, M. Y. Choi (1994), Optimization by Multicanonical Annealing and the traveling salesman problem, Phys. Rev. E **50**, R651–R654
128. D. H. Lehmer (1949), Mathematical methods in large-scale computing units, in: Proc. 2nd Symp. on Large-Scale Digital Calculating Machinery, Cambridge, MA, pp. 141–146, Harvard University Press, Cambridge, MA
129. S. Lin (1965), Computer solutions of the Traveling Salesman Problem, Bell Syst. Tech. J. **44**, 2245–2269
130. S. Lin, B. W. Kernighan (1973), An Effective Heuristic Algorithm for the Traveling Salesman Problem, Oper. Res. **21**, 498–516
131. Th. R. Malthus (1798), *An Essay on the Principle of Population, As It Affects the Future Improvement of Society with Remarks on the Speculations of Mr. Godwin, M. Condorcet, and Other Writers*, printed for J. Johnson, in St. Paul's Church-Yard, London
132. E. Marinari, G. Parisi (1992), Simulated Tempering: a new Monte Carlo scheme, Europhys. Lett. **19**, 451
133. G. Marsaglia (1968), Random numbers fall mainly in the planes, Proc. Natl. Acad. Sci. USA **61**, 25–28
134. G. Marsaglia, T. A. Bray (1964), A convenient method for generating normal variables, SIAM Rev. **6**, 260–264
135. D. McAllester, B. Selman, H. Kautz (1997), Evidence for Invariants in Local Search, in: Proc. 14th AAAI-97
136. R. Mennicken, E. (1977) Wagenführer, *Numerische Mathematik 1*, Rowohlt, Reinbek, Hamburg
137. N. Metropolis, A. W. Rosenbluth, M. N. Rosenbluth, A. H. Teller, E. Teller (1953), Equation of state calculations for fast computing machines, J. Chem. Phys. **21**, 1087–1092
138. M. Mézard, G. Parisi, R. Zecchina (2002), Analytic and algorithmic solutions of random satisfiability problems, Science **297**, 812–815
139. M. Mézard, R. Zecchina (2002), The random K-satisfiability problem: from an analytic solution to an efficient algorithm, Phys. Rev. E **66**, 056126
140. M. L. Minsky, S. Papert (1969), *Perceptrons: An Introduction to Computational Geometry*, MIT Press, Cambridge, MA
141. A. Misevičius (2004), Using Iterated Tabu Search for the Traveling Salesman Problem, Informacines Technologijos Ir Valdymas **3**, 29–40
142. D. Mitra, F. Romeo, A. Sangiovanni-Vincentelli (1985), Convergence and finite-time behavior of Simulated Annealing, in: Proc. 24th Conf. on Decision and Control, 761–767
143. A. Mitsutake, Y. Okamoto (2000), Replica-exchange simulated tempering method for simulations of frustrated systems, arXiv:cond-mat/0009387
144. A. Mitsutake, Y. Sugita, Y. Okamoto (2003), Replica-exchange multicanonical and multicanonical replica-exchange Monte Carlo simulations of peptides. I. Formulation and benchmark test, J. Chem. Phys. **118**, 6664–6675

145. R. Monasson, R. Zecchina, S. Kirkpatrick, B. Selman, L. Troyansky (1996), Phase transition and search cost in the $2+p$-sat problem, in: T. Toffoli, M. Biafore, J. Leao (eds.), Proc. PhysComp 96, Boston
146. R. Monasson, R. Zecchina, S. Kirkpatrick, B. Selman, L. Troyansky (1999), Determining computational complexity from characteristic phase transitions, Nature **400**, 133–137
147. R. Monasson, R. Zecchina, S. Kirkpatrick, B. Selman, L. Troyansky (1999), 2+pSAT: relation of typical-case complexity to the nature of the phase transition, Random Struct. Algor. **15**, 414–440
148. A. Montanari, G. Parisi, F. Ricci-Tersenghi (2004), Instability of one-step replica-symmetry-broken phase in satisfiability problems, J. Phys. A Math. Gen. **37**, 2073–2091
149. I. Morgenstern (1993), *Simulationsmethoden in der Physik*, script, University of Regensburg, Germany
150. P. Moscato, J. F. Fontanari (1990), Stochastic vs. deterministic update in Simulated Annealing, Phys. Lett. A **146**, 204–208
151. R. Motvani, P. Raghavan (1995), *Randomized Algorithms*, Cambridge University Press, Cambridge, UK
152. N. Mousseau, G. T. Barkema (1997), Traveling through potential energy landscapes of disordered materials: the activation-relaxation technique, Phys. Rev. E **57**, 2419–2424
153. N. Mousseau, G. T. Barkema (1999), arXiv:cond-mat/9905410
154. H. Müller-Merbach (1983), Zweimal travelling salesman, DGOR Bull. **25**, 12–13
155. K. Neumann, M. Morlock (1993), *Operations Research*, Hanser, Munich
156. M. Ninio, J. J. Schneider (2005), Weight Annealing, Phys. A **349**, 649–666
157. K. J. Nurmela (1993), Constructing Combinatorial designs by local search, Research Report A27, Helsinki University of Technology
158. M. Padberg (1995), *Linear Optimization and Extensions*, Springer, Berlin Heidelberg New York
159. M. W. Padberg, G. Rinaldi (1987), Optimization of a 532-city symmetric travelling salesman problem by branch and cut, Oper. Res. Lett. **6**, 1–7
160. C. Papadimitriou (1991), On selecting a satisfying truth assignment, in: Proc. Foundations of Computer Science, 163–169
161. G. Parisi (2003), On the probabilistic approach to the random satisfiability problem, in: Proc. SAT 2003 and arXiv:cs:CC/0308010v1
162. J. Pearl (1988), *Probabilistic Reasoning in Intelligent Systems*, 2nd edn., Kauffmann, San Francisco
163. T. J. P. Penna (1995), Traveling salesman problem and Tsallis statistics, Phys. Rev. E **51**, R1–R3
164. D. Polani (1996), *Adaption der Topologie von Kohonen-Karten durch Genetische Algorithmen*, Ph.D. thesis, University of Mainz, Germany
165. Th. Pongratz (1999), *Das Konzept lokaler Temperaturen angewandt auf das Problem des Handlungsreisenden*, Diploma thesis, University of Regensburg, Germany
166. W. H. Press, S. A. Teukolsky, W. T. Vetterling, B. P. Flannery (1992), *Numerical Recipes in Fortran*, Cambridge University Press, Cambridge, UK
167. M. Puchta (2004), Optimierung von Problemstellungen aus der diskreten und der Prozess-Industrie unter Verwendung physikalischer Verfahren, Ph.D. thesis, University of Regensburg, Germany

168. R. E. Randelman, G. S. Grest (1986), N-City Traveling Salesman Problem – Optimization by Simulated Annealings, J. Stat. Phys. **45**, 885–890
169. I. Rechenberg (1975), *Evolutionsstrategie*, Friedrich Frommann, Stuttgart
170. G. Reinelt (1994), *The Traveling Salesman*, Springer, Berlin Heidelberg New York
171. F. Ricci-Tersenghi, M. Weigt, R. Zecchina (2001), Simplest random K-Satisfiability Problem, Phys. Rev. **E63**, 026702
172. H. J. Ritter, T. M. Martinetz, K. J. Schulten (1991), *Neuronale Netze*, Addison-Wesley, Munich
173. F. Romeo, A. Sangiovanni-Vincentelli (1985), in: Proc. Chapel Hill Conf. on VLSI, 393
174. F. Rosenblatt (1958), The perceptron: a probabilistic model for information storage and organization in the brain, Psychol. Rev. **65**, 386–408
175. F. Rosenblatt (1962), *Principles of Neurodynamics*, Spartan, New York
176. D. J. Rosenkrantz, R. E. Stearns, P. M. Lewis II (1977), An analysis of several heuristics for the Traveling Salesman Problem, SIAM J. Comput. **6**, 563–581
177. J. O'Rourke (1993), *Computational Geometry in C*, Cambridge University Press, Cambridge, UK
178. H. A. Salkin (1975), *Integer Programming*, Addison-Wesley, Reading, MA
179. A. Scherer (1997), *Neuronale Netze – Grundlagen und Anwendungen*, Vieweg, Braunschweig, Wiesbaden, Germany
180. T. Scheuer (1995), pers. commun.
181. J. Schneider (1995), *Parallelisierung physikalischer Optimierungsverfahren*, Diploma thesis, University of Regensburg, Germany
182. J. Schneider (1999), *Effiziente parallelisierbare physikalische Optimierungsverfahren*, Ph.D. thesis, University of Regensburg, Germany
183. J. Schneider (2003), Searching for Backbones – a high-performance parallel algorithm for solving combinatorial optimization problems, Future Generat. Comput. Syst. **19**, 121–131
184. J. J. Schneider (2004), Searching for backbones – An Efficient Parallel Algorithm for Finding Groundstates in Spin Glass models, in: M. Tokuyama, I. Oppenheim, *3rd International Symposium on Slow Dynamics in Complex Systems, Sendai, Japan, 3-8 November 2003*, AIP Conference Proceedings **708**, 426–429
185. J. Schneider, J. Britze, A. Ebersbach, I. Morgenstern, M. Puchta (2000), Optimization of production planning problems – a case study for assembly lines, Int. J. Mod. Phys. C **11**, 949–972
186. J. Schneider, M. Dankesreiter, W. Fettes, I. Morgenstern, M. Schmid, J. M. Singer (1997), Search space smoothing for combinatorial optimization problems, Phys. A **243**, 77–112
187. J. Schneider, Ch. Froschhammer, I. Morgenstern, Th. Husslein, J. M. Singer (1996), Searching for Backbones – an efficient parallel algorithm for the Traveling Salesman Problem, Comput. Phys. Comm. **96**, 173–188
188. J. J. Schneider, S. Kirkpatrick (2005), Selfish vs. unselfish optimization of network creation, J. Stat. Mech. **P08007**
189. J. Schneider, I. Morgenstern (2001), Optimization of production lines by methods from statistical physics, in: K.-H. Hoffmann, M. Schreiber (eds.), *Computational Statistical Physics: From Billiards to Monte-Carlo*, in: Proc. Heräus meeting 2000 in Chemnitz, 77–96, Springer, Berlin Heidelberg New York

190. J. Schneider, I. Morgenstern, J. M. Singer (1998), Bouncing towards the optimum: improving the results of Monte Carlo optimization algorithms, Phys. Rev. E **58**, 5085
191. E. Schöneburg, N. Hansen, A. Gawelczyk (1990), *Neuronale Netzwerke*, Markt und Technik, Munich
192. E. Schöneburg, F. Heinzmann (1994), S. Feddersen, *Genetische Algorithmen und Evolutionsstrategien*, Addison-Wesley, Bonn
193. A. Schrijver (2005), On the history of combinatorial optimization (till 1960), in: K. Aardal, G. L. Nemhauser, R. Weismantel (eds.), *Handbook of Discrete Optimization*, Elsevier, Amsterdam, pp. 1–68
194. G. Schrimpf, J. Schneider, H. Stamm-Wilbrandt, G. Dueck (1999), Record breaking optimization results using the Ruin and Recreate principle, J. Comput. Phys. **159**, 139–171
195. J. Schwabe (1998), Team-Probleme, c't **26**, 80
196. B. Selman, H. A. Kautz (1993), An empirical study of greedy local search strategies for satisfiability testing, in: Proc. AAAI-93, 46–51
197. B. Selman, H. Kautz, B. Cohen (1994), Noise strategies for local search, in: Proc. AAAI-94, 337–343
198. B. Selman, S. Kirkpatrick (1996), Critical Behavior in the Computational Cost of Satisfiability Testing, Artif. Intell. J. **81**, 273–295
199. B. Selman, D. G. Mitchell, H. J. Levesque (1996), Generating hard satisfiability problems, Artif. Intell. J. **81**, 17–29
200. G. Semerjian, R. Monasson (2003), Relaxation and metasability in a local search procedure for the random satisfiability problem, Phys. Rev. E **67**, 066103-1-10
201. P. F. Stadler, W. Schnabl (1992), The Landscape of the Traveling Salesman Problem, Phys. Lett. A **161**, 337–344
202. H. Stamm-Wilbrandt (1995), pers. commun.
203. G. Stattenberger, M. Dankesreiter, F. Baumgartner, J. J. Schneider (2005), On the Neighborhood Size of the Lin-k-Opts, preprint, arXiv:physics/0608269
204. D. Stauffer, A. Aharony (1992), *Introduction to Percolation Theory*, 2nd edn., Taylor and Francis, London
205. F. H. Stillinger, T. A. Weber (1988), Nonlinear optimization simplified by hypersurface deformation, J. Stat. Phys. **52**, 1429–1445
206. R. H. Storer, S. D. Wu, R. Vaccari (1992), New search spaces for sequencing problems with application to job shop scheduling, Manage. Sci. **38**, 1495–1509
207. P. N. Strenski, S. Kirkpatrick (1991), Analysis of finite length annealing schedules, Algorithmica **6**, 346–366
208. T. Stühn (2000), *Methoden zur Equilibrierung unterkühlter Flüssigkeiten*, Diploma thesis, University of Mainz, Germany
209. taboo (2006), Encyclopaedia Britannica, retrieved 1 April 2006 from Encyclopaedia Britannica Premium Service, `http://www.britannica.com/eb/article?tocId=9070845`
210. C. Tsallis, D. A. Stariolo (1995), Generalized Simulated Annealing, arXiv:cond-mat/9501047
211. S. Tschernikow (1971), *Lineare Ungleichungen*. VEB Deutscher Verlag der Wissenschaften, Berlin
212. Ch. Voudouris, E. Tsang (1995), Guided Local Search, Technical Report CSM-247, University of Essex, UK

213. Ch. Voudouris, E. Tsang (1998), Guided local search, Eur. J. Oper. Res. **113**, 80119
214. Ch. Voudouris, E. Tsang (1999), Guided local search and its application to the traveling salesman problem, Eur. J. Oper. Res. **113**, 469–499
215. W. Wenzel, K. Hamacher (1999), Stochastic Tunneling Approach for Globel Minimization of Complex Potential Energy Landscapes, Phys. Rev. Lett. **82**, 3003–3007
216. A. Zell (1994), *Simulation Neuronaler Netze*, Addison-Wesley, Bonn

Index

Scientific Computation

A Computational Method in Plasma Physics
F. Bauer, O. Betancourt, P. Garabechan

Implementation of Finite Element Methods for Navier-Stokes Equations
F. Thomasset

Finite-Different Techniques for Vectorized Fluid Dynamics Calculations
Edited by D. Book

Unsteady Viscous Flows
D. P. Telionis

Computational Methods for Fluid Flow
R. Peyret, T. D. Taylor

Computational Methods in Bifurcation Theory and Dissipative Structures
M. Kubicek, M. Marek

Optimal Shape Design for Elliptic Systems
O. Pironneau

The Method of Differential Approximation
Yu. I. Shokin

Computational Galerkin Methods
C. A. J. Fletcher

Numerical Methods for Nonlinear Variational Problems
R. Glowinski

Numerical Methods in Fluid Dynamics
Second Edition M. Holt

Computer Studies of Phase Transitions and Critical Phenomena O. G. Mouritsen

Finite Element Methods in Linear Ideal Magnetohydrodynamics
R. Gruber, J. Rappaz

Numerical Simulation of Plasmas
Y. N. Dnestrovskii, D. P. Kostomarov

Computational Methods for Kinetic Models of Magnetically Confined Plasmas
J. Killeen, G. D. Kerbel, M. C. McCoy, A. A. Mirin

Spectral Methods in Fluid Dynamics
Second Edition
C. Canuto, M. Y. Hussaini, A. Quarteroni, T. A. Zang

Computational Techniques for Fluid Dynamics 1 Fundamental and General Techniques Second Edition
C. A. J. Fletcher

Computational Techniques for Fluid Dynamics 2 Specific Techniques for Different Flow Categories Second Edition
C. A. J. Fletcher

Methods for the Localization of Singularities in Numerical Solutions of Gas Dynamics Problems
E. V. Vorozhtsov, N. N. Yanenko

Classical Orthogonal Polynomials of a Discrete Variable
A. F. Nikiforov, S. K. Suslov, V. B. Uvarov

Flux Coordinates and Magnetic Filed Structure: A Guide to a Fundamental Tool of Plasma Theory
W. D. D'haeseleer, W. N. G. Hitchon, J. D. Callen, J. L. Shohet

Monte Carlo Methods in Boundary Value Problems
K. K. Sabelfeld

The Least-Squares Finite Element Method Theory and Applications in Computational Fluid Dynamics and Electromagnetics
Bo-nan Jiang

Computer Simulation of Dynamic Phenomena
M. L. Wilkins

Grid Generation Methods
V. D. Liseikin

Radiation in Enclosures
A. Mbiock, R. Weber

Higher-Order Numerical Methods for Transient Wave Equations
G. C. Cohen

Fundamentals of Computational Fluid Dynamics
H. Lomax, T. H. Pulliam, D. W. Zingg

The Hybrid Multiscale Simulation Technology An Introduction with Application to Astrophysical and Laboratory Plasmas A. S. Lipatov

Scientific Computation

GPSR Compliance

The European Union's (EU) General Product Safety Regulation (GPSR) is a set of rules that requires consumer products to be safe and our obligations to ensure this.

If you have any concerns about our products, you can contact us on ProductSafety@springernature.com

In case Publisher is established outside the EU, the EU authorized representative is:

Springer Nature Customer Service Center GmbH
Europaplatz 3
69115 Heidelberg, Germany

Batch number: 10755310

Printed by Printforce, the Netherlands